HEMOCYTIC AND HUMORAL
IMMUNITY
IN ARTHROPODS

HEMOCYTIC AND HUMORAL IMMUNITY IN ARTHROPODS

Edited by

AYODHYA P. GUPTA
Professor of Entomology
Rutgers University
New Brunswick, New Jersey

A WILEY-INTERSCIENCE PUBLICATION
JOHN WILEY & SONS
New York   Chichester   Brisbane   Toronto   Singapore

Library of Congress Cataloging in Publication Data:

Hemocytic and humoral immunity in arthropods.

  "A Wiley-Interscience publication."
  Includes bibliographies and index.
  1. Arthropoda—Immunology.   I. Gupta, A. P.,
1928–
QL434.72.H46   1986   595.2′0429   86-15913
ISBN 0-471-82812-2

Printed in the United States of America

10 9 8 7 6 5 4 3 2 1

Contributors

Gianni A. Amirante
Department of Biology
University of Trieste
Trieste, Italy

Robert S. Anderson
Chesapeake Biological Laboratory
Center for Environmental and
  Estuarine Studies
University of Maryland
Solomons, Maryland

June S. Chadwick
Department of Microbiology and
  Immunology
Queens University
Kingston, Canada

Benjamin M. Chain
Tumor Immunology Unit
Department of Zoology
University College of London
London, England

Elias Cohen
Department of Laboratory
  Medicine
Roswell Park Memorial Institute
Departments of Pathology and
  Microbiology
School of Medicine
State University of New York at
  Buffalo
Buffalo, New York

Gary B. Dunphy
Department of Biosciences
Simon Fraser University
Burnaby, Canada

Dieter Ehlers
Section of Biology
Ernst-Moritz-Arndt University
Greifswald, East Germany

Peter Götz
Division of Biology
Institute for General Zoology
Free University of Berlin
Berlin, West Germany

Sandra A. Graf
School of Life and Health
  Sciences
University of Delaware
Newark, Delaware

Ayodhya P. Gupta
Department of Entomology and
  Economic Zoology
Rutgers University
New Brunswick, New Jersey

Kenneth D. Hapner
Department of Chemistry
Montana State University
Bozeman, Montana

J. Michael Kehoe
Department of Microbiology and
  Immunology
Northeastern Ohio Universities
  College of Medicine
Rootstown, Ohio

Ann M. Lackie
Department of Zoology
The University
Glasgow, Scotland

Catherine M. Leonard
School of Biological Sciences
University College of Swansea
Swansea, Wales

Werner Mohrig
Section of Biology
Ernst-Moritz-Arndt University
Greifswald, East Germany

Herold Müller
Institute for Plant Physiology and
  Cell Biology
Free University of Berlin
Berlin, West Germany

Thomas G. Pistole
Department of Microbiology
University of New Hampshire
Durham, New Hampshire

Norman A. Ratcliffe
School of Biological Sciences
University College of Swansea
Swansea, Wales

Lothar Renwrantz
Zoological Institute and
  Zoological Museum
University of Hamburg
Hamburg, West Germany

Elaine H. Richards
School of Biological Sciences
University College of Swansea
Swansea, Wales

Rose M. Rizki
Division of Biological Sciences
University of Michigan
Ann Arbor, Michigan

Tahir M. Rizki
Division of Biological Sciences
University of Michigan
Ann Arbor, Michigan

Andrew F. Rowley
School of Biological Sciences

University College of Swansea
Swansea, Wales

Dietmar Schittek
Section of Biology
Ernst-Moritz-Arndt University
Greifswald, East Germany

Werner Schwemmler
Institute for Plant Physiology and
  Cell Biology
Free University of Berlin
Berlin, West Germany

Rochelle K. Seide
Brumbaugh, Graves, Donohue
  and Raymond
New York, New York

Valerie J. Smith
University Marine Biological
  Station
Millport, Scotland

Kenneth Söderhäll
Institute of Physiological Botany
University of Uppsala
Uppsala, Sweden

Mark R. Stebbins
Genetic Systems Corporation
Seattle, Washington

John G. Stoffolano, Jr.
Department of Entomology
University of Massachusetts
Amherst, Massachusetts

Alain Vey
Research Station of Comparative
  Pathology
Institute of Agronomical Research
St. Chistol, France

Robin W. Yeaton
Department of Microbiology
Molecular Biology Institute
University of California
Los Angeles, California

Preface

Compared with the comprehensive body of knowledge currently available on mammalian immunology, that on the arthropod immune system is truly minuscule. Indeed, it would be no exaggeration to say that in terms of its lack of both the breadth and depth of our understanding of its modus operandi, arthropod immunology trails so far behind mammalian immunology that it seems to have advanced no farther than the stone age! Furthermore, most of the old and new information on this system remains scattered. The raison d'être of this book is to bring together in a single publication most, if not all, available information on the various components of the arthropod immune system and their functions. As I began to plan the book some three years ago, it became apparent that the many gaps in our knowledge of the various aspects of this system would make the task of synthesizing and systematizing the available information difficult. Thus, perhaps the single most important contribution of this book might be the uncovering of the very gaps in our knowledge of the arthropod immune system for future research.

Although arthropod hemocytes have been investigated for nearly 150 years, we still find ourselves debating over their nomenclature and identity (types). Any one venturing into the study of arthropod hemocytes knows only too well how discouraging it is to have to contend with the bewildering array of terminologies that exists in the literature, and the tendency—or perhaps the irresistible urge—on the part of hematologists to introduce new terms for variant forms of morphologically distinct types of hemocytes further confounds the situation. Unfortunately, this seemingly controversial status of hemocyte classification leads many arthropod hematologists to ascribe immune functions to "hemocytes" rather than to well-defined and morphologically distinct hemocyte types, especially the plasmatocytes and the granulocytes (referred to as immunocytes elsewhere). This practice, I am afraid, is tantamount to saying that in mammals nonself tissue is recognized by "lymphocytes" rather than by the B- and T-lymphocytes! Needless to say, unless a consensus on the identities and nomenclature of the most basic hemocyte types is reached among arthropod hematologists and adhered to in practice, the functional interpretations ascribed to those types will continue to be confusing and unreliable, a situation that will hinder the progress of our understanding of the functions of the arthropod immune system. Although the two major arthropod immunocytes and other hemocyte types can generally be clearly recognized and identified, more reliable methods of characterizing them must be found.

The most basic ultrastructure and chemistry of the immunocyte plasma membrane—especially of the glycoproteins and the glycolipids and their monosaccharide moieties—remain unknown. Without the knowledge of the nature, arrangement, and mode of formation of the intramembranous particles, it is difficult to understand and interpret their locations and/or movement (e.g., their patching or capping properties) and functions as plasma membrane receptors. Similarly, without the basic information on the nature and mode of formation of the membrane gap junctions, especially during encapsulation and nodule formation, it is difficult to decipher the cell-to-cell communication that must occur among the plasmatocytes that form the layers of capsules and nodules.

Among the humoral factors, the origins and precise roles of hemagglutinins (lectins) in several major arthropod groups remain uncertain, and we have only vague ideas about the origins and modes of action of the complementlike molecules and opsonins. Although recent works on hemolymph coagulation and the prophenoloxidase-activating system (PAS) in some arthropods have increased our understanding of the roles of these mechanisms in the immune system, the origins, biochemical nature, and modes of action of the proteins released both during coagulation and PAS activation—as well as those of lysozymes and antibacterial and antiviral molecules—remain unresolved. We now know that both the activation of the PAS and the conversion of the cellular coagulogen into coagulin are mediated by a serine protease enzyme or enzymes; however, whether the same enzyme participates in both processes remains to be confirmed. Attention must also be paid to the contributions (in addition to that of hemocyte production) of the hemopoietic organs to the effectiveness of the immune system.

Thus, as a whole, comparative studies of all the components of the immune systems in all major arthropod groups must remain the primary goal of hematologists, because such studies alone would provide us with the insight to understand the functions of those components and also shed light on their evolutionary history and commonality with the immune systems of vertebrates. Furthermore, among the arthropods themselves, any phylogenetic or taxonomic differences in immune systems will be difficult to delineate until detailed studies of these systems in both the immature and adult stages of many more representatives are available. Finally, the possible role of land colonization and the exigencies of a terrestrial mode of life in the evolution of the immune system of the terrestrial arthropods should be investigated.

Biomedical applications of arthropod lysate to characterize several human diseases are presently confined almost exclusively to the lysate of the most primitive arthropods: the horseshoe crabs. Considering the rich faunal diversity of arthropods, it would not surprise me if future research would uncover many other uses of the lysates of higher arthropods.

Clearly, a great deal remains to be done, and I hope this book will provide the incentive and leads for much-needed research in the above-mentioned areas.

The book is organized into three parts: "Hemocytic Immunity," "Humoral Immunity," and "Techniques and Biomedical Applications." In any multiauthored book some overlap is inevitable, and this book is no exception. However, overlap has been kept to a minimum and retained only where necessary for understanding the discussion at hand. Wherever relevant, overlaps and divergences of opinions in various chapters have been cross-referenced. Most authors have pointed out the dearth of information on their respective topics, and many have suggested areas of further research. Because taxonomic ranking of major arthropod groups is highly controversial, each contributor has used his or her preferred taxonomic categories. Furthermore, each contributor has had complete freedom to develop, interpret, and present his or her views. Each chapter attempts to present an in-depth review of the topics it covers. In the chapters of the non–English-speaking contributors, editing has been confined to removing obvious infelicities in order to retain the original style and contents.

Organizing a multiauthored book is in many ways comparable to, but more difficult than, conducting an orchestra or directing a complicated movie. While the conductor or the director may, of necessity, compel endless rehearsals and retakes to achieve the desired goal, an editor, alas, does not enjoy such professional luxuries! Thus, an endeavor such as this book could not have been successfully completed without the cooperation and assistance of all the authors who responded to my invitation to contribute and thus made the book possible. To them, I am most grateful.

The following individuals, journals, societies, and publishers gave their permissions to reproduce published and/or unpublished materials: C. A. Abel, D. D. Dorai, J. D. Jamieson, H. Komano, S. Natori, Academic Press, Agricultural Chemical Society of Japan, American Society of Biological Chemists, *Biochimica et Biophysica Acta, Developmental and Comparative Immunology*, Elsevier North Holland, Faculty Press, *Insect Biochemistry*, Japanese Biochemical Society, *Journal of Biochemistry, Journal of Biological Chemistry, Journal of Histochemistry and Cytochemistry, Journal of Molecular Biology*, and Pergamon Press.

In addition, I am grateful to my wife and children for their understanding, ungrudging support, and enthusiasm during the preparation of the book.

Finally, I hope the book will stimulate further research to fill the gaps that it reveals in our knowledge of the arthropod immune system.

AYODHYA P. GUPTA

New Brunswick, New Jersey
September, 1986

Contents

HEMOCYTIC AND HUMORAL
IMMUNITY
IN ARTHROPODS

PART I
HEMOCYTIC IMMUNITY

CHAPTER 1

Arthropod Immunocytes
Identification, Structure, Functions, and Analogies to the Functions of Vertebrate B- and T-Lymphocytes

Ayodhya P. Gupta

Department of Entomology and Economic Zoology
Rutgers University
New Brunswick, New Jersey

1.1. INTRODUCTION

This chapter is intended to provide the reader, especially one not familiar with the arthropod immune system, with an overview of the subject. Details of the two major arthropod immunocytes and their functional analogies with those of vertebrate B- and T-lymphocytes are emphasized in the hope that they will prompt a thorough search of the functions of these immunocytes that would be comparable in precision and detail to the functions of the vertebrate B- and T-lymphocytes.

The most important measure of the competence of an immune system is its ability to distinguish between self and nonself tissues. In arthropods, this is accomplished by the combined actions of two types of immunocytes (blood cells), several humoral (hemolymph) factors (agglutinins, lysozymes, complement factors, and antibacterial factors, and antiviral factors), and the polyphenoloxidase system.

It is well known that in arthropods the most important immunologic functions of phagocytosis and encapsulation of foreign antigen are generally performed by two blood cell types, called *granulocytes* (GRs) and *plasmatocytes* (PLs), which I have previously (Gupta, 1985b) referred to as *immunocytes* in the cockroach, *Gromphadorhina portentosa*, because of their commonly known capacity to distinguish between self and nonself tissues. In both phagocytosis and encapsulation, once the nonself tissue, or target cell (TC), is recognized by the surface receptors of the two immunocytes and contact with it is established, either ingestion (phagocytosis) or encapsulation begins, depending on the size of the TC.

1.2. ARTHROPOD HEMOCYTES

Classifications of arthropod hemocytes have been variously based on their morphology, functions, and staining or histochemical reactions. Thus, it is not unusual to find the same hemocyte type or its variant forms being referred

TABLE 1.1 SUMMARY OF HEMOCYTE TYPES IN VARIOUS GROUPS OF ARTHROPODS AND ONYCHOPHORA

Groups	Prohemocyte (PR)	Plasmatocyte (PL)	Granulocyte (GR)	Spherulocyte (SP)	Adipohemocyte (AD)	Coagulocyte (CO)	Oenocytoid (OE)	Others[a]
Aquatic								
Chelicerata (Xiphosura)	—	—	GR	—	—	—	—	Cyanoblast
Crustacea	PR	PL	GR	SP	AD	CO	OE	Cyanocyte
Terrestrial								Cyanocyte
Chelicerata	PR	PL	GR	SP	AD	CO	OE	Cyanocyte
Myriapoda	PR	PL	GR	SP	AD	CO	OE	Crystal cell
Insecta	PR	PL	GR	SP	AD	CO	OE	For numerous other terms see Section 2.3.3.
Onychophora	PR	PL	GR	SP	—	—	OE	—

Source: From Gupta, 1985a.

[a] Includes cells that could not be homologized.

5

to by different names in various arthropods by different authors—a situation that has inevitably resulted in a confusing mass of terminologies. Consequently, it is often difficult to compare hemocytes of one species with those of others. Furthermore, arthropod hematologists often disagree as to the actual number of hemocyte types in various arthropods. Indeed, from one to as many as nine or more types have been described, particularly by light microscopy (see reviews by Gupta, 1979a–d, 1985a). Ultrastructurally, however, only seven types are generally recognized (Table 1.1): prohemocytes (PRs), plasmatocytes, granulocytes, spherulocytes (SPs), adipohemocytes (ADs), coagulocytes (COs), and oenocytoids (OEs). Note that only the GR has been reported from all major arthropod groups as well as from the Onychophora. I have therefore regarded the GR as the most primitive (plesiomorphic hemocyte in arthropods (Gupta, 1979a, 1985a), from which the other types seem to have evolved (differentiated) during the course of arthropod evolution. Of these seven, the true identities of COs and ADs, as separate types, are doubtful, and the SPs are generally regarded as more mature forms of GRs (Gupta and Sutherland, 1966). By contrast, the morphological identity of the two immunocytes (GRs and PLs) is less controversial, and they can ordinarily be recognized by light microscopy.

Among the Aquatic Chelicerata, the hemocytes of only one species of one group (Xiphosura)—namely, the most primitive of all arthropods, *Limulus*— have so far been described (see the review by Gupta, 1979a). Only two types of hemocytes have been described in *Limulus*: the GR (called "amoebocytes" by other authors) and the cyanocyte (CYN), the latter producing hemocyanin (Fahrenbach, 1968, 1970).

For a general survey of hemocyte types in Crustacea, Aquatic and Terrestrial Chelicerata, Myriapoda, Insecta, and Onychophora, see reviews by Gupta (1979a,b,e, 1985a), Bauchau (1981), Sherman (1981), Rowley and Ratcliffe (1981), and Ravindranath (1981).

1.3. IDENTIFICATION OF ARTHROPOD IMMUNOCYTES

The cytoplasm of the GR is prominently and characteristically granular (Fig. 1.1D); the granules may or may not be numerous, depending on the developmental (differentiation) stage of the hemocyte; the nucleus is often smaller than that of the PL, round or elongate, and generally centrally located. By contrast, the polymorphic PL often shows agranular or slightly granular cytoplasm (Fig. 1.1C) and a round or elongate nucleus that may or may not appear punctate but is often centrally located. For identification keys of all the seven types of arthropod hemocytes, see Gupta (1979a–d, 1985a).

1.4. ULTRASTRUCTURE AND SYNONYMS OF ARTHROPOD IMMUNOCYTES

The following account is based on my previous works and reviews (Gupta, 1979a–e, 1985a).

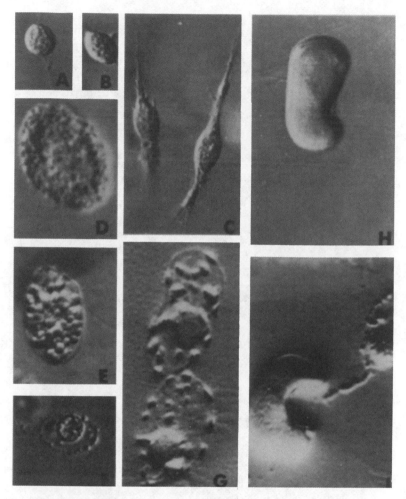

Figure 1.1. Various hemocyte types in the cockroach *Gromphadorhina portentosa*. (A) Prohemocyte (×4,333). (B) Prohemocyte (×5,633). (C) Plasmatocytes (×6,500). (D) Granulocyte (×5,687). (E) Spherulocyte (×5,687). (F) Coagulocyte (×3,250). (G) Coagulocytes (×6,314). (H) Enucleate oenocytoid (×5,525). (I) Oenocytoid with extruded nucleus (×7,150). (From Gupta, 1985a.)

1.4.1. Granulocytes

Spherical or oval cells (Fig. 1.1D), granulocytes vary in size from small to large (10–45 μm long and 4–32 μm wide, rarely larger). The plasma membrane may or may not have micropapillae or other irregular processes. The nucleus may be relatively small compared to that of the PL, is round or elongate, and is generally centrally located. Nuclear size is variable (2–8 μm long and 2–7 μm wide).

The laminar nature of the plasma membrane may not be visible (see Sec-

AYODHYA P. GUPTA

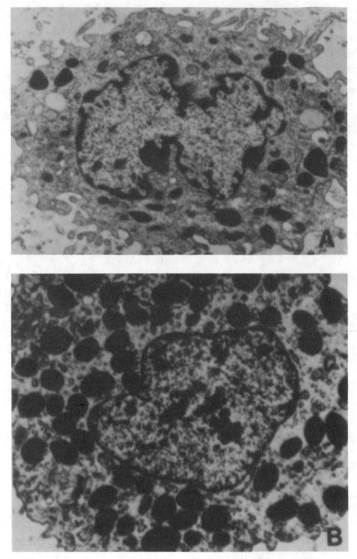

Figure 1.2. Ultrastructures of plasmatocytes and granulocytes. (A) Plasmatocyte of the stick insect *Carausius morosus* showing micropapillae and lobate nucleus (×6,570). (B) Granulocyte of the beetle *Melolontha melolontha* (×6,300). (From Gupta, 1985a.)

tion 1.4.3.1). The cytoplasm is characteristically granular (Fig. 1.2B). Several types of membrane-bound granules have been described in the GRs of various insects and Crustacea: (1) *structureless, electron-dense granules* (Fig. 1.4) (unstructured inclusion [type 1] of Baerwald and Boush, 1970; melanosomelike granules of Hagopian, 1971; opaque body of Moran, 1971;

type 2 bodies of Scharrer, 1972; and electron-dense granules of Raina, 1976, and others); (2) *structureless, thinly granular bodies* (Type 3 bodies of Scharrer, 1972; and electron-lucent granules of Raina, 1976); and (3) *structured granules* (*globules* or *granules multibullaires* of Beaulaton, 1968; *grains denses structures*, in the AD, of Devauchelle, 1971, and Landureau and Grellet, 1975; *corpus fibrillaires* of Hoffmann and coworkers, 1968b, 1970; cylinder inclusions [type 2], regular-packed inclusions [type 3], and inclusions with bandlike units [type 4] of Baerwald and Boush, 1970; *Granula mit tubularer Binnenstruktur* of Stang-Voss, 1970; premelanosomelike granules of Hagopian, 1971; tubule-containing bodies, or TCB, of Moran, 1971; type 1 bodies of Scharrer, 1972; and granules with a microtubular structure of Ratcliffe and Price, 1974). The length or the diameter of the structureless granules varies from 0.15 to 3 μm or more in various insects, while that of the structured granules varies from 0.5 to 2 μm. The shape of the granules may be spherical, ovoid, elongate, or irregularly polygonal. The diameter of the microtubules within the structured granules varies from 15 to 80 nm in various insects. Internally, the microtubules may show micro-microtubules about 5 nm in diameter (Hagopian, 1971), Akai and Sato (1973) also have described "subunits of fibrils" in what they call secretory vesicles. The number of microtubules per granule may vary from 9 to 80. From the accounts provided by Hagopian (1971), Scharrer (1972), Akai and Sato (1973), and François (1975), it appears that the granules are derived from the Golgi bodies (Fig. 1.3A), the microtubules developing during the later stages of morphogenesis. It is conceivable that the structureless, electron-dense granules represent the final stages of development of these granules, in which the structured nature becomes obliterated. Histochemically, most of the granules contain sulfated, periodate-reactive sialomucin and other glycoproteins or neutral mucopolysaccharides (Costin, 1975; François, 1974, 1975). That these granules contain the factors that cause coagulation has been reported by several authors (Howell, 1885; Loeb, 1904, 1910; Maluf, 1939; Dumont et al., 1966; Murer et al., 1975); and Stang-Voss (1971) has suggested that they are the sites of hemagglutinin storage in *Astacus*. Occasionally, lipid droplets may be present, especially in older GRs. In addition to the structureless and structured granules, the cytoplasm is rich in free ribosomes (polysomes), lysosomes (membrane-bounded, electron-dense bodies, 0.1–1.30 μm in size), Golgi bodies, and both smooth (ER) and rough endoplasmic reticulum (RER). Mitochondria are generally few in number. Marginal bundles of microtubules also may be present (Fig. 1.3A).

The GRs of hemimetabolous insects are generally larger and have more prominent granules than those of holometabolous insects (Arnold, 1979), and Akai (1969) even found two types of GRs in the same insect (*Philosamia cynthia ricini*).

I have proposed (see Section 1.10.3) that arthropod GRs perform the functions of vertebrate B-lymphocytes as well as the helper and suppressor functions of T-lymphocytes. In other words, the GRs are functional hybrids

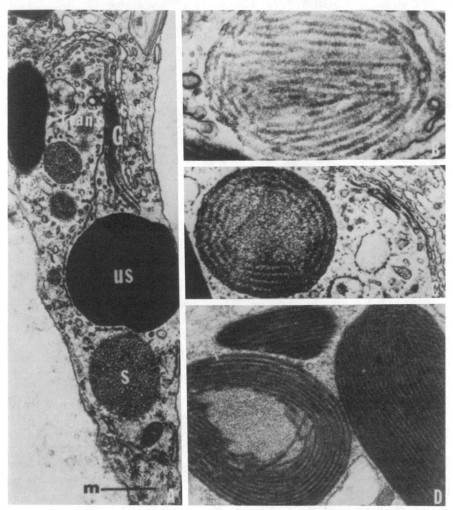

Figure 1.3. Various organelles of a granulocyte of the cockroach *Leucophaea maderae.*
(A) Derivation of structured (premelanosomelike) granules from Golgi bodies (G), structureless or unstructured (melaninlike) granules (US), and intracytoplasmic microtubules
(m). Note *cis* and *trans* faces of Golgi bodies (×50,000). (B) Structured granule showing
internal microtubules (×40,000). (C) An earlier stage of development of internal microtubules (×50,000). (D) Section of a structured granule showing concentric arrangement
of internal microtubules (×38,000). (From Gupta, 1985a.)

of both B- and T- lymphocytes as well as of the macro- and microphages (polymorphonuclear leucocytes) in some primitive arthropods.

1.4.1.1. Synonyms of GRs. Jones (1846) was apparently the first to establish the category of *granular cells*. Later, Cuénot (1896) described *amoebocytes* with finely granular cytoplasm. Since then, GRs have been described by several other names, such as *adipohemocytes, cystocytes (coagulocytes), hyaline cells, phagocytes, pycnoleucocytes,* and *spherulocytes.* Indeed, GRs have been widely misidentified and confused with spherulocytes, adiphemocytes, and coagulocytes.

1.4.2. Plasmatocytes

The polymorphic plasmatocytes also vary in size (3.3–5 μm wide and 3.3–40 μm long). The plasma membrane may have micropapillae or other irregular processes, as well as pinocytotic or vesicular invaginations (Fig. 1.2A). The nucleus may be round, elongate, or even lobate (Fig. 1.2A) and is generally centrally located. It may vary in size (3–9 μm wide and 4–10 μm long) in various arthropods and may appear punctate. Scattered chromatin masses may be present along with the nucleolus (Fig. 1.2A). Occasionally, a binucleate PL may be found.

The laminar nature of the plasma membrane may or may not be visible. The cytoplasm is generally abundant and may be agranular or slightly granular; it is basophilic and rich in organelles. Generally, there is well-developed and extensive RER, which may form greatly distended cisternae or a vacuolar system. Golgi bodies (referred to as dictyosomes, golgiosomes, or internal reticular apparatus) and lysosomes may be numerous. Lysosomes can be identified by the presence in them of the reaction products of the hydrolytic marker enzymes acid phosphatase and thiamine pyrophosphatase (Scharrer, 1972); and lysosomes are often associated with the RER or the vacuolar system. Microsomes and cisternae of the ER (called *ergastoplasme* by French authors) may be present. Free ribosomes (polysomes or polyribosomes) or ribosomes attached to microsomes or RER may be present; microtubules, sometimes arranged in bundles, are present.

The PLs are generally abundant and in some species may be indistinguishable from PRs and GRs. Several types of PLs (most often transitional forms) have been described on the basis of size and shape (see Section 1.4.2.1).

I have proposed (see Section 1.10.4) that arthropod PLs are functionally analogous to the killer T-lymphocytes and macrophages of vertebrates.

1.4.2.1. Synonyms of PLs. Since Yeager and Munson (1941) introduced the term *plasmatocyte*, these cells or their variant forms have been described

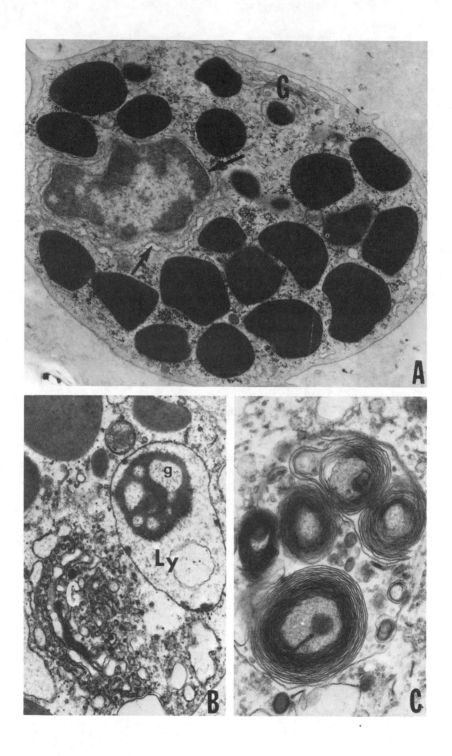

under such names as *amoebocytes*, *giant fusiform cells*, *hyaline cells*, *lamellocytes*, *leucocytes*, *lymphocytes*, *micronucleocytes*, *nematocytes*, *phagocytes*, *podocytes*, *radiate cells*, *star-shaped amoebocytes*, and *vermicytes* or *vermiform cells* (see the review of Gupta, 1985a).

1.4.3. Plasma Membrane of Immunocytes

1.4.3.1. Basic Organization of Cell Membranes. The following accounts are adopted from the work of Alberts and associates (1983). The cell plasma membranes of nucleated cells generally consist of lipids (phospholipids, cholesterol, and glycolipids) and proteins (e.g., spectrin, glycophorin, and band III in the red blood cell [RBC] membrane), often in a 50:1 ratio. The phospho- and glycolipids together form the lipid bilayer, responsible for the laminar nature of the plasma membrane (see Sections 1.4.1 and 1.4.2), and the cholesterol molecules orient themselves in the lipid bilayer with their OH groups close to the polar (hydrophilic) head groups of the phospholipid molecules.

Compared to the lipid molecules, the protein molecules are larger and generally extend across the lipid bilayer (and are therefore called transmembrane protein molecules); their polar ends are thus exposed to the aqueous environments on both the extracellular and cytoplasmic sides of the membrane. Like the lipid molecules, however, they also possess hydrophobic (nonpolar) and hydrophilic (polar) regions, the former interacting with the hydrophobic tails of the lipid molecules.

Both the glycoprotein and the glycolipid molecules have covalently bound oligosaccharide side chains, generally only one side chain on the glycolipid molecule and more than one on the glycoprotein molecule. The oligosaccharide side chains of the glycoprotein and glycolipid molecules constitute the so-called glycocalyx, or cell coat, on the extracellular surface of the plasma membrane, which can be visualized with certain dyes (e.g., ruthenium red).

The monosaccharides commonly found on the glycoprotein and glycolipid molecules are galactose, mannose, fucose, galactosamine, glucose, and sialic acid (*N*-acetylneuraminic acid, or NANA); NANA residues are usually located at the ends of the carbohydrate side chains. Glycolipids containing one or more NANA residues are called gangliosides, one of which, called GM1 (i.e., with one NANA residue), acts as surface receptor for the bacterial toxin that causes diarrhea of cholera. These gangliosides may also serve as receptors for normal cell-to-cell recognition. Because antibodies for many

---

Figure 1.4. Granules and organelles of a granulocyte of the crustacean *Eriocheir sinensis.* (A) Electron-dense (structureless) granules, Golgi bodies (G), and perinuclear and endoplasmic reticular cisternae (arrows) (×15,500). (B) Portion of granulocyte showing a lysosome (Ly) with a granule (g) in the process of resorption and Golgi bodies (G) (×17,500). (C) Granules with concentric filaments (myelin) (×23,400). (From Gupta, 1979a, after Bauchau and De Brouwer, 1972.)

gangliosides are available, it should be possible to study the specific function of gangliosides by binding antibodies to them and studying stimulation or inhibition of specific functions (e.g., during capsule formation in arthropods).

Like the glycolipids, the plasma membrane glycoproteins also serve as surface receptors. Perhaps the best-known plasma membrane glycoprotein is the band III protein component of the human RBC membrane; this band III protein forms the intramembranous particles.

Kaplan (1981) has categorized mammalian membrane receptors for soluble polypeptides into two groups on the basis of their functions. Class I receptors, whose main function is to transmit information and thus cause changes in cell behavior or metabolism, bind to the ligand but most often do not internalize it. The primary function of class II receptors is to internalize the ligand and thus remove it from the extracellular environment. Class I receptors include hormonal and nonhormonal (e.g., IgE of mast cells) humoral agents and are independent of divalent cations (e.g., Ca^{2+}) for ligand binding and catabolized after it, whereas class II receptors are dependent on divalent cations and are not catabolized after ligand binding; apparently they are used, since their number on the plasma membrane remains unchanged.

The carbohydrate residues of both the glycoproteins and the glycolipids can bind with bivalent antibodies (or hemagglutinins or lectins) and form complexes on the plasma membrane surface; the residues of the glycoproteins move laterally in the fluid bilayer, thus producing clustering or patching. This clustering or patching, produced by antibodies or lectins when coupled with ferritin or hemocyanin, may be used to map receptor sites or macromolecules on the membrane surface by immobilizing the glycoprotein molecules by fixing the cell.

Ingestion of the receptor-mediated macromolecules or foreign microorganisms (e.g., during phagocytosis) is called *adsorptive endocytosis* and occurs via coated pits. According to Kaplan (1981), endocytosis terminates, rather than initiates, information transfer by the membrane receptor (which Kaplan calls class II receptors).

1.4.3.2. Status of Knowledge of Arthropod Immunocyte Membranes. Although the plasma membrane and its receptors actively participate in nonself tissue recognition (phagocytosis and encapsulation), coagulation, wound repair, and release of humoral immunologic factors into the hemolymph, virtually nothing is known of the ultrastructure and molecular structure of the arthropod immunocyte membrane itself, especially of the intramembranous particles present on one or both of its characteristic and complementary fracture faces: the internal, convex *cytoplasmic* (also called A, P, or plus) and the external, concave *extracellular* (also called B, E, or minus) faces, obtained after freeze-fracturing the membrane along the internal hydrophobic phospholipid bilayer (Branton, 1966). The precise roles of these particles, or fibrils (composed of glycoproteins), or antigenic or receptor sites (e.g., for the ABO determinants, or epitopes [components of the antigens that recognizes the ABO blood groups as foreign], of the human erythrocytes, bacteria, fungi, or other proteins, such as hemagglutinins or lectins) and in cell-to-cell communication via gap junctions (e.g., during capsule or nodule formation) is unknown in arthropods. Furthermore, it is likely that, as in the ghost RBC (Pinto da Silva et al., 1971; Alberts et al., 1983), these intramembranous particles in the plasma membranes of arthropod immunocytes traverse the phospholipid-glycoprotein matrix of the membrane and protrude on its surface, where they carry the heterosaccharide ABO determinants. In addition, because arthropod hemagglutinins (lectins) are known to act as receptors on the surface of the immunocytes that synthesize them (see Section 1.8.4.1), it is reasonable to suggest that they are probably attached to the protruding intramembranous particles that function as receptor sites. And because the locations of antibodies or the hemaglutinins on the membrane surface can be pinpointed by labeling them (after fixing the cell) with markers, such as hemocyanin, ferritin, or fluorescein, they should most likely correspond to the precise distribution of the protruding ends of the particles. Furthermore, because the binding sites or receptors are known to be mobile, it appears that the particles also must move through the phospholipid-glycoprotein matrix of the membrane. All these aspects of arthropod immunocytes are currently unknown and need to be studied and confirmed. It should be mentioned that what we map or measure is not actually the receptor molecule, but receptor function, that is, ligand-binding; and this provides us the insight into receptor function.

During encapsulation, nodule formation, and wound repair, cell-to-cell communication between plasmatocytes supposedly occurs through gap junctions; the latter are formed in other animals when the plasma membranes of two adjacent cells not more than 2–3 nm apart (Giese, 1979) form hexagonal or tetragonal connecting junctions between them; these junctions, composed of polypeptides, have central pores through which ions and molecules of less than 1000 MW pass between the two cells, thus establishing cell-to-cell communication. In mammalian cells, the outer diameter of the pore is 8 nm and the inner diameter about 1.4 nm; several hundred such pores are arranged in quasicrystalline structures (Brümmer and Hulser, 1982).

In the very few arthropods in which they had been studied by the early 1970s, the gap junction particles were generally reported to be located on the extracellular (B or E) fracture face, and therefore Flower (1972) designated these gap junctions "inverted type," with reference to the mammalian gap junctions, in which the location of the particles is reversed (i.e., on the cytoplasmic, or A or P, face). Thus, Staehelin (1974) and Baerwald (1975) suggested that the extracellular or E-type gap junctions may be characteristic of arthropods, although such junctions have also been found in the primitive coelenterate *Hydra* (Wood, 1977). The nematode *Trichinella spiralis* (Wright and Lee, 1984) and some mollusks (Flower, 1971; Gilula and Satir, 1971), on the other hand, have P-type gap junctions. In view of this, it appears that probably the E type represents the most primitive stage, while the P type is the most advanced type that originated before the evolution of the arthropods, in which both E and P types are present. Indeed, P-type gap junctions are now known to be present in several arthropods (Peracchia, 1974; Shaw and Stowe, 1982; St. Marie and Carlson, 1982).

According to Baerwald (1979), gap junctions are more numerous in the deeper layers of *Periplaneta americana* capsules. Brümmer and Hulser (1982) reported that in cultured multicell spheroids of mammary tumor cells of the marshall rat, two days after the formation of the spheroids, the pores in the gap junctions apparently close (indicated by the fact that injected Lucifer yellow dye was retained in the injected cell and did not spread to other cells of the spheroid). If a similar cessation of cell-to-cell communication occurs during capsule formation in arthropods, it would explain the progressive retardation of plasmatocyte attachment to the capsule as its thickness increases over a period of time. The existence of such a mechanism needs to be comfirmed, however.

1.5. INTERRELATIONSHIP BETWEEN GRs AND PLs

Authors disagree as to how the various types of hemocytes differentiate from the prohemocyte, or stem cell; it appears that probably several differentiation pathways exist in various arthropods (Gupta and Sutherland, 1966). I have discussed elsewhere (Gupta, 1979a,b, 1985a) that during evolution the primitive and ubiquitous GR (see Sections 1.2 and 1.10.3) differentiated (evolved) into other hemocyte types. It can be postulated that the GR originated from the so-called PR and passed through the evanescent PL stage before becoming a distinct GR type. In taxa in which only GRs have been reported (e.g., *Limulus*), PRs and PLs are merely evanescent stages and have not achieved distinctness as types. However, in taxa that are reported to possess other types besides GRs, the latter perhaps further differentiated into the SP, AD, CO, and OE, not necessarily in that order (Fig. 1.5). Post-GR differentiation is generally accompanied by distinct PRs and PLs. Furthermore, in the more highly evolved taxa, any of the types may be suppressed and its functions shifted to other types.

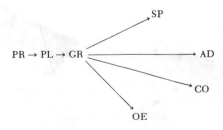

Figure 1.5. Differentiation pathways of various hemocyte types (AD, adipohemocyte; CO, coagulocyte; GR, granulocyte; PL, plasmatocyte; PR, prohemocyte; OE, oenocytoid; SP, spherulocyte). (From Gupta, 1979b.)

1.6. CONTINUOUS LINEAR INCREASE IN IMMUNOCYTE POPULATION DURING ONTOGENY

Given the fact that immunocytes are important components of the immune system, one would expect a gradual increase in their population as the animal progresses from the early immature stages toward the adult stage. Such an increase in fact occurs. Several authors have reported both qualitative and quantitative changes in hemocyte populations in varius insects (Gupta, 1979c). During our studies of the effects of insect growth regulators on hemocyte populations (see Section 1.7.1) in various developmental stages of the German cockroach, *Blattella germanica*, we found a progressively linear increase not only in the total hemocyte counts (Fig. 1.6A) in the controls but also in the differential hemocyte counts (DHCs) of the two immunocytes (Fig. 1.6B,C, Hazarika and Gupta, 1986), suggesting a strong correlation between the THCs and the DHCs of the PLs and GRs. These two immunocytes represent more than 70%–80% of the THCs in all the developmental stages, thus determining the quantitative picture of the hemocyte populations in any of the seven developmental stages of this roach. The continuous and linear increases of PLs and GRs seem to be commensurate with the growing demand for cellular immunity as the roach progresses toward the adult stage. Because the effectiveness of phagocytosis, encapsulation, and other related defense mechanisms depends largely on the available circulating immunocytes (PLs and GRs, Gupta, 1985a), it is reasonable to suggest that the larger the PL and GR populations, the stronger the cellular defense (see Section 1.8.1). By implication, the early immature instars would be immunologically less competent than the adult and thus more susceptible to pathogen attack.

1.7. EFFECTS OF INSECT GROWTH REGULATORS ON IMMUNOCYTES

1.7.1. Depression of Immunocyte Populations

For more than a decade, investigations in my laboratory on the effects of insect growth regulators (IGRs) have shown many morphogenetic and behavioral abnormalities in IGR-treated insects. Our most recent studies of the effects of injected doses of juvenile hormone I (JH I) on the THCs and immunocyte DHCs in *B. germanica* adultoids (adults emerging from sixth-instar nymphs injected with 0.33 μl of JH I) showed that the PLs decreased significantly in the adultoids (Table 1.2); surprisingly, there was no significant decrease in the GRs compared with those in the untreated adults and a more than 40% increase in the coagulocytes; the latter increase accounts for the excessive melanization of IGR-treated adultoids (Das and Gupta, 1977) because GRs and COs are thought to contain both the phenols and the phenoloxidase necessary for melanization (see Section 1.8.2). Because

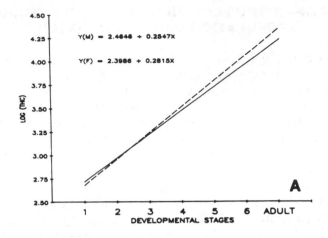

$Y(M) = 2.4646 + 0.2547X$

$Y(F) = 2.3986 + 0.2815X$

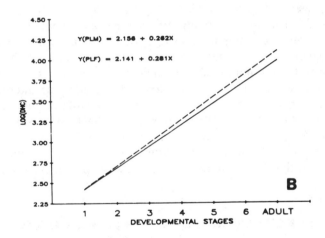

$Y(PLM) = 2.156 + 0.262X$

$Y(PLF) = 2.141 + 0.281X$

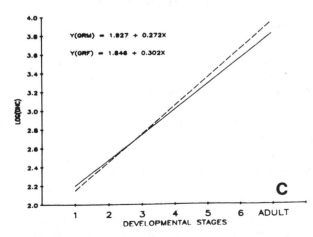

$Y(GRM) = 1.927 + 0.272X$

$Y(GRF) = 1.846 + 0.302X$

TABLE 1.2. HEMOCYTE COUNTS OF JH I–TREATED ADULTOIDS AND UNTREATED ADULTS (CONTROL) OF *BLATTELLA GERMANICA* (L.)

Type	Sex	Mean (SE)[a]		Percent Increase or Decrease
		Control	Treated	
THC	F	23352 (1831.6)	16360** (1549)	− 29.9
	M	19236 (1499.9)	14983* (906.1)	− 22.1
CO	F	414.7 (76)	781.2 (333.2)	+ 46.9
	M	431.1 (51.1)	712.7 (344.1)	+ 39.5
GR	F	8050.4 (688.6)	7382.8 (349.6)	− 8.2
	M	8314.3 (584.1)	7982.5 (11.2)	− 3.9
OE	F	362 (91.1)	287.6 (88.9)	− 20.6
	M	357.7 (137.7)	287.6 (88.9)	− 19.5
PL	F	13383.6 (1908.5)	4453.2** (613)	− 66.7
	M	9188.6 (733.7)	3759.2** (473.4)	− 59.1
PR	F	1344.9 (156.7)	2525.6** (195.6)	+ 48.0
	M	1014.8 (364.7)	2515** (297.1)	+ 59.7

Key: CO, coagulocyte, GR, granulocyte; OE, oenocytoid; PL, plasmatocyte; PR, prohemocyte; SE, standard error; THC, total hemocyte count. (From Hazarika and Gupta, 1986)
[a] Means with *'s are significantly different between treatment groups; * = 0.05, ** = 0.01 (Kruskal-Wallis Test); no comparisons should be made between M (male) and F (female).

we observed widespread degeneration (probably as a result of the interaction of the exogenous JH I and endogenous ecdysteroids) in the PLs in general in the adultoids (which might account for the significant decrease in these immunocytes), the conclusion is inescapable that the degeneration would result in severe induced immune deficiency in the adultoids. This study clearly shows the important role of arthropod growth regulators (AGRs) in

Figure 1.6. Linear increases in the total hemocyte counts and differential counts of plasmatocytes and granulocytes during various developmental stages of the German cockroach, *Blattella germanica* (1–6, six nymphal stages). (A) Total hemocyte counts of male (Y (M), ———; $r = 0.96$) and female (Y (F), – – –; $r = 0.95$). (B) Differential plasmatocyte counts of male (Y (PLM), ———; $r = 0.95$) and female (Y (PLF), – – –; $r = 0.96$). (C) Differential granulocyte counts of male (Y (GRM), ———; $r = 0.96$) and female (Y (GRF), – – –; $r = 0.96$).

the regulation of the immune system of arthropods and hence their potential as pest control agents by inducing immune deficiency.

1.7.2. Effect on Immunocyte Membrane Receptors and Gap Junctions

Virtually nothing is known of the effects of AGRs on the immunocyte plasma membrane, its receptors, and gap junctions. The only account available is that by Johnson and coworkers (1974), who reported that ecdysterone (when injected at the rate of 10 μg/g body weight) caused an increase in the formation of gap junctions in the midgut cells in *Limulus polyphemus*. Obviously, the effects of IGRs and/or AGRs on the ultrastructure of immunocyte membranes, its receptors, and gap junctions need to be investigated.

1.8. IMMUNOLOGIC FUNCTIONS OF GRs AND PLs

Most of the following account has been adopted from my previous reviews (Gupta, 1979a, 1985a). The hemocytic immunity in arthropods is accomplished by phagocytosis, encapsulation, nodule formation, secretion of immunologic factors, coagulation, and poison detoxification mechanisms; these processes are triggered in nature as defense reactions against foreign biologic agents (immunogens) and toxic substances. Artificially injected inanimate objects also induce some of these reactions.

1.8.1. Phagocytosis

Metchnikoff (1892) was probably the first to recognize the existence of phagocytosis in invertebrates. Among the extant arthropods, *Limulus* is unique in that it possesses only the most primitive and functionally versatile GR, which performs phagocytosis in this animal. Apparently, in some Crustacea the GRs are also phagocytic (see Chapter 2); in other higher arthropods, however, phagocytosis is generally performed by PLs. Oenocytoids, spherulocytes, and adipohemocytes also have been occasionally reported as phagocytes, but it is likely that some of these hemocytes were incorrectly identified. In arthropods in which PLs, SPs, and ADs are present as distinct types, it appears that the phagocytic function has shifted from GRs to PLs (see Section 1.10.3). It is also likely that some of these hemocytes were incorrectly identified.

Phagocytosis is accomplished in three stages: (1) recognition of the foreign body, or immunogen, (2) its ingestion, and (3) its final disposal, or clearance, from the body (Bang, 1975; Ratcliffe and Rowley, 1979). For details of these stages, see Gupta (1985a) and Chapters 2–4 and 16.

Opsonins are known to facilitate phagocytosis; however, the nature of these molecules remains controversial (see Section 1.8.4.1). Brewer and

Vinson (1971) suggested that tyrosine-containing proteins or polyphenols may be involved in opsonization of alien surfaces. According to Rizki and Rizki (1982), in in vitro phagocytosis of bacteria by *Drosophila melanogaster* hemocytes, opsonins are not required. Dularay and Lackie (1985) reported that phenoloxidases do not act as opsonins. Apparently, the prophenoloxidase-activating system (PAS) in arthropods produces a number of sticky proteins that might function as opsonins (see Chapter 9); the phenoloxidase itself, although sticky, does not function as an opsonin.

The effectiveness of phagocytosis depends on, among other factors, the phagocytic index (number of phagocytizing immunocytes), the nature of the microorganisms (immunogens), and the frequency of attack (Gupta, 1985a). It has been reported that when the number of phagocytizing cells is high, ingestion and hence removal of the microorganism from the hemolymph is hastened (Wheeler, 1962; Wittig, 1966; Leutenegger, 1967). The availability of circulating hemocytes is important for the effectiveness of phagocytosis, for it is known that insects become susceptible to infection if the activity of the phagocytizing cells is blocked (Bettini et al., 1951; Stairs, 1964). However, Hoffmann and colleagues (1979) reported that in *Locusta migratoria*, circulating hemocytes do not play a major role in taking up bacteria. Wago and Ichikawa (1979) reported that the rate of phagocytosis of injected goose erythrocytes in *Bombyx mori* larvae was highest in the fifth larval instar, probably because of the increased number of circulating hemocytes in this instar. Unfortunately, little or no precise information regarding the phagocytic rate at any given time in arthropods is available. Recently, Steinkam and associates (1982), using a new technique, measured the percentage of phagocytizing cells in rats and suggested that this technique could be used in many cell systems. Apparently not all microorganisms invading arthropods are equally affected by the phagocytizing cells. Phagocytosis appears to be ineffective against the bacteria *Bacillus thuringiensis* and *B. cereus*, microsporidia, *Nosema* sp., some viruses, and certain fungi (*Sorosporella uvella* and *Coelomomyces* sp., see Arnold, 1974). By contrast, a single hemocyte can ingest a number of particles (Harshbarger and Heimpel, 1968). The β-1,3-glucans (the carbohydrate moieties on the cell walls of fungi) enhance the rate of phagocytosis of bacteria in vitro by causing degranulation of the GRs, which in turn causes stimulation of the PLs (see Chapter 9).

The role of the so-called phagocytic organs or tissues in the effectiveness of phagocytosis (and perhaps also in encapsulation) deserves mention. To the extent that these organs in many insects are also hemopoietic in nature and thus produce hemocytes (e.g., PLs, GRs, and ADs in *L. migratoria*, Hoffmann et al., 1968a, 1979), they surely play a role in phagocytosis, a fact that was first observed by Cuénot (1895, 1896, 1897). He reported (1896) that the phagocytic tissue of the orthopteroid insects takes up bacteria. However, according to Hoffmann and colleagues (1979), it is the reticular cells of the hemopoietic organ in *L. migratoria*, not the circulating hemocytes, that engulf the bacteria. Furthermore, according to these authors, these reticular cells can be induced to produce large amounts of proteinaceous antibacterial substance by the injection of immunizing doses of pathogens. These authors have suggested the possibility that, as in vertebrates, some of the reticular cells may be differentiated into immunocompetent cells and that it is these cells that synthesize the antibacterial factor, as evidenced by marked dilations of the cisternae of the RER and the presence of crystalloid inclusions in them.

1.8.2. Encapsulation

Hemocytic encapsulation is a mechanism that isolates the foreign microorganism if it is too large to be phagocytized. Encapsulation is evoked by both biologic and nonbiologic (inert) materials. As in phagocytosis, the primary hemocyte types involved are the two immunocytes, although SPs, COs, and OEs, or the so-called thrombocytoids in Diptera (Zachary et al., 1975), also have been reported to take part in encapsulation.

Encapsulation is accomplished in two stages: recognition of the foreign body by the GRs, and capsule formation and melanization by the GRs and PLs. In most cases of encapsulation, generally the GRs first establish contact with the target cell (TC) and then lyse (Lackie, 1976; Levin, 1979; Ratcliffe and Rowley, 1979; unlike the vertebrate cytotoxic T-lymphocyte [CTL]–TC binding, in which the TC lyses [Berke, 1983]) and release the so-called recognition factors into the hemolymph (see also Section 1.10.3).

The discharge of material from the GRs (or COs of other authors) during the initial phase of encapsulation apparently occurs (Poinar et al., 1968; Crossley, 1975; Rizki and Rizki, 1976; Ratcliffe and Gagen, 1976, 1977; Schmit and Ratcliffe, 1977). Ratcliffe and Rowley (1979) stated that during encapsulation in *Galleria mellonella*, sticky acid mucopolysaccharidelike substances "appear to be discharged onto the foreign surfaces by GRs, . . . and, independently of or in combination with material(s) from the foreign bodies, apparently specifically attracts PLs to the alien surface." Parish (1977) hypothesized that 5-glycosyltransferases secreted by hemocytes form subunits of recognition factors and polymerize, after being secreted into the hemolymph, into hexamers that in turn react with the foreign body. Ratner and Vinson (1983) suggested that the so-called encapsulation factors originate in hemocytes in *Heliothis virescens*. For details, see the review by Gupta (1985a) and Chapters 2, 4, and 6.

Cellular capsules vary in terms of the structural changes in the PLs that form the capsule and the number of cell layers (regions) in the completed capsule. After the release of the recognition factor or factors from the lysing GRs (or COs), PLs (GRs in some primitive arthropods) are attracted to the foreign body and form an inner layer around it. In many insects, this layer shows somewhat flattened and necrotic PLs. As more PLs adhere to the developing capsule, they undergo more intense flattening but less necrosis and form the so-called middle layer of the capsule. Apparently, both the flattening and necrosis of the PLs cease after the middle layer has attained a certain thickness until finally the outermost layer consists of only loosely attached, normal PLs. The degree of flattening of PLs and their layered arrangements vary among insects. Apart from the flattening of the PLs, several other changes in them have been observed. For example, PLs surrounding the foreign body may (I.S. Misko, 1972, cited by Ratcliffe and Rowley, 1979) or may not (Poinar et al., 1968; I.S. Misko, 1972, cited by Ratcliffe and Rowley, 1979) form a syncytium. The formation of membrane-bound, electron-dense vesicles in the PLs of the inner layer has been reported in many insects; the vesicles seem to contain melanin, which causes melanization in the capsules of many insects. The PLs of the inner layer of the G. mellonella capsule contain cytolysosomes and a dense intercellular substance, which in Ephestia kühniella is composed of acid mucopolysaccharide fused with a protein secreted by the PLs (L. Reik, 1968, cited by Ratcliffe and Rowley, 1979). The PLs of the middle and the outer layers contain numerous microtubules, mitochondria, and ribosomes (François, 1974; Brehélin et al., 1975). According to Baerwald (1979), these microtubules may be associated with desmosomes. In addition to the intracellular structural changes in the PLs, several kinds of intercellular cell junctions, such as desmosomes (local "weld" spots without channels for intercellular ionic communication) or gap junctions (with cell-to-cell communicating channels that allow molecules of low molecular weight, 300–500, to pass freely). The structural changes in the PLs after they contact the foreign body apparently do not occur in all insects, for in the larvae of D. melanogaster parasitized by the wasps Pseudeucoila bochei and P. mellipes, structural and behavioral changes occurred in the free hemocytes before capsule formation (Walker, 1959; Nappi and Streams, 1969).

According to Baerwald (1979), interspecific variation in the cellular makeup of capsules is to be expected owing to the variation in the free, circulating hemocytes in various species. Generally, three layers have been reported in the capsules of many insects. However, no layers are present in the capsule in the phasmids Carausius morosus and C. extradentatus (Ratcliffe and Rowley, 1979).

Melanization does not necessarily occur in all cellular capsules, however. Those around inanimate objects are always without melanization and so are some around animate objects. More often than not, however, melanization occurs during encapsulation and may even be essential for capsule formation in certain cases (Nappi, 1973). Melanization always occurs in humoral capsules, however. The question whether melanization originates in the animate nonself tissue or in the host's hemocytes is still debated, although it has been documented in many cases that the hemocytes provide both the phenolic substrates and phenoloxidizing enzymes, with the GRs and/or PLs (Ferron, 1978) supposedly providing the phenols and the OEs supplying the enzymes (Nappi and Stoffolano, 1971, 1972; Crossley, 1979). It should be noted, however, that the hemocytic origin of both the phenols and the phenoloxidases is controversial and that the GRs, COs, and OEs have all been reported to provide both (Neuwirth, 1973; Schmit et al., 1977). Mazzone (1976) reported that cultured gypsy moth cells produced melanin for long periods.

In addition to cellular capsules, capsules of nonhemocytic origin (humoral) have been reported (Schell, 1952; Burns, 1961; Coutourier, 1963; Götz, 1969, 1973; Götz and Vey, 1974; Götz et al., 1977). These humoral capsules are formed in insects that have few circulating hemocytes (e.g., Diptera) and thus apparently without the latter's participation. According to Ratcliffe and Rowley (1979), humoral capsules are formed only against living parasites, such as nematodes and fungi, and not against inanimate objects (for details on humoral encapsulation, see Chapter 15).

1.8.3. Nodule Formation

Nodule formation is similar to encapsulation in several respects and occurs in response to both animate and inanimate particulate material that cannot normally be phagocytized. Both encapsulation and nodule formation may occur within the same insect in response to the same parasite (Ratcliffe and Rowley, 1979), and it may be difficult to distinguish between the two defense reactions (Salt, 1970). Nodules, or granuloma, can be described as multicellular sheaths formed variously by the two immunocytes (Vey et al., 1968; Gagen and Ratcliffe, 1976; Ratcliffe and Gagen, 1976, 1977; see also Chapters 2 and 4).

Typically, a nodule is formed around a melanized core of coagulum produced by the degranulating GRs and/or COs and that entraps the particulate material and the GRs. As in encapsulation, the PLs are soon attracted to this core and form a sheath, indicating a chemotactic response (Ratcliffe and Rowley, 1979) to the recognition factors released by the lysing GRs (or COs). According to these authors, the nodules in *G. mellonella* and *Pieris brassicae*, in response to certain bacteria, contained the same three distinct layers as in encapsulation: the inner layer, composed of flattened PLs; the middle layer, also composed of flattened PLs; and the outer layer, made up of normal PLs. The melanization of the nodules seems to occur in the same way as in encapsulation.

1.8.4. Secretion of Immunologic Factors

It is well known that mammals possess serum proteins (γ-globulins) called immunoglobulins (Igs), or antibodies, that act as humoral recognition factors in removing foreign antigens. Although arthropods do not possess Igs, they do produce humoral recognition factors, called hemagglutinins, that are secreted by the immunocytes (see Section 1.8.4.1). In addition, hemolymph complement factors, antibacterial and antiviral factors, and lysozymes are believed to originate in the GRs.

1.8.4.1. Hemagglutinins. Since annelid hemocytes are capable of distinguishing between self, closely related nonself, and distantly related nonself (Cooper, 1969a,b), it is not surprising that those of their descendants, the arthropods, also possess this capability. This recognition is accomplished by the surface receptors on the plasma membranes of the immunocytes and the humoral recognition factors, variously called hemagglutinins, heteroagglutinins (structurally different agglutinins in the same organism), or agglutinins (lectins). These humoral factors may be likened to guided missles that are meant to deal with the invading foreign antigens. Thus, they are an important component of the humoral immune system.

They are called agglutinins because of their capacity to agglutinate vertebrate erythrocytes and certain microorganisms as a result of the presence of certain polysaccharides on the surface of these foreign cells. It should be noted, however, that other substances also agglutinate erythrocytes (Tsivion and Sharon, 1981; Ravindranath and Cooper, 1984).

Agglutinins are not unique to arthropods but are, rather, ubiquitous proteins or glycoproteins with multiple carbohydrate-binding sites; they are found widely in plants (first observed by Renkonen, 1948, and Boyd and Reguera, 1949), microorganisms (Goldstein et al., 1980), and other animals and are inhibited by carbohydrates and chelating agents. As a matter of fact, the carbohydrate-binding specificity of an agglutinin is defined by the chemical structure of its inhibitors, (e.g., mono- or oligosaccharide or glycoprotein), which reduce or obliterate its agglutinating property. Furthermore, the agglutination specificity and inhibition of agglutinins depend on both the erythrocyte type and agglutinin purity (see Chapters 8, 18, and 19).

Perhaps the most fully characterized and studied arthropod heteroagglutinins are those of the most primitive arthropod, *L. polyphemus* (see Chapters 8, 11, 12, and 19). These heteroagglutinins agglutinate vertebrate erythrocytes and have electrophoretic mobility similar to that of the γ- and β-globulins (Cohen et al., 1965).

It should be noted that, according to Yeaton (1981), variation in lectin concentration unrelated to age and sex occurs among individuals of the same population in both mollusks and arthropods and that lectins may be absent temporarily. For example, the spider crab, *Maia squinado*, shows no lectin before molting (Bang, 1967), and in the coconut crab, *Birgus latro*, lectin is apparently absent in the young. However, lectin is present even in the egg of the silk moth *Hyalophora cecropia*. It appears three days before hatching and steadily increases in concentration until pupation and adult development (Yeaton, 1981).

Hemocytic Origin. According to Alberts and coworkers (1983), in vertebrates the first antibodies (known as *membrane-bound antibodies*) are made by the B-lymphocytes and, instead of being secreted into the blood, are inserted in the plasma membrane and serve as receptors. Each B-lymphocyte has about 10^5 such receptors on its plasma membrane. The *soluble* antibodies, with the same antigenic-binding sites as the membrane-bound antibodies, are produced by stimulated plasma cells (B-lymphocytes) and secreted into the blood.

In fact, the various receptors or antibodies on the plasma surfaces are produced by different *families* or *clones* or B- or T-lymphocytes, each consisting of B- or T-lymphocytes descended from a common stem cell. These receptors or antibodies are predestined to specifically bind to a specific foreign antigen or its determinants; the latter stimulate the production of antibodies by B-lymphocytes or reactions in T-lymphocytes.

During their early development, B- and T-lymphocytes with receptors for self tissue are suppressed or eliminated. This mechanism enables the immune system, which is capable of recognizing both self and nonself, to respond only to nonself or foreign antigens. However, if a nonself antigen is continuously present, the immune system also learns not to respond to it, in other words, it develops an *acquired immunologic tolerance*. When the tolerance to self tissue breaks down, the B- and T-lymphocytes or both destroy their own tissue (a reaction called *autoimmune reaction*).

The hemolytic plaque technique (Jerne et al., 1974) may be used to demonstrate that the B-lymphocytes (or other cells) produce antibodies.

From the standpoint of the origin of hemagglutinins in insects, the most interesting works are those of Amirante (1976), Valvassori and Amirante (1976), and Amirante and Mazzalai (1978), who demonstrated for the first time that in the cockroach *Leucophaea maderae* the hemagglutinins are present in the vacuoles and the plasma membranes of the GRs. Amirante (1976) also showed that in this cockroach the hemagglutinins are in the β-globulin range. It should be mentioned here that Marek (1970) reported an "immunoglobulin" in *G. mellonella* that is induced by cooling and is within the "range of vertebrate γ-globulin." That the arthropod hemagglutinins are synthesized by GRs (and possibly also by PLs) has been reported by several authors. For example, Stang-Voss (1971) reported that the granules of the so-called clotting cells in *Astacus* are the sites of hemagglutinin. Yeaton (1981, 1983) reported that in the American silk moth, *H. cecropia*, the GRs synthesize and contain hemagglutinin and release it into the culture medium (see also Chapter 20). In the lobster *Homarus americanus*, the agglutinins are also synthesized by hemocytes and released into the hemolymph (Cornick and Stewart, 1973, 1978). Furman and Pistole (1976) reported that *Limulus* plasma from which GRs were removed did not show any bactericidal action, but the serum (obtained after hemolymph clotting with GRs) did. Thus, in this arthropod, clotting is an important factor in defense (see Chapter 11).

Nelstrop and colleagues (1970) have cautioned that, although proteins with electrophoretic mobility comparable to that of vertebrate γ-globulin may be detected in insects (e.g., *Locusta*), they should not be characterized as immunoglobulins just on the basis of their electrophoretic mobility in the γ-globulin range. Considering the lower phylogenetic position of arthropods, one would not expect their hemagglutinins to be comparable to the γ-globulins of the most highly evolved vertebrates! It is not surprising, therefore, that attempts (Scott, 1971; Anderson et al., 1973) to demonstrate properties of hemagglutinins comparable to those of vertebrate immunoglobulins have been unsuccessful.

At any rate, one may draw several conclusions regarding hemagglutinins from the works of Amirante and associates and others:

1. Hemolymph hemagglutinins—at least in insects and crustaceans and probably in other arthropods as well—seem to be of hemocytic origin. Note that the mammalian immunoglobulins also are produced by plasma cells (B-lymphocytes, see also Section 1.10.1).
2. Because they have been reported to be present on the plasma membranes of GRs, hemagglutinins are thought to function as surface receptors in foreign body recognition, in addition to their opsonin role as humoral recognition factors (see "Functions," below). This proposal is supported by the fact that GRs are the first to be attracted to the foreign body during phagocytosis, encapsulation, and nodule formation.
3. The hemolymph-independent attachment of vertebrate erythrocytes to hemocytes (Ratcliffe and Rowley, 1979; Wago, 1981) is most likely due to these hemagglutinins, and insects, like vertebrates, do possess serum-independent surface recognition factors on their immunocytes.

Nonhemocytic Origin. Although the presence of agglutinins has been reported in several tissues other than hemocytes, we do not know whether these tissues synthesize them; the function of the agglutinins in these tissues is also unknown.

According to Hall and Rowlands (1974a,b), agglutinins (lectins) are present in several tissues in the lobster *H. americanus*, such as the lymphatic gland, pericardium, ovaries and testes, and hepatopancreas. Among insects, agglutinins have been reported in the alimentary canal of *Rhodnius prolixus* and *Glossina austeri* (Pereira et al., 1981; Ibrahim et al., 1984), the pupal wing of *P. brassicae* (Mauchamp, 1982), the coxal depressor muscle of the cockroach *P. americana* (Denburg, 1980), the fat body of *Sarcophaga peregrina* (Komano et al., 1983), and the peritrophic membrane of *Calliphora erythrocephala* (Peters et al., 1983, see also Chapter 20).

Agglutinating Property. Arthropod hemagglutinins generally possess multiple binding sites that are responsible for their binding specificity with certain polysaccharides present on the plasma membranes of microorganisms or vertebrate erythrocytes. This binding specificity is variable and inhibited by several mono- and oligosaccharides and glycoproteins. Ca^{2+} is necessary for carbohydrate binding (Marchalonis and Edelman, 1968; see also Chapter 8). According to Yeaton (1981), vertebrate erythrocytes contain only 7 (L-fucose, D-galactose, *N*-acetyl-D-galactosamine [NAgalNH], D-glucose, *N*-acetyl-D-glucosamine, D-mannose, and *N* acetyl [or glycolyl]-neuraminic acid) of the 100 known monosaccharides in nature.

According to Pistole and Rostam-Abadi (1979), Rostam-Abadi and Pistole (1979), and Pistole (1982), *limulin* reacts with bacterial lipopolysaccharides (LPS), and the specific site responsible for this reaction is 2-keto-3-deoxyoctonate (KDO), which is a key component of the LPS in gram-negative bacteria. According to O. Luderitz and associates (C. Galanos, V. Lehmann, M. Nurminen, E. T. Rosenfelder, M. Simon, O. Westjahal, 1973, cited by Pistole, 1982), the KDO directly links the polysaccharide side chain with the glucosamine backbone of lipid A. NANA and KDO are structurally similar (Pistole, 1982). The reaction of the lectin on the NANA residues on vertebrate erythrocytes is responsible for agglutinating the erythrocytes. *Limulus* serum has three recognized heteroagglutinins (Cohen et al., 1965).

Pistole (1982) has suggested two models to explain the dual binding of the lectin with the GR and the foreign microbe. According to the first model, the lectin molecule is not a part of the surface membrane of the GR but, rather, a humoral factor; and it has two sites, one binding with the GR and the other with the microbe. According to Pistole, this is comparable to the dual binding of certain mammalian immunoglobulins that bind to specific antigens via antigen-specific sites in the *Fab* region of the molecule and to the phagocyte via the immunoglobulin-specific sites in the *Fc* region. Alternatively, the second model postulates that the lectin is an integral part of the surface membrane of the GR (the so-called amoeboid cell) and thus the GR directly binds with the microbe in vivo. Thus, according to Pistole, the in vitro binding of the serum lectin to the pathogen is an artifact.

Note that this property of lectins to agglutinate cells has been found to vary with antigenic type, state of infection, stage of cell cycle, cell density, and contact during embryonic development (Nicolson, 1978). Rabinovitch and De Stefano (1970) have shown that the attractiveness of the sheep erythrocytes to phagocytizing cells is altered by various modifications of the surface charge by iron salt; altering the protein structure by tanning or formalin fixation stimulated increased phagocytosis.

Inhibition of Agglutinating Property. The binding capacity of arthropod agglutinins is inhibited by a wide variety of compounds, including NANA-containing carbohydrates, glycoproteins, and glycolipids. Of the various carbohydrate inhibitors, NAgalNH and NANA seem to be the most common. It seems that the effectiveness of certain inhibitors varies among arthropod groups. For example, according to Hapner and Stebbins (see Chapters 8 and 18), although agglutinins from Aquatic Chelicerata and Crustacea are inhibited by hexuronic acid and NANA, the latter is not a prominent inhibitor of insect agglutinins. The agglutinin of *Teleogryllus commodis* is an exception (Hapner and Jermyn, 1981) in that it "shows merostom-crustacean type of carbohydrate inhibition," and that of *B. mori* is inhibited by hexuronic acids and acidic oligosaccharides. Insect agglutinins are generally inhibited by galactosidic and glucosidic carbohydrates. Most recently, Stynen and associates (1985) reported that the most effective saccharide inhibitors for lectins in the adult female, *Sarcophaga bullata* were α-D(+)-melibiose (0.25 mM) and D-galactose (0.50 mM). On the whole, however, the NANA-binding characteristic of arthropod agglutinins seems to be widespread among various arthropod groups, a conclusion based on the fact that vertebrate erythrocytes, when treated with neuraminidase, are not agglutinated by NANA–binding agglutinins.

Functions. Lectins were originally used to characterize blood group specificities (Boyd, 1963) and are now commonly used in cell biology as structural probes to study cell surface and membrane structure and function, especially to determine the number of lectin receptors and their affinities. Fluorescent lectins (e.g., Con A) are often used to study localization of lectin receptors by light and electron microscopy.

Because of their capacity to bind to a variety of carbohydrates (e.g., D-galactose, KDO, glucuronic acid, *N*-acetylmuramic acid, and colominic acid) present on foreign cell membranes, agglutinins show multiple specificities and thus may function as a carbohydrate-based recognition system for nonself tissue. From what we know of these molecules, the following functions are or can be attributed to them.

OPSONIZATION OF NONSELF TISSUE. Agglutinins cause opsonization of the nonself tissue, or target cell, and thus act as antibodylike surveillance components of the immune system (Komano et al., 1980, 1981; Suzuki and Natori, 1983; Sharon, 1984; Stynen et al., 1985). These authors suggest that lectins act as opsonins. According to Weir (1980), lectins constitute the main portion of the serum opsonins in many invertebrates. Tyson and Jenkin (1973, 1974) reported that in crayfish, hemolymph hemagglutinins are responsible for opsonic activity. Apparently, the same occurs in *H. americanus* (Paterson and Stewart, 1974; Paterson et al., 1976). Tripp (1966) and Anderson and Good (1976) suggested that in mollusks hemagglutinins may be involved in opsonization. Rowley and Ratcliffe (1981) believe that in insects agglutinins function as opsonins, and they have found a nonagglutinating factor in the hemolymph of *P. americana* that causes uptake of *B. cereus* isolates and thus, according to them, functions as an opsonin (Rowley and Ratcliffe, 1980). Most recently, Stynen and colleagues (1985) suggested that *S. bullata* lectins act as opsonins. It is reasonable to assume that the hemagglutinin first opsonizes the foreign bodies, which are then recognized and phagocytized by PLs and/or GRs.

Apparently, not all agglutinins function as opsonins. Of the two hemagglutinins (HA-1 and HA-2) in the tunicate *Bottrylloides leachii*, HA-1 does not have opsonin function but HA-2 does (Coombe et al., 1982, see also Chapter 14).

ATTRACTION OF PLs AND OTHER GRs TO THE GR-TC COMPLEX. The role of agglutinins in attracting PLs and other GRs to the GR-TC complex is based on the assumption that the GR, upon lysis, releases granules that in fact represent hemagglutinins. According to Chain and Anderson (1983, Chapter 3 in this volume), following exposure to foreign bodies, GRs become sticky, and these hemocytes contain alcian blue–positive granules (absent in PLs) that are secreted in abundance following infection with *B. cereus*.

Apparently, the attraction of PLs to the GR-TC complex occurs because of local stickiness caused by the lysis of the GRs (Schmit and Ratcliffe, 1977) or local clotting of the plasma (Ratcliffe and Gagen, 1976) and release of the recognition factors into the hemolymph (Ratcliffe and Rowley, 1979). Note, however, that Lackie (1981) commented that because insects have open circulatory systems, a chemotactic gradient (as implied by recognition factors) exists only in the tissues. Thus, according to her, the question of the release of a chemotactic, chemokinetic, or sticky substance should be carefully examined. Note that Söderhäll and coworkers (1979, Söderhäll and Smith, 1983, Chapter 9 in this volume) reported that phenoloxidase also is sticky.

PROLIFERATION OF PLs AND POSSIBLY GRs. Because they cause proliferation of PLs and possibly GRs, agglutinins may be regarded as modulators of phagocytosis, encapsulation, and perhaps also nodule formation. This modulation is most likely accomplished by inducing mitosis. In fact, Majerus and Brodie (1972) reported that lectins induce mitosis and platelet release in vertebrates. If these compounds indeed cause the proliferation of the immunocytes (PLs and GRs), it should be possible to demonstrate the cessation of proliferation by inactivating the hemagglutinins on the surface of the immunocytes or those in the hemolymph. Such a demonstration could have important implications because the failure of PLs and GRs to release these compounds could result in immunologic tolerance of the foreign antigen. In vertebrates, the helper T-lymphocytes induce other white blood cells, called killer T-lymphocytes (cytotoxic T-lymphocytes) to destroy the foreign antigen. Widmer and Bach (1981) discovered that some killer T-lymphocytes can cause self-proliferation and, independently of the helper T-lymphocytes, destroy the foreign antigen. These cells are referred to as helper cell–independent cytotoxic T-lymphocytes (HIT cells, see also Section 1.10.2).

MEMBRANE SURFACE RECEPTORS. As stated earlier, in both vertebrates and arthropods the antibodies and hemagglutinins are produced by blood cells (B-lymphocytes and GRs, respectively) and act as membrane surface receptors. On the basis of erythrocyte-rosette formation, it is known that molluscan hemocytes have agglutinin receptors on their surfaces (Renwrantz and Chang, 1977a,b). In insects, Amirante and Mazzalai (1978) have shown that molecules that agglutinate rabbit erythrocytes are associated with the hemocyte membrane. In other words, these hemagglutinins act as antibodies on the surface of the immunocytes. It is this property of these compounds that enables us to pinpoint the location of surface receptors by labeling them with markers, such as hemocyanin, ferritin, or fluorescein. The latter has recently been used by Rizki and Rizki (1983) and Nappi and Silvers (1984) to demonstrate blood cell surface changes in two hemocyte populations in *Drosophila* mutants with melanotic tumors (see also Chapter 6).

If hemagglutinins are indeed the counterparts of vertebrate immunglobulins, it would be interesting to find out if they possess similar structures.

In addition to hemagglutinins, several types of chemical surface receptors have been reported in insects, primarily on the basis of the affinity of hemocytes for attaching to vertebrate erythrocytes. For example, Scott (1971) reported that in *P. americana*, the attachment of the hemocytes to mammalian erythrocytes is due to trypsin-labile surface receptors. Anderson (1976a,b) suggested the presence of nonspecific surface receptors on the hemocytes of *Spodoptera eridania* and *Estigmene acrea*. Ratcliffe and Rowley (1979) reported that in *P. brassicae*, the SPs (specialized GRs in this species) have a layer of sticky acid mucopolysaccharide that probably is responsible for binding sheep erythrocytes and the bacteria *Escherichia coli* and *Staphylococcus aureus* to these hemocytes. It should be noted, however, that Wago and Ichikawa (1979) reported that in *B. mori* larva, the initial reaction of the GRs to injected goose erythrocytes was due to random contact and not to any chemotaxis.

ROLE IN METAMORPHOSIS (HISTOGENESIS) AND WOUND HEALING. Komano and coworkers (1980, 1981), Suzuki and Natori (1983), and Stynen and coworkers (1985) suggest that lectins act as opsonins not only against foreign microorganisms but also against disintegrating pieces of tissue in pupae and that they also mediate cell-to-cell communication during wound healing.

POSSIBLE ROLE AS VITELLOGENINS. Stynen and De Loof (1982) reported that in the adult Colorado potato beetle, *Leptinotarsa decemlineata*, hemagglutinins were identical to vitellogenins. It is interesting to note that these authors found that egg homogenates of this beetle showed considerable hemagglutinin activity. However, Stynen and colleagues (1985) reported that the vitellogenin of *S. bullata* is not involved in hemagglutination.

OTHER FUNCTIONS. In addition to the functions described above, lectins may mimic hormone action (Nicolson, 1978) and inhibit cell growth, cell movement, and phagocytosis (Berlin, 1972), and delayed hypersensitivity, allograft rejection, and fertilization (Nicolson, 1978, see also Chapter 8).

Finally, according to Pistole (1982), it is possible that lectins play no role in defense, and because they bind with sugars, they may be involved in sugar transport. According to Hankins and Shannon (1978), lectins have enzymatic activity.

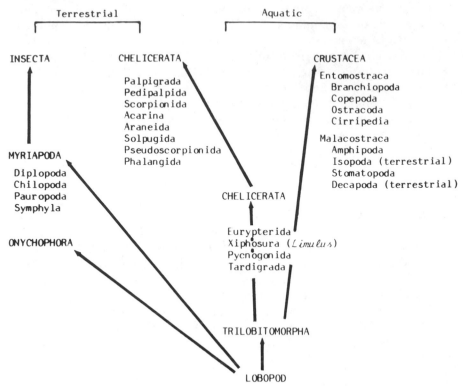

Figure 1.7. Monophyletic arrangement of major arthropod groups. (Modified from Gupta, 1979a.)

Evolution in Various Arthropod Groups. Let us examine lectins in an evolutionary context in an attempt to gain some understanding of the evolution of the hemagglutinin repertoire in various groups of arthropods.

Figure 1.7 shows a monophyletic arrangement of the major aquatic and terrestrial arthropod groups. It is generally acknowledged that arthropods originated from an annelidlike lobopod ancestor (see Gupta, 1979a, 1983). Note that, according to this scheme, the lobopod ancestor gave rise to both the Myriapoda and the Trilobitomorpha; the latter were the progenitors of both the Crustacea and the Aquatic Chelicerata. Exactly when the Aquatic Chelicerata and the ancestors of Myriapoda colonized land is controversial and shrouded in mystery (Gupta, 1979b,c). According to Størmer (1977), the arthropod invasion of land occurred during late Silurian and Devonian times, some 400 million years ago. It is also believed that aquatic arthropods colonized land independently several times. All controversies nothwithstanding, land colonization brought about important anatomic and physiologic changes, some of which may have had important bearing on the evolution of the hemagglutinin repertoire and its specificities.

Lectins have been reported in some annelid species—*Chaetopterus variopedatus*, *Lumbricus terrestris*, and *Sabellastarte magnifica* (Yeaton, 1981)—so it is not surprising that arthropods that evolved from annelidlike ancestors also possess lectins. However, lectins have been reported in only few major arthropod groups so far (see Chapters 8 and 18). Note also that even in the most primitive arthropods, the Aquatic Chelicerata, including *Limulus*, *Tachypleus*, and *Carcinoscorpius*, three or more agglutinins have been reported, indicating the existence of a hemagglutinin repertoire at the very beginning of the arthropod evolution. Further detailed studies of the hemagglutinins in many more species of the major terrestrial arthropod groups should reveal more extensive hemagglutinin repertoires.

Vasta and Cohen (1984a–d, 1985a,b) have studied lectins in several terrestrial chelicerates (scorpions) and found that lectins in these arthropods bind to NANA or sialoconjugates on the erythrocyte, and according to them (Vasta and Cohen, 1984a), Chelicerata "is probably the only group of high taxonomic rank that exhibits certain homogeneity in the specificity of their humoral lectins." They found that after their NANA cleavage by neuraminidase (from *Vibrio cholerae*), the erythrocytes are no longer agglutinated. Crustacean lectins also bind to sialic acid, and the same might be true in other arthropods.

According to Vasta and Cohen (1984b), human ABH blood groups were not discriminated by the three chelicerate sera (two scorpions and *Limulus*). However, Jurenka and associates (1982) reported that grasshopper (*Melanoplus*) sera agglutinate ABO types. This finding may be indicative of further evolution of multiple specificity among higher, and perhaps in terrestrial, arthropods.

1.8.4.2. Secretion of Complement Factors. According to Alberts and colleagues (1983), the vertebrate complement system consists of a complex of soluble proteins (MW 24,000–400,000), most of which are inactive unless stimulated by an immune response caused by a foreign antigen. There are about 20 interacting components in the complement system, including components designated C1–C9, factors B and D, and a variety of other proteins. Components C1–C4 are called *early complement components* and C5–C9 *late complement components*. The latter form a large complex that participates in cell lysis by making the plasma membrane leaky and thus causing it to swell and burst. The aggregation of C5–C9 is brought about by a cascade of proteolytic activation reactions that sequentially cleave the early components (most of which are proenzymes) to generate a series of proteolytic enzymes, each of which cleaves the next proenzyme. The reactions occur on the plasma membrane, and C3 is the most important component of the proteolytic cascade. An antigen-antibody complex is generally necessary to activate the sequential reaction of the complement complex, although cobra venom factor (CVF) and bacterial endotoxin are also known to activate it (Anderson et al., 1972).

It is also interesting to note that arthropods possess at least some components of the vertebrate complement system necessary for the operation of the many immunologic reactions, such as antigen-antibody reactions and agglutination of cells. Day and coworkers (1970) demonstrated the presence of a proactivator of the vertebrate complement component 3 (C3) in *L. polyphemus* and several other invertebrates; apparently a similar factor is present in *Blaberus craniifer* (Anderson et al., 1972) and *Galleria* (Aston et al., 1976). Because *Limulus* has only one of the two arthropod immunocytes, GRs (Gupta, 1979a), it is reasonable to suggest that this complement factor is probably secreted by the GRs in this animal and probably in higher arthropods as well. Because immunoglobulins activate the complement system, it is likely that the hemagglutinins in *Limulus* activate the C3 proactivator in this arthropod. Note, however, that the hemocytic origin of both the hemagglutinins and the C3 proactivator in *Limulus* has not yet been confirmed.

D'Cruz (1983) reported three peptide components of the larval hemolymph of *B. mori* that reacted with human complement-mediated lysis of rabbit erythrocytes (RaRBC). One of these hemolymph complement activators (HCA, MW < 1400) enhanced the lytic activity of the human serum, while another of the three peptides prevented this lysis, the activity of the HCA being modulated by a major hemolymph protein (MHP, which D'Cruz called LSP, MW 500,000). The MHP is induced during pupation and on injury or immunization. The HCA peptide exists both in bound and free form in the larval stages, and complete release occurs in vivo during onset of pupation and in vitro by heat treatment (at 56°C for 10 min). More important, the HCA acts as a specific opsonin for protein A–positive *S. aureus*. D'Cruz and coworkers (1984) isolated an MHP (which they called LSP, MW 580,000–600,000) with two subunits (MW 90,000 and 84,000) from the larval hemolymph of *Spodoptera*, which enhanced fourfold the lysis rate of sheep erythrocytes by human complement (C). According to them, "the interaction of *Spodoptera* LSP with mammalian C implicates a new function for the highly conserved, developmentally regulated, neuroendocrine-dependent storage protein of insects."

1.8.4.3. Antibacterial and Antiviral Factors.

Various antibacterial and antiviral toxins have been reported in arthropods, but the direct role of immunocytes in the secretion of these toxins is unexplored. Hoffmann and colleagues (1979) speculated that some of the reticular cells of the hemopoietic organs in *L. migratoria* may differentiate into immunocompetent cells and synthesize the antibacterial factor.

According to Götz and Boman (1985), a group of induced proteins called cecropins have broad antibacterial activity; these authors have reported that both cecropins and lysozymes are necessary for complete destruction of the bacterial cell wall (for details, see Chapters 10 and 11).

1.8.4.4. Lysozymes. Lysozymes are mucolytic enzymes located in the ly-sosomes, which are acid phosphatase–positive organelles found in hemo-cytes, pericardial cells (Crossley, 1979), and ecdysial (prothoracic) glands (Gupta, 1985a). It is very likely that hemolymph lysozymes originate from these three sources, probably from the immunocytes. In fact, Chain and Anderson (1983) found that both immunocytes are rich in lysosomes, and according to Zachary and Hoffmann (1984), lysozymes are stored in GRs and/or COs. Lysozymes, along with hemagglutinins and other complementlike substances in the hemolymph, play important roles in arthropod immunity. According to Houk and Griffiths (1980), hemolymph lysozymes may be im-portant in eliminating any symbionts that escape from the mycetocyts. Moh-rig and Messner (1968), Chadwick (1970), and Powning and Davidson (1973) considered lysozymes as possible antibacterial factors in several insects. Many studies have demonstrated that during phagocytosis, the lysosome fuses with the phagosome that contains the ingested microorganism (Rat-cliffe and Rowley, 1979). According to Crossley (1975), lysozymes are read-ily identifiable in the hemolymph of injured *Galleria* and rise from 25–500 $\mu g/ml^{-1}$ hemolymph to 9000 $\mu g/ml^{-1}$ in insects stressed by injection of gram-positive bacteria. Lysozymes are also found in *C. erythrocephala* (Crossley, 1972) and *Locusta* (Hoffmann et al., 1977, see also Chapters 10 and 17).

Apparently, the lytic effect of lysozymes is due to their capacity to cleave the β-1,4-glycosidic bond between *N*-acetylmuramic acid and *N*-acetylglu-cosamine of the murein in the cell wall of intracellular symbionts. For more information on lysozymes, see Chapter 17 and Dunn (1986).

1.8.5. Coagulation

In general, hemolymph coagulation in arthropods is caused by (1) hemocyte agglutination (in *Calliphora*, Åkesson, 1975; Crossley, 1975), (2) hemolymph gelation (in *Locusta*; Brehélin, 1972; in *Periplaneta*; Franke, 1960a,b; Yeager and Knight, 1933), or (3) a combination of both (Gupta, 1979a, 1985a). Al-though most often the granules of GRs are reported to induce coagulation, the role of GRs in hemocytic or humoral immunity appears to be restricted to forming a localized coagulum around the target cell. Durliat (1985) pro-vides the most recent review of the clotting processes in Xiphosura and Crustacea.

The coagulation function of the GR in *Limulus* is well documented. Du-mont and associates (1966) studied the clotting process in this arthropod by electron microscopy and described the ultrastructural changes in normal and clotted GRs (i.e., after clotting sets in). According to them, in this animal the hemolymph by itselt is not capable of coagulation, and it is the granules of the GR that provide the material that participates actively in the formation of the clot. It should be noted that Howell (1885), Loeb (1904, 1910), and Maluf (1939) also suggested much earlier that the material that actually forms the clot is contained within the hemocyte and not in the hemolymph. These observations have been supported by Murer and coworkers (1975), who claim that the granules of GRs (which they call amoebocytes) contain all the factors, including the clottable protein (coagulogen), and that these factors are released when the granules rupture during cell aggregation. The GRs in *Limulus* also aggregate in response to injuries (Kenny et al., 1972). White (1976) compared the microtubules within the granules of the GR in this animal to the "polymer" (a term that appears to refer to the microtubules) of sickled human erythrocytes. According to Durliat (1985), in the horseshoe crab, the cellular coagulogen (MW 20,000) is converted into an insoluble coagulin by a serine protease enzyme (the clotting enzyme) located in the GRs. The *L. polyphemus* coagulogen contains 220 amino acids with a high half-cystine content of 18 residues (Liu et al., 1979). The clotting enzyme (a serine protease enzyme) is remarkably similar to the enzyme that activates the prophenoloxidase for conversion into phenoloxidase (Liu et al., 1979; Harada et al., 1979, see also Chapter 9). The clotting enzyme exists in two active forms with MWs of 78,000 and 40,000, respectively and very similar amino acid composition (Liu et al., 1979). The clotting enzyme originates from an inactive proclotting enzyme that can be activated either by the lipopolysaccharide moiety of the endotoxin of gram-positive bacteria or by β-1,3-glucans from the cell walls of certain fungi and algae (Durliat, 1985). Endotoxin causes degranulation of the GRs in *Limulus* (Armstrong and Rickles, 1982), resulting in the release of their contents into the hemolymph.

The role of the crustacean GR (or its differentiated form, the CO) in coagulation has been directly reported or indirectly interpreted by several authors (see reviews by Gupta, 1979a, 1985a). Durliat and Vranckx (1976) reported three soluble proteins in the hemocytes of *Astacus*, one of which they referred to as fibrinogen, which is the origin of the clottable hemolymph protein. If this is true, it indicates that in Crustacea also the clottable protein (coagulogen) originates in the hemocytes, as in *Limulus* (Murer et al., 1975; Durliat, 1985). Apparently, a multienzyme system, including both hemocytic and humoral factor, is involved in coagulation in Crustacea (Durliat, 1985).

The literature on coagulation in Terrestrial Chelicerata is meager; coagulation by COs (differentiated forms of GRs) has been reported by Deevey (1941) and Ravindranath (1974).

In Myriapoda, the presence of COs (in addition to GRs) has been reported by Gupta (1968) and Ravindranath (1973). Apparently, coagulation in this group is brought about by COs, not GRs, an indication of a shift of function.

Grégoire (1955) described the coagulation patterns in the Onychophora, and since COs are not reported in *Peripatus* and other onychophoran species, GRs might be responsible for causing coagulation in this group of animals.

Among insects, coagulation has been extensively studied. For a review of COs and coagulation, see Grégoire and Goffinet (1979). It is generally accepted that in insects, COs cause coagulation. However, Rowley and Ratcliffe (1976) have demonstrated that in *G. mellonella*, hemolymph coagulation is initiated by granules from GRs. This finding revives the controversy over whether GRs and COs are in reality two separate categories in insects as well as in other higher arthropods.

1.8.6. Poison Detoxification Mechanisms

Virtually nothing is known about the direct role of immunocytes in the detoxification of poisons (Gupta, 1985a). However, an indirect role of hemocytes in detoxification mechanisms via esterases is possible (Patton, 1961). Patton suggested that nonspecific esterases in hemocytes might be involved in detoxifying parathion. Whitten (1968) also suggested an active role for hemocytes in detoxification of contact insecticides that penetrate the tarsal cuticle in the fly, *S. bullata*. Yeager and coworkers (1942) have also reported that hemocytes detoxify poisons. And we have reported that exposure to sublethal doses of the insecticide chlordane caused increases in the total and differential hemocyte counts of PLs, GRs, COs, and SPs in *P. americana* (Gupta and Sutherland, 1968). Yeager and Munson (1941), however, demonstrated some pathologic effects on PLs and COs in *Prodenia eridania* larvae after feeding them turnip leaf–cornstarch "sandwiches" with sodium fluoride and mercuric chloride.

1.9. ROLE OF IMMUNOCYTES IN THE PROPHENOLOXIDASE-ACTIVATING SYSTEM

Let us briefly examine the role of the prophenoloxidase-activating system in the recognition of foreign antigens. It is well known that parasites that invade insects and other arthropods are melanized by the host's polyphenoloxidase (tyrosinase or phenolase); this enzyme also plays a role in wound healing. Phenoloxidase is produced as a result of the activation of its precursor proenzyme, prophenoloxidase, by a serine protease enzyme. The PAS has been extensively studied in *B. mori* (Ashida and Ohnishi, 1967; Ashida and Dohke, 1980) and in the crayfish *Astacus astacus* (Söderhäll, 1981; Ashida and Söderhäll, 1984; see also Chapter 9). Taylor (1969) apparently was the first to suggest not only that the activation of phenoloxidase produces "disinfectants" in the form of quinones but also that this enzyme is a prerequisite for hemocytic adhesion to nonself tissue, suggesting regulation of the immunologic behavior of immunocytes by the phenoloxidase system. The quinones eventually produce melanin as a result of a series of nonenzymatic reactions.

1.9.1. Site of PAS

Unestam and Nylund (1972) reported that prophenoloxidase is stored in the granules of the GRs of the crayfish *A. astacus*. Later, Söderhäll and Smith (1983) also concluded that phenoloxidase is present only in the GRs of the four decapod species that they studied. According to these authors, in the crayfishes *A. astacus* and *Pacifastacus tenuiculus*, the PAS contains opsoninlike factors and can lyse crayfish hemocytes and produce fungitoxic compounds. In crustaceans, phagocytosis, encapsulation, and microbial killing are influenced by the PAS present within the hemocytes (Söderhäll, 1981, 1982; Söderhäll and Ajaxon, 1982; Söderhäll and Smith, 1983; see also Chapter 9). We have reported (Gupta and Sutherland, 1967) that SPs do not show this activity, and I have recently confirmed that the OEs of *G. portentosa* also do not have phenoloxidase (Gupta, 1985b). Phenoloxidase extracted from *A. astacus* hemocytes attaches to the fungal wall, with Ca^{2+} being necessary (Söderhäll et al., 1979) for attachment, as is the case for agglutinins (see Section 1.8.4.1).

1.9.2. Activation of PAS

The PAS is reported to be activated by the carbohydrate moieties (e.g., β-1,3-glucans) of the plasma membranes of bacteria and fungi and bacterial endotoxin or lamarin. Apparently, these compounds activate the prophenoloxidase via a proteolytic activator, serine protease, in the hemolymph (Ashida and Söderhäll, 1984), which itself needs to be activated by another proteolytic factor or factors. The end result of this proteolytic action is the production of a sticky substance or substances, with Ca^{2+} being necessary for the completion of the reaction. Serine protease is present in the hemocyte lysate (70,000 g supernatant of homogenized hemolymph) in an inactive form, and its inhibitors completely prevent activation of the proenzyme prophenoloxidase (Söderhäll, 1981). Prophenoloxidase may also be activated by other agents (for details, see Chapter 9).

1.10. FUNCTIONAL ANALOGIES BETWEEN IMMUNOCYTES AND VERTEBRATE B- AND T-LYMPHOCYTES

1.10.1. Morphology of B- and T-Lymphocytes

According to Alberts and coworkers (1983), both B- and T-lymphocytes are usually small, with the nucleus filling the cell; and both cells, when activated by antigens, proliferate and differentiate. More important, the T- and B-lymphocytes show morphologic differences only after they are stimulated by an antigen, the resting or unstimulated T- and B-lymphocytes appearing very similar except that the latter have extensive rough endoplasmic reticulum, while the former have little RER.

Some of the glycoproteins present on the plasma membranes of these lymphocytes can serve as cell surface antigenic markers to distinguish and separate the lymphocytes from each other. For example, in mice, Thy-1 glycoprotein is present on the T-lymphocyte plasma membrane but absent

TABLE 1.3. MARKERS OF HUMAN PERIPHERAL BLOOD LYMPHOCYTES

Markers	B Cells	T Cells
Surface Ig	+	−
Human B-lymphocyte antigen (HBLA)	+	−
Sheep erythrocyte receptors (E rosette)	−	+
Human T-lymphocyte antigen (HTLA)	−	+
Fc receptors (aggregated Igs or EA rosette)	+	−
Complement receptors (EAC rosette)	+	−

Source: Modified from Kersey and Gajl-Peczalska, 1975.

on that of B-lymphocyte. Thus, by producing Thy-1 antibodies, one can remove or purify T-lymphocytes from a mixed population of mouse lymphocytes (Alberts et al., 1983). T-lymphocytes are generally recognized by the presence on their plasma membranes of receptors for spontaneous binding of sheep erythrocytes (Brunning and Parkin, 1980). Table 1.3 lists the markers on human peripheral B- and T-lymphocytes (Kersey and Gajl-Peczalsak, 1975).

1.10.2. Major Functions of Vertebrate B- and T-Lymphocytes

It is both possible and useful to establish functional analogies between the arthropod immunocytes and the vertebrate B- and T-lymphocytes on the basis of their known functions.

The two primary B-lymphocyte functions are production of immunoglobulins or antibodies upon stimulation by an antigen, and recognition of nonself tissue.

T-lymphocytes do not secrete antibodies (Alberts et al., 1983). When stimulated by antigens, they divide and differentiate into three subsets of effector cells, distinguishable from one another by the cell surface antigens they express. The three subsets of T-lymphocytes each perform different functions:

1. Helper-inducer T-lymphocytes cause the proliferation and/or activation of B-lymphocytes; cytotoxic, or killer, T-lymphocytes; and macrophages (by secreting such factors as lymphokines and interleukins) and promote immunoglobulin production following recognition of nonself tissue.

2. Suppressor T-lymphocytes suppress the immune response by inactivating the B-lymphocytes and the cytotoxic T-lymphocytes and thus maintain a balance between the two cells. The helper and suppressor lymphocytes are called regulatory T-lymphocytes because they act as regulators of the immune system.

3. Cytotoxic T-lymphocytes, also called effector cells, attack nonself tissue when activated.

These three subsets of T-lymphocytes modulate the cellular immune response in mammals.

In addition, transplantation and graft reactions are mediated by T-lymphocytes, probably by both helper and cytotoxic subsets (Alberts et al., 1983). These reactions occur against foreign versions of cell surface antigens called *transplantation or histocompatibility antigens*, the most important of which is a group encoded by a complex of genes called the *major histocompatibility complex* (MHC). T-lymphocytes have special affinity for the antigens of the MHC, which are present on the plasma membranes of higher vertebrates (Table 1.3). Two classes of MHC antigens are known: H-2 antigens in mice and human-leucocyte–associated antigens (HLA) in human beings. The genes for the two complexes are located on chromosomes 17 and 6 of mouse and human beings, respectively. Furthermore, the HLA complex consists of class I and class II antigens, each representing a set of cell-surface glycoproteins. Since the glycoproteins of class I are found on virtually all nucleated cells (Alberts et al., 1983), it is possible that they exist on the membranes of the nucleated arthropod immunocytes as well. According to these authors, the class I glycoproteins are encoded by at least three separate gene loci in both mice and human beings, each locus encoding a single polypeptide chain (MW 45,000, with about 345 amino acid residues), which is inserted in the plasma membrane. Each glycoprotein molecule consists of a short cytoplasmic COOH-terminal, a middle hydrophobic segment traversing the lipid bilayer, and a large extracellular NH_2 terminal segment, which represents about 80% of the total mass and is folded into three separate domains. The class I glycoproteins constitute about 1% of the plasma membrane protein. Alberts and colleagues state that "the MHC glycoproteins help to channel each class of antigen into the appropriate immune response pathway," thus functioning "as antigen recognition molecules distinguishing between classes of antigens."

The three T-lymphocyte subsets can be categorized by monoclonal antibodies. For example, the helper-inducer T cells can be labeled with OKT4 and the suppressor and cytotoxic T cells (and their precursors) with OKT8 (Valdiserri, 1984). More important, it is now known that the ratio of helper to suppressor cells (H/S ratio), as defined by the surface antigens OKT4 and OKT8, is characteristically altered in some human diseases. For example, the H/S ratio increases in patients with active multiple sclerosis and decreases in autoimmune hemolytic anemia, autoimmune thyroid disease, and acquired immune deficiency syndrome (AIDS, Valdiserri, 1984). According to Valdiserri, this H/S ratio is not specific for a particular disease, such as AIDS, and may be altered by other factors.

Both B and T cells are capable of recognizing nonself tissue, as, of course, are the arthropod immunocytes, GRs and PLs.

1.10.3. The Primitive Immunocyte

On the basis of the known functions of the two arthropod immunocytes and comparisons with the functions of the B- and T-lymphocytes, it can be hypothesized that the primitive GR, as is found in the msot primitive arthropod, *Limulus*, performs the functions of both the B- and T-lymphocytes (Fig. 1.8A) as well as of the macro- and microphages (polymorphonuclear leucocytes).

Thus, the primitive arthropod GR can be regarded as producing and releasing into the hemolymph at least five factors that are functionally analogous to those of B- and T-lymphocytes (Fig. 1.8A):

- Factor I represents the hemagglutinins and/or opsonins, which aid in the recognition of nonself tissue. This role is comparable to the two major functions of the B-lymphocytes (see Section 1.10.2).
- Factor II, referred to as the recognition or encapsulation factor in insects, causes proliferation and/or activation of PLs or GRs, as occurs in phagocytosis, encapsulation, and nodule formation. These functions are comparable to those of the helper-inducer T-lymphocytes (see Section 1.10.2).
- Factor III, so far unidentified in arthropods, maintains the balance between factors I and II by suppressing their overactivity. This function is comparable to that of the suppressor T-lymphocytes.
- Factor IV, so far unidentified in arthropods (PAS may be a likely candidate)) activates PLs and/or GRs to attack nonself tissue. This function is comparable to that of the cytotoxic T-lymphocytes.
- Factor V, so far unidentified in arthropods, regulates the levels of hemolymph hemagglutinin and/or opsonins, the occurrence of coagulation and melanin synthesis; the activation of polyphenoloxidase, complement factors, and proteolytic enzymes; and other humoral activities.

According to Söderhäll and Smith (1983), recognition and phagocytosis of bacteria by the hemocytes of the decapod *Carcinus maenas* is mediated by the phenoloxidase-activating system, residing exclusively in the GRs.

1.10.4. The Differentiated PL and Its Functions

It is generally recognized that other types of hemocytes evolved or differentiated from the primitive GR (Gupta, 1985a). Now, what changes occurred in the functions of the primitive GR when it gave rise to other hemocyte types in higher arthropods? Let us consider two rather simple alternatives.

First, let us suppose that when the PL evolved and became a distinct type, it acquired all the functions of the T-lymphocyte (Fig. 1.8B), while the GR retained the functions of the B-lymphocyte (Fig. 1.8C, see Section

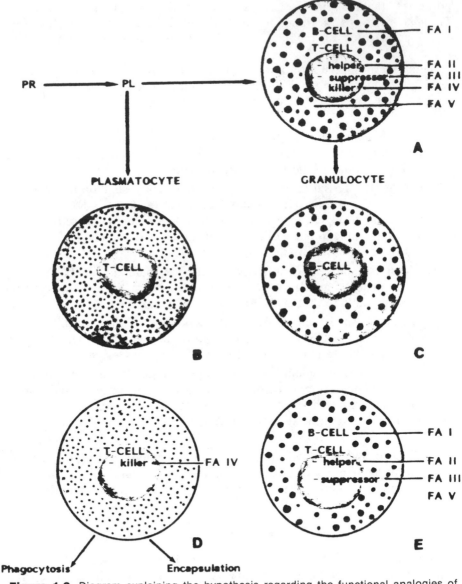

Figure 1.8. Diagram explaining the hypothesis regarding the functional analogies of arthropod immunocytes and those of vertebrate B- and T-lymphocytes. (A) The most primitive granulocyte (e.g., in *Limulus*) that is functionally analogous to both B- and T-lymphocytes, as well as to the macro- and microphages, and produces at least five (FA I–V) factors. At this evolutionary stage, the plasmatocyte is evanescent and has not differentiated into a distinct hemocyte type. (B,C) The first alternative of the hypothesis, according to which the plasmatocyte (B) evolved into a distinct type and acquired all the functions

1.10.2). However, this alternative is inconsistent with the known functions of both GRs and PLs. From what we know of the functions of PLs in arthropods, it seems that they did not acquire all three functions (helper, suppressor, and killer) of the T-lymphocytes, as this alternative would suggest. For example, PLs are not known to cause proliferation of GRs; instead, the latter are known to cause proliferation of PLs upon lysis. Furthermore, PLs, rather than GRs, are predominantly involved in phagocytosis and encapsulation in higher arthropods.

Alternatively, it seems more reasonable to suppose that only the killer, or cytotoxic, function was shifted to the PL (Fig. 1.8D) and that all the functions of the B-lymphocyte and the helper and suppressor functions of the T cell were retained by the GR (Fig. 1.8E). Thus, functionally the GR is a hybrid of B- and T-lymphocytes as well as of macro- and microphages.

According to the second alternative, the GRs in most arthropods perform the two important functions of the vertebrate B-lymphocytes: the production of antibodies or immunoglobulins (arthropod hemagglutinins) and recognition of nonself tissue. In addition, they perform the helper and suppressor functions of the T-lymphocytes. For example, we know (Gupta, 1985a) that following recognition of a nonself, or target, cell and after forming a complex with it (GR-TC), the GRs lyse and release into the hemolymph the so-called recognition factors that cause the proliferation of PLs and attract them to the GR-TC complex (comparable helper T cell function of GR) to encapsulate the nonself tissue (comparable killer T-cell function of PL). And the suppressor function of keeping the production of hemagglutinins and proliferation of PLs in balance by suppressing their overactivity would be necessary to maintain the effectiveness of the arthropod immune system.

Thus, the PLs (Fig. 1.8D) perform only the killer function of the cytotoxic T-lymphocyte and of the helper cell–independent T cells by phagocytosis, thus acting as sort of double-duty cells. According to Lackie (1981), PLs are "capable of recognizing and phagocytosing foreign particles without the help of granulocytes." In addition, they also act as macrophages (Anderson, 1976a,b). When they become abnormal (as do killer T cells in some diseases,

---

of the vertebrate T-lymphocyte, with the granulocyte (C) retaining the functions of the B-lymphocyte. (D,E) The second alternative of the hypothesis, according to which only the killer function of the primitive granulocyte (A) was shifted to the plasmatocyte (D) and all the functions of the B-lymphocyte and the helper and suppressor functions of the T-lymphocyte were retained by the granulocyte (E); thus, functionally, the granulocyte is a hybrid of the B- and T-lymphocytes and of the macro- and microphages. (FA I, functions of nonself recognition and hemagglutinin synthesis and release; FA II, function of causing proliferation and/or activation of plasmatocytes and/or granulocytes; FA III, function of maintaining balance between FA I and FA II by suppressing their overactivities; FA IV, function of activating plasmatocytes and/or granulocytes to attach to nonself tissue; FA V, functions of regulating the titer of hemolymph hemagglutinins, complement factors, and opsonins; coagulation; and activation of the prophenoloxidase system, melanin production, and proteolytic enzymes, etc.)

e.g., rheumatoid arthritis), they turn against their self tissue, as in case of the mutant malignant PLs in *Drosophila* (Gateff, 1984). According to Gateff, mutant genes cause malignant transformation of larval PLs, which then fail to recognize self tissue and thus invade it.

Thus, in summary,

1. To some degree arthropod GRs and PLs are functionally comparable to vertebrate (mammalian) B and T cells. GRs are a functional hybrid of both B- and T-lymphocytes.
2. In arthropods (e.g., *Limulus*) that possess only GRs, these cells normally perform the functions of both B- and T-lymphocytes.
3. In arthropods that possess both GRs and PLs, only the killer function of the T-lymphocytes has been shifted to the PLs, and the GRs retain the functions of the B-lymphocytes and the helper and suppressor functions of the T-lymphocytes.
4. Under certain conditions (e.g., in response to a native tumor), the PLs assume the functions of helper cell–independent cytotoxic T-lymphocytes, as reported by Widmer and Bach (1981) in mice. The PLs also perform the functions of vertebrate macrophages.

In a preliminary test, we found that a hemocyte sample from the cockroach *G. portentosa*, when treated with the monoclonal antibody OKT 11, specific for the peripheral human T-lymphocytes, showed 2%–10% positivity of the hemocytes. One important reason for the low positivity could be the fact that short homologues of the amino acid sequences of the human T-lymphocyte antibodies are present in the surface receptors of the arthropod immunocytes. This proposal is not farfetched if we accept the fact that immunoglobulins of higher vertebrates and mammals evolved as modifications of a common ancestral protein. If this is true, the amino acid homology would suggest at the molecular level the evolutionary antiquity of the origin of the defense mechanism in all animals. In fact, Alberts and colleagues (1983), pointing out the homology between domains of immunoglobulins, have suggested that in evolution a primordial gene coded for a single 110 amino acid domain of unknown function, and it is from this chain that the immunoglobulin chains of higher animals evolved by a series of gene duplications. Recently, Shepherd and associates (1984) showed that the homeotic genes of the bithorax complex (BX-C), which determine the identity of the thoracic and abdominal segments, and those of the antennapedia complex (ANT-C) in *D. melanogaster* share brief sequences of 60 amino acids (of the "homeo domain") with those of a frog, *Xenopus laevis*; and, according to them, this domain may also be present in earthworm, beetle, chicken, mouse, and human genomes. Vasta and coworkers (1984) have reported similar sharing of short stretches of amino acid sequences between immunoglobulin and nonimmunoglobulin molecules.

1.11. SUMMARY

This chapter provides an overview of the components of the immune system and their functions in arthropods. Details of the two major arthropod immunocytes—plasmatocytes (PLs) and granulocytes (GRs)—and their functional analogies to those of vertebrate B- and T-lymphocytes have been emphasized in the hope of prompting a thorough search of the functions of these immunocytes comparable in precision and detail to studies of B- and T-lymphocytes. The arthropod immune system recognizes nonself tissue by the combined actions of the two immunocytes, several humoral factors (hemagglutinins, lysozymes, complement factors, antibacterial factors, and antiviral factors), and the polyphenoloxidase system. Immunocyte population increases linearly throughout development, reaching its peak in the adult. Insect growth regulators depress the immunocyte population and are hence potential pest control agents. The major functions of the immunocytes are phagocytosis, encapsulation, nodule formation, secretion of immunologic factors, coagulation, and poison detoxification. In the most primitive arthropods (Xiphosura) and in some Crustacea, GRs perform phagocytosis; in higher arthropods, this function has been shifted to the PLs. Both immunocytes participate in encapsulation and nodule formation. Apparently, in most arthropods, the GRs synthesize hemagglutinins, complement factors, and other humoral factors and release them into the hemolymph. GRs are also involved in coagulation and the prophenoloxidase-activating system. Unfortunately, virtually nothing is known about the ultrastructure and molecular composition of the plasma membrane of the two immunocytes, especially of the intramembranous particles present on one or both of its characteristic and complementary fracture faces, or of the gap junctions. The same is true of the detoxification mechanisms.

On the basis of the known functions of the two immunocytes, it has been suggested that

1. To some degree, arthropod GRs and PLs are functionally comparable to vertebrate (mammalian) B- and T-lymphocytes, and GRs are a functional hybrid of both B- and T-lymphocytes.
2. In arthropods (e.g., *Limulus*) that possess only GRs, these cells normally perform the functions of both B- and T-lympohcytes.
3. In arthropods that possess both GRs and PLs, only the killer function of the T-lymphocytes has been shifted to the PLs, and the GRs retain the functions of the B-lymphocytes and the helper and suppressor functions of the T-lymphocytes.
4. Under certain conditions (e.g., in response to a native tumor), PLs assume the functions of helper cell–independent cytotoxic T-lymphocytes. PLs also perform the functions of vertebrate macrophages.

Furthermore, it has been suggested that short homologues of the amino acid sequences of vertebrate T-lymphocyte antibodies are present in the surface receptors (antibodies) of arthropod immunocytes.

ACKNOWLEDGMENTS

I thank Elias Cohen and Thomas Pistole for reading portions of the initial draft of this manuscript. This is New Jersey Agricultural Experiment Station Publication No. F-08112-O1-86, supported by state funds and U.S. Hatch Act funds.

REFERENCES

Akai, H. 1969. Ultrastructure of haemocytes observed on the fat-body cells in *Philosamia* during metamorphosis. *Jpn. J. Appl. Entomol. Zool.* **13:** 17–21.

Akai, H. and S. Sato. 1973. Ultrastructure of the larval hemocytes of the silkworm, *Bombyx mori* L. (Lepidoptera: Bombycidae). *Int. J. Insect Morphol. Embryol.* **2**(3): 207–231.

Åkesson, B. 1976. Observations on the haemocytes during the metamorphosis of *Calliphora erythrocephala* (Meig.) *Ark. Zool.* **6**(12): 203–211.

Alberts, B., D. Bray, J. Lewis, M. Raff, K. Roberts, and J.D. Watson. 1983. *Molecular Biology of the Cell.* Garland Publishing, New York.

Amirante, G.A. 1976. Production of heteroagglutinins in haemocytes of *Leucophaea maderae*. *Experientia* **32:** 526–528.

Amirante, G.A. and F.G. Mazzalai, 1978. Synthesis and localization of haemagglutinins in haemocytes of the cockroach *Leucophaea maderae*. *Dev. Comp. Immunol.* **2:** 735–740.

Anderson, R.S. 1976a. Expression of receptors by insect macrophages, pp. 27–34. In R.W. Wright and E.L. Cooper (eds.) *Physiology of Thymus and Bone Marrow-Bursa Cells.* Elsevier North Holland, Amsterdam.

Anderson, R.S. 1976b. Macrophage functions in insects, pp. 215–219. In T.A. Angus, P. Faulkner, and A. Rosenfield (eds.) *Proc. 1st International Colloqium of Invertebrate Pathology.* Queens University, Kingston, Canada.

Anderson, R.S., N.K.B. Day, and R.A. Good. 1972. Specific hemagglutinin and a modulator of complement in cockroach hemolymph. *Infect. Immun.* **5:** 55–59.

Anderson, R.S. and R.A. Good. 1976. Opsonic involvement in phagocytosis by mollusk hemocytes. *J. Invertebr. Pathol.* **27:** 57–64.

Anderson, R.S., B. Holmes, and R.A. Good. 1973. *In vitro* bactericidal capacity of *Blaberus craniifer* hemocytes. *J. Invertebr. Pathol.* **22**(1): 127–135.

Armstrong, P.B. and F.F. Rickles. 1982. Endotoxin induced degranulation of the *Limulus* amoebocyte. *Exp. Cell Res.* **140:** 15–24.

Arnold, J.W. 1974. The hemocytes of insects, pp. 201–254. In M. Rockstein (ed.) *The Physiology of Insecta*, vol. 5, 2d ed., Academic Press, New York.

Arnold, J.W. 1979. Controversies about hemocyte types in insects, pp. 231–258. In A.P. Gupta (ed.) *Insect Hemocytes*. Cambridge University Press, Cambridge.

Ashida, M. and K. Dohke. 1980. Activation of prophenoloxidase by the activating enzyme of the silkworm, *Bombyx mori*. *Insect Biochem*. **10**: 37–47.

Ashida, M. and E. Ohnishi. 1967. Activation of prophenoloxidase in the hemolymph of the silkworm, *Bombyx mori*. *Arch. Biochem. Biophys*. **122**: 411–416.

Ashida, M. and K. Söderhäll. 1984. The prophenoloxidase activating system in crayfish. *Comp. Biochem. Physiol*. **77**B: 21–26.

Aston, W.P., J.S. Chadwick, and M.J. Henderson. 1976. Effects of cobra venom factor on the *in vivo* immune response in *Galleria mellonella* to *Pseudomonas aeruginosa*. *J. Invertebr. Pathol*. **27**: 171–176.

Baerwald, R.J. 1975. Inverted gap and other cell junctions in cockroach hemocyte capsules: A thin section and freeze-fracture study. *Tissue Cell* **7**(3): 575–585.

Baerwald, R.J. 1979. Fine structure of hemocyte membranes and intercellular junctions formed during hemocyte encapsulation, pp. 155–188. In A.P. Gupta (ed.) *Insect Hemocytes*. Cambridge University Press, Cambridge.

Baerwald, R.J. and G.M. Boush. 1970. Fine structure of the hemocytes of *Periplaneta americana* (Orthoptera: Blattidae) with particular reference to marginal bundles. *J. Ultrastruct. Res*. **31**: 151–161.

Bang, F.B. 1967. Serological responses among invertebrates other than insects. *Fed. Proc*. **26**: 1680–1684.

Bang, F.B. 1975. Phagocytosis in invertebrates, pp. 137–151. In K. Maramorosch and R. E. Shore (eds.) *Invertebrate Immunity*. Academic Press, New York.

Bauchau, A.G. 1981. Crustaceans, pp. 385–420. In N.A. Ratcliffe and A.F. Rowley (eds.) *Invertebrate Blood Cells*, vol. 2. Academic Press, New York.

Bauchau, A.C. and M.B. De Brouwer. 1972. Ultrastructure des hémocytes d'*Eriocheir sinensis*, Crustacé Décapode Brachyoures. *J. Microsc*. (Paris) **15**: 171–180.

Beaulaton, J. 1968. Étude ultrastructurale et cytochimique des glandes prothoraciques de vers à soie auz quatrième et cinquième âges larvaires. 1. La tunica propria et ses relations avec les fibres conjonctives et les hémocytes. *J. Ultrastruct. Res*. **23**: 474–498.

Berke, G. 1983. Cytotoxic T-lymphocytes: How do they function? *Immunol. Rev*. **72**: 5–42.

Berlin, R.D. 1972. Effect of Con A on phagocytosis. *Nature* (London) **235**:44–45.

Bettini, S., D.S. Sakaria, and R.L. Patton. 1951. Observations on the fate of vertebrate erythrocytes and hemoglobin injected into the blood of the American cockroach (*Periplaneta americana*). *Science* (Washington, DC) **113**: 9–10.

Boyd, W.C. 1963. The lectins: Their present status. *Vox Sang*. **8**: 1–32.

Boyd, W.C. and R.M. Reguera. 1949. Hemagglutinating substances in various plants. *J. Immunol*. **62**: 333–339.

Branton, D. 1966. Fracture faces of frozen membranes. *Proc. Nat. Acad. Sci. U.S.A.* **55**: 1048–1056.

Brehélin, M.M. 1972. Étude du mécnaniseme de la coagulation de l'hémolymph d'un acridien, *Locusta migratoria migratorioides* (R. and F.). *Acrida* **1**: 167–175.

Brehélin, M., J.A. Hoffmann, G. Matz, and A. Porte. 1975. Encapsulation of implanted foreign bodies by hemocytes in *Locusta migratoria* and *Melolontha melolontha*. *Cell Tissue Res.* **160**: 283–289.

Brewer, F.D. and S.B. Vinson. 1971. Chemicals affecting the encapsulation of foreign material in an insect. *J. Invertebr. Pathol.* **18**: 287–289.

Brümmer, F. and D.F. Hulser. 1982. Gap junctions in multicell spheroids. *Zeiss Inf.* **1982**(1): 34–38.

Brunning, R.D. and J.L. Parkin. 1980. The leukocyte: Morphogenesis and ultrastructure, pp. 145–176. In R.M. Schmidt (ed.) *CRC Handbook Series in Clinical Laboratory. Section I. Hematology*, vol. 2. CRC Press, Boca Raton, Fla.

Burns, W.C. 1961. Penetration and development of *Allassogonoporus vespertilionis* and *Acanthatrium oregonense* (Trematoda: Lecithodendriidae) cercariae in caddisfly larvae. *J. Parasitol.* **47**: 927–938.

Chadwick, J.S. 1970. Relation of lysozyme concentration to acquired immunity against *Pseudomonas areuginosa*, in *Galleria mellonella*. *J. Invertebr. Pathol.* **15**: 455–456.

Chain, B.M. and R.S. Anderson. 1983. Observations on the cytochemistry of the hemocytes of an insect, *Galleria mellonella*. *J. Histochem. Cytochem.* **31**: 601–607.

Cohen, E., A.W. Rose, and F.C. Wissler. 1965. Heteroagglutinins of the horseshoe crab *Limulus polyphemus*. *Life Sci.* **4**: 2009–2016.

Coombe, D.R., S.F. Schluter, P.L. Ey, and C.R. Jenkin. 1982. Identification of the HA-2 agglutinin in the haemolymph of the ascidian *Botrylloides leachii* as the factor promoting adhesion of sheep erythrocytes to mouse macrophages. *Dev. Comp. Immunol.* **6**(1): 65–74.

Cooper, E.L. 1969a. Chronic allograft rejection in *Lumbricus terrestris*. *J. Exp. Zool.* **171**:69–74.

Cooper, E.L. 1969b. Allograft rejection in *Eisenia foetida*. *Transplantation* **8**: 220–223.

Cornick, J.W. and J.E. Stewart, 1973. Partial characterization of a natural agglutinin in the hemolymph of the lobster *Homarus americanus*. *J. Invertebr. Pathol.* **21**: 255–262.

Cornick, J.W. and J.E. Stewart. 1978. Lobster (*Homarus americanus*) hemocytes: Classification, differential counts, and associated agglutinin activity. *J. Invertebr. Pathol.* **31**: 194–203.

Costin, N.M. 1975. Histochemical observations of the haemocytes of *Locusta migratoria*. *Histochem. J.* **7**: 21–43.

Coutourier, A. 1963. Recherches sur des Mermithidae, Nematodes parasites du Hanneton commun (*Melolontha melolontha* L. Coleop. Scarab.). *Ann. Epiphyt.* **14**: 203–267.

Crossley, A.C. 1972. The ultrastructure and function of pericardial cells and other nephrocytes in an insect: *Calliphora erythrocephala*. *Tissue Cell* **4**: 529–560.

Crossley, A.C.S. 1975. The cytophysiology of insect blood. *Adv. Insect Physiol.* **11**: 117–222.

Crossley, A.C.S. 1979. Biochemical and ultrastructural aspects of synthesis, storage, and secretion in hemocytes, pp. 423–473. In A.P. Gupta (ed.) *Insect Hemocytes.* Cambridge University Press, Cambridge.

Cuénot, L. 1895. Études physiologiques sur les crustacés décapodes. *Arch. Biol. Liège* **13:** 245–303.

Cuénot, L. 1896. Études physiologiques sur les Orthoptères. *Arch. Biol.* **14:** 293–341.

Cuénot, L. 1897. Les globules sanguins et les organes lymphoides des invertebrès: Revue critiques et nouvelles recherches. *Arch. Anat. Microsc.* **1:** 153–192.

Das, Y.T. and A.P. Gupta. 1977. Nature and precursors of juvenile hormone-induced excessive cuticular melanization in German cockroach. *Nature* (London) **268:** 139–140.

Day, N.K.B., H. Gewurz, R. Johannsen, J. Finstad, and R.A. Good. 1970. Complement and complement-like activity in lower vertebrates and invertebrates. *J. Exp. Med.* **132:** 941–950.

D'Cruz, O.J.M. 1983. Interaction of silkworm hemolymph components with human complement. *Fed. Proc.* **42:** 1237.

D'Cruz, O.J.M., N.K. Burton, and N.K. Day. 1984. A new role for the major larval serum protein of *Spodoptera* in modulating mammalian complement. *Fed. Proc.* **43:** 1764.

Deevey, G.B. 1941. The blood cells of the Haitian tarantula and their relation to the molting cycle. *J. Morphol.* **68:** 457–491.

Denburg, J.L. 1980. Cockroach muscle hemagglutinins-candidate recognition macromolecules. *Biochem. Biophys. Res. Commun.* **97:** 33–40.

Devauchelle, G. 1971. Étude ultrastructurale des hémocytes du Coléoptère *Melolontha melolontha* (L.). *J. Ultrastruct. Res.* **34:** 492–516.

Dularay, B. and A.M. Lackie, 1985. Haemocytic encapsulation and the phenoloxidase-activation pathway in the locust, *Schistocerca gregaria.* Forsk. *Insect Biochem.* **15**(6): 827–834.

Dumont, J.N., E. Anderson, and G. Winner. 1966. Some cytologic characteristics of the hemocytes of *Limulus* during clotting. *J. Morphol.* **119:** 181–208.

Dunn, P.E. 1986. Biochemical aspects of insect immunity. *Annu. Rev. Entomol.* **31:** 321–339.

Durliat, M. 1985. Clotting processes in Crustacea Decapoda. *Biol. Rev.* **60:** 473–498.

Durliat, M. and R. Vranckx. 1976. Analysis of the hemocyte proteins from *Astacus leptodactylus. C. R. Séances Acad. Sci.* (III) **282**(24)D: 2215–2218.

Fahrenbach, W.H. 1968. The cyanoblast: Hemocyanin formation in *Limulus polyphemus. J. Cell Biol.* **39:** 43a.

Fahrenbach, W.H. 1970. The cyanoblast: Hemocyanin formation in *Limulus polyphemus. J. Cell. Biol.* **44:** 445–453.

Ferron, P. 1978. Biological control of insect pests by entomogenous fungi. *Annu. Rev. Entomol.* **23:** 409–442.

Flower, N.E. 1971. Septate and gap junctions between the epithelial cells of an invertebrate, the mollusk *Cominella maculosa. J. Ultrastruct. Res.* **37:** 259–268.

Flower, N.E. 1972. A new junctional structure in the epithelia of insects of the order Dictyoptera. *J. Cell Sci.* **10:** 683–691.

François, J. 1974. Étude ultrastructurale des hémocytes du Thysanoure *Thermobia domestica* (Insecte, Aptérygote). *Pedobiologia* **14**: 157–162.

François, J. 1975. Hemocyte et organe hématopoiètique de *Thermobia domestica* (Packard) (Thysanura: Lepismatidae). *Int. J. Insect Morphol. Embryol.* **4**(6): 477–494.

Franke, H. 1960a. Licht- und elektronenmikroskopische Untersuchungen über die Blutgerinnung bei *Periplaneta orientalis. Zool. Jahrb. Abt. Allg. Zool. Physiol. Tiere* **68**: 499–518.

Franke, H. 1960b. Licht- und elektronenmikroskopische Untersuchungen über die Blutgerinnung bei *Periplaneta orientalis. Zool. Abt. Allg. Zool. Physiol. Tiere* **69**: 131–132.

Furman, R.M. and T.G. Pistole. 1976. Bactericidal activity of hemolymph from the horseshoe crab, *Limulus polyphemus. J. Invertebr. Pathol.* **28**: 239–244.

Gagen, S.J. and N.A. Ratcliffe. 1976. Studies on the *in vivo* cellular reactions and fate of injected bacteria in *Galleria mellonella* and *Pieris brassicae* larvae. *J. Invertebr. Pathol.* **28**(1): 17–22.

Gateff, E. 1984. Hematopoietic malignancies of genetic origin in *Drosophila melanogaster. 17th Int. Congr. Entomol.* **S18.1**(5): 760, abstract.

Giese, A.C. 1979. *Cell Physiology*, 5th ed., W.B. Saunders, Philadelphia.

Gilula, N.P. and P. Satir. 1971. Septate and gap junctions in molluscan gill epithelium. *J. Cell Biol.* **51**: 869–872.

Goldstein, I.J., R.C. Hughes, M. Monsigny, T. Osawa, and N. Sharon. 1980. What should be called a lectin? *Nature* (London) **285**: 66.

Götz, P. 1969. Die Einkapselung von Parasiten in der Hämolymphe von *Chironomus*-Larven (Diptera). *Zool. Anz. Verh. Zool. Ges.* **33**: 610–617, supplement.

Götz, P. 1973. Immunreaktionen bei Insekten. *Naturwiss. Rundsch.* **26**: 367–375.

Götz, P. and H.G. Boman. 1985. Cellular and humoral immunity in insects, pp. 453–485. In C.A. Kerkut and L.I. Gilbert (eds.) *Comprehensive Insect Physiology, Biochemistry and Pharmacology*, vol. 3. Pergamon Press, Oxford.

Götz, P. and A. Vey. 1974. Humoral encapsulation in Diptera (Insecta): Defence reaction of *Chironomus* larvae against fungi. *Parasitology* **68**: 193–205.

Götz, P., J. Roettgen and W. Lingg. 1977. Encapsulement humoral en tant que réaction de défense chez les Diptères. *Ann. Parasitol. Hum. Comp.* **52**: 95–97.

Grégoire, C. 1955. Blood coagulation in arthropods. 6. A study of phase contrast microscopy of blood reactions *in vitro* in Onychophora and in various groups of arthropods. *Arch. Biol.* **66**: 489–508.

Grégoire, C. and G. Goffinet. 1979. Controversies about the coagulocyte, pp. 189–229. In A.P. Gupta (ed.) *Insect Hemocytes*. Cambridge University Press, Cambridge.

Gupta, A.P. 1968. Hemocytes of *Scutigerella immaculata* and ancestry of Insecta. *Ann. Entomol. Soc. Am.* **61**(4): 1028–1029.

Gupta, A.P. 1979a. Arthropod hemocytes and phylogeny, pp. 669–735. In A.P. Gupta (ed.) *Arthropod Phylogeny*. Van Nostrand Reinhold, New York.

Gupta, A.P. 1979b. Hemocyte types: Their structures, synonymies, interrelationships, and taxonomic significance, pp. 85–127. In A.P. Gupta (ed.) *Insect Hemocytes*. Cambridge University Press, Cambridge.

Gupta, A.P. 1979c. Identification key for hemocyte types in hanging-drop preparations, pp. 527–529. In A.P. Gupta (ed.) *Insect Hemocytes*. Cambridge University Press, Cambridge.

Gupta, A.P. (editor). 1979d. *Insect Hemocytes*. Cambridge University Press, Cambridge.

Gupta, A.P. 1979e. Origin and affinities of Myriapoda, pp. 373–390. In M. Camatini (ed.) *Myriapod Biology*. Academic Press, New York.

Gupta, A.P. 1983. Neurohemal and neurohemal-endocrine organs and their evolution in arthropods, pp. 17–50. In A.P. Gupta (ed.) *Neurohemal Organs of Arthropods*. Charles C Thomas, Springfield, IL.

Gupta, A.P. 1985a. Cellular elements in the hemolymph, pp. 401–451. In G.A. Kerkut and L.I. Gilbert (eds.) *Comprehensive Insect Physiology, Biochemistry and Pharmacology*, vol. 3. Pergamon Press, Oxford.

Gupta, A.P. 1985b. The identity of the so-called crescent cell in the hemolymph of the cockroach, *Gromphadorhina portentosa* (Schaum) (Dictyoptera: Blaberidae). *Cytologia* **50:** 739–746.

Gupta, A.P. and D.J. Sutherland. 1966. *In vitro* transformations of the insect plasmatocyte in some insects. *J. Insect Physiol.* **12:** 1369–1375.

Gupta, A.P. and D.J. Sutherland. 1967. Phase contrast and histochemical studies of spherule cells in cockroaches. *Ann. Entomol. Soc. Am.* **60**(3): 557–565.

Gupta, A.P. and D.J. Sutherland. 1968. Effects of sublethal doses of chlordane on hemocytes and midgut epithelium of *Periplaneta americana*. *Ann. Entomol. Soc. Am.* **61**(4): 910–918.

Hagopian, M. 1971. Unique structures in the insect granular hemocytes. *J. Ultrastruct. Res.* **36:** 646–568.

Hall, J.L. and D.T. Rowlands, Jr. 1974a. Heterogeneity of lobster agglutinins. 1. Purification and physiochemical characterization. *Biochemistry* **13:** 821–827.

Hall, J.L. and D.T. Rowlands, Jr. 1974b. Heterogeneity of lobster agglutinins. 2. Specificity of agglutinin-erythrocyte binding. *Biochemistry* **13:** 828–832.

Hankins, C.N. and L.M. Shannon. 1978. The physical and enzymatic properties of a phytohemagglutinin from mung beans. *J. Biol. Chem.* **253:** 7791–7797.

Hapner, K.D. and M.A. Jermyn. 1981. Haemagglutinin activity in the haemolymph of *Teleogryllus commodus* (Walker). *Insect Biochem.* **11:** 287–295.

Harada, T., T. Morita, S. Iwanaga, S. Nakamura, and M. Niwa. 1979. A new chromogenic substrate method for assay of bacterial endotoxins using *Limulus* hemocyte lysate, pp. 209–220. In E. Cohen (ed.) *Biomedical Applications of the Horseshoe Crab (Limulidae)*. Alan R. Liss, New York.

Harshbarger, J.C. and A.M. Heimpel. 1968. Effects of zymosan on phagocytosis in the larvae of the greater wax moth, *Galleria mellonella*. *J. Invertebr. Pathol.* **10:** 176–179.

Hazarika, L.K. and A.P. Gupta. 1986. Variation in hemocyte populations during various developmental stages of *Blattella germanica* (L.) (Dictyoptera: Blattelildae). *Zool. Sci.* (in press).

Hoffmann, D., M. Brehélin, and J.A. Hoffmann. 1977. Premiers résultats sur les réactions de défense antibactériennes de larves et d'imagos de *Locusta migratoria*. *Ann. Parasitol. Hum. Comp.* **52:** 87–88.

Hoffmann, J.A., A. Porte, and P. Joly. 1968a. Présence d'un tissue hématopoïètique qu niveau du diaphragme dorsal de *Locusta migratoria* (Orthoptère). *C. R. Séances Acad. Sci.* (III) **266D**: 1882–1883.

Hoffmann, J.A., A. Porte, and P. Joly. 1970. On the localization of phenoloxidase activity in coagulation of *Locusta migratoria* (L.) (Orthoptera). *C. R. Séances Acad. Sci.* (III) **270D**: 629–631.

Hoffmann, J.A., M.E. Stoekel, A. Porte, and P. Joly. 1968b. Ultrastructure des hémocytes de *Locusta migratoria* (Orthoptère). *C. R. Séances Acad. Sci.* (III) **266D**: 503–505.

Hoffmann, J.A., D. Zachary, D. Hoffmann, and M. Brehélin. 1979. Postembryonic development and differentiation: Hemopoietic tissues and their functions in some insects, pp. 29–66. In A.P. Gupta (ed.) *Insect Hemocytes*, Cambridge University Press, Cambridge.

Houk, E.J. and G.W. Griffiths. 1980. Intracellular symbiotes of the Homoptera. *Annu. Rev. Entomol.* **25**: 161–187.

Howell, W.H. 1885. Observations upon the chemical composition and coagulation of the blood of *Limulus polyphemus, Callinectes hastatus,* and *Cucumaria* sp. *Johns Hopkins Univ. Cir.* **5**: 4–5.

Ibrahim, E.A.R., G.A. Ingram, and D.H. Molyneau. 1984. Haemagglutinins and parasite agglutinins in haemolymph and gut of *Glossina* Tropenmed. *Parasitology* **35**: 151–156.

Jerne, N.K., C. Henry, A.A. Nordin, H. Fuji, A.M. Koros, and I. Lefkovits. 1974. Plaque forming cells: Methodology and theory. *Transplant. Rev.* **18**: 130–191.

Johnson, G., D. Quick, R. Johnson, and W. Herman. 1974. Influence of hormones on gap junctions in horseshoe crabs. *J. Cell Biol.* **63**: 157a.

Jones, T.W. 1846. The blood corpuscle considered in its different phases of development in the animal series. *Philos. Trans. R. Soc. Lond.* (Biol.) **136**: 1–106.

Jurenka, R., K. Manfredi, and K.D. Hapner. 1982. Haemagglutinin activity in Acrididae (grasshoppers) haemolymph. *J. Insect Physiol.* **28**: 177–181.

Kaplan, J. 1981. Polypeptide-binding membrane receptors: Analysis and classification. *Science* (Washington, DC) **212**: 14–20.

Kenny, D.M., F.A. Belamarich, and D. Shepro. 1972. Aggregation of horseshoe crab (*Limuluss polyphemus*) amoebocytes and reversible inhibition of aggregation by EDTA. *Biol. Bull.* (Woods Hole) **143**: 548–567.

Kersey, J.H. and K.J. Gajl-Peczalska. 1975. T and B lymphocytes in humans: A review. *Am. J. Pathol.* **81**: 446–458.

Komano, H., D. Mizuno, and S. Natori. 1980. Purification of lectin induced in the hemolymph of *Sarcophaga peregrina* larvae on injury. *J. Biol. Chem.* **255**: 2919–2924.

Komano, H., D. Mizuno, and S. Natori. 1981. A possible mechanism of induction of insect lectin. *J. Biol. Chem.* **256**: 7087–7089.

Komano, H., R. Nozawa, D. Mizuno, and S. Natori. 1983. Measurement of *Sarcophaga peregrina* lectin under various physiological conditions by radioimmunoassay. *J. Biol. Chem.* **258**: 2143–2147.

Lackie, A.M. 1976. Evasion of the haemocytic defence reaction of insects by larvae of *Hymnolepis diminuta* (Cestoda). *Parasitology* **73**: 97–104.

Lackie, A.M. 1981. Immune recognition in insects. *Dev. Comp. Immunol.* **5:** 191–204.

Landureau, J.C. and P. Grellet. 1975. New permanent cell lines from cockroach hemocytes: Physiological and ultrastructural characteristics. *J. Insect Physiol.* **21:** 137–152.

Leutenegger, R. 1967. Early events of *Sericesthis* iridescent virus infection in hemocytes of *Galleria mellonella* (L.). *Virology* **32:** 109–116.

Levin, J. 1979. The reaction between bacterial endotoxin and amoebocyte lysate. *Prog. Clin. Biol. Res.* **29:** 131–146.

Liu, T.Y., R.C. Seid, J.P. Tai, S.M. Liang, T.P. Sakmar, and J.B. Robbins, 1979. Studies on *Limulus* lysate coagulating system, pp. 147–158. In E. Cohen (ed.) *Biomedical Applications of the Horseshoe Crab (Limulidae).* Alan R. Liss, New York.

Loeb, L. 1904. On the spontaneous agglutination of blood cells of arthropods. *Univ. Penn. Med. Bull.* **16:** 441–443.

Loeb, L. 1910. Über die Blutgerinnung bei Wirbellosen. *Biochem. Z.* **24:** 478–495.

Majerus, P.W. and G.N. Brodie. 1972. The binding of phytohemagglutinins to human platelet plasma membranes. *J. Biol. Chem.* **247:** 4253–4273.

Maluf, N.S.R. 1939. The blood of arthropods. *Q. Rev. Biol.* **14:** 149–191.

Marchalonis, J.J. and G.M. Edelman. 1968. Isolation and characterization of a hemagglutinin from *Limulus polyphemus. J. Mol. Biol.* **32:** 453–465.

Marek, M. 1970. Effect of stress of cooling on the synthesis of immunoglobulin in haemolymph of *Galleria mellonella* L. *Comp. Biochem. Physiol.* **35:** 615–622.

Mauchamp, B. 1982. Purification of an *N*-acetyl-D-glucosamine specific lectin (P.B.A.) from epidermal cell membranes of *Pieris brassicae* L. *Biochemie* **64:** 1001–1008.

Mazzone, H.M. 1976. Influence of polyphenol oxidase on hemocyte cultures of the gypsy moth, pp. 275–278. In E. Kurstak and K. Maramorosch (eds.) *Invertebrate Tissue Culture: Applications in Medicine, Biology and Agriculture.* Academic Press, New York.

Metchnikoff, E. 1892. *Leçons sur la Pathologie Comparée de l'Inflammations.* Masson et Cie, Paris. (Reprinted by Dover Publications, New York, 1968, as *Lectures on the Comparative Pathology of Inflammation.*)

Mohrig, W. and B. Messner. 1968. Immunreaktionen bei Insekten. I. Lysozym als grundlegender antibakterieller Faktor im humoralen Abwehrmechanismus der Insekten. *Sonderb. Biol. Zentralbl.* **87:** 439–470.

Moran, D.T. 1971. The fine structure of cockroach blood cells. *Tissue Cell* **3:** 413–422.

Murer, E.H., J. Levin, and R. Holme, 1975. Isolation and studies of the granules of the amoebocytes of *Limulus polyphemus*, the horseshoe crab. *J. Cell Physiol.* **86:** 533–542.

Nappi, A.J. 1973. The role of melanization in the immune reaction of larvae of *Drosophila algonquin* against *Pseudeucoila bochei. Parasitology* **66:** 23–32.

Nappi, A.J. and M. Silvers. 1984. Cell surface changes associated with cellular immune reactions in *Drosophila. Science* (Washington, DC) **225:** 1166–1168.

Nappi, A.J. and J.G. Stoffolano, Jr. 1971. *Heterotylenchus autumnalis*: Hemocytic reaction and capsule formation in the host, *Musca domestica*. *Exp. Parasitol.* **29:** 116–125.

Nappi, A.J. and J.G. Stoffolano, Jr. 1972. Haemocytic changes associated with the immune reaction of nematode-infected larvae of *Orthellia caesarion*. *Parasitology* **65:** 295–302.

Nappi, A.J. and F.A. Streams. 1969. Hemocytic reactions of *Drosophila melanogaster* to the parasites *Pseudeucoila mellipes* and *P. bochei. J. Insect Physiol.* **15:** 1551–1566.

Nelstrop, A.E., G. Taylor, and P. Collard. 1970. Comparative studies of serum and haemolymph proteins. *Comp. Biochem. Physiol.* **35:** 191–196.

Neuwirth, M. 1973. The structure of the hemocytes of *Galleria mellonella* (Lepidoptera). *J. Morphol.* **139**(1): 105–124.

Nicolson, G.L. 1978. Ultrastructural localization of lectin receptors, pp. 1–38. In J.K. Koehler (ed.) *Advanced Techniques, Biological Electron Microscopy* Vol. II. Springer-Verlag, Berlin.

Parish, C.R. 1977. Simple model for self–non-self discrimination in invertebrates. *Nature* (London) **267:** 711–713.

Paterson, W.D. and J.E. Stewart. 1974. *In vitro* phagocytosis by hemocytes of the American lobster (*Homarus americanus*). *J. Fish. Res. Bd. Can.* **31:** 1051–1056.

Paterson, W.D., J.E. Stewart, and B.M. Zwicker. 1976. Phagocytosis as a cellular immune response mechanism in the American lobster, *Homarus americanus. J. Invertebr. Pathol.* **27:** 95–104.

Patton, R.L. 1961. The detoxification function of insect hemocytes. *Ann. Entomol. Soc. Am.* **54:** 696–698.

Peracchia, C. 1974. Excitable membrane ultrastructure. 1. Freeze-fracture of crayfish axons. *J. Cell Biol.* **61:** 107–122.

Pereira, M.E.A., F.B. Andradea, and T.M.C. Ribeiro. 1981. Lectins of distinct specificity in *Rhodnius prolixus* interact selectively with *Trypanosoma cruzi. Science* (Washington, DC) **211:** 597–600.

Peters, W., H. Kolb, and V. Kolb-Bachofen. 1983. Evidence for a sugar receptor (lectin) in the peritrophic membrane of the blowfly larva, *Calliphora erythrocephala* Mg. (Diptera). *J. Insect Physiol.* **29:** 275–280.

Pinto da Silva, P., S.D. Douglas, and D. Branton. 1971. Localization of A antigen sites on human erythrocyte ghosts. *Nature* (London) **232:** 194–196.

Pistole, T.G. 1982. *Limulus* lectin: Analogues of vertebrate immunoglobulins, pp. 283–288. In *Physiology and Biology of Horseshoe Crabs: Studies on Normal and Environmentally Stressed Animals*. Alan R. Liss, New York.

Pistole, T.G. and H. Rostam-Abadi. 1979. Lectins from the horseshoe crab, *Limulus polyphemus*, reactive with bacterial lipopolysaccharides, pp. 423–426. In H. Peters (ed.) *Protides of Biological Fluids*. Pergamon Press, Oxford.

Poinar, G.O., Jr., R. Leutenegger, and P. Götz. 1968. Ultrastructure of the formation of a melanotic capsule in *Diabrotica* (Coleoptera) in response to a parasitic nematode (Mermithidae). *J. Ultrastruct. Res.* **25:** 293–306.

Powning, R.F. and W.J. Davidson. 1973. Studies on insect bacteriolytic enzymes.

1. Lysozyme in hemolymph of *Galleria mellonella* and *Bombyx mori*. *Comp. Biochem. Physiol.* **45**B: 669–686.

Rabinovitch, M. and M.J. De Stefano. 1970. Interactions of red cells with phagocytes of the wax moth (*Galleria mellonella* L.) and mouse. *Exp. Cell. Res.* **59**: 272–282.

Raina, A.K. 1976. Ultrastructure of the larval hemocytes of the pink bollworm, *Pectinophora gossypiella* (Saunders) *Lepidoptera: Gelechiidae*). *Int. J. Insect Morphol. Embryol.* **5**(3): 187–195.

Ratcliffe, N.A. and S.J. Gagen. 1976. Cellular defense reactions of insect hemocytes *in vivo*: Nodule formation and development in *Galleria mellonella* and *Pieris brassicae* larvae. *J. Invertebr. Pathol.* **28**(3): 373–382.

Ratcliffe, N.A. and S.J. Gagen. 1977. Studies on the *in vivo* cellular reactions of insects: An ultrastructural analysis of nodule formation in *Galleria mellonella*. *Tissue Cell* **9**(1): 73–85.

Ratcliffe, N.A. and C.D. Price. 1974. Correlation of light and electron microscope hemocyte structure in the Dictyoptera. *J. Morphol.* **144**: 485–497.

Ratcliffe, N.A. and A.F. Rowley. 1979. Role of hemocytes in defense against biological agents, pp. 331–414. In A.P. Gupta (ed.) *Insect Hemocytes*. Cambridge University Press, Cambridge.

Ratner, S. and S.B. Vinson. 1983. Encapsulation reactions *in vitro* by haemocytes of *Heliothis virescens*. *J. Insect Physiol.* **29**: 855–863.

Ravindranath, M.H. 1973. The hemocytes of a millipede, *Thyropygus poseidon*. *J. Morphol.* **141**: 257–268.

Ravindranath. M.H. 1974. The hemocytes of a scorpion *Palamnaeus swammerdami*. *J. Morphol.* **144**: 1–10.

Ravindranath, M.H. 1981. Onychophorans and myriapods, pp. 328–354. In N.A. Ratcliffe and A.F. Rowley (eds.) *Invertebrate Blood Cells*, vol. 2. Academic Press, New York.

Ravindranath, M.H. and E.L. Cooper. 1984. Crab lectins: Receptor specificity and biomedical applications, pp. 83–96. In E. Cohen (ed.) *Recognition Proteins, Receptors and Probes: Invertebrates*. Alan R. Liss, New York.

Renkonen, K.O. 1948. Studies on hemagglutinins present in seeds of some representatives of Leguminosae. *Ann. Med. Exp. Fenn.* (Helsinki) **26**: 66–72.

Renwrantz, L.R. and T.C. Chang. 1977a. Agglutinin mediated attachment of erythrocytes to hemocytes of *Helix pomatia*. *J. Invertebr. Pathol.* **29**: 97–100.

Renwrantz, L.R. and T.C. Chang. 1977b. Identification of agglutinin receptors on hemocytes of *Helix pomatia*. *J. Invertebr. Pathol.* **29**: 88–96.

Rizki, T.M. and R.M. Rizki. 1976. Cell interactions in hereditary melanotic tumor formation in *Drosophila*, pp. 137–141. In T.A. Angus, F. Faulkner, and A. Rosenfield (eds.) *Proc. 1st International Colloquium of Invertebrate Pathology* Queens University, Kingston, Canada.

Rizki, T.M. and R.M. Rizki. 1982. The cellular defense reactions of *Drosophila melanogaster*, pp. 173–176. In H. Akai, R.C. King, and S. Morohoshi (eds.) *The Ultrastructure and Functioning of Insect Cells*. Plenum Press, New York.

Rizki, T.M. and R.M. Rizki. 1983. Blood cell surface changes in *Drosophila* mutants with melanotic tumors. *Science* (Washington, DC) **220**: 73–75.

Rostam-Abadi, H. and T.G. Pistole. 1979. Sites on the lipopolysaccharide molecule reactive with *Limulus* agglutinins. *Proc. Clin. Biol Res.* **29:** 525–535.

Rowley, A.F. and N.A. Ratcliffe. 1976. The granular cells of *Galleria mellonella* during clotting and phagocytic reactions *in vitro*. *Tissue Cell* **8:** 437–446.

Rowley, A.F. and N.A. Ratcliffe. 1980. Insect erythrocyte agglutinins: *In vitro* opsonization experiments with *Clitumnus extradentatus* and *Periplaneta americana* haemocytes. *Immunology* **40:** 483–492.

Rowley, A.F. and N.A. Ratcliffe. 1981. Insects, pp. 421–488. In N.A. Ratcliffe and A.F. Rowley (eds.) *Invertebrate Blood Cells*, vol 2. Academic Press, New York.

St. Marie, R.L. and S.D. Carlson. 1982. Synaptic vesicle activity in stimulated and unstimulated photoreceptor axons in the housefly: A freeze-fracture study. *J. Neurocytol.* **11:** 747–761.

Salt, G. 1970. *The Cellular Defense Reactions in Insects*. Cambridge University Press, Cambridge.

Scharrer, B. 1972. Cytophysiological features of hemocytes in cockroaches. *Z. Zellforsch. Mikrosk. Anat.* **129:** 301–313.

Schell, S.C. 1952. Tissue reactions of *Blattella germanica* L. to the developing larva of *Physaloptera hispida* Schell, 1950 (Nematoda: Spiruroidea). *Trans. Am. Microsc. Soc.* **71:** 293–302.

Schmit, A.R. and N.A. Ratcliffe. 1977. The encapsulation of foreign tissue implants in *Galleria mellonella* larvae. *J. Insect Physiol.* **23:** 175–184.

Schmit, A.R., A.F. Rowley, and N.A. Ratcliffe. 1977. The role of *Galleria mellonella* hemocytes in melanin formation. *J. Invertebr. Pathol.* **29:** 232–234.

Scott, M.T. 1971. A naturally occurring haemagglutinin in the haemolymph of the American cockroach. *Arch. Zool. Exp. Gen.* **112:** 73–80.

Sharon, N. 1984. Carbohydrates as recognition determinants in phagocytosis and in lectin-mediated killing target cells. *Biol. Cell.* **51:** 239–246.

Shaw, S.R. and S. Stowe. 1982. Freeze-fracture evidence for gap junctions connecting the axon terminals of dipteran photoreceptors. *J. Cell Sci.* **53:** 115–141.

Shepherd, J.C.W., W. McGinnis, A.E. Carrasco, E.M. De Robertis, and W.J. Gehring. 1984. Fly and frog homeo domains show homologues with yeast mating types regulatory protein. *Nature* (London) **310:** 70–71.

Sherman, R.G. 1981. Chelicerates, pp. 356–384. In N.A. Ratcliffe and A.F. Rowley (eds.) *Invertebrate Blood Cells*, vol. 2. Academic Press, New York.

Söderhäll, K. 1981. Fungal cell wall β-1,3-glucans induce clotting and phenoloxidase attachment to foreign surfaces of crayfish hemocyte lysate. *Dev. Comp. Immunol.* **5:** 565–573.

Söderhäll, K. 1982. Prophenoloxidase activation system and melanization—a recognition mechanism of arthropods? A review. *Dev. Comp. Immunol.* **6:** 601–611.

Söderhäll, K. and R. Ajaxon. 1982. Effect of quinones and melanin on mycelial growth of *Aphanomyces* spp. and extracellular protease of *Aphanomyces astaci*, a parasite on crayfish. *J. Invertebr. Pathol.* **39:** 105–109.

Söderhäll, K. and V.J. Smith. 1983. Separation of the haemocyte populations of *Carcinus maenas* and other marine decapods, and phenoloxidase distribution. *Dev. Comp. Immunol.* **7:** 229–239.

Söderhäll, K., L. Häll, T. Unestam, and L. Nyhlen. 1979. Attachment of pheno-loxidase to fungal cell walls in arthropod immunity. *J. Invertebr. Pathol.* **34:** 285–294.

Staehelin, L.A. 1974. Structure and function of intercellular junctions. *Int. Rev. Cytol.* **39:** 191–283.

Stairs, G.R. 1964. Changes in the susceptibility of *Galleria mellonella* (Linnaeus) larvae to nuclear-polyhedrosis virus following blockage of phagocytes with India ink. *J. Insect Pathol.* **6:** 373–376.

Stang-Voss, C. 1970. Zur Ultrastruktur der Blutzellen wirbelloser Tiere. 1. Über die Haemocyten der Karve des Mehlkafers *Tenebrio molitor*. *Z. Zellforsch. Mikrosk. Anat.* **103:** 589–605.

Stang-Voss, C. 1971. Zur Ultrastruktur der Blutzellen wirbelloser Tiere. V. Über die Haemocyten von *Astacus astacus* (Crustacea). *Z. Zellforsch. Mikrosk. Anat.* **122:** 68–75.

Steinkamp, J.A., J.S. Wilson, G.C. Saunders, and C.C. Stewart. 1982. Phagocy-tosiss: flow cytometric quantitation with fluorescent microspheres. *Science* (Washington, DC) **215:** 64–66.

Størmer, L. 1977. Arthropod invasion of land during Silurian and Devonian times. *Science* (Washington, DC) **197:** 1362–1364.

Stynen, D., M. Peferson, and A. De Loof. 1982. Proteins with hemagglutinin activity in larvae of the Colorado potato beetle *Leptinotarsa decemlineata*. *J. Insect Phys-iol.* **28:** 465–470.

Stynen, D., K. Vansteenwegen, and A. De Loof. 1985. Anti-galactose lectins in the haemolymph of *Sarcophaga bullata* and three other calliphorid flies. *Comp. Biochem. Physiol.* **81**B(1): 171–175.

Suzuki, T. and S. Natori. 1983. Identification of a protein having hemagglutinating activity in the hemolymph of the silkworm, *Bombyx mori*. *J. Biochem.* **93:** 583–590.

Taylor, R.L. 1969. A suggested role for polyphenol-phenoloxidase system in inver-tebrate immunity. *J. Invertebr. Pathol.* **14:** 427–428.

Tripp, M.R. 1966. Hemagglutinin in the blood of oyster *Cassostraca virginica*). *J. Invertebr. Pathol.* **8:** 478–484.

Tsivion, Y. and N. Sharon. 1981. Lipid-mediated hemagglutination and its relevance to lectin-mediated agglutination. *Biochim. Biophys. Acta* **642:** 336–344.

Tyson, C.J. and C.R. Jenkin. 1973. The importance of opsonic factors in the removal of bacteria from the circulation of the crayfish (*Parachaeraps bicarinatus*). *Aust. J. Exp. Biol. Med. Sci.* **51:** 609–615.

Tyson, C.J. and C.R. Jenkin. 1974. Phagocytosis of bacteria *in vitro* by crayfish (*Parachaeraps bicarinatus*). *Aust. J. Exp. Biol. Med. Sci.* **52:** 341–348.

Unestam, T. and J.E. Nylund. 1972. Blood reactions *in vitro* in crayfish against a fungal parasite, *Aphanomyces astaci*. *J. Invertebr. Pathol.* **19:** 94–106.

Valdiserri, R.O. 1984. Decreased T-helper/T-suppressor ratios in homosexual men: A non-specific ratio. *Lab. Med.* **15**(7): 472–474.

Valvassori, R. and G.A. Amirante. 1976. Caracteristiques immunochimiques et ul-trastructurales des hémocytes de *Leucophaea maderae* L. (Insecta Blattoidea). *Monit. Zool. Ital.* (N.S.) **10:** 403–412.

Vasta, G.R. and E. Cohen. 1984a. Carbohydrate specificities of *Birgus latro* serum lectins. *Dev. Comp. Immunol.* **8**: 197–202.

Vasta, G.R. and E. Cohen. 1984b. Characterization of the carbohydrate specificity of serum lectins from the scorpion *Hadrurus arizonensis* Stahnke. *Comp. Biochem. Physiol.* **77**B(4): 721–727.

Vasta, G.R. and E. Cohen. 1984c. Humoral lectins in the scorpion *Vaejovus confucius:* A serological characterization. *J. Invertebr. Pathol.* **43**: 226–233.

Vasta, G.R. and E. Cohen. 1984d. Sialic acid–binding lectins in the "Whip Scorpion" (*Mastigoproctus giganteus*) serum. *J. Invertebr. Pathol.* **43**: 333–342.

Vasta, G.R. and E. Cohen. 1985a. Naturally occurring hemagglutinins in the hemolymph of the scorpion *Paurauronctonus mesanensis* Stahnke. *Experientia*, **39**: 721–722.

Vasta, G.R. and E. Cohen. 1985b. Serum lectins from the scorpion *Vaejovus spinigerus* bind sialic acids. *Experientia*, **40**: 485.

Vasta, G.R., J.J. Marchalonis, and H. Kohler. 1984. Invertebrate recognition protein cross-reacts with an immunoglobulin idiotype. *J. Exp. Med.* **159**: 1270–1276.

Vey, A., J.M. Quiot, and C. Vago. 1968. Formation *in vitro* de réaction d'immunité cellulaire chez les insectes, pp. 254–263. In C. Barigozzi (ed.) *Proc. 2nd International Colloquim of Invertebrate Tissue Culture*. Instituto Lombardo di Scienze e Lettere, Milan.

Wago, H. 1981. The role of hemolymph in the initial cellular attachment to foreign cells by the hemocytes of the silkworm, *Bombyx mori. Dev. Comp. Immunol.* **5**: 217–227.

Wago, H. and Y. Ichikawa. 1979. Changes in the phagocytic rate during the larval development and manner of hemocytic reactions to foreign cells in *Bombyx mori. Appl. Entomol. Zool.* **14**(4): 397–403.

Walker, I. 1959. Die Abwehrreaktion des Wirtes *Drosophila melanogaster* gegen die Zoophage Cynipidae *Pseudeucoila bochei* Weld. *Rev. Suisse Zool.* **68**: 569–632.

Weir, D.M. 1980. Surface carbohydrates and lectins in cellular recognition. *Immunol. Today* **1**: 45–51.

Wheeler, R.E. 1962. Studies on the total haemocyte count and haemolymph volume in *Periplaneta americana* (L.) with special reference to the last moulting cycle. *J. Insect Physiol.* **9**: 223–235.

White, J.G. 1976. A comment on the ultrastructure of amoebocytes from the horseshoe crab (*Limulus polyphemus*), pp. 97–101. In *Animal Models of Thrombosis and Hemorrhagic Diseases*. National Institutes of Health, Bethesda, MD.

Whitten, J.M. 1968. Haemocytic activity in relation to epidermal cell growth, cuticle secretion and cell death in metamorphosing cyclorrhaphan pupa. *J. Insect Physiol.* **15**: 673–778.

Widmer, M.B. and F.H. Bach. 1981. Antigen-driven helper cell–independent cloned cytotoxic T lymphocyte. *Nature* (London) **294**: 750–752.

Wittig, G. 1966. Phagocytosis by blood cells in healthy and diseased caterpillars. 2. A consideration of the method of making hemocyte counts. *J. Invertebr. Pathol.* **8**: 461–477.

Wood, R.L. 1977. The cell junctions of *Hydra* as viewed by freeze-fracture replication. *J. Ultrastruct. Res.* **58**: 299–315.

Wright, K.A. and D.L. Lee. 1984. Membrane specializations in a nematode revealed by freeze-fracture technique. *Zeiss Inf.* **1984**(3): 35–39.

Yeager, J.F. and H.H. Knight. 1933. Microscopic observation on blood coagulation in several different species of insects. *Ann. Entomol. Soc. Am.* **26:** 591–602.

Yeager, J.F. and S.C. Munson. 1941. Histochemical detection of glycogen in the blood cells of the southern armyworm (*Prodenia eridania*) and in other tissues, especially midgut epithelium. *J. Agric. Res.* **63**(5): 257–294.

Yeager, J.F., E.R. McGovran, S.C. Munson, and E.L. Mayer. 1942. Effects of blocking hemocytes with Chinese ink and staining nephrocytes with trypan blue upon resistance of the cockroach *Periplaneta americana* (L.) to sodium arsenite and nicotine. *Ann. Entomol. Soc. Am.* **35**(1): 23–40.

Yeaton, R.W. 1981. Invertebrate lectins. 1. Occurrence. *Dev. Comp. Immunol.* **5:** 391–402.

Yeaton, R.W. 1983. Wound responses in insects. *Am. Zool.* **23:** 195–203.

Zachary, D. and D. Hoffman. 1984. Lysozyme is stored on the granules of certain haemocyte types of *Locusta*. *J. Insect Physiol.* **30:** 405–413.

Zachary, D., M. Brehélin, and J.A. Hoffmann. 1975. Role of the "Thrombocytoids" in the capsule formation in the dipteran *Calliphora erythrocephala*. *Cell Tissue Res.* **162**(3): 343–348.

CHAPTER 2

Cellular Immune Responses in Crustaceans

Gianni A. Amirante
Department of Biology
University of Trieste
Trieste, Italy

2.1. INTRODUCTION

Like other groups of arthropods, crustaceans exhibit various types of hemocytic reactions against foreign bodies. Usually, when an animal is invaded by foreign bodies of varying dimensions (bacteria, free cells, fungi, or protozoan or metazoan parasites), the hemocytes react against them by means of phagocytosis in small organisms and multicellular reactions (encapsulation or nodule formation) in larger corpuscles (mostly > 10 μm) (Ratner and

Vinson, 1983). In these reactions, only certain classes of hemocytes are involved.

2.2. STRUCTURE AND CLASSIFICATION OF HEMOCYTES INVOLVED IN IMMUNOLOGIC CELL REACTIONS

There is still much confusion with regard to the classification of hemocytes in crustaceans, especially because many authors classify them on a morphologic basis only, while others classify them on the basis of their physiologic activities. Hemocytes are very reactive cells and display many transformations in vitro. Although described by different names by various authors, three or more classes of hemocytes have been detected and described in crustaceans by light and electron microscopy and cytochemical methods. Of the various types, the phagocytic hemocytes are very interesting from an immunologic point of view. They vary morphologically (Bauchau et al., 1975), and although they have been classified as hyaline cells, semigranular cells, and granulocytes, the first two types are actually plasmatocytes (Gupta, 1979, 1985).

2.2.1. Hyaline Cells

The hyaline cells are the smallest cells, with a large central nucleus surrounded by basophilic cytoplasm. The cytoplasm exhibits scarce endoplasmic reticulum and ribosomes; Golgi bodies may be absent, and the granules may be absent or present in small numbers. In vitro, these cells display pseudopodia (Fig. 2.1). Hyaline cells are in fact plasmatocytes (see Chapters 1, 9, and 13).

2.2.2 Semigranular Cells

The semigranular cells are transitional cells between the granulocytes and hyaline cells. They have a spherical, bilobate, and central or eccentric nucleus. They have free ribosomes, endoplasmic reticulum, two or more Golgi bodies, and numerous eosinophilic granules. They may contain many tubular or microfibril-containing formations (Fig. 2.2).

2.2.3. Granulocytes

The granulocytes are the largest hemocytes and have a small, eccentric, bilobate nucleus. A Golgi body is almost always present. Around the nucleus and along the cell periphery is rough as well as smooth endoplasmic reticulum. There are free ribosomes in the cytoplasm, which contains large, membrane-bound granules. The granules are electron-dense and usually strongly acidophilic. Hearing and Vernick (1967) divided the granules of

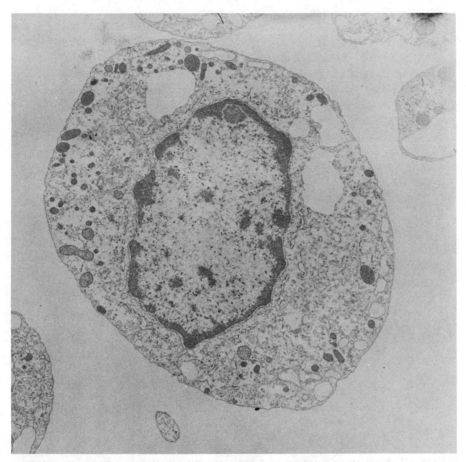

Figure 2.1. Electron micrograph of a hyaline cell (plasmatocyte) of *Eriocheir sinensis*. Some small granules begin to appear at the periphery of cytoplasm (×11,880). (Courtesy of A.G. Bauchau.)

Homarus americanus into two classes: eosinophils and ovoid basophils. These classes vary remarkably with the age of the specimens (Fig. 2.3; for details about these granules, see Chapter 1).

2.2.4. Biochemical Characteristics of Granules

By means of specific cytochemical techniques, various authors have reported different kinds of granules. Williams and Lutz (1975) divided the *Carcinus maenas* hemocytes into two types: granulocytes with granules containing glycogen and granulocytes without glycogen. Glycogen was also detected in the cytoplasm of the plasmatocytes of *Orconectes virilis* (Wood and Visentin, 1967), *Eriocheir sinensis* (Bauchau et al., 1975), *Pachygrapsus*

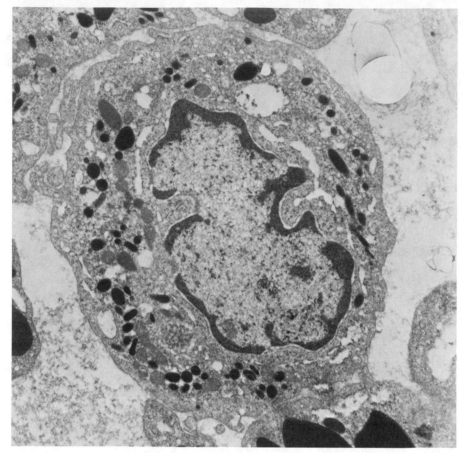

Figure 2.2. Electron micrograph of a semigranular cell of *E. sinensis* (plasmatocyte) (×17,820). (Courtesy of A.G. Bauchau.)

marmoratus (Arvy, 1954), and *Artemia salina* (Lockhead and Lockhead, 1941); in the latter, the cytoplasmic glycogen was never present in the granules. Bauchau and coworkers (1975), treating the *E. sinensis* plasmatocytes with proteolytic enzymes (e.g., pepsin and trypsin), proved that the granules contained large amounts of basic proteins and carbohydrates but no glycogen.

Mengeot and colleagues (1976, 1977) detected mucopolysaccharides in the granules of certain species. By means of electrophoretic analysis of the granule content, these authors demonstrated that the granules, instead of being of a homogeneous class, are highly heterogeneous. This finding supports the assumption that plasmatocytes and/or granulocytes play an essential role in various physiologic activities. Thus, these cells would be important in immunologic defenses, synthesis of lectins and agglutinins

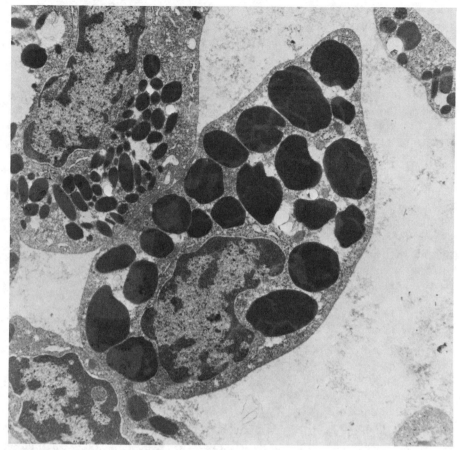

Figure 2.3. Electron micrograph of a granulocyte of *E. sinensis*. The granules and bilobate nucleus are evident. On the left, a portion of a semigranular cell can be seen (×11,880). (Courtesy of A.G. Bauchau.)

(Amirante and Basso, 1984), synthesis of chitin (Johnston et al., 1973), and synthesis of hemocyanins (Ghiretti-Magaldi et al., 1977).

2.3. PHAGOCYTOSIS

Inert substances (e.g., carmine and carbon) or living cells (e.g., yeast, bacteria, protozoa, and erythrocytes) are quickly phagocytized by plasmatocytes both in vivo and in vitro.

The two plasmatocyte subcategories (hyaline cells and semigranular cells) and the granulocytes (see Section 2.2) perform phagocytosis; however, the most active phagocytes are the hyaline cells and the semigranular cells.

Johnson (1977) indicated that in *Callinectes sapidus* the protozoa (amoebas) are phagocytized only by hyaline cells, while the bacteria are phagocytized only by granulocytes.

It seems that before phagocytosis begins, it is necessary that the phagocyte membrane receptors (probably lectins; see Chapter 13) recognize the foreign object. In some species, the opsonin action of certain serum factors has been demonstrated. In vertebrates, the immunoglobulins (Igs) and the third component of complement (C3) act as opsonins (see also Chapter 1). Arthropods, and especially crustaceans, have neither vertebrate complement nor immunoglobulins. But do they have opsonins?

2.3.1. The Role of Opsonins in Phagocytosis

The standard method of demonstrating the presence of serum opsonins consists of the addition of serum-treated or serum-untreated test particles (often mammalian erythrocytes) to hemocyte suspensions or monolayers. If an opsonin is present, serum pretreatment increases the binding or uptake of particles by the hemocytes.

While in several insects attempts to show the presence of opsonins have been unsuccessful (Rabinovitch and De Stefano, 1970; Scott, 1971; Anderson et al., 1972; Wago and Ichikawa, 1979; Rowley and Ratcliffe, 1980), in certain decapods the presence of nonclassified opsonins has been demonstrated by the abovementioned method (McKay et al., 1969; McKay and Jenkin, 1970c; Paterson and Stewart, 1974; Schapiro et al., 1977).

It has also been shown that a component of the *Parachaeraps bicarinatus* serum contributes to the elimination of bacteria in vivo (Tyson and Jenkins, 1973), although it has not been conclusively determined that phagocytosis is the cause of their elimination. In many cases, certain stages of phagocytosis or binding may take place even in the absence of serum.

The invagination of the phagocyte plasma membrane and the subsequent phagocytosis of the foreign cell are not only activated by a receptor, and sometimes facilitated by opsonins, but also depend on the localized contraction of the cytoplasmic microfilament network under the membrane. Moreover, it seems that Ca^{2+} and basic pH facilitate this mechanism.

In *C. sapidus*, the protozoan *Paramoeba perniciosa* is phagocytized only by hyaline cells (Johnson, 1977). This phagocytosis is inhibited by acid pH and low temperatures. The successive stages of phagocytosis, such as the displacement of phagosomes in the cytoplasm, fusion with lysosomes, rejection or storage or digestion wastes, and the possible degeneration of the phagocyte, have not yet been studied in crustaceans (Bauchau and Mengeot, 1978). Furthermore, the plasmatocytes in crustaceans phagocytize not only foreign cells but also the organism's own degenerating cells (Chassard-Bauchaud and Hubert, 1975) and damaged cells (Shivers, 1977). Reade (1968) studied phagocytosis of carbon particles in vitro in *P. bicarinatus*. These particles were injected into the ventral abdominal sinus and accumulated in the gills and the hepatopancreas. Within the gills, the particles were not associated with any cells; however, in the hepatopancreas, large mononuclear cells (probably hemocytes), forming a cyst around the aggregated carbon particles, phagocytized them actively. It has also been shown in this animal that the plasmatocytes adhered to vertebrate red blood cells (RBCs) injected into the hemocoel and subsequently phagocytized them (McKay et al., 1969; McKay and Jenkin, 1970a). Furthermore, McKay and Jenkin (1970a) demonstrated in vitro that the plasmatocytes phagocytized sheep red blood cells (SRBCs) whenever these cells had been previously treated with the animal's hemolymph; serum-untreated bacteria, horse red blood cells (HRBCs), and SRBCs are not phagocytized (McKay and Jenkin, 1970b,c; Tyson and Jenkin, 1973, 1974). The same phenomenon occurs in *H. americanus* (Paterson and Stewart, 1974; Paterson et al., 1976). In *C. maenas*, Smith and Ratcliffe (1980) noted that after 3 hr only 3% of the *Bacillus cereus* was phagocytized, compared with 15% of *Moraxella* sp.; however, during the same time, the RBCs were not phagocytized. Schapiro and coworkers (1977) showed that bacteria were phagocytized when opsonized with plasma but not when opsonized with serum. Certain microorganisms, however, may be immune to phagocytosis. For example, in *H. americanus, Aerococcus virideus* is not destroyed by phagocytosis and continues to multiply, inducing the death of the crustacean due to septicemia (Cornick and Stewart, 1968). Other viruses, rickettsiae, bacteria, and protozoa can also live in the hemocoel without being phagocytized (e.g., *Anophis* sp. in *C. maenas*; Bang, 1967).

2.4. FORMATION OF CAPSULES AND NODULES

Phagocytosis is possible only when the corpuscle to be phagocytized is smaller or equal in volume to the plasmatocyte; when it is larger than the plasmatocyte, encapsulation or nodule formation occurs. This phenomenon is widespread in crustaceans. The first studies of the formation of nodules or capsules were performed by Cuénot (1895, 1902) and Poisson (1930) in *Carcinus*.

2.4.1. Cellular Encapsulation

Most crustaceans have the ability to form a multilayered hemocytic capsule around macroscopic parasites and pathogens. This process often kills the offending organism or at least reduces its movements and stunts its growth. The theories put forward to explain the death of the encapsulated organism include asphyxia, the buildup of wastes, and the toxic action of quinones (precursors of melanin; Salt, 1963; Nappi, 1977; Poinar et al., 1979).

A typical capsule consists of 5–30 layers of tightly crowded hemocytes with scarce or no intercellular space. As with phagocytosis, the formation of capsules occurs when the plasmatocytes come into contact both with the biologic object and, in certain cases, with inert or synthetic substances. Bauchau and Mengeot (1978) demonstrated this mechanism in *C. maenas* by introducing nylon or keratin filaments into the hemocoel. After 5 hr, three concentric layers of cells formed around the foreign body. The cells forming the capsule were mainly ameboid phagocytic cells, which resemble (and may actually be) plasmatocytes. Ultrastructural studies of the early stages of encapsulation show, nevertheless, that different classes of hemocytes are responsible for capsule formation. This holds true at least for the recognition of the foreign cell and the initiation of encapsulation. These labile hemocytes have been termed hyaline cells, cystocytes, spherule cells, or granulocytes.

The inner layer of the capsule is made up of approximately a dozen layers of flat plasmatocytes adhering to one another and completely adherent to the surface of the foreign body. Some promptly degenerate and become small plates of waste. This necrosis becomes more pronounced after 48 hr. The middle sheath is always the thickest and displays three types of flat and tiled hemocytes without any sign of necrosis. The granules of the hemocytes gradually disappear, while the cytoplasm is invaded by numerous formations similar to but larger than microtubules. Furthermore, many bundles of microfilaments fill the intercellular spaces, obscuring the profiles of the cellular membranes. These bundles look like collagen fibers and strengthen the capsule. The external sheath of the capsule is made up of only five or six hemocyte layers, which become flatter as they become progressively peripheral.

2.4.2. Metabolic Mechanisms of Encapsulation

As the capsule grows in size, it is probably subjected to repeated compression; at the same time, a melanin mass is formed around the foreign object. For example, in *Pacifastacus lemisculus* and *Astacus astacus*, it has been shown that the hemocytes form cysts when *Aphamomyces astaci* hypha is injected into the animals (Unestam and Nylund, 1972). The granules, freed by the granulocytes, surround the hyphae and help the encapsulation process. It has often been noted both in vivo and in vitro that after a few hours an intense melanization occurs around the hyphae (Unestam and Weiss, 1970). These authors did not morphologically identify the hemocytes involved in encapsulation. Besides the degranulation of the granulocytes due to exocytosis, they noted the outcropping of a thin refractive surface of the hypha, apparently formed by a material similar to that of the granules. The intense melanization is the result of the activation of hemocytic prophenoloxidase in the presence of a substrate probably supplied by the same hemocytes.

Apparently, melanin is not formed when the foreign body is cellulose, chitin, or a synthetic substance (Unestam and Ajaxon, 1976; Unestam and Beskow, 1976). Similar reactions (i.e., no melanin formation) were observed in shrimps by De Backer (1961) and Solanji and Lightner (1976). Even the most successful parasite of *C. maenas, Sacculina carcini*, elicits a hemocytic reaction. Granulocytes stick to the sacculin roots and cover them with a collagenlike film originating from the granules. However, the parasite does not seem to be affected and remains healthy (Hubert et al., 1976).

Certain species of shrimps (*Prochristianella penaei, Penaeus aztecus*, and *Penaeus setiferus*), although infected with trypanorhynchid cestodes, survive. The hemocytes around the hepatopancreas form cysts, or capsules, around the parasites (Aldrich, 1965). These cysts are made up of humerous hemocytes, collagen fibers, and fibroblasts and, after their necrosis, kill the parasite. Similar cysts also form in the hemocoel, although the parasite is not destroyed (Sparks and Fontaine, 1973). Ultrastructural evidence indicates that granulocytes rupture upon encountering a foreign surface. The released granular and/or electron-dense flocculent material seems to cause the plasmatocytes to adhere to the foreign surface and may contain the chemical signal that induces them to flatten and conjoin to form capsules (Nappi and Streams, 1969; Salt, 1970; Unestam and Nylund, 1972, Brehélin et al., 1975; Schmit and Ratcliffe, 1977; Ratner and Vinson, 1983). This material is referred to as encapsulation promoting factor (EPF).

What is the nature of the EPF? When foreign objects are added to short-term insect hemolymph cultures, hemocytes promptly aggregate around them, and after 2 hr the capsule begins to form. If no further foreign objects are added, the hemocytes aggregate spontaneously, sometimes forming smooth, rounded nodules. This response can be triggered by the surface of the culture capsule or by factors freed from the damaged tissues during the bleeding (Cherbas, 1973). If the addition of foreign objects is delayed until after nodule formation is well under way (about 30 min), the object is not encapsulated. The formation of capsules (and of nodules) can be inhibited by a short treatment with trypsin. These formations can be restored by the addition of lysate of fresh whole hemolymph. This observation indicates that the trypsin digests a protein released by the same hemocytes (e.g., EPF).

The inhibitors of glycolysis and of oxidative phosphorylation do not inhibit the initial aggregation of hemocytes, but they do inhibit the maturation of the capsule. The recognition is passive, but the consolidation of the capsule requires metabolic energy. Thus, there could be a link between the phenoloxidase and encapsulation, because the inhibitors of melanin formation inhibit encapsulation (Nappi, 1973; Beresky and Hall, 1977). Indeed, hemolymph of many species of arthropods displays a prophenoloxidase, which can be activated (Ashida and Dohke, 1980); the activated prophenoloxidase adheres to the surface of foreign objects.

2.4.3. Immunologic Mechanisms of Encapsulation

Lectins seem to play a role in capsule formation (Amirante and Mazzalai, 1978; Lackie, 1981a,b). These lectins, acting as membrane receptors, bind to certain oligosaccharides present on the surface of the foreign objects.

This, however, does not account for the continuing layering of hemocytes after the first hemocyte, or layers of hemocytes, have completely enveloped and isolated the foreign object. Evidently, their receptors can no longer react with the oligosaccharides of the foreign body! There may be two explanations. First, remember that this stage is an active phenomenon and that the first granulocytes release the contents of their granules. Thus, it is possible that as these substances are released and cover the hemocyte layers, they are recognized by the free hemocytes or, alternatively, that the adhesion phenomenon triggers in the hemocytes the neoformation of new acceptors on the membrane that in turn are recognized by the lectins of the hemocytes. In any case, the recognition by hemocytes of a foreign cell can be demonstrated in vitro and analyzed in its various stages by means of a classic reaction known as immunocytoadherence (ICA), which implies the formation of "rosettes" by the binding of hemocyte with heterologous cells (see Chapter 13).

In 1976, we were the first to demonstrate the following immunologic phenomena in a group of insects and subsequently in crustaceans (Amirante, 1976; Amirante and Guidali-Mazzalai, 1978; Amirante and Basso, 1984):

1. The hemocytes synthesize specific lectins against mammalian red blood cells.
2. Some of these lectins are present not only in the circulating fluids but also on the membranes of two classes of hemocytes: hyaline cells and granulocytes.
3. There are at least two different lectins, specific to different oligosaccharide radicals of the same erythrocyte.
4. Of these two lectins, only one is present on the hemocyte membrane; the other is in the hemocyte cytoplasm and is secreted by the hemocyte.

In these hemocytes, specific cell recognition and successive adhesion can be observed, and this recognition can be demonstrated in vitro by ICA. It is noteworthy that the rosette also forms in the presence of metabolic inhibitors but not in the presence of specific sugars of the erythrocyte. Once the rosette is formed, a second stage can be observed: on the first layer of erythrocytes, many other layers are formed that eventually completely hide the hemocyte. This second stage is conversely inhibited, for example, by cycloheximide (CHI); furthermore, it can be observed by means of immunofluorescence that, while membrane lectins of the hemocyte remain, those present in the cytoplasm disappear (unpublished data). We failed to provide an explanation for this phenomenon (Mazzalai et al., 1978). Recently, however, Wright and Cooper (1981), resuming our experiments on tunicates, provided an answer to this question. Following the formation of the rosette (i.e., after the adhesion of the hemocyte), a metabolic process is triggered that leads to the synthesis of neolectins that are different from those of the membrane and that, once secreted, recognize another sugar present on the erythrocyte membrane. In this case, these lectins act as a bridge between the first layer of erythrocytes, already linked to the hemocyte, and other, free erythrocytes, thus favoring the formation of a progressively growing large mass.

2.5. FORMATION OF NODULES

When bacterial masses form in the hemocoel (probably agglutinated by lectins or serum heteroagglutinins), the hemocytes adjust by forming generally melanized nodules. These nodules are small capsules from which certain hemocytes detach themselves and enter the mass of agglutinated bacteria trying to phagocytize them. Nodule formation represents a hybrid of the encapsulation and phagocytic reactions.

Some parasites often present in crustaceans induce a similar hemocytic reaction. For example, the "roots" of sacculin attract numberous hemocytes, mostly granulocytes, whose granules possibly give rise to the collagen pockets that surround the roots of the parasite, the surface of granulocyte being linked to them (Hubert et al., 1976).

2.6. SUMMARY

The cell immuno-defenses in crustaceans can be classified into three types of reactions: phagocytosis, encapsulation, and nodule formation. All three often involve a class of hemocytes, namely the plasmatocytes, as well as granulocytes. The former is a very heterogeneous class made up of two cell types: hyaline cells, small cells with pseudopods and few granules; and semigranular cells, which are cells with an intermediate morphology between the hyaline cells and the granulocytes, which are larger and have numerous eosinophilic and basophilic granules. The phagocytosis takes place both in vivo and in vitro when the dimension of the foreign object (bacteria, erythrocytes and often inert substances), recognized by the plasmatocytes, does not exceed the dimension of the hemocyte. In the endocytic process, the membrane receptors are involved (lectins). This process is assisted in crustaceans by the presence of serum or plasma opsonins, calcium ions, and basic pH.

When the foreign object is larger than the hemocyte, then encapsulation takes place. This process involves membrane receptors, metabolic mechanisms that entail the formation of melanin after the triggering of the prophenoloxidase, degranulation of granulocytes, and an ex novo synthesis of lectins. The triggering of this mechanism is possibly due to a factor known as encapsulation promoting factor.

The formation of nodules is an intermediate mechanism between phagocytosis and encapsulation. It takes place whenever the organism comes into contact with agglutinated masses of bacteria, which are probably linked together by lectins or heteroagglutinins.

REFERENCES

Aldrich, D.V. 1965. Observations on the ecology and life cycle of *Prochristianella penaei* Kruse (Cestoda: Tripanoryncha). *J. Parasitol.* **51**: 370–376.

Amirante, G.A. 1976. Production of heteroagglutinins in haemocytes of *Leucophaea maderae*. *Experientia* **32**: 526–528.

Amirante, G.A. and V. Basso. 1984. Analytical study of lectins in *Squilla mantis* L. (Crustacea, Stomatopoda) using monoclonal antibodies. *Dev. Comp. Immunol.* **8**: 721–726.

Amirante, G.A. and F. Mazzalai. 1978. Synthesis and localization of heemagglutinins in haemocytes of the cockroach *Leucophaea maderae*. *Dev. Comp. Immunol.* **2**: 735–740.

Anderson, R.S., N.K.B. Day, and R.A. Good. 1972. Specific hemagglutinin and a modulator of complement in cockroach hemolymph. *Infect. Immun.* **5**: 55–59.

Arvy, L. 1954. Donées sur la leucopoiése chez *Musca domestica* L. *Proc. R. Entomol. Soc.* (*London*) **29**: 39–41.

Ashida, M. and K. Dohke. 1980. Activation of prophenoloxidase by the activating enzyme of the silkworm, *Bombyx mori*. *Insect Biochem.* **10**: 37–47.

Bang, F.B. 1967. Serological responses among invertebrates other than insects. *Fed. Proc.* **26**: 1680–1684.

Bauchau, A.G. and J.C. Mengeot. 1978. Structure et function des hémocytes chez les Crustacés. *Arch. Zool. Exp.* **119**: 227–248.

Bauchau, A.G., M.B. DeBrouwer, E. Passalaq-Guerin, and J.C. Mengeot. 1975. Étude cytochimique des hémocytes des Crustacés décapodes brachyoures. *Histochemistry* **45**: 101–113.

Beresky, M.A. and D.W. Hall. 1977. The influence of phenylthiourea on encapsulation, melanization and survival in larvae of the mosquito *Aedes aegypti* parasitized by the nematode *Neoplectana carpocapsae*. *J. Invertebr. Pathol.* **29**: 74–80.

Brehélin, M., J. A. Hoffmann, G. Matz, and A. Porte. 1975. Encapsulation of implanted foreign bodies in *Locusta migratoria* and *Melolontha melolontha*. *Cell Tissue Res.* **160**: 283–289.

Chassard-Bouchaud, C. and M. Hubert. 1975. Étude ultrastructurale des hémocytes présents dans l'organe Y de *Carcinus maenas* L. (Crutacé Décapode). *C. R. Séances Acad. Sci.* (*III*) **281**D: 807–810.

Cherbas, L. 1973. The induction of an injury reaction in cultured haemocytes from saturniid pupae. *J. Insect Physiol.* **19**: 2011–2033.

Cornick, J.W. and J.E. Stewart. 1968. Interaction of the pathogen *Gaffkya homari* with natural defense mechanism of *Homarus americanus*. *J. Fish. Res. Bd. Can.* **25**: 695–709.

Cuénot, L. 1895. Étude physiologiques sur les crustacés décapodes. *Arch. Biol. Liè ge.* **13**: 245–303.

Cuénot, L. 1902. Organes agglutinants et organes cytophagocytaires. *Arch. Zool. Exp. Gen.* **3**:10–79.

De Backer, J. 1961. Role joué par les hémocytes dans les réactions tissulaires de défense chez les crustacés. *Ann. Soc. R. Zool. Belg.* **92**:141–151.

Ghiretti-Magaldi, A.F. Ghiretti, and B. Salvato, 1977. The evolution of hemocyanin. *Symp. Zool. Soc. Lond.* **38**: 513–523.

Gupta, A. P. 1979. Arthropod hemocytes and phylogeny, pp. 669–735. In A.P. Gupta (ed.) *Arthropod Phylogeny.* Van Nostrand Reinhold, New York.

Gupta, A.P. 1985. Cellular elements in the hemolymph, pp. 401–451. In G.A. Kerkut and L.I. Gilbert (eds.) *Comprehensive Insect Physiology, Biochemistry and Pharmacology*, vol. 3. Pergamon Press, Oxford.

Hearing, V. and S.H. Vernick. 1967. Fine structure of the blood cells of the lobster *Homarus americanus. Chesapeake Sci.* **8**: 170–186.

Hubert, M., C. Chassard-Bouchaud, and J. Bocquet-Vèrdine. 1976. Aspects ultrastructuraux des hémocytes de *Carcinus maenas* L. (Crustacé Décapode), parasitè par *Sacculina carcini*: Activitè réactionelle, gènese du collagène *C. R. Séances Acad. Sci.* (III) **283**D: 789–792.

Johnson, P.T. 1977. Paramoebiasis in the blue crab *Callinectes sapidus. J. Invertebr. Pathol.* **29**: 303—320.

Johnston, M.A., H.Y. Elder, and P. Spencer Davies. 1973. Cytology of *Carcinus* hemocytes and their function in carbohydrate metabolism. *Comp. Biochem. Physiol.* **46**: 569–581.

Lackie, A.M. 1981a. Immune recognition in insects. *Dev. Comp. Immunol.* **5**: 191–204.

Lackie, A.M. 1981b. Immune recognition mechanisms in two insect species. *Dev. Comp. Immunol.* **5**: 99–104.

Lockhead, J.H. and M.S. Lockhead. 1941. Studies on the blood and related tissues in *Artemia* (Crustacea Anacostraca). *J. Morphol.* **68**: 593–632.

Mazzalai, F., P. Bergamo, and G.A. Amirante, 1978. Studi sulla sintesi di emagglutinine da parte di colture di emociti di Insetti Blattoidei (*Nauphoeta cinerea* Oliv. e *Leucophaea maderae* Fabr.). *Rend. Accad. Naz. Lincei.* **65**: 338–342.

McKay, D. and C.R. Jenkin. 1970a. Immunity in the invertebrates: Correlation of the phagocytic activity of hemocytes with resistance to infection in the crayfish. *Aust. J. Exp. Biol. Med. Sci.* **48**: 609–616.

McKay, D. and C.R. Jenkin. 1970b. Immunity in the invertebrates: The fate and distribution of bacteria in normal and immunized crayfish (*Parachaeraps bicarinatus*). *Aust. J. Exp. Biol. Med. Sci.* **48**: 599–607.

McKay, D. and C.R. Jenkin. 1970c. Immunity in the invertebrates: The role of serum factors in phagocytosis of erythrocytes by hemocytes of the freshwater crayfish (*Parachaeraps bicarinatus*). *Aust. J. Exp. Biol. Med. Sci.* **48**: 139–147.

McKay, D., C.R. Jenkin, and D. Rowley. 1969. Immunity in the invertebrates. 1. Studies on the naturally occurring haemagglutinins in the fluid from invertebrates. *Aust. J. Exp. Biol. Med. Sci.* **48**: 139–150.

Mengeot, J.C., A. Bauchau, M.B. De Brouwer, and E. Passalaq-Guerin. 1976. Séparation des granules présents dans les hémocytes des crustacés par exocytose provoqué. *Comp. Biochem. Physiol.* **54:** 145–148.

Mengeot, J.C., A.G. Bauchau, M.B. De Brouwer, and E. Passalaq-Guerin. 1977. Isolement des granules des hémocytes de *Homarus vulgaris*: Examens éléctrophoretiques du contenu protéique des granules. *Comp. Biochem. Physiol.* **58:** 393–403.

Nappi, A.J. 1973. The role of melanization in the immune reaction of larvae of *Drosophila algonquin* against *Pseudeucoila bochei*. *Parasitology* **66:** 23–32.

Nappi, A.J. 1977. Comparative ultrastructural studies of cellular immune reactions and tumorigenesis in *Drosophila*, pp. 155–188. In L.A. Bulla, Jr., and T.C. Cheng (eds.) *Comparative Pathology*, vol. 3. Plenum Press, New York.

Nappi, A.J. and F.A. Streams. 1969. Hemocytic reactions of *Drosophila melanogaster* to the parasites *Pseudeucoila mellipes* and *P. bochei*. *J. Insect Physiol.* **15:** 1551–1566.

Paterson, W.D. and J.E. Stewart. 1974. In vitro phagocytosis by hemocytes of the American lobster (*Homarus americanus*). *J. Fish. Res. Bd. Can.* **31:** 1051–1056.

Paterson, W.D., J.E. Stewart, and B.M. Zwicker. 1976. Phagocytosis as a cellular immune response mechanism in the American lobster, *Homarus americanus*. *J. Invertebr. Pathol.* **27:** 95–104.

Poinar, G.O., R.T. Hess, and J.J. Petersen. 1979. Immune responses of mosquitoes against *R. culiensifax* (Nematoda). *J. Nematol.* **11:** 110–116.

Poisson, R. 1930. Observations sur *Anophrys sargophaga* (Cohn). *Bull. Biol. Fr. Belg.* **64:** 288–331.

Rabinovitch, M. and M.J. De Stafano. 1970. Interactions of red cells with phagocytes of the wax moth (*Galleria mellonella* L.) and mouse. *Exp. Cell Res.* **59:** 272–282.

Ratner, S. and S.B. Vinson, 1983. Phagocytosis and encapsulation: Cellular immune response in Arthropoda. *Am. Zool.* **23:** 185–194.

Reade, P.C. 1968. Phagocytosis in invertebrates. *Aust. J. Exp. Biol. Acad. Sci.* **46:** 219–229.

Rowley, A.F. and N.A. Ratcliffe. 1980. Insect erythrocyte agglutinins: *In vitro* opsonization experiments with *Clitumnus extradentatus* and *Periplaneta americana* hemocytes. *Immunology* **40:** 483–492.

Salt, G. 1963. The defense reactions of insects of metazoan parasites. *Parasitology* **53:** 527–642.

Salt, G. 1970. *The Cellular Defense Reactions in Insects*. Cambridge University Press, Cambridge.

Schapiro, H.C., J.F. Steenburgen, and Z.A. Fitzgerald. 1977. Hemocytes and phagocytosis in the American lobster *Homarus americanus* L. pp. 128–133. In L.A. Bulla, Jr., and T.C. Chang (eds.) *Comparative Pathology*, vol. 3, Plenum Press, New York.

Schmit, A.R. and N.A. Ratcliffe. 1977. The encapsulation of foreign tissue implants in *Galleria mellonella* larvae. *J. Insect Physiol.* **23:** 175–184.

Scott, M.T. 1971. Recognition of foreignness in invertebrates. 2. *In vitro* studies of cockroach hemocytes. *Immunology* **21:** 817–828.

Shivers, R.R. 1977. Formation of functional complexes at site of contact of hemocytes with tissue elements in degenerating nerves of the crayfish *Orconectes virilis*. *Tissue Cell* **9:** 43–56.

Smith, V.J. and N.A. Ratcliffe. 1980. Cellular defense reactions of the shore crab *Carcinus maenas*: In vivo hemocytic and histopathologic responses to injected bacteria. *J. Invertebr. Pathol.* **35:** 65–74.

Solanji, M.A. and D.V. Lightner. 1976. Cellular inflammatory response of *Penaeus aztecus* and *P. setiferus* to the pathogenic fungus *Fusarium* sp. isolated from the California brown shrimp, *P. californiensis*. *J. Invertebr. Pathol.* **27:** 77–86.

Sparks, A.K. and C.T. Fontaine. 1973. Host response in the white shrimp, *Penaeus setiferus*, to infection by the larval trypanorhynchid cestode, *Prochristianella penaei*. *J. Invertebr. Pathol.* **22:** 213–219.

Tyson, C.J. and C.R. Jenkin. 1973. The importance of opsonic factors in the removal of bacteria from the circulation of the crayfish (*Parachaeraps bicarinatus*). *Aust. J. Exp. Biol. Med. Sci.* **51:** 609–615.

Tyson, C.J. and C.R. Jenkin. 1974. Phagocytosis of bacteria *in vitro* by crayfish (*Parachaeraps bicarinatus*). *Aust. J. Exp. Biol. Med. Sci.* **52:** 341–348.

Unestam, T. and R. Ajaxon. 1976. Phenoloxidation in soft cuticle and blood of crayfish compared with that in other arthropods and activation of the phenoloxidase by fungal and other cell walls. *J. Invertebr. Pathol.* **27:** 287–295.

Unestam, T. and S. Beskow. 1976. Phenoloxidase in crayfish blood: Activation by and attachment on cells of other organisms. *J. Invertebr. Pathol.* **27:** 297–305.

Unestam, T. and J.E. Nylund. 1972. Blood reactions *in vitro* in crayfish against a fungal parasite, *Aphanomyces astaci*. *J. Invertebr. Pathol.* **19:** 94–106.

Unestam, T. and D.W. Weiss, 1970. The host-parasite relationship between freshwater crayfish and the crayfish disease fungus *Aphanomyces astaci*: Responses to infection by a susceptible and a resistant species. *J. Gen. Microbiol.* **60:** 77–90.

Wago, H. and Y. Ichikawa. 1979. *In vitro* analysis of cyto-adherence phenomenon between *Bombyx mori* hemocytes and goose erythrocytes in relation to the larval development. *Appl. Entomol. Zool.* **14:** 256–263.

Williams, A. J. and P. L. Lutz. 1975. Blood cell types in *Carcinus maenas* and their physiological role. *J. Mar. Biol. Assoc.* **55:** 671–674.

Wood, P.J. and L.P. Visentin. 1967. Histological and histochemical observations of the hemolymph cells in the crayfish *Orconectes virilis, J. Morphol.* **123:** 559–567.

Wright, R.K. and E.L. Cooper, 1981. Agglutinin-producing hemocytes in the ascidian *Styela clava*. *Am. Zool.* **21:** 974–981.

CHAPTER 3

Macrophage Functions in Insects
Responses of *Galleria mellonella* Hemocytes to Bacterial Infection

Robert S. Anderson*
Benjamin M. Chain**
Walker Laboratory
Sloan-Kettering Institute for Cancer Research
Rye, New York

3.1. INTRODUCTION

Resistance to microbes and parasites in insects, as well as many other invertebrates, tends to rely more heavily on inherent mechanisms than on

* Current address: Chesapeake Biological Laboratory, University of Maryland, Box 38, Solomons, MD 20688-0038.
** Current address: Tumor Immunology Unit, Department of Zoology, University College of London, London WC1E 6BT, England.

forms of adaptive immunity. This reliance is a result of the absence of immunoglobulins and immune memory cells, such as lymphocytes. Nevertheless, insects can certainly differentiate between self and nonself, are capable of distinguishing degrees of foreignness, and can show increased resistance to infections after appropriate experimental treatment. Central to their ability to recognize and respond to foreign organisms are the hemocytes that can kill or isolate invading microbes by intracellular mechanisms or by encapsulation. Hemocytes can also release a variety of molecules that react with foreign material and/or influence the behavior of the hemocytes. Immune recognition in insects and the role of hemocytes in this process have been reviewed by Ratcliffe and Rowley (1979) and Lackie (1981).

This chapter presents data on several responses observed in larvae of the wax moth, *Galleria mellonella*, following injection of the pathogen *Bacillus cereus*. These responses include the release from hemocytes of chemical signals modifying the behavior of other hemocytes as well as the release of adhesive acid mucopolysaccharides. The possible roles of these reactions in immunologic mechanisms of insects are discussed.

3.2. HEMOCYTE PROFILES AFTER INJECTION OF THE PATHOGEN *BACILLUS CEREUS*

3.2.1. Differential Hemocyte Count

Throughout our studies of differential hemocyte count, we routinely used last-instar *Galleria* larvae. *B. cereus* were from 18-hr broth cultures, washed and resuspended to give approximately 10^8 bacteria/ml Grace's medium. Experimental animals were injected with 10 μl of bacterial suspension at predetermined intervals before collection of hemolymph via a proleg. Ten microliters of hemolymph were mixed with an equal volume of Grace's medium containing a trace of phenylthiourea to inhibit melanization and placed on a clean coverslip in a moist chamber. After 15 min of adhesion, a further 50 μl of Grace's medium containing 100 U/ml penicillin and 100 μg/ml streptomycin were added. The resultant cell monolayer preparations were incubated at 30°C for 3 hr, washed with Grace's medium to remove nonadherent cells and the fluid portion of the hemolymph, fixed with methanol, stained with Papanicolaou's hematoxylin, and counterstained with Papanicolaou's stain.

Two types of hemocytes predominated the preparations: plasmatocytes and granulocytes. Other cell types—spherulocytes, oenocytoids, and prohemocytes—were present in small numbers in the hemolymph upon collection but did not adhere to the coverslips under the conditions of the experiment. It was easy to differentiate plasmatocytes from granulocytes based on the more extensive spreading behavior and relative lack of intracellular granules in the former (Chain and Anderson, 1983b). The identification of

**TABLE 3.1. DIFFERENTIAL HEMOCYTE COUNT ONE HOUR
AFTER INJECTION**

Injected Material (10 μl)	Percent Plasmatocytes (± SEM)	Number of Larvae
None	46 ± 3	12
Bacillus cereus	<1	12
Grace's medium	46 ± 2	7
Trypticase soy broth	42 ± 4	12
Broth from 18-hr *B. cereus* culture (bacteria free)	30 ± 3	8

these cell types is in agreement with hemocyte classification of Price and Ratcliffe (1974).

The total hemocyte count in control animals was 25,020 ± 1,890 cells/μl ($n = 6$). As previously reported by Gagen and Ratcliffe (1976), injection of *B. cereus* was followed by a rapid decrease in the numbers of circulating blood cells to 13,190 ± 1,107/μl. This effect persisted for at least 6 hr. It is interesting to note that this decrease in the total hemocyte count was produced by the selective depletion of plasmatocytes from the circulation. Plasmatocytes usually made up 45%–50% of the total hemocytes; this number could be reduced to less than 1% by the injection of *B. cereus* (Table 3.1). Plasmatocyte depletion was not a nonspecific effect of injury or of the injection of foreign fluids into the insects. Wounding with a hypodermic needle, injection of Grace's medium, or injection of the bacterial growth medium had little effect on plasmatocyte numbers. Injection of spent bacterial broth did produce a slight reduction in plasmatocyte numbers, probably mediated by bacterial products liberated during incubation.

3.2.2. Speed of Reaction and Dose Dependency

The speed of plasmatocyte depletion following injection of *B. cereus* was remarkable (Chain and Anderson, 1982). The maximal response developed within 5 min and persisted for at least 6 hr. Furthermore, the response was dose dependent. Injections of 9×10^3 bacteria reduced the number of plasmatocytes to 30.5% ± 4.0% of the total number hemocytes; a tenfold-increased dose further reduced this percentage to 5.0% ± 3.8%, and essentially no plasmatocytes were recovered after injection of 9×10^5 bacteria. Based on estimates of hemolymph volume, this dose was equivalent to about 1.25×10^4 *B. cereus*/μl hemolymph, close to the dose required for maximal reduction of total circulating hemocytes according to Gagen and Ratcliffe (1976). These authors also reported that about 95% of a 10^5/μl dose of bacteria was removed by 5 min after injection by phagocytosis and nodule formation. However, pathogens, such as *B. cereus*, are not killed during the

TABLE 3.2. PLASMATOCYTE DEPLETION AFTER INJECTION OF COMPARABLE DOSES OF VARIOUS BACTERIA

Bacterial Species	Percent Plasmatocytes	Number of Larvae	Mortality (% dead at 24 hr)
Bacillus cereus	< 1	16	100
B. cereus (heat killed)	7 ± 3	5	0
B. subtilus	12 ± 2	8	0
B. megaterium	20 ± 2	8	0
Micrococcus luteus	22 ± 3	9	0
Staphylococcus aureus	28 ± 5	9	0
Escherichia coli B	24 ± 2	7	0
Serratia marcescens	34 ± 3	6	100
Proteus vulgaris	50 ± 3	8	0
None	46 ± 3	12	0

process of nodule formation but actually multipy to produce a fatal septicemia (Walters and Ratcliffe, 1983). It is tempting to speculate that the plasmatocyte depletion is an immediate result of nodule formation; however, early nodules contain mainly granulocytes, and only later are plasmatocytes recruited in large numbers. It is possible that the plasmatocytes undergo a change in affinity for the membranes lining the hemocoel and probable that the free, circulating hemocytes and a population of hemocytes adhering to the hemal linings are in dynamic equilibrium.

3.2.3. Response to Other Bacteria

Plasmatocyte depletion was not a general response to injection of bacterial suspensions. Table 3.2 shows that the response varied from total depletion after *B. cereus* injection to no response after injection of *Proteus vulgaris*. Clearly, pathogenicity was not the determining factor, since heat-killed *B. cereus* was about as effective as viable *B. cereus* in producing the effect; also highly pathogenic *Serratia marcescens* induced only slight plasmatocyte depletion. The data indicated that some degree of plasmatocyte depletion is seen during infection with many bacterial species. However, the response is particularly strong after exposure to some antigenic determinant shared by *B. cereus* and *B. subtilis*, which are systematically closely related.

3.2.4. Plasmatocyte Depletion Factor, Hemolymph Transfer Studies

The speed of plasmatocyte depletion suggested the possibility that initial bacteria-hemocyte interactions might trigger the release of some soluble factor that could subsequently affect the behavior of the plasmatocytes. Indeed this appeared to be the case, and plasmatocyte depletion factor (PDF) was identified in the cell-free hemolymph of *Galleria* injected with *B. cereus* (Chain and Anderson, 1983a). As in the previous study, *Galleria* were in-

TABLE 3.3. HEMOLYMPH TRANSFER STUDIES

Treatment of Recipient	Percent Circulating Plasmatocytes	Number of Larvae
Untreated control	46 ± 3	12
10 µl viable *Bacillus cereus* (bacteria injected 5 min before hemolymph withdrawal)	< 1	16
10 µl hemolymph from *B. cereus*–injected donor (bacteria injected 5 min before hemolymph withdrawal from donor)	18 ± 5	14
10 µl hemolymph from *B. cereus*–injected donor (bacteria injected 3 hr before hemolymph withdrawal from donor)	20 ± 4	14

jected with about 10^6 viable *B. cereus*, and hemolymph was collected at various times thereafter. Hemolymph was mixed (2.5:1) with Grace's medium containing a trace of phenylthiourea and centrifuged for 10 min at 800 g. The cell-free hemolymph was ultrafiltered to remove any bacteria. Differential hemocyte counts were made on naive larvae 1 hr after injection of 10 µl of sterile, cell-free hemolymph from *B. cereus*–injected animals.

The data from these hemolymph transfer studies are given in Table 3.3. It was apparent that injection of cell-free hemolymph from untreated donors produced significant plasmatocyte depletion in recipient larvae. This depletion probably resulted from the release of PDF from the hemocytes during hemolymph collection and centrifugation. The hemocytes of *Galleria* are relatively unstable during handling, and interaction with foreign surfaces can lead to disruption and degranulation. However, when cell-free hemolymph from *B. cereus*–injected donors was injected, the percentage of circulating plasmatocytes was significantly ($P < .01$) further reduced. The reduction was not as marked as that produced by direct injection of *B. cereus*, but it must be remembered that the donor hemolymph is diluted about eightfold upon injection into a naive larva's hemocoel, based on the hemolymph volume estimate of Gagen and Ratcliffe (1976). It was interesting that maximal release of PDF could be detected by hemolymph transfer studies 5 min after bacterial injection when plasmatocyte depletion was also marked. Comparable PDF levels were seen in hemolymph samples taken 3 hr after injection of *B. cereus*. The most likely source of PDF was thought to be the hemocytes themselves, since they are the first cells to make contact with injected bacteria. To test this hypothesis, we designed the following method of measuring in vitro release of PDF.

3.2.5. PDF Production In Vitro

Whole hemolymph (25 µl) mixed with 10 µl of Grace's medium containing a trace of phenylthiourea was incubated 15 min in 200-µl wells of a microtiter

TABLE 3.4. PDF SECRETION BY HEMOCYTES IN VITRO

Incubation Mixture	Plasmatocytes (as % total hemocytes) after Injection of 10 μl Sterile Supernatant from Incubation Mixture
Hemocytes alone (no bacteria)	42 ± 3 (16)
Bacillus cereus alone (no hemocytes)	36 ± 6 (5)
Hemocytes + *B. cereus*	17 ± 3 (14)
Hemocytes + *Proteus vulgaris*	37 ± 2 (8)

plate. An additional 50 μl of Grace's medium were added and the plates incubated at 30°C for 45 min to permit cellular adhesion. The cells were washed three times with Grace's medium and 25 μl of a *B. cereus* suspension (approximately 10^8/ml) were added and incubated for 15 min at 30°C. The supernatant medium was then removed, centrifuged at 1,600 g for 10 min to remove suspended bacteria, and tested for plasmatocyte depletion activity, as described above. Bacterial counts in 10-μl samples of these supernatants were never more than 100 and usually less than 30, a concentration already shown not to produce plasmatocyte depletion.

The idea that the hemocytes themselves are a likely source of PDF is supported by the data in Table 3.4. Supernatants in which *B. cereus* or hemocytes alone had been incubated had little effect on plasmatocyte numbers when injected into *Galleria* larvae. However, when they were incubated together, a significant release of PDF was consistently recorded. PDF release in vitro was shown to depend on the numbers of bacteria in the incubation mixture; a half-maximal reaction was obtained with 5×10^7 *B. cereus*/ml, corresponding to a hemocyte-bacteria ratio of about 1:2. It was interesting to note that *P. vulgaris*, which caused no plasmatocyte depletion following direct injection, also failed to trigger PDF secretion from hemocytes in vitro. Although bacteria-cell incubations were routinely carried out for 15 min, considerable PDF release was noted after only 5 min, similar to the in vitro situation. In view of the rapidity of its release, it seems unlikely that de novo synthesis is involved. Probably, stimulation of particular cell surface receptors by bacterial binding mediates the release of PDF. Once cells are stimulated to release PDF by in vitro contact with *B. cereus*, no additional PDF can be detected in the medium upon subsequent 15-min exposures.

3.3. HYPERSECRETION OF ACID MUCOPOLYSACCHARIDE

3.3.1. Prior Investigations and Background

In our studies of the responses of *G. mellonella* to injections of *B. cereus*, we had occasion to examine many hemolymph samples from injected and control larvae. It soon became apparent that hemocytes taken from injected

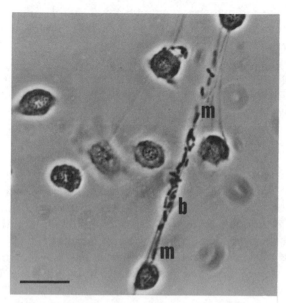

Figure 3.1. Hemocyte monolayer preparation from larvae injected 1 hr previously with 10^6 viable *Bacillus cereus* in 10 μl Grace's insect tissue culture medium. Granulocytes have secreted acid mucopolysaccharide strands (m) to which numerous bacteria (b) have become affixed. (Scale bar = 12 μm.)

larvae often had long strands of mucuslike material attached to them and that few such strands were usually seen on cells from control animals. Furthermore, other hemocytes and/or bacteria were often attached to the strands (Fig.3.1).

Increased "stickiness" of insect hemocytes following exposure to foreign bodies can be mediated by prophenoloxidase activation and/or acid mucopolysaccharide discharge. Melanization, mediated by phenoloxidase, is recognized as an important component of defense reactions in several invertebrate groups (Taylor, 1969). The role of the prophenoloxidase-activating system in the immune responses of arthropods has been the subject of several recent reviews (Söderhäll, 1982; Söderhäll and Smith, 1983; see also Chapter 9). Zymosan, a yeast cell wall preparation, was shown to activate phenoloxidase in immunized *G. mellonella* larvae (Pye, 1974). In crustaceans, β-1,3-glucans found in fungal cell walls and lipopolysaccharides (LPSs) from bacterial cell walls specifically activate prophenoloxidase. The reaction is very sensitive, being triggered by glucans and LPS concentrations as low as 10^{-10} g/ml. After activation, phenoloxidase becomes sticky and attaches strongly to foreign surfaces. Phenoloxidase activation has been suggested to play a role in opsonization, coagulation, and antimicrobial activity. Prophenoloxidase activation in insects is currently under study, particularly in *Bombyx mori* (Ashida and Dohke, 1980) and *G. mellonella* (Ratcliffe et al., 1984).

Galleria hemocyte monolayers phagocytized *B. cereus* more avidly in the presence of laminarin (a β-1,3-D-glucan extracted from fungal cell walls) or *Escherichia coli* endotoxin. In this study, the prophenoloxidase system was activated by laminarin but not by endotoxin. Activation is concomitant with release of the enzyme from granulocytes upon contact with foreign surfaces, which may represent the recognition stage in the insect's immune response. The second stage involves the phagocytosis or encapsulation of foreign materials coated with factors derived from activation of the prophenoloxidase system.

The sticky strands that we observed are probably identical to the adhesive matrix formed around the granulocytes of *Galleria* after injection of *B. cereus* (Ratcliffe and Gagen, 1976). Melanization was not seen in our preparations due to the presence of phenylthiourea. Clearly, these strands may play roles in coagulation and nodule formation as well as in removal of bacteria from the fluid portion of the hemolymph. The presence of acid mucopolysaccharides in the adhesive extracullular fibrillar substance around stressed spherulocytes and granulocytes was established by Ratcliffe (1975).

3.3.2. Acid Mucopolysaccharide as a Granulocyte Marker Enzyme

G. mellonella hemolymph contains five cell types—prohemocytes, plasmatocytes, granulocytes, spherulocytes, and oenocytoids—according to the classification scheme of Price and Ratcliffe (1974). However, in our preparations the majority of the cells could be identified as either plasmatocytes or granulocytes, based on their morphology and spreading characteristics. The other three types constituted only a small percentage of the population and/or were weakly adherent and were lost during washing of the cell monolayers. Upon histochemical examination (Chain and Anderson, 1983b), it was found that both plasmatocytes and granulocytes were rich in lysosomes containing acid phosphatase and nonspecific esterase. However, the granulocytes contained a population of alcian blue–positive granules not present in the plasmatocytes. This dye, used at pH 3.0, can be used as a specific stain for acid mucopolysaccharides (Pearse, 1968). Acid mucopolysaccharide–rich granules had been previously described in insect hemocytes (Ashhurst and Richards, 1964; Lea and Gilbert, 1966; Gupta and Sutherland, 1967; Neuwirth, 1973; Ratcliffe, 1975). In our preparations, these granules were restricted to the granulocyte class; however, occasionally some cells with typical granulocytic features were seen with few or no alcian blue–staining granules. It is possible that these cells represent an immature stage in granulocyte development. The results shown in Table 3.5 indicate that a series of developmental changes in granulocyte maturation can be observed in terms of a positive correlation between the number and size of alcian blue–staining granules and cell size. Therefore, it is possible that the granule-

TABLE 3.5. CORRELATION BETWEEN GRANULOCYTE SIZE AND INTENSITY OF ALCIAN BLUE–STAINING GRANULES

Cell Diameter (μm)	Staining Intensity[a]	Percent Granulocytes Scored[b]
4.9–6.3	0.8 ± 0.7	24.7
6.4–7.6	1.6 ± 0.9	36.7
7.7–9.0	2.4 ± 1.0	28.0
9.1–10.4	3.0 ± 0.0	10.6

[a] Scoring scale: 0, no staining; 1, a few positive granules; 2, many positive granules; 3, many large positive granules.
[b] Percentage of granulocytes in each cell size group, based on counts of 150 granulocytes from 8 different animals.

associated acid mucopolysaccharide content of insect granulocytes, as well as the numbers of lysosomes, increases with maturation. The numbers of lysosomes and the activity of lysosomal enzymes increases with developmental stage and level of activation in the macrophages of higher animals.

In a recent paper, Ashhurst (1982) reported that, under acidic conditions, the granules of *Galleria* spherulocytes also stain intensely with alcian blue. These results of a battery of histochemical tests indicate the presence of a sulphated, glycosaminoglycanlike polymer in the granules of granulocytes, a probable subclass of granulocytes.

3.3.3. Extracellular Acid Mucopolysaccharide

Following experimental infection with *B. cereus*, there is an obvious hypersecretion of alcian blue–positive material by the circulating granulocytes of *G. mellonella*. The cell-associated, adhesive fibrils were seen to entrap and bind large numbers of injected bacteria as well as other cells and extracellular debris (Fig. 3.1). This sticky cell coat probably also participates in nodule and/or capsule formation. Apparently, secretion of mucopolysaccharide strands proceeds under normal conditions; 21% ± 3% of the granulocytes from untreated larvae were engaged in this activity. However, the response was augmented after *B. cereus* injection, when 56% ± 3% of the granulocytes had alcian blue–positive material associated with their surfaces. In many cells, alcian blue–positive granules were concentrated in the area of cytoplasm from which the strands were secreted and appeared to be involved in the secretory process (Fig. 3.2).

Mucus-secreting blood cells that hypersecrete during infection have been demonstrated in another invertebrate phylum by Bang and Bang (1965, 1975, 1980); such cells may represent a more widespread phenomenon of nonspecific cellular immune systems than previously suspected.

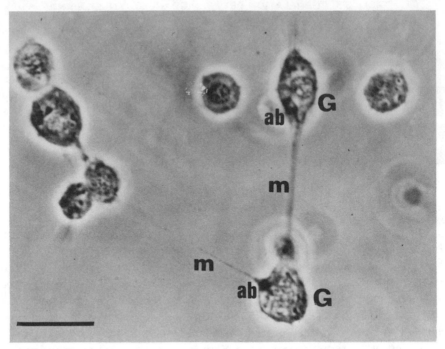

Figure 3.2. Granulocytes from *B. cereus*–injected *Galleria*, as in Figure 3.1. Many of the granulocytes (G) are connected by mucopolysaccharide strands (m) apparently secreted by clusters of alcian blue–stained granules (ab). (Scale bar = 10 μm.)

3.4 SUMMARY

The responses of *G. mellonella* to pathogenic bacteria, such as *B. cereus*, are varied and complex. Two immediate effects of experimental infection have been summarized in this chapter: plasmatocyte depletion and acid mucopolysaccharide hypersecretion. Within minutes of contact with *B. cereus*, the hemocytes release a humoral factor that mediates the rapid margination of circulating plasmatocytes. Presumably, these cells attach to the hemal lining in order to reach a substrate suitable to support their movement to inflammatory foci, hemocyte-bacteria clumps, and/or nodules. It is suggested that both plasmatocyte depletion factor and plasmatocyte chemotactic factor are released by granulocytes following binding of bacteria to particular cell surface receptors. It appears that the receptors are somewhat specific, since not all bacteria—indeed, not all foreign particulates—have equal ability to trigger plasmatocyte depletion. In addition to possibly releasing plasmatocyte regulatory factors, the granulocytes have been shown to rapidly become sticky due to the release of acid mucopolysaccharide and/or phenoloxidase. Bacteria are trapped in this material, melanin deposition occurs, the granulocytes undergo physical deterioration, and plasmatocytes arrive to surround the mass to make a typical nodule. Important details concerning characterization of the humoral factors and cell surface receptors involved have yet to be elucidated.

REFERENCES

Ashhurst, D.E. 1982. Histochemical properties of the spherulocytes of *Galleria mellonella* L. (Lepidopteria: Pyralidae). *Int. J. Insect Morphol. Embryol.* **11:** 285–292.

Ashhurst, D.E. and A.G. Richards. 1964. Some histochemical observations on the wax moth *Galleria mellonella*. *J. Morphol.* **114:** 247–254.

Ashida, M. and K. Dohke. 1980. Activation of prophenoloxidase by the activating enzyme of the silkworm, *Bombyx mori*. *Insect Biochem.* **10:** 37–47.

Bang, B.G. and F.B. Bang. 1965. The mucus secretory apparatus of the free urn cell of *Sipunculus nudus*. *Cah. Biol. Mar.* **17:** 423–432.

Bang, B.G. and F.B. Bang. 1975. Cell recognition by mucus secreted by the urn cell of *Sipunculus nudus*. *Nature* (London) **253:** 634–635.

Bang, B.G. and F.B. Bang. 1980. The urn cell complex of *Sipunculus nudus*: A model for study of mucus-stimulating substances. *Biol. Bull.* (Woods Hole) **159:** 571–581.

Chain, B.M. and R.S. Anderson. 1982. Selective depletion of the plasmatocytes in *Galleria mellonella* following injection of bacteria. *J. Insect Physiol.* **28:** 377–384.

Chain B.M. and R.S. Anderson. 1983a. Inflammation in insects: The release of a plasmatocyte depletion factor following interaction between bacteria and haemocytes. *J. Insect Physiol.* **29:** 1–4.

Chain, B.M. and R.S. Anderson. 1983b. Observations on the cytochemistry of the hemocytes of an insect, *Galleria mellonella*. *J. Histochem. Cytochem.* **31:** 601–607.

Gagen, S.J. and N.A. Ratcliffe. 1976. Studies on the *in vivo* cellular reactions and fate of injected bacteria in *Galleria mellonella* and *Pieris brassicae* larvae. *J. Invertebr. Pathol.* **28:** 17–24.

Gupta, A.P. and D.J. Sutherland. 1967. Phase contrast and histochemical studies of spherule cells in cockroaches (Dictyoptera). *Ann. Entomol. Soc. Am.* **60:** 557–564.

Lackie, A.M. 1981. Immune recognition in insects. *Dev. Comp. Immunol.* **5:** 191–204.

Lea, M.S. and L.E. Gilbert. 1966. The hemocytes of *Hyalophora cecropia* (Lepidoptera). *J. Morphol.* **118:** 197–215.

Neuwirth, M. 1973. The structure of the hemocytes of *Galleria mellonella* (Lepidoptera). *J. Morphol.* **139:** 105–124.

Pearse, A.G.E. 1968. *Histochemistry: Theoretical and Applied*, pp. 345–346, vol. 1, 3d ed. Little, Brown, Boston.

Price, C.D. and N.A. Ratcliffe. 1974. A reappraisal of insect haemocyte classification by the examination of blood from fifteen insect orders. *Z Zellforch.* **147:** 537–549.

Pye A.E. 1974. Microbial activation of prophenoloxidase from immune insect larvae. *Nature* (London) **251:** 610–613.

Ratcliffe, N.A. 1975. Spherule cell–test particle interactions in monolayer cultures of *Pieris brassicae* hemocytes. *J. Invertebr. Pathol.* **26:** 217–223.

Ratcliffe, N.A. and S.J. Gagen. 1976. Cellular defense reactions of insect hemocytes in vivo: Nodule formation and development in *Galleria mellonella* and *Pieris brassicae* larvae. *J. Invertebr. Pathol.* **28:** 373–382.

Ratcliffe, N.A. and A.F. Rowley, 1979. Role of hemocytes in defense against biological agents, pp. 331–414. In A.P. Gupta (ed.) *Insect Hemocytes.* Cambridge University Press, Cambridge.

Ratcliffe, N.A., Leonard, C. and A.F. Rowley. 1984. Prophenoloxidase activation: Nonself recognition and cell cooperation in insect immunity. *Science* (Washington, DC) **226:** 557–559.

Söderhäll, K. 1982. Prophenoloxidase activating system and melanization—A recognition mechanism of arthropods? A review. *Dev. Comp. Immunol.* **6:** 601–611.

Söderhäll, K. and V.J. Smith. 1983. The prophenoloxidase activating system: A complement-like pathway in arthropods? pp. 1–6. In J. Aist and D.W. Roberts (eds.) *Infection Processes of Fungi.* Rockefeller Foundation, New York.

Taylor, R.L. 1969. A suggested role for the polyphenol-phenoloxidase system in invertebrate immunity. *J. Invertebr. Pathol.* **14:** 427–428.

Walters, J.B. and N.A. Ratcliffe. 1983. Variable cellular and humoral defense reactivity of *Galleria mellonella* larvae to bacteria of differing pathogenicities. *Dev. Comp. Immunol.* **7:** 661–664.

CHAPTER 4

Antifungal Cellular Defense Mechanisms in Insects

Alain Vey

Research Station of Comparative Pathology
St. Christol-les-Ales, France

Peter Götz

Division of Biology
Institute for General Biology
Free University of Berlin
Berlin, West Germany

4.1. INTRODUCTION

Insects exhibit various types of cellular and humoral immune reactions, as recently reviewed by Ratcliffe and Rowley (1979) and Götz and Boman (1985). While data concerning antibacterial immunity have often been reported, only a few papers have been devoted to antifungal defense mechanisms, in spite of the importance of fungi as parasites. The major reaction observed in insects against fungal parasites is the formation of multicellular hemocytic capsules, or granuloma (Flandre et al., 1968; Vey, 1984; (see Fig. 4.1). Because of its significance, this type of antifungal immune mechanism is discussed in detail in this chapter. Phagocytosis also commonly occurs during development of fungal infections, but it does not seem to be very

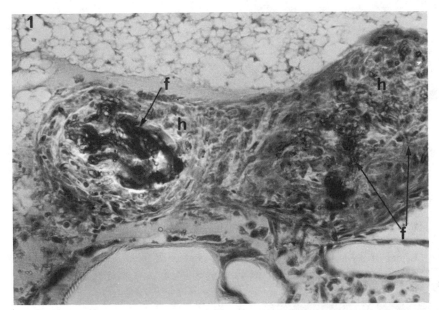

Figure 4.1. Hemocytic granuloma formed in a larva of *Galleria mellonella* injected with a suspension of conidia of *Aspergillus niger* 7 days after inoculation (h, hemocytic envelope; f, melanized fungal filaments) (×280).

efficient. Humoral reactions probably intervene with cellular mechanisms, but information concerning them is limited (Boczkowska, 1935; Müller-Kögler, 1965). New data based on actual biochemical and immunologic techniques are necessary for further discussion of this aspect of antifungal defense.

4.2. PHAGOCYTOSIS

Phagocytosis occurs during mycoses of insects when the fungus produces free-floating blastospores. Although phagocytosis is generally not efficient against virulent fungi, since the ingested cryptogamic cells are able to attack the surrounding hemocyte (Speare, 1920; Paillot, 1930), it plays a defensive role against weakly pathogenic fungi. Sussman (1952a) has reported that in the cecropia moth, phagocytosis is more intense with a nonpathogen, *Aspergillus niger*, than with *A. flavus*, which is pathogenic for this moth. Kawakami (1965) also considered the phagocytic activity of hemocytes to play an important role in the defense of *Bombyx mori* larvae against nonvirulent hyphomycetes.

Successes have been reported in the study of phagocytosis in insects under in vitro conditions (Rabinovitch and De Stefano, 1970; Anderson et al., 1973; Anderson 1974, 1975, 1976; Ratcliffe and Rowley, 1975) and in the observation of the dynamics of the reaction in the host by means of microcinematography (Marschall, 1966). Fungal cells, however, have not been used as test particles in such investigations, except by Vago and Vey (1972). These authors have shown that various methods of uptake of fungal particles occur in invertebrate hemocytes. After contact with a pseudopodium, small unicellular blastospores are very rapidly attracted to and incorporated into the cytoplasm without the formation of a phagocytic vacuole. With longer mycelial filaments, a slow process of attachment and spreading of the cell in close contact with the fungal wall, followed by the binding of the mycelium, are observed; the cells thus succeed in the uptake of branched elements of a length much greater than their diameter.

4.3. ENCAPSULATION

The formation of concentric layers of densely packed hemocytes around fungal elements has been reported by Speare (1920), Boczkowska (1935), Sussman (1951a), and Amargier (1967). Certain authors have described melanotic structures of an undetermined nature (Lepesme, 1938; Weber, 1939; Smirnoff, 1968) that probably originate from hemocytic aggregations. The efficiency of hemocyte aggregation around a pathogen and the melanization of the cellular capsule vary, depending on the type of fungus involved and its mode of penetration.

4.3.1. Induction by Injection

We have studied the encapsulation mechanism in detail in some Lepidoptera, particularly *Galleria mellonella* L., by injecting them with conidia of *A. niger* (Vey, 1971a,b, 1985). Attachment of the first hemocytes (granulocytes) to the spores and further recruitment of others occur rapidly, in less than 30 min. The attached hemocytes discharge part of their cytoplasmic content or disintegrate. Coincidently, melanization occurs near the surface of the conidia. The granuloma, or capsule, progressively acquires a tight consistency, owing to the close packing of the blood cells around the conidia. After 24 hr, the hemocytic envelope is composed of numerous concentric layers of flattened blood cells (plasmatocytes), and the conidia are covered with a thick but irregular deposit of melanin (Fig. 4.2). Although enclosed in a tough pseudotissue, the spores of *Aspergillus* start germination.

Attraction of additional hemocytes leads, within a few days, to the formation of voluminous capsules, sometimes reaching more than 1 mm in size (Fig. 4.1). The newly attached hemocytes show a fibroblastlike appearance, as described by Flandre and coworkers (1968). A material, presumed to be polysaccharide or collagen, occurs in the intercellular spaces of the flattened plasmatocytes. Four layers can be distinguished in the capsule. At the periphery, slightly modified hemocytes are visible. The middle layer is composed of concentric layers of flattened cells that contain abundant microtubules and membranes of endoplasmic reticulum, which are arranged parallel to the surface of the flattened cells (Fig. 4.3). The hemocytes of the internal region of the capsule are slightly elongated or isodiametric, and large inclusions similar to those described by Grimstone and colleagues (1967) in capsules formed around foreign implants and to the autophagic vacuoles studied by Beaulaton (1967a,b), Fain-Maurel and Cassier (1969), and Gharagozlou (1970) are visible in their cytoplasm (Fig. 4.4). The presence of such bodies may result from an autolysis of the aggregated hemocytes that may be provoked by lack of oxygen and nutrients or due to the effect of noxious compounds produced by the fungus. A fourth zone is found directly adjacent to the fungal body. The contents of these hemocytes are relatively electron-dense, almost homogeneous, with rarely recognizable cytoplasmic organelles.

In cases in which larvae survived fungal injection, the capsules continued to persist up to the adult stage. Such older capsules showed little change except for increased melanization within their cellular envelopes. Some granuloma are enveloped by hypodermal cells and separated from the hemocoel

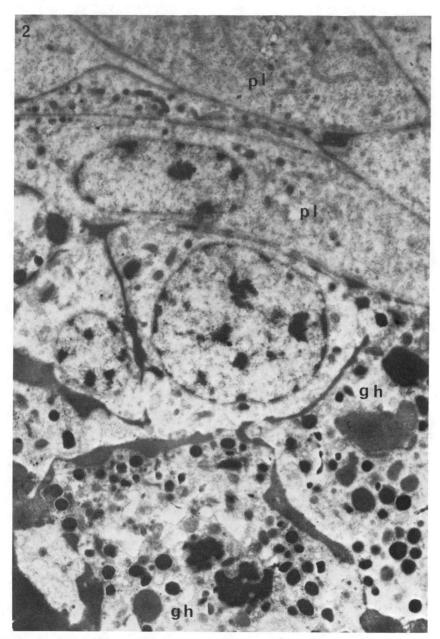

Figure 4.2. Types of hemocytes participating in the encapsulation of fungi in Lepidoptera (gh, granulocytes, partly disintegrated; pl, plasmatocytes in a more external position) (×8000).

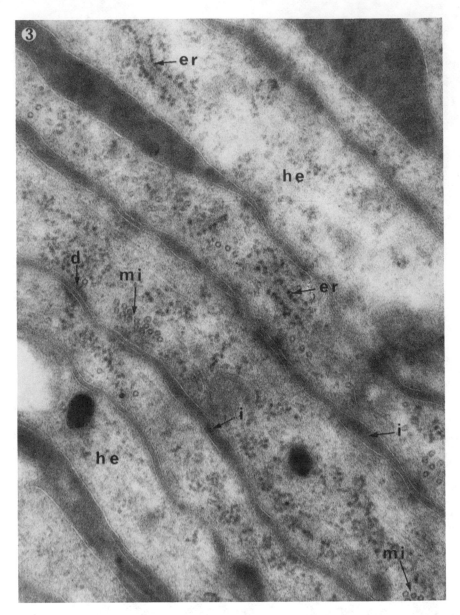

Figure 4.3. Ultrastructure of the envelope of a granuloma formed around fungal elements (he, elongated hemocytes; er, endoplasmic reticulum; mi, microtubules; i, intercellular material; d, desmosomes) (×55,000).

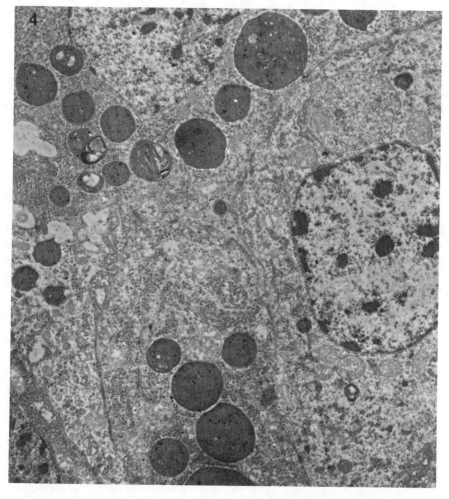

Figure 4.4. Hemocytes of the internal part of the envelope of a 7-day capsule showing large, dense cytoplasmic granules ($\times$62,500).

by a newly formed cuticle. They eventually degenerate into brownish masses that are eliminated at larval ecdysis.

4.3.2. Fungal Invasion through Wounds

Certain fungi that are unable to penetrate the integument or gut lining can attack insects via cuticular wounds. One of these fungi is the zygomycete *Mucor hiemalis* (Hurpin and Vago, 1958). The mechanism of such infection has been discussed by Vago (1959). Heitor (1962) demonstrated experimentally the capacity of *M. hiemalis* to settle at the clot area of a wound and penetrate the integumental barrier. Vey (1968) reported that the hyphae of *Mucor* penetrate the pseudotissue formed at the wound site and reach the body cavity, where they are trapped in a hemocytic envelope. Nevertheless, the hyphae continue to grow and invade the adipose tissue of *Galleria*, and the progression of the mycosis becomes fatal in spite of the fact that hemocytic reactions continue and decrease in intensity only just before the death of the larva. It is obvious that the efficiency of encapsulation in protecting the insect is very low, due to the fact that *Mucor* has rapidly growing hyphae of relatively strong mechanical power and produces compounds that are toxic to the host (Vey and Quiot, 1976).

4.3.3. Invasion through the Integument

Highly pathogenic fungi often penetrate the intact integument. *Beauveria bassiana*, ubiquitious hyphomycete, and *Metarhizium anisopliae* are among the best known of such pathogenic fungi (Ferron, 1985). Their hyphae trigger a flux of hemocytes at the area of penetration, as observed by Paillot (1933). The blood cells moving to this zone form a dense pseudotissue, accumulate between the hyphae, and phagocytize some of the blastosphores. Later, recruitment and aggregation of hemocytes becomes inadequate. As a result, the filaments and spores of the fungus are only partially enveloped by a few blood cells, which show little tendency to adhere. Typical encapsulation of such highly pathogenic fungi occurs only after weak local infections (Vey and Fargues, 1977).

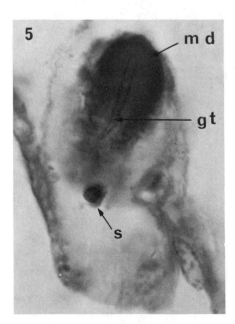

Figure 4.5. Germination of a spore of *Aspergillus flavus* in a trachaea of *Locusta migratoria* and penetration of the germ tube. Strong melanization occurs before cellular encapsulation (s, spore; gt, germ tube; md, melatonic deposit) (×450).

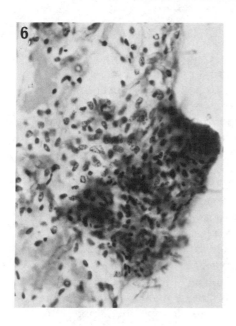

Figure 4.6. Infection through the respiratory system. Aggregation of blood cells around the penetrating hyphae in a late stage of the reaction ($\times$350).

4.3.4. Invasion through the Respiratory System

Beauveria, Metarhizium, and weaker pathogenic fungi, such as *Aspergillus* sp., sometimes infect the insect via the respiratory system, as in the Orthoptera. Defense reactions against such infections have been studied under experimental conditions by Ogloblin and Jauch (1943) and Michel (1981). The conidia of *Beauveria* and *Aspergillus* germinate in the tracheae, and the developing hyphae penetrate the taenidium. Their apexes provoke a rapid and strong melanization reaction, which occurs without any visible involvement of blood cells and gives rise to large brown areas at the inner surface of the respiratory organs (Figs. 4.5, 4.6). The aggregation of blood cells around the penetrating fungal elements starts later between the intima and the tracheal cells. The mechanism by which the hemocytes travel through the tracheal epithelium is still unknown. The hemocytic granuloma increases in size, but the growing mycelium, while provoking the degradation of surrounding cells, finally reaches the hemocoel (Michel, 1981).

4.3.5. Reactions to Different Species and Pathotypes

Reactions to diverse species of hyphomycetes have been studied comparatively by injecting the same insect with deuteromycetes showing different degrees of virulence: *A. niger, A. flavus, A. tamarii, M. anisopliae, B. bassiana, Penicillium granulatum,* and *Trichothecium roseum.* When injected into the body cavity of *G. mellonella* larvae, all of these entomogenous fungi triggered granuloma formation (Vey, 1971b). With all the pathogens tested, the initial encapsulation reaction was identical. However, differences were observed from one fungal species to another regarding the number of aggregating hemocytes, the extent of their transformation, and the intensity of melanization. These differences were important in determining how quickly the mycelial filaments were able to penetrate the cellular envelope. Thus, the outgrowth of fungal elements was particularly rapid in the case of the highly pathogenic fungus *B. bassiana* (Vey and Vago, 1971).

A possible specificity of the encapsulation process has been investigated at an infraspecific level using different pathotypes of *M. anisopliae* whose immunologic and enzymatic characteristics have been extensively described (Duriez et al., 1981; Fargues et al., 1981). The specificity of the parasitic action of these isolates on scarabeid larvae does not seem to be related to their capacity to induce in a particular host a weak or strong cellular reaction, in contrast to the case of implantation of eggs of insect parasitoids (Carton, 1976). However, the long-term development of the granuloma reaction varies depending on the isolate tested. The aggregate of hemocytes is pierced by the specific pathotype but not by the nonspecific one. An antihemocytic toxic effect of the specific fungus appears to be one essential factor in explaining this situation. A probable intervention of specific inhibitors or the absence of necessary nutritional factors was also postulated (Vey et al., 1982).

4.3.6. Melanization

If a fungus penetrates the cuticle of an insect, it is not generally melanized inside the exoskeleton, even if phenoloxidase is in the vicinity of the mycelial filaments (Michel, 1981). However, certain dipteran species develop noncellular humoral capsules at this level (Götz and Vey, 1974). When hyphae develop in the hemocoel, they trigger a melanization reaction. The yellowish to deep brown deposit attached to the surface of fungal propagules shows all the physical and biochemical characteristics of melanins (Vey, 1971a; see Fig. 4.7). Melanization is the result of phenoloxidase activity that transforms phenolic substrates into melanins. The prophenoloxidase-activating system has been analyzed and described in crustaceans (Söderhäll and Smith, 1984; Chapter 9). Nonself molecules (e.g., β-1, 3-glucans) activate prophenoloxidase (proPO) through a complex chain of reactions including a serine protease and other still unknown factors (Söderhäll, 1981, 1982). A similar pathway also occurs in insects. In this group, the prophenoloxidase is considered to be present in the blood plasma (Ashida and Dohke, 1980; Ashida, 1981) or in the hemocytes (Leonard et al., 1985).

Söderhäll and associates (1984) have shown that the hemocytes of *Astacus astacus* exhibit a stronger reaction to blastospores of *B. bassiana* coated with an activated hemocyte lysate than to spores treated with plasma or buffer. These results constitute the first evidence that hemocyte lysate contains activated phenoloxidase, which attaches to fungal cell walls, together

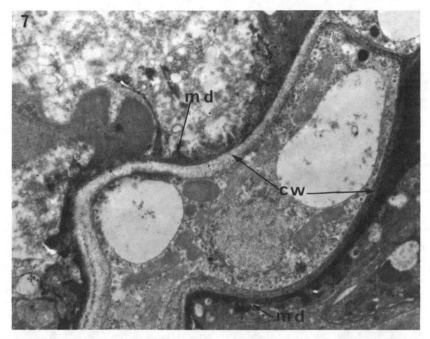

Figure 4.7. Irregular melanotic deposit covering the surface of an encapsulated hypha of *Aspergillus* in *Galleria* (md, melanotic deposit; cw, cell wall of the fungal filament) (×9800).

with four unidentified proteins of the proPO system and that this enzyme system may be considered to have an opsonic effect on arthropod blood cells.

4.4. IN VITRO STUDIES OF ENCAPSULATION

Our knowledge of multicellular defense reactions in invertebrates is mainly based on histologic and cytologic studies. However, some experiments have also been carried out in vitro, allowing continuous observation of the progress of these mechanisms (Vey et al., 1968; Vey, 1969b; Unestam and Nylund, 1972). The most favorable conditions for the experimental production of granuloma have been obtained by tissue culture, particularly of lepidopteran and coleopteran blood cells (Vago and Chastang, 1958, 1960; Vago and Quiot, 1969).

Encapsulation of the conidia of *A. niger* and *A. flavus* by *Galleria* hemocytes has been studied in Vago and Chastang's BM 22 medium (1960). Glutathion, an inhibitor of melanization (Wyatt, 1956), was added to the medium. The gathering of cells around the fungal elements starts very early,

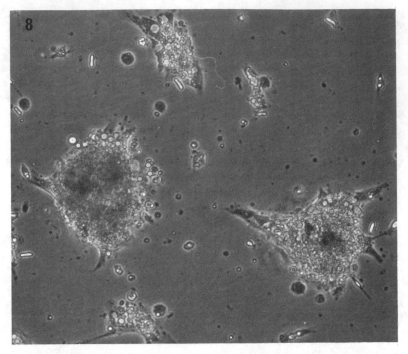

Figure 4.8. Voluminous granuloma formed in in vitro conditions by the aggregation of hemocytes of *Oryctes* around the spores of Metarhizium partly inactivated (×300).

some hemocytes attaching to the particles within 5 min. The hemocytic envelope undergoes a reduction in volume owing to the tendency of the cells to flatten, and the shape of the granuloma becomes more regular. This initial stage of encapsulation is followed by a progressive participation of other hemocytes during the next 24 hr. Among the hemocytic types present in the preparation, only coagulocytes (granulocytes) and plasmatocytes participate in encapsulating *Aspergillus* spores (Vey et al., 1968; Vey 1969b). Conidia of *Metarhizium* treated by heating at 45°C have been used to induce encapsulation. The ability of the spores to trigger the reaction is not altered, but the development of the fungus is postponed. Hemocytes of the coleopteran *Oryctes rhinoceros*, previously cultivated by Quiot and coworkers (1972), have been used as a source of insect blood cells without the addition of melanization inhibitor. After 24 hr of incubation with *Metarhizium*, small groups of hemocytes gathered around the spores, and within 3 days, voluminous aggregates had formed. A few conidia produced germ tubes (Fig. 4.8). The central areas of these capsules showed an abundant melanotic deposit comparable to that occurring in vivo. Thereafter, the disintegration of granuloma started under the influence of the growing pathogen.

The encapsulation of fungi in insects has not been analyzed in detail by means of time-lapse cinematography. However, Vey and colleagues (1975), using mollusk hemocytes, demonstrated that the application of microcinematographic techniques is particularly useful for experimental studies of the dynamics of immune processes of invertebrates at the cellular level.

Granuloma formed in experimentally infected larvae can be removed aseptically and transferred into fresh tissue culture medium (Vey, 1969b). Experiments carried out under these conditions demonstrate that most of the cells that participate in the formation of 2- or 3-day old capsules do not detach and disperse, probably as a consequence of severe modifications induced by their intimate association. Only the most peripheral hemocytes manage to detach themselves, appear mobile, and not show any apparent alteration. The influence of encapsulation on the vitality of the participating blood cells is even better demonstrated by the fact that in older granuloma (e.g., 14 days) very few hemocytes are able to regain their mobility. Explantation of granuloma has also provided information about the vitality of the encapsulated fungus (see Section 4.5.2.).

4.5. EFFECTIVENESS OF ENCAPSULATION

4.5.1. Resistance and Recovery of the Host

Although a large number of investigations have been devoted to acute fungal infections in insects, studies of fungal infection recovery are rare (Müller-Kögler, 1965). Encapsulation protects the host against weakly virulent fungi, such as *Aspergillus luchuensis* (Sussman, 1951a). In a population of *Galleria* larvae injected with a given dose of *A. niger* spores, the infection is not lethal for a fraction of the animals (Vey and Vago, 1969). Among these insects, some do not show any visible sign of infection, while in others a weakening, followed by an improvement, occurs. Studies of such recovering larvae show an initial development of the mycelium of *Aspergillus* and a strong hemocytic reaction, resulting in a complete encapsulation of the hyphae.

Studies have also been undertaken on wound infections, using *M. hiemalis* (Vey, 1968). Some of the infected hosts recover within a week because the hemocytes in these larvae form large aggregates in the area around the wound. The hyphae can grow within these aggregates without damaging other tissues of the larvae. Besides this, the formation of a neohypodermis isolates the largest part of the fungal elements from the internal tissues of the host. This structure is eliminated during the next molt.

4.5.2. Effect of Encapsulation on the Fungus

The effect of encapsulation on the pathogen can be followed by comparing the structures of fungal elements growing on an artificial medium and of hyphae trapped in granuloma for various periods of time. Two weeks after

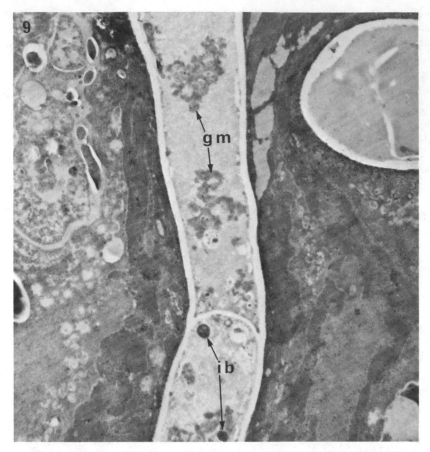

Figure 4.9. Altered structure of a filament of *Aspergillus* in a 5-week-old granuloma. Only dense inclusion bodies (ib) and some granular masses (gm) can be recognized in the cytoplasm of the hypha (×7500).

injection, alterations begin to appear in the filaments of *Aspergillus*. The disappearance of the internal structures of the fungal cells becomes more pronounced, and in 4- to 5-week-old capsules these cells appear structureless (Fig. 4.9). Their contents consist of granular masses and a few dense inclusion bodies that are sometimes associated with membrane remnants. Cryptogamic elements showing these types of lesions have obviously lost their vitality (Vey, 1985). The fungus reacts to the unfavorable conditions created by encapsulation by producing intrahyphal hyphae. Inside these small filaments, the normal structures of a living hypha can be recognized. Filaments of *Aspergillus* displaying such unaltered structures are therefore still alive, even in older granuloma.

The efficiency of encapsulation has also been characterized by in vitro

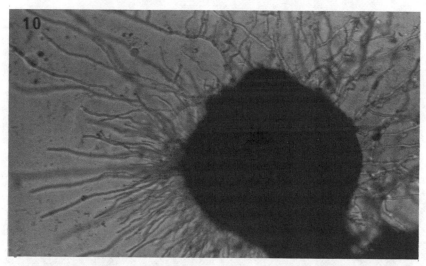

Figure 4.10. Abundant mycelium originating from a 1-week-old granuloma, 24 hr after transplantation in a mycologic medium ($\times$110).

experiments. The vitality of the encapsulated pathogen were checked by sampling granuloma and placing them in liquid medium. All 4- to 7-day-old capsules produced abundant mycelium within 24 hr (Fig. 4.10). Fewer filaments grew out from 2-week-old capsules, and with 5-week-old granuloma, the final mycelium growth rate was only 55%. When the infection dose is lower, the effect of encapsulation on the vitality of the microorganism is even stronger.

4.6. EFFECT OF VARIOUS FACTORS

4.6.1. Factors Produced by the Fungus

4.6.1.1. Toxins. It has been known for some time that fungi produce toxins during mycosis and that the death of the host is often principally due to these toxic substances. The production of such fungal toxins has been demonstrated in artificial cultures and to a lesser extent in vivo. The principal class of toxins produced are cyclodepsipeptides, as shown by studies of *M. anisopliae* (Kodaira, 1961; Roberts, 1966; Païs et al., 1981), *B. bassiana* (Hamill et al., 1973; Vey et al., 1973; Elsworth and Grove, 1977, 1980a,b see Fig. 4.11), and *B. brongnartii* (Frappier et al., 1975).

One possible role of toxins in the development of pathogenesis is their inhibitory effect on cellular defense reactions. This proposal is supported by histologic and ultrastructural observations, revealing early alterations in the hemocytes and retardation of blood cell aggregation. These events occur

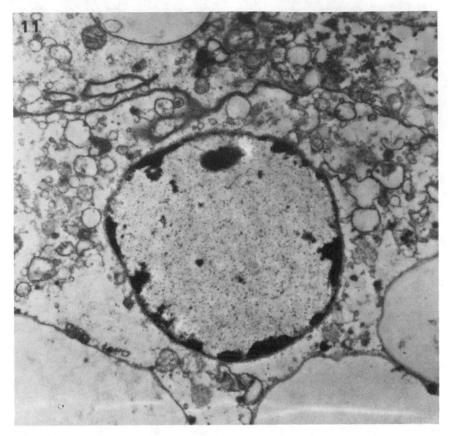

Figure 4.11. Cytotoxic effect of beauvericin on the cells of *Leucophaea maderae* cultivated in vitro (×9000).

during infection by various types of pathogens: (1) highly virulent hypho-mycetes (Vey and Vago, 1971), (2) certain pathotypes that are well adapted to their specific hosts (Vey et al., 1982), and (3) efficient wound parasites (Vey, 1968). Further evidence supporting the inhibitory role of toxins comes from an in vitro study of *Metarhizium* culture filtrates that contain com-pounds toxic for the cells of *Oryctes* and their inhibitory effect on encap-sulation (Vey and Quiot, 1975).

The immunodepressive effect of toxins has been conclusively established and characterized by experiments on insects by using purified compounds of the family of "destruxins." These are cyclodepsipeptides produced by the entomogenous deuteromycete *M. anisopliae*. These fungal peptides have strong cytotoxic effect, depress the metabolic activity of cells, and show insecticidal properties (Kodaira, 1961; Roberts, 1966; Fargues, 1981; Far-gues et al., 1985; Quiot et al., 1985; Vey et al., 1985). The effect of these

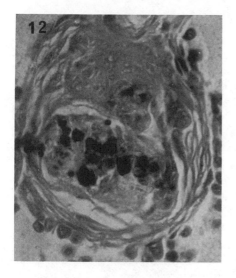

Figure 4.12. Immunodepressive effect of oligotoxicosis provoked by destruxin E. Fully developed reaction in a control insect (×420).

destruxins (DA, DB, and DE) has been studied at low dosage levels, which create conditions of oligotoxicosis in *Galleria* larvae. When caused by DE, these weakly toxic actions have a favorable effect on the development of infections provoked by *A. niger*. This phenomenon is linked to the influence of the toxin on the hemocytic reaction of the host and is due to an inhibitory effect on the movement of blood cells and to a disturbance in the mode of hemocyte association (Vey et al., 1985; see Figs. 4.12 and 4.13). This immunodepressive effect varies from one toxin to another. Thus, if the action

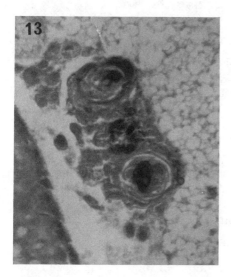

Figure 4.13. Immunodepressive effect of oligotoxicosis provoked by destruxin E. Partial inhibition of this process in a larva treated with the toxin (×420).

of DE is very strong, the effect of DA is reduced, and DB is completely inefficient.

4.6.1.2. Compounds of the Fungal Cell Wall. Certain compounds of the fungal cell wall (e.g., β-1, 3-glucans) play an important role in defense reactions of crustaceans as well as of insects because these compounds activate the prophenoloxidase through a cascade of reactions (Söderhäll and Unestam, 1979; Ashida and Dohke, 1980; Ashida, 1981; Söderhäll and Smith, 1984). The activated phenoloxidase becomes sticky and attaches firmly to foreign substrates, such as the surface of fungal filaments. This activation occurs even at a concentration of 10^{-10} and is very specific, since other polysaccharides do not trigger the reaction (Söderhäll and Smith, 1984; Chapter 9).

Certain strains of Entomophthorales are able to produce spontaneous protoplasts and provide good experimental models for studying the differences in the reactions of insects to mycelial elements and to protoplasts enveloped only by a membrane. Dunphy and Nolan (1980b) have reported that the plasmatocytes, granulocytes, and spherulocytes of *Lambdina fiscellaria* do not adhere to *Entomophtora egressa* protoplasts both in vivo and in vitro. Spherulocytes do, however, attach to the hyphal bodies and hyphae of *E. egressa* in vitro. The nonrecognition of the protoplast stage of the fungus by the insect's cellular immune system indicates a high degree of adaptation of the pathogen to the host. Protoplasts of the same *Entomophtora* species are similarly not encapsulated in another lepidopteran host, *Choristoneura fumiferana*. While the hyphal bodies do not react with the granulocytes of this insect, the hyphae are readily encapsulated (Dunphy and Nolan, 1980a, 1982).

4.6.2. Factors Concerning the Host

4.6.2.1. Developmental Stage. The sensitivity of the same insect species to a particular pathogen may vary during various developmental stages. The possibility that variations in the resistance of the host are related to changes in the cellular defense mechanisms has been investigated by Vey (1971b). In Lepidoptera, insects in all developmental stages react to *Aspergillus* spores by aggregating hemocytes and forming a melanotic deposit. The encapsulation process of pupa, however, is slightly different from that of larvae, melanization of the fungus being particularly intense (Fig. 4.14). In imagoes, the granuloma reaction is characterized by a reduced recruitment of cells. The development, often rapid and fatal, of the *Aspergillus* mycosis in adult insects is certainly related to this peculiarity.

4.6.2.2. Molting. The mode of infection by entomogenous fungi during the molting period in insects has rarely been studied (Zakaruk, 1973; Fargues and Vey, 1974). The penetration of *B. bassiana* in *Leptinotarsa decemlineata* larvae during molting and the cellular reactions of the host have been ana-

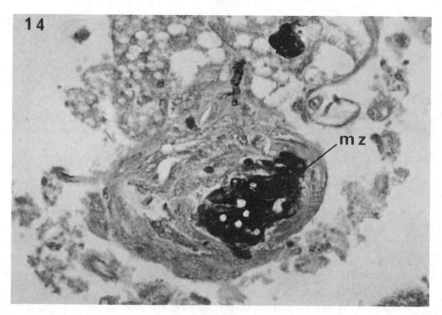

Figure 4.14. Granuloma formed in a *Galleria* pupa 48 hr after injection of *Aspergillus*. Note the large melanized area (mz) (×420).

lyzed by Vey and Fargues (1977) to determine the process by which these animals survive infections over long periods. A large part of the fungus is eliminated with the exuvium during ecdysis, leaving only a few hyphae to attack the new integument, pass through it, and reach the body cavity. The rare intrusive filaments are trapped inside well-formed granuloma. Thus, the rejection of the contaminated exoskeleton plays the role of a complementary mechanism to cellular defense, and the combination of these two processes is fairly effective in the protection of the host.

4.6.3. Physical, Chemical, and Pathologic Factors

4.6.3.1. Temperature It is well known that the intensity of phagocytosis decreases when the temperature falls (Paillot, 1933). Temperature also exerts a strong effect on the germination and growth of fungi. For these two reasons, temperature influences multicellular defense reactions to fungal pathogens. Sussman (1952b) has reported that pupae of *Platysamia cecropia* resist injected conidia of *A. flavus* if the temperature reaches 10–15°C. A similar situation has been described in *Galleria* larvae infected with *A. niger* (Vey, 1971b). In the range of 15–35°C, the process of encapsulation of *Aspergillus* does not show considerable variation. However, the transfer of infected animals from 15 to 35°C accelerates pathogenesis. According to Sussmann (1952b) and Vey (1971b), the restricted development of acute mycoses at low temperatures is principally due to the inhibition of fungal growth under such conditions.

4.6.3.2. Radiations and Chemical Substances Exposure of *Mamestra brassicae* larvae to γ rays at a dose of 7000 rad weakens encapsulation of *A. flavus*. The defense mechanism of irradiated hosts gives rise to granuloma that are less voluminous and well organized than those formed in control insects. *A. flavus* escapes easily from these structures, and the death of the larvae, occurring quickly, is essentially due to the action of the pathogen. Such a mechanism, which can be explained by the reduction of the number of circulating hemocytes (El Brady, 1964; Hoffmann, 1971), is significant for the control of insect pests by radiation (Vey and Causse, 1979).

The encapsulation reaction is also susceptible to anti-inflammatory drugs. Repeated injections of hydrocortisone hemisuccinate in *Mamestra* have a depressive effect on the encapsulation of spores of *A. niger*. The weakening of the cellular mechanism, which seems to result from an effect of the chemical on hemocyte migration, has a favorable influence on the development of mycoses (as also known in vertebrates; Vey, 1978).

The possible effect of inhibitors on phenoloxidases has also been investigated (Vey, 1979). Injection of glutathion in *Galleria* results in a strong reduction of the melanotic deposit at the surface of the pathogen, whose formation is, however, not completely inhibited. On the contrary, the association of hemocytes around *Aspergillus* seems to be normal. This observation agrees with those of Brewer and Vinson (1971) concerning the nonmodification of cellular aspects of encapsulation of parasitoids by melanization inhibitors. Compared with nontreated insects, the germination and growth of the fungus under conditions of reduced melanization are more rapid, and a faster and higher mortality is observed (Vey and Quiot, 1979).

4.6.3.3. Development of Other Pathogens in the Host Different organisms can develop simultaneously or successively in the same insect host. During such multiple infections, diverse interactions can arise between pathogens: coexistence, antagonism, or synergism (Vago, 1959, 1963). In these different cases, the influence that a pathologic factor can exert on the cellular defense system of the insect has been often neglected. However, some research has been carried out in the field of virus infection. For example, the role of the parvovirus of *Galleria* in provoking the "densonucleosis" disease in this moth has been investigated. Vey and Vago (1967) and Vey (1969a) have demonstrated that the densonucleosis virus progressively undermines cellular defense of the larvae of *Galleria*. At an advanced stage of the viral disease, the hemocytes continue to participate in the formation of capsules but are less regular in their aggregation due to abnormalities in the nucleus. Finally, when the virosis has developed further, the hemocytic envelope becomes extremely reduced, and the capacity of the host to form granuloma is severely restricted.

4.7. SUMMARY

During fungal infections, insects exhibit a hemocytic reaction that leads to the formation of capsules, or granuloma; phagocytosis does not seem to play an important role. The dynamics of encapsulation has been described in living insects by histologic and cytologic studies that permit one to follow the different steps of progressive aggregation of cells around the melanized pathogen. The types of hemocytes involved are granulocytes and plasmatocytes. Differences in the modalities of the reaction are observed, depending on the degree of pathogenicity of the fungus, the different pathotypes of the pathogen, and the mode of penetration in the host.

Infections by weakly pathogenic fungi and wound parasites may give rise to cases of immediate resistance and of recovery, essentially owing to the intervention of encapsulation. This efficiency of the cellular defense mechanism is related to a progressive fungicidal action.

Certain factors concerning the host, such as molting and developmental stage, have an effect on the progression and efficiency of the defense reactions. Factors associated with the pathogen, such as compounds of the fungal cell wall and toxins with immunodepressive properties, also exert an effect.

Low temperatures reduce the intensity of encapsulation, and melanization inhibitors, corticoids, and γ rays have an inhibitory effect, thus making the host more susceptible to the fungus. The development of another pathogen, such as a virus, in the host also undermines the antifungal defense.

A better understanding of cellular reactions depends on further studies concerning the biochemical mechanisms involved in these defense processes and using purified populations of hemocytes under in vitro conditions, microcinematography, and computer analysis.

REFERENCES

Amargier, A. 1967. Étude des lésions des glandes séricigènes chez les Lépidoptères atteints de différents types de maladies. Doctoral dissertation, Fac. Sci. Aix-Marseille, France.

Anderson, R.S. 1974. Metabolism of insect hemocytes during phagocytosis, pp. 47–54. In E.L. Cooper (ed.) *Contemporary Topics in Immunobiology*, vol. 4, *Invertebrate Immunology*. Plenum Press, New York.

Anderson, R.S. 1975. Phagocytosis by invertebrate cells *in vitro*: Biochemical events and other characteristics compared with vertebrate phagocytic systems, pp. 152–180. In K. Maramorosch and R.E. Shope (eds.) *Invertebrate Immunity*. Academic Press, New York.

Anderson, R.S. 1976. Macrophage function in insects, pp. 215–219. In T.A. Angus, P. Faulkner, and A. Rosenfield (eds.) *Proc. First International Colloquium Invertebrate Pathology*, Queens University, Kingston, Canada.

Anderson, R.S., B. Holmes, and R.A. Good. 1973. *In vitro* bactericidal capacity of *Blaberus craniifer* hemocytes. *J. Invertebr. Pathol.* **22:** 127–235.

Ashida, M. 1981. A cane sugar factor suppressing activation of prophenoloxidase in hemolymph of the silkworm, *Bombyx mori. Insect Biochem.* **11:** 57–65.

Ashida, M. and M. Dohke. 1980. Activation of prophenoloxidase by the activating enzyme of the silkworm, *Bombyx mori* L. *Insect Biochem.* **10:** 37–47.

Beaulaton, J. 1967a. Localisation d'activités lytiques dans les glandes du ver à soie du chêne (*Antheraea pernyi* Guér.) au stade prénymphal. 1. Structures lysosomiques, appareil de Golgi et ergastoplasme. *J. Microsc.* (Paris) **6:** 179–200.

Beaulaton, J. 1967b. Localisation d'activités lytiques dans les glandes du ver à soie du chêne (*Antheraea pernyi* Guér.) au stade prénymphal. 2. Les vacuoles autolytiques (cytolysomes). *J. Microsc.* (Paris) **6:** 349–370.

Boczkowska, M. 1935. Contribution à l'immunité chez les chenilles de *Galleria mellonella* L. contre les champignons entomophytes. *C. R. Soc. Biol.* (Paris) **119:** 39–40.

Brewer, F.D. and S.B. Vinson. 1971. Chemicals affecting the encapsulation of foreign material in an insect. *J. Invertebr. Pathol.* **18:** 287–289.

Carton, Y. 1976. Isogenic, allogenic and xenogenic transplants in an insect species. *Transplantation* **21:** 17–22.

Dunphy, G.B. and R.A. Nolan. 1980a. Hemograms of selected stages of the spruce budworm, *Chroistoneura fumiferana* (Lepidoptera: Tortricidae). *Can. Entomol.* **112:** 443–450.

Dunphy, G.B. and R.A. Nolan. 1980b. Response of eastern hemlock looper to selected stages of *Entomophthora egressa* and other foreign particles. *J. Invertebr. Pathol.* **36:** 71–84.

Dunphy, G.B. and R.A. Nolan. 1982. Cellular immune responses of spruce budworm larvae to *Entomophthora egressa* protoplats and other test particles. *J. Invertebr. Pathol.* **39:** 81–92.

Duriez, T., J. Fargues, P. Robert, and R. Popeye. 1981. Étude enzymatique comparée de champignons entomopathogènes: *Beauveria* et *Metarhizium. Mycopathologia* **75:** 109–126.

El Brady, E. 1964. The effect of irradiation on the haemocyte counts of larvae of the potato tuberworm, *Gnorimoschema operculella* (Zeller). *J. Insect. Pathol.* **6:** 327–330.

Elsworth, J.F. and J.F. Grove. 1977. Cyclodepsipeptides from *Beauveria bassiana* Bals. 1. Beauverolides H and I. *J. Chem. Soc.* **1:** 270–273.

Elsworth, J. and J.F. Grove. 1980a. Cyclodepsipeptides from *Beauveria bassiana* Bals. 2. Beauverolides A to F and their relationship to Isarolide. *J. Chem. Soc.* **1:** 1795–1799.

Elsworth, J.F. and J.F. Grove. 1980b. Cyclodepsipeptides from *Beauveria bassiana* Bals. 3. The isolation of Beauverolides Ba, Ca, Ja and Ka. *J. Chem. Soc.* **1:** 2878–2880.

Fain-Maurel, M.A. and P. Cassier. 1969. Étude infrastructurale des glandes de mue de *Locusta migratoria migratorioides* (R. et F.). 2. Analyse des étapes de la dégénérescence chez les imagos grégaires. *Arch. Zool. Exp. Gén.* **110:** 91–126.

Fargues, J. 1981. Spécificité des hyphomycètes entomopathogènes et résistance interspécifique des larves d'insectes. Doctoral dissertation, University of Paris.

Fargues, J. and A. Vey. 1974. Modalités d'infection des larves de *Leptinotarsa decemlineata* par *Beauveria bassiana* au cours de la mue. *Entomophaga* **19**: 311–323.

Fargues, J., P. Robert, and A. Vey. 1985. Effet des destruxines A, B et E dans la pathogénèse de *Metarhizium anisopliae* chez les larves de coléoptères Scarabeidae. *Entomophaga* **30**, in press.

Fargues, J., T. Duriez, R. Popeye, P. Robert, and J. Biguet. 1981. Serological characterization of the entomopathogenic hyphomycetes *Beauveria* and *Metarhizium*: Comparison of strains. *Mycopathologia* **75**: 101–108.

Ferron, P. 1985. Fungal control, pp. 313–346. In G.A. Kerkut and L.I. Gilbert (eds.) *Comprehensive Insect Physiology, Biochemistry and Pharmacology*, vol. 12. Pergamon Press, Oxford.

Flandre, O., C. Vago, J. Secchi, and A. Vey. 1968. Les réactions hémocytaires chez les insectes. *Rev. Pathol. Comp. Med. Exp.* **5**: 101–106.

Frappier, F., P. Ferron, and M. Païs. 1975. Chimie des champignons entomopathogènes: Le Beauvellide, nouveau cyclodepsipeptide isolé d'un *Beauveria tenella*. *Phytochemistry* **14**: 2703–2705.

Gharagozlou, I.D. 1970. Infrastructure et activité lytique des lysosomes du tissu adipeux de *Periplaneta americana* (Blattidae). *J. Microsc.* (Paris) **9**: 563–566.

Götz, P. and H.G. Boman. 1985. Insect immunity, pp. 454–485. In G.A. Kerkut and L.I. Gilbert (eds.) *Comprehensive Insect Physiology, Biochemistry and Pharmacology*, vol. 3. Pergamon Press, Oxford.

Götz, P. and A. Vey. 1974. Humoral encapsulation in Diptera (Insecta): Defence reactions of *Chironomus* larvae against fungi. *Parasitology* **68**: 1–13.

Grimstone, A.V., S. Rotheram, and G. Salt. 1967. An electron-microscope study of capsule formation by insect blood cells. *J. Cell Sci.* **2**: 281–292.

Hamill, R.L., C.E. Higgens, and M. Gorman. 1968. Beauvericin, a new depsipeptide antibiotic toxic to *Artemia salina*, p. 18, abstract. *8th International Conference on Antimicrobial Chemotherapy*.

Heitor, F. 1962. Parasitisme de blessure par le champignon *Mucor hiemalis* Wehmer chez les insectes. *Ann. Epiphyt.* **13**: 179–203.

Hoffmann, J.A. 1971. Effet d'une irradiation sélective du tissu hématopoétique sur l'hémogramme d'imagos mâles de *Locusta migratoria* (Orthoptère). *C. R. Séances Acad. Sci.* (III) **273D**: 1604–1607.

Hurpin, B. and C. Vago 1958. Les maladies du hanneton commun (*Melolontha melolontha*) (Col. Scarabaeidae). *Entomophaga* **3**: 285–330.

Kawakami, K. 1965. Phagocytosis in muscardine diseased larvae of the silkworm, *Bombyx mori* L. *J. Invertebr. Pathol.* **7**: 203–208.

Kodaira, Y. 1961. Biochemical studies on the muscardine fungi in the silkworm, *Bombyx mori* L. *J. Facult. Text. Sci. Technol. Shinshu Univ.*, Nr 29, (Ser. E) *Agric* **5**: 1–69.

Leonard, C. K., Söderhäll, and N. A. Ratcliffe. 1985. Studies on prophenoloxidase and protease activity of *Blaberus craniifer* haemocytes. *Insect Biochem.*, in press.

Lepesme, P. 1938. Recherches sur une aspergillose des acridiens. *Bull. Soc. Hist. Nat. Afr. N.* **29:** 373–381.

Marschall, K.J. 1966. Bau und Funktionen der Blutzellen des Mehlkäfers *Tenebrio molitor*. *Z. Morphol. Oekol. Tiere* **58:** 182–246.

Michel, B. 1981. Recherches expérimentales sur la pénétration des champignons pathogènes chez les insectes. Doctoral dissertation, Spec. Univ. Sci. Techn., Montpellier, France.

Müller-Kögler, E. 1965. *Pilzkrankheiten bei Insekten*. P. Parey, Berlin.

Ogloblin, A. and C. Jauch. 1943. Reaciones patologicas de los acridos atacados por *Aspergillus parasiticus, Rev. Argent. Agric.* (Buenos Aires) **10:** 256–267.

Paillot, A. 1930. *Traité des Maladies du ver à soie*. Doin, Paris.

Paillot, A. 1933. *L'infection Chez les Insectes: Immunité et Symbiose*. Patissier, Trévoux, France.

Païs, M., B.C. Das, and P. Ferron. 1981. New depsipeptides from *Metarhizium anisopliae*. *Phytochemistry* **20:** 715–723.

Quiot, J.M., A. Vey, and C. Vago. 1985. Effect of mycotoxins on invertebrate cells *in vitro*. *Adv. Cell Cult.* **4:** 199–212.

Quiot, J.M., P. Montsarrat, G. Meynadier, G. Croizier, and C. Vago. 1972. Infection des cultures cellulaires de Coléoptères par le virus "Oryctes." *C. R. Séances Acad. Sci.* (III) **278**D: 3229–3231.

Rabinovitch, M. and M.J. De Stefano. 1970. Interactions of red cells with phagocytes of the wax moth (*Galleria mellonella* L.) and mouse. *Exp. Cell Res.* **59:** 272–282.

Ractliffe, N.A. and A.F. Rowley. 1975. Cellular defense reactions of insects *in vitro*: Phagocytosis in a new suspension culture system. *J. Invertebr. Pathol.* **26:** 225–233.

Roberts, D.W. 1966. Toxins from the entomogenous fungus *Metarhizium anisopliae*. *J. Invertebr. Pathol.* **8:** 212–227.

Roberts, D.W. 1981. Toxins of entomopathogenic fungi, pp. 441–464. In H.D. Burgess (ed.) *Microbial Control of Pests and Plant Diseases, 1970–1980*. Academic Press, New York.

Smirnoff, W.A. 1968. The nature of cysts found in pupae and adults of *Neodiprion swainei*. *Can. Entomol.* **100:** 313–318.

Söderhäll, K. 1981. Fungal cell wall β-1, 3-glucans induce clotting and phenoloxydase attachment to foreign surfaces of crayfish hemocyte lysate. *Dev. Comp. Immunol.* **5:** 565–573.

Söderhäll, K. 1982. Prophenoloxydase activating system and melanization—a recognition mechanism of arthropods? A review. *Dev. Comp. Immunol.* **6:** 601–611.

Söderhäll, K. and V. Smith. 1984. The prophenoloxidase activating system: A complement-like pathway in arthropods? pp. 160–167. In J. Arst and D.W. Roberts (eds.) *Infection Processes of Fungi*. Rockefeller Foundation, New York.

Söderhäll, K. and T. Unestam. 1979. Activation of crayfish serum prophenoloxidase in arthropod immunity: The specificity of cell wall glucan activation and activation by purified fungal glycoproteins. *Can. J. Microbiol.* **25:** 404–416.

Söderhäll, K., A. Vey, and M. Ramstedt. 1984. Hemocyte lysate enhancement of fungal spore encapsulation by crayfish hemocytes. *Dev. Comp. Immunol.* **8:** 23–29.

Speare, A.T. 1920. Further studies on *Sorosporella uvella*, a fungus parasite of Noctuid larvae. *J. Agric. Res.* **18**: 399–439.

Sussman, A.S. 1951. Studies of an insect mycosis. 1. Etiology of the disease. *Mycologia* **43**: 423–429.

Sussman, A.S. 1952a. Studies of an insect mycosis. 3. Histopathology of an aspergillosis of *Platysamia cecropia* L. *Ann. Entomol. Soc. Am.* **45**: 243–245.

Sussman, A.S. 1952b. Studies of an insect mycosis. 4. The physiology of host-parasite relationship of *Platysamia cecropia* and *Aspergillus flavus*. *Mycologia* **44**: 493–505.

Unestam, T. and J.E. Nylund. 1972. Blood reactions *in vitro* in crayfish against a fungal parasite, *Aphanomyces astaci*. *J. Invertebr. Pathol.* **19**: 94–106.

Vago, C. 1959. L'enchaînement des maladies chez les insectes: Recherches en pathologie comparée. *Ann. Epiphyt.* **10**: 1–181.

Vago, C. 1963. Predispositions and interrelations in insect diseases, pp. 339–379. In F. A. Steinhaus (ed.) *Insect Pathology: An Advanced Treatise*, vol. 1. Academic Press, New York.

Vago, C. and S. Chastang. 1958. Obtention de lignées cellulaires en culture de tissus d'invertébrés. *Experientia* **14**: 110–113.

Vago, C. and S. Chastang. 1960. Culture de Borrelinavirus dans les organes d'insectes en survie. *C. R. Séances Acad. Sci.* (III) **241**D: 903–905.

Vago, C. and J.M. Quiot. 1969. Recherches sur la composition des milieux pour cultures des cellules d'invertébrés. *Ann. Zool. Ecol. Anim.* **1**: 281–288.

Vago, C. and A. Vey. 1972. *Mycoses d'Invertébrés*, film. Service du Film de Recherche Scientifique, Paris.

Vago, C., G. Meynadier, and J.L. Duthoit. 1964. Étude d'un nouveau type de maladie à virus chez les Lépidoptères. *Ann. Epiphyt* **15**: 475–479.

Vey, A. 1968. Réactions de défense cellulaire dans les infections de blessures à *Mucos hiemalis* Wehmer. *Ann. Epiphyt.* **19**: 695–702.

Vey, A. 1969a. Action de la densonucléose sur les réactions de type granulome des larves de *Galleria mellonella* vis à vis du champignon *Aspergillus niger* v. Tiegh. *Ann. Zool. Ecol. Anim.* **1**: 113–120.

Vey, A. 1969b. Étude *in vitro* des réactions hémocytaires anticryptogamiques des larves de lépidoptères. *Ann. Zool. Ecol. Anim.* **1**: 93–100.

Vey, A. 1971a. Étude des réactions cellulaires anticryptogamiques chez *Galleria mellonella* L.: Structure et ultrastructure des granulomes à *Aspergillus niger*. v. Tiegh. *Ann. Zool. Ecol. Anim.* **3**: 17–30.

Vey, A. 1971b. Recherches sur la réaction hémocytaire anticryptogamique de type granulome chez les insectes. Doctoral dissertation, University of Toulouse, France.

Vey, A. 1978. Sensibilité de la réaction hémocytaire multicellulaire de l'insecte *Galleria mellonella* à un corticoide anti-inflammatoire. *C. R. Séances Acad. Sci.* (III) **287**D: 337–340.

Vey, A. 1979. Effect of melanization inhibitors, anti-inflammatory drugs and irradiations on insect multicellular reactions, pp. 229–231. In J. Weiser (ed.) *Progresses in Invertebrate Pathology, Proceedings of Intern. Colloquium of Invertebrate Pathology*, Prague.

Vey, A. 1984. Cellular antifungal reactions in invertebrates, pp. 168–172. In J. Aist and D.W. Robert (ed.) *Infection Processes of Fungi*. Rockefeller Foundation, New York.

Vey, A. 1985. Efficacité de là réaction d'encapsulement chez les insectes: Effet sur la structure et la vitalité du champignon encapsulé. In *Proc. I.S.D.C.I. Invertebra. Immunol. Conf.*, Montpellier, France.

Vey, A. and R. Causse. 1979. Effet de l'exposition aux rayons γ sur la réaction multicellulaire des larves de *Mamestra brassicae* (Lep.: Noctuidae). *Entomophaga* **24:** 41–47.

Vey, A. and J. Fargues. 1977. Histological and ultrastructural studies of *Beauveria bassiana* infection of *Leptinotarsa decemlineata* larvae during ecdysis. *J. Invertebr. Pathol.* **30:** 207–215.

Vey, A. and J.M. Quiot. 1975. Effet *in vitro* de substances toxiques produites par le champignon *Metarhizium anisopliae* (Metsch.) Sorok. sur la réaction hémocytaire du coléoptère *Oryctes rhinoceros* L. *C. R. Séances Acad. Sci.* (III) **280**D: 931–934.

Vey, A. and J.M. Quiot. 1976. Action toxique du champignon *Mucor hiemalis* sur les cellules d'insectes en culture *in vitro*. *Entomophaga* **21:** 275–279.

Vey, A. and J.M. Quiot. 1979. Toxic effects of entomogenous fungi on *in vitro* cultured insect cells, pp. 232–234. In J. Weiser (ed.) *Progresses in Invertebrate Pathology, Proceeding International Colloquium Invertebrate Pathology*, Prague.

Vey, A. and C. Vago. 1967. Influence d'une maladie virale sur les réactions hémocytaires anticryptogamiques chez les insectes. *C. R. Séances Acad. Sci.* (III) **265**D: 1568–1570.

Vey, A. and C. Vago. 1969. Recherches sur la guérison dans les infections cryptogamiques d'invertébrés: Infection à *Aspergillus niger* v. Tiegh. chez *Galleria mellonella* L. *Ann. Zool. Ecol. Anim.* **1:** 121–126.

Vey, A. and C. Vago. 1971. Réactions anticryptogamiques de type granulome chez les insectes. *Ann. Inst. Pasteur* **121:** 527–532.

Vey, A., J. Fragues, and P. Robert. 1982. Histological and ultrastructural studies of factors determining the specificity of pathotypes of the fungus *Metarhizium anisopliae* for scarabeid larvae. *Entomophaga* **27:** 387–397.

Vey, A., J.M. Quiot, and C. Vago. 1968. Formation *in vitro* de réactions d'immunité cellulaire chez les insectes, pp. 254–263. In C. Barigozzi (ed.) *Proc. 2nd. Intern. Coll. Invert. Tissue Cult.* Succ. Fusi, Pavia, Italy.

Vey, A., J.M. Quiot, and C. Vago. 1973. Mise en évidence et étude de l'action d'une mycotoxine, la beauvericine, sur des cellules d'insectes cultivées *in vitro*. *C. R. Séances Acad. Sci.* (III) **276**D: 2489–2492.

Vey, A., M. Bouletreau, J.M. Quiot, and C. Vago. 1975. Étude *in vitro* en microcinématographie des réactions cellulaires d'invertébrés vis à vis d'agents bactériens et cryptogamiques. *Entomophaga* **20:** 337–351.

Vey, A., J.M. Quiot, C. Vago, and J. Fargues. 1985. Effet immunodépresseur des toxines fongiques: Inhibition de la réaction d'encapsulement multicellulaire par les destruxines. *C. R. Séances Acad. Sci.* (III) **300**D: 647–651.

Weber, H. 1939. Beobachtungen über die wirkung von *Empusa weberi* Lacon auf die Larve von *Raphidia ophiopsis* L. *Z. Ang. Entomol.* **26:** 522–535.

Wyatt, S.S. 1956. Culture *in vitro* of tissue from the silkworm, *Bombyx mori* L. *J. Gen. Physiol.* **39:** 841–852.

Zakaruk, R.Y. 1973. Penetration of the cuticular layers of elaterid larvae by the fungus *Metarhizium anisopliae*, and notes on a bacterial invasion. *J. Invertebr. Pathol.* **21:** 101–106.

CHAPTER 5

Nematode-Induced Host Responses

John G. Stoffolano, Jr.
Department of Entomology
University of Massachusetts
Amherst, Massachusetts

5.1. INTRODUCTION

The major objective of research on cellular defensive reactions of insects
has been to gain a more complete understanding of how the insect distin-

guishes self from nonself. Anderson (1981) stated that the "blood cells of all animals are capable of recognizing foreign materials." Since insect hemocytes constitute the major line of defense once the parasite or pathogen has gained access to the hemocoel, numerous studies have been designed to determine the types of hemocytes involved in these reactions, to define their surface properties and organelle composition, and, more recently, to use in vitro systems to help in clearly defining the biophysical and biochemical properties involved in hemocytic recognition. The various responses of insects to nematode parasites provide researchers with a unique opportunity for investigating the various aspects of nematode-induced host responses. This chapter reviews developments in this area since Poinar's review of the topic (1974).

5.1.1. Model Systems

The demonstrated diversity of insect responses to entomogenous nematodes and their economic importance provides investigators with a host-parasite relationship that is ideal for investigating the current problems concerning recognition of foreignness. Throughout this chapter, the utility of nematodes in answering various questions concerning host recognition is identified. The potential implications of genetic engineering and other biotechnologies for insect control has recently been addressed (Kirschbaum, 1985). One of the areas designated as important is better understanding by researchers of the molecular basis of the cellular defenses of insects and how various parasites and pathogens have dealt with these host defenses. The insect-nematode model proves very useful in answering many of these questions.

Before launching into the specific details of nematode-induced host responses, a brief look at a model describing the dynamic nature of the host-parasite interactions is appropriate (Fig. 5.1). The insect host responds to damage of the cuticle and/or midgut, the major sites of entry for nematode parasites. Penetration of the cuticle and/or epidermal cells may elicit a tissue response by the host. A hemocytic and/or humoral response is usually not involved at this time. Generally, the initial recognition process of identifying a foreign object, in this case a nematode, begins when the parasite breaks through the basement membrane and enters the hemocoel. At this time, the host recognition of nonself can take one of two pathways or both, depending on the host. A humoral response is produced in insects, usually the nematocerous Diptera, having few hemocytes (see also Chapter 15). This response usually results in melanization and death of the parasite. Insects having large numbers of hemocytes respond by hemocytic encapsulation, which in some cases is accompanied by melanization. Nematode parasites have evolved various tactics for dealing with these host responses. These mechanisms are either countertactics against the encapsulation and melanization responses or evasive tactics that involve three possible strategies: entering host tissues or cells, molecular mimicry, and antigen sharing. The host responses and

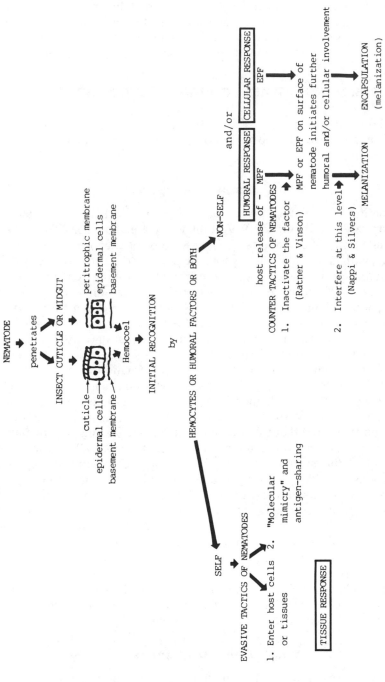

Figure 5.1. Model showing nematode-induced host responses in insects. Whether they use insects as definitive hosts or intermediate hosts, nematodes must actively evade or counter the insect host defensive response. This relationship is dynamic, and selective pressures are placed on the host to improve its defensive responses and on the nematode to evade or counter the host response. (EPF, encapsulation-promoting factor; MPF, melanization-promoting factor.)

the tactics used by the nematodes to avoid being killed are discussed later in greater detail. An attempt is made not only to describe what is known about the proposed model but to use the model to reveal areas in need of research and to help generate new hypotheses.

5.2. INSECT RESPONSES TO NEMATODES

Nematode-induced responses of insects can be categorized as three basic types of reactions by the host: humoral response, cellular response, and tissue response.

5.2.1. Humoral Responses

Encapsulation of nematodes without participation of hemocytes has been previously reported by Götz (1969), Poinar and Leutenegger (1971), Poinar (1974), and Götz and coworkers (1977). This type of encapsulation is usually associated with insects having low total hemocyte counts (Götz *et al.*, 1977). In fact, some species of chironomids respond to invading nematodes exclusively by humoral encapsulation (Götz, 1964). Andreadis and Hall (1976), using ligation techniques, concluded that both the encapsulation and melanization responses of *Aedes aegypti* larvae to the nematode *Neoaplectana carpocapsae* could still take place without participation of the hemocytes in other parts of the body; but since mosquitoes, like other Nematocera, have few circulating hemocytes (Hall, 1983), formation of melanized capsules is believed to proceed without participation of host hemocytes. Götz (1984) suggested that the humoral encapsulation response of *Chironomus* may follow a pathway similar to that proposed by Ashida and Söderhäll (1984), who suggested that the prophenoloxidase-activating system may operate as a recognition system in crayfish (see also Chapter 9). These authors suggested that this system may also function as a complement like system in arthropods. Bartlett (1984) suggested that the host response to *Dirofilaria scapiceps* (i.e., melanization within the fat body of the host) represented a humoral response, since hemocytes were never associated with larvae within the fat body; but in some cases melanization outside the fat body represented hemocytic melanization. The humoral response of insects to nematode parasites without the involvement of hemocytes is discussed at greater length in Chapter 15.

Chironomids and their nematode parasites should constitute ideal models for investigating the mechanisms of both recognition and melanin encapsulation by the humoral response system in insects.

5.2.2. Cellular Responses

Cellular responses of insects to nematodes fall into four major areas: wound healing, hemolymph coagulation, intracellular melanization, and hemocytic encapsulation and melanization.

5.2.2.1. Wound Healing. Unlike the hymenopterous parasites, which circumvent the cuticle or gut barrier by being injected directly into the host via the female's ovipositor, most nematodes must, on their own, actively penetrate either the external cuticle or the lining of the alimentary tract. Also, unlike the hymenopterous parasites, for which the female seeks out and recognizes the host for her offspring, nematodes must do this on their own.

Nematodes that penetrate the external cuticle are either Tylenchida or Mermithoidea, while those that enter via the digestive tract include the Spiruroidea, Rhabditoidea, Subuluroidea, Filarioidea, and Mermithoidea that do not penetrate the external cuticle (Poinar, 1969).

Information concerning the mechanisms used by nematodes to actually penetrate either the cuticle or the digestive tract of insects is scarce. Esslinger (1962) stated that "detailed information regarding the events occurring in the early phase of infection of arthropods by the Filarioidea is lacking." This statement also applies to the other nematode groups. In addition, little is known concerning the mechanisms used to recognize either the host or the specific site of the host for penetration. It is highly possible that both the recognition and the penetration mechanisms used by nematodes to gain access to the internal environment through the external cuticle differ from those involving gut penetration, especially since the cuticular linings of the fore- and hindgut are identical to the external cuticle, while the midgut lacks a cuticular lining and almost all nematodes that enter the host via the digestive tract do so mainly through the midgut (Poinar, 1969). The excellent review of plant nematode chemotaxis and host recognition mechanisms by Zuckerman and Jansson (1984) clearly illustrates how little we know about these processes in entomophilic nematodes; their paper should provide new ideas for future research using nematode parasites of insects. Nordbring-Hertz and colleagues (1982) demonstrated that nematophagous fungi recognize the plant nematode by means of a recognition mechanism that is mediated by a lectin on the fungi that specifically binds to a carbohydrate on the nematode. In addition, Jansson and associates (1984) and Jeyaprakash and associates (1985) were able to inhibit chemotactic responses of *Caenorhabditis elegans* by enzyme-mediated behavioral modification that affected the nematode's receptors. Thus, the mechanism of molecular recognition used by plant nematodes in their plant hosts (Zuckerman and Jansson, 1984) may be similar to that in nematodes that use insects either as their primary host or as intermediate hosts.

How do nematodes that enter the insect host through the external cuticle recognize the host? To date, little information exists concerning the sensilla of the free-living stage of entomophilic nematodes or the host cues used for recognition. The study of Dickinson (1959) and the comments of Welch (1963) suggest that infective-stage larval nematodes may be attracted to and attempt penetration of any hydrophobic membrane. Since the cuticle of insects is a hydrophobic structure, this hydrophobicity may serve as the stimulus. Once they recognize the cuticle, how do these nematodes penetrate it, and do they elicit a host response during the act of penetration? Poinar and Doncaster (1965) provide the only known study of how the nematode penetrates the external cuticle of an insect. They showed that the nematode *Tripius sciarae* penetrated the cuticle of *Bradysia paupera* by means of a digestive secretion and with the aid of a stylet. These authors made no mention of any host response to penetration. Keilin and Robinson (1933), however, reported that mycetophilid larvae that contained nematodes had scars on the cuticle and assumed that these indicated where the parasite had gained entrance. Stoffolano and Streams (1971) confirmed that in the larvae of *Musca domestica* exposed to nematodes the dark spotting on the outer cuticle was the site of either actual or attempted penetration (Figs. 5.2 and 5.3). They also suggested that since some of the spots were considerably larger than the diameter of the nematode, the response may also represent a reaction by the host to digestive enzymes used by the stylet-bearing nematode to penetrate the cuticle (Fig. 5.2). The cuticle and the underlying hypodermal or epidermal cells of the host are usually disrupted during penetration (Fig. 5.3); however, the opening is always closed, and both hemolymph loss and exposure to other pathogens entering are prevented. The darkening of the external cuticle was believed by Stoffolano and Streams (1971) to be melanin. Specific details concerning hemocytic involvement to counter nematode penetration of the external cuticle and subsequent scar formation and melanization is needed.

The fact that the majority of nematodes that enter their insect hosts through the digestive tract do so mainly in the midgut, which differs in chemical composition from the fore- and hindgut, strongly suggests some sort of recognition of this region by the nematode. Bronskill (1962) noted that juvenile DD-136, which failed to penetrate the gut in the proventriculus region, appeared unable to enter elsewhere, while Poinar (1969) reported that a host response resulting in death of the nematode usually occurred in cases in which the nematode entered the wrong region of the digestive tract. Is there anything special about the midgut of insects? Recently, Peters and coworkers (1983) demonstrated a sugar receptor (lectin) in the peritrophic membrane of a fly larva. Since most insects produce this special structure (i.e., peritrophic membrane) and since certain nematodes that penetrate the midgut have been shown to possess surface carbohydrates (Furman and Ash, 1983), a molecular recognition binding system similar to that proposed by Zuckerman and Jansson (1984) for phytophagous nematodes may exist for entomophilic nematodes and their hosts.

As with our understanding of how nematodes penetrate the external cuticle, we also know little about their passage through the midgut. To date, many believe (Barr and Shope, 1975) that the gut and/or the peritrophic membrane may provide a barrier to the passage of certain nematode parasites (Orihel, 1975). Esslinger (1962) suggested that a motile cephalic hook aids the *Brugia* microfilaria in gaining access to the mosquito hemocoel. Figure

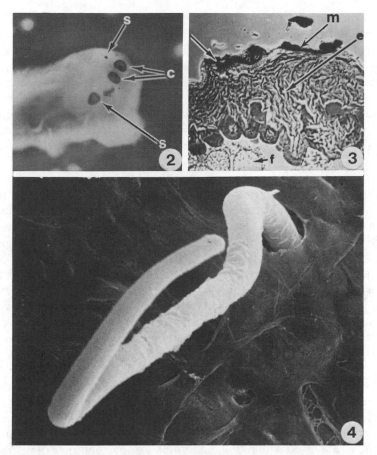

Figure 5.2. Housefly larva showing caudal spiracles (c) and melanic spotting (s) of the cuticle due to nematode penetration or attempted penetration. (From Stoffolano and Streams, 1971.)

Figure 5.3. Cross section through a melanic spot (arrow), similar to that in Figure 5.2 on the cuticle of a housefly larva, showing the distorted endocuticle (e) and melanin (m) in the epicuticle. Note that the underlying hypodermal cells are distorted and that fat body cells (f) are present. (From Stoffolano and Streams, 1971.)

Figure 5.4. Anterior end of a microfilaria of *Brugia pahangi* emerging from the midgut of an adult *Aedes aegypti* mosquito. (From Christensen and Sutherland, 1984.)

5.4 shows a microfilaria of *Brugia* emerging from the midgut of an adult mosquito (Christensen and Sutherland, 1984). Do nematodes that enter the host via the midgut produce a host response? The question remains unanswered. Day and Bennetts (1953) showed that mechanically wounding the gut of an adult mosquito failed to elicit a hemocytic response and that the wound was sealed by contraction of gut muscles, followed by a response of the hypodermal cells. An eschar frequently formed of necrotic tissue, and they suggested it was composed of coagulated hemolymph. Esslinger (1962) noted that the epithelial cells of the midgut were damaged in a manner similar to that shown in Figure 5.3 when the microfilariae entered the mosquito. Bronskill (1962) also reported distortion of the caudal epithelium as a result of juvenile DD-136 penetration. One major difference in host response to nematodes that penetrate the external cuticle as opposed to the midgut is the lack of a dark (melanic) scar in the midgut where the nematode has penetrated. As with cuticular penetration, we need more information concerning how nematodes recognize the site of midgut penetration, the mechanism used for penetration, and the specific host responses to the act of penetration.

Once the nematode has broken through the external cuticle or midgut and crossed the hypodermal cells and the basement membrane, it is recognized as either foreign or self. If it is recognized as foreign, either a humoral response and/or a hemocytic encapsulation response results (Fig. 5.1). Yeaton (1983) presents the most current information of the wounding response of insects as it relates to various aspects of insect cellular immunity. More information is needed concerning the possible presence of an injury factor, or hemokinin, as reported by Cherbas (1973), who showed that when tissue of *Hyalophora cecropia* was injured, a hemokinin was released, activating the hemocytes, thus causing them to increase in mobility and adhesiveness and to aggregate at the sites of the wound (see also Chapter 16).

5.2.2.2. Hemolymph Coagulation. In his two reviews of insect immunity to nematodes, Poinar (1969, 1974) made no mention of the role of hemolymph coagulation; yet reports exist in the literature that suggest that hemolymph coagulation is important not only in wound healing but possibly in hemocytic encapsulation (Rowley and Ratcliffe, 1981). Insects, possessing an open vascular system, are faced with both hemolymph loss and entrance of pathogens through any opening in the cuticle or lining of the midgut. Consequently, hemolymph coagulation is an important adaptive mechanism. The literature concerning the role of this process in insect-nematode relationships is meager. Too often, investigators have overlooked this response in search of the more obvious: hemocytic encapsulation and melanization.

In experimental studies to determine whether dipterous larvae other than the natural host (face fly, *Musca autumnalis*) would accept the nematode *Heterotylenchus autumnalis*, three different fly larvae (*M. domestica, Orthellia caesarion*, and *Ravinia l'herminieri*) were exposed to infective nematode larvae. The nematodes successfully penetrated all the larvae and elicited a host response in all species. Close examination under phase contrast and oil immersion microscopy often revealed small ameboid cytoplasmic structures (Figs. 5.5 and 5.6) attached to and often moving over the surface of the encapsulated nematode. These structures started out as hemocytes of undetermined origin that sent out cytoplasmic extensions (Figs. 5.5 and 5.6) and eventually formed massive networks (Figs. 5.7 and 5.8) very similar to those reported by Grégoire (1971) and Crossley (1975), who iden-

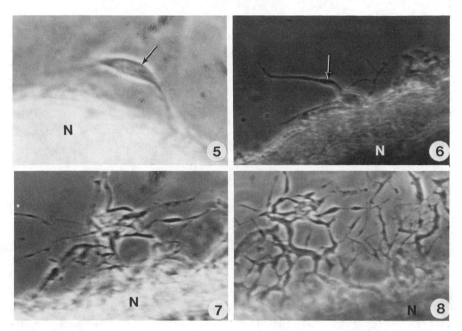

Figures 5.5, 5.6. Phase contrast micrographs of ameboid cytoplasmic structure (plasmatocytes, indicated by arrows) of undetermined hemocytic origin on the surface of encapsulated nematodes (N).
Figures 5.7, 5.8. Phase contrast micrographs of cytoplasmic extensions originating from hemocytes (elongated plasmatocytes) of undetermined origin and eventually forming such cytoplasmic networks on the surface of the nematode.

tified these structures as important components of the cellular agglutination process of blood clotting. These observations of a nematode infection have identified structures comparable to those already reported by Grégoire (1971) in an insect larva (*M. domestica*), for which coagulation of the hemolymph has already been reported (Hewitt, 1914).

It remains for researchers to give equal time to the study of the roles of both cellular agglutination (Lackie, 1981) and plasma coagulation in the initial phases of recognition of foreignness. Whether substances (hemagglutinins) act as encapsulation-promoting factors (opsoninlike), as proposed by Ratner and Vinson (1983), remains to be proved. Reviews of the literature on hemagglutinins in invertebrate immunology are presented by Cooper (1974), Ratcliffe and Rowley (1984), and Amirante (see Chapter 13).

5.2.2.3. Intracellular Melanization. As an immune response of insects to nematodes, intracellular melanization is briefly covered by Poinar (1974). Since nematodes usually do not enter host cells, this response often goes unnoticed; however, microfilariae often develop within specific host cells or tissues. In some cases, these nematodes die and become darkened, presumably through a melanization response (Poinar, 1974). It is important to determine whether the nematode dies and is then darkened or whether the melanization response causes the death and darkening of the parasite. Whether intracellular melanization involves the same biochemical pathway as extracellular melanization of parasites remains to be proved.

5.2.2.4. Hemocytic Encapsulation and Hemocytic Melanization. Salt (1963) reported that the earliest reference he found to encapsulation in any insect was by Deslongchamps (1824), who described the nematode *Spirura (Filaria) rytipleurites* encapsulated in the nymph of the cockroach *Blatta (Periplaneta) orientalis*. However, based on the work of Seureau (1973), one questions whether this response to *S. rytipleurites* really represented hemocytic encapsulation. It is more likely that what Deslongchamps observed was a tissue response (see Section 5.2.3). The spirurids and some other nematode groups evade the host's defensive response by developing inside cells or tissues of the host. Other nematodes develop free in the hemocoel, thus exposing themselves to a humoral and/or hemocytic response. Since Deslongchamps' report, a hemocytic encapsulation response to most nematode groups has been observed (Salt, 1963; Poinar, 1969, 1974).

An in-depth study was conducted by Stoffolano and Streams (1971) on the different kinds of host responses three dipterous larvae produced to *H. autumnalis,* a natural allantonematid parasite of the face fly, *M. autumnalis*. In addition to the spotting of the external cuticle (Fig. 5.2), specific internal changes occurred that are typical of most insect responses to nematodes developing free within the hemocoel. No obvious host response of *M. domestica* larvae was evident during the first 24 hr postinfection; however, 2–3 days following infection, host hemocytes were observed adhering to the nematode's cuticle. Four days postinfection, many of the gamogenetic female nematodes were partially or completely encapsulated. A cephalic cap, dark in color and presumably containing melanin, often covered the nematode's mouth and presumably interfered with feeding (Figs. 5.9 and 5.10). One general host response to nematodes is to encapsulate natural openings of the parasite. Schacher and Khalil (1968) and Christensen (1981) also reported cephalic caps covering the mouth of different nematode species, while these same investigators, along with Bartlett (1984), reported plugs covering the execretory pore. It was proposed (Schacher and Khalil, 1968) that this response to natural openings was a host response to secretions produced by the nematode. Encapsulated nematodes often have tracheoles (Fig. 5.11) attached to their surface. The next obvious host response is partial or complete hemocytic encapsulation, usually accompanied by hemocytic melanization. Viewed under the light microscope, hemocytic capsules are seen to contain an outer, unpigmented layer (Fig. 5.12) and an inner, crusty and darkened layer (Fig. 5.13). Bronskill (1962) reported a similar structure for the capsule surrounding DD-136 in *A. aegypti* larvae. Oftentimes, nematodes are partially encapsulated and fail to die, and the unencapsulated end con-

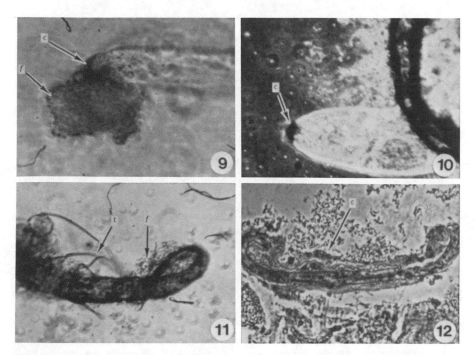

Figure 5.9, 5.10. Two live gamogenetic nematodes (*Heterotylenchus autumnalis*) shown here with cephalic caps (c) covering the mouth and globules of fatlike material (f) stuck to the nematode. (From Stoffolano and Streams, 1971.)

Figure 5.11. Posterior end of an encapsulated gamogenetic female nematode (*H. autumnalis*) showing proliferation of tracheoles (t) to the nematode's surface and globules of fatlike material (f) stuck to the cuticle. (From Stoffolano and Streams, 1971.)

Figure 5.12. Longitudinal section of an encapsulated female nematode removed from a housefly larva showing the unpigmented outer cellular capsule (c) surrounding the nematode. (From Stoffolano and Streams, 1971.)

tinues to live and grow. In such cases, the hemocytic capsule restricts growth and results in a girdle around the nematode (Fig. 5.14). Encapsulated nematodes are usually not found freely floating in the hemocoel but are usually attached to other tissues or body wall by strands or filaments. When pulled away, these structures fray at the ends (Fig. 5.15). It is suggested that these strands are connective tissue (Ashhurst, 1982). Bronskill (1962) reported a most interesting response (parasitic expulsion) of a mosquito to a nematode parasite. Little attention has been given to the mechanism whereby encapsulated parasites are restrained in the body cavity (i.e., prevented from floating free, thus causing damage to other tissues by obstructing growth, etc.) or to the host's ability to expel the encapsulated parasite from its body cavity. Bronskill (1962) reported the expulsion of an encapsulated DD-136 nematode from *A. aegypti* at the time of pupation. Christensen (1981) noted that melanized *Dirofilaria immitis* microfilariae were voided from the malpighian

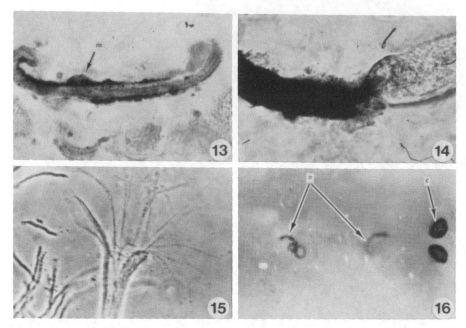

Figure 5.13. Same longitudinal section shown in Figure 5.12 but with different lighting, which reveals the crusty melanin (m) surrounding and in contact with the encapsulated nematode. (From Stoffolano and Streams, 1971.)

Figure 5.14. An encapsulated, small gamogenetic female nematode, *H. autumnalis,* that had its posterior end encapsulated and did not die. The front-end continued to grow, resulting in a girdling effect. (From Stoffolano and Streams, 1971.)

Figure 5.15. A strand of what is believed to be connective tissue that was attached to the encapsulated nematode. When pulled away, the end frayed.

Figure 5.16. Two encapsulated nematodes (n) just underneath the cuticle of an infected 3-day-old housefly larva. The two black structures to the right are the caudal spiracles (c) of the fly larva.

tubules into the gut and ultimately out the anus. Failure to find encapsulated *H. autumnalis* in adult house flies that were known to be infected as larvae suggested a similar mechanism of parasitic expulsion (Stoffolano and Streams, 1971). Closer examination revealed that encapsulated nematodes somehow become closely appressed to the cuticle of third-instar fly larvae (Fig. 5.16) and during pupation are eliminated with the last larval cuticle, which in the Diptera becomes the puparium. Poinar and coworkers (1968) also reported that encapsulated mermithids were readily visible through the host's cuticle but did not report whether they were expelled with the cuticle during molting. The benefit of parasitic expulsion is the prevention of further potential damage to the host. Bronskill (1962) reported that for *A. aegypti* during histogenesis from larva to adult, displacement or distortion of tissues and organs was often evident and presumably was caused by the encapsulated nematodes.

In general, besides being a humoral response of the insect host to nematode parasites free within the hemocoel, the host defense represents a cellular response in the form of hemocytic encapsulation and melanization (see Fig. 5.1). Poinar and Leutenegger (1971) and Gray and Anderson (1983) provided evidence that the dark pigment associated with encapsuled nematodes is melanin.

What do we know concerning hemocytic recognition of nematodes that have invaded the host's hemocoel? It has been established that some strains of mosquitoes are refractory, while others are susceptible (see Section 5.3.2), and that certain groups of nematodes (i.e., spirurids), may avoid a hemocytic response by entering different tissues (see Section 5.2.3.4). The ideas of Salt (1970) that nonself is equivalent to anything other than intact host basement membrane and of Ratner and Vinson (1983) that granuloyctes (hyaline cells, cystocytes, or spherulocytes) rupture upon contacting a foreign surface, thus releasing encapsulation-promoting factors, should be tested using one of the insect-nematode model systems (see Section 5.3.3).

Since hemocytes are involved in the majority of nematode-induced host responses, one should see evidence of both hemocyte chemotaxis to the nematode and changes in the differential hemocyte counts. Shapiro (1979) reviewed the literature concerning the influence of various disease agents on hemocyte population changes but did not discuss the effects of nematodes.

Few studies using the insect-nematode model have examined the importance of hemocyte chemotaxis as part of the hemocytic encapsulation response (Nappi and Stoffolano, 1972). If certain insects have few circulating hemocytes and these insects respond to a nematode parasite via the humoral response (see Chapter 15), one does not have to regard chemotaxis of hemocytes as important. However, in most insect hosts, nematode encapsulation and melanization involve the participation of the host's hemocytes. No definitive evidence exists demonstrating chemotaxis in an insect-nematode model. Nappi and Stoffolano (1972) suggested that in *M. domestica* larvae parasitized by *H. autumnalis*, chemotaxis of hemocytes may explain how this nematode is encapsulated and melanized in an insect reported to have few circulating hemocytes. In their study, Nappi and Stoffolano (1972) found most hemocytes in the last three abdominal segments of nonparasitized larvae. In parasitized larvae, however, hemocytes were found in greater numbers elsewhere. At the time, Nappi and Stoffolano (1972) did not know the location of the hemopoietic tissues in Diptera, but since then Hoffmann and colleagues (1979) have reported the location in *Calliphora erythrocephala* larvae as surrounding the dorsal blood vessel in abdominal segments 5–8. Whether the hemocytes in the study by Nappi and Stoffolano (1972) came from the aggregations in the last three segments or arose from the hemopoietic tissue remains to be demonstrated. There is evidence suggesting that chemotaxis is important in the encapsulation response of insects to other pathogens and parasites (Rowley and Ratcliffe, 1981), but confirming evi-

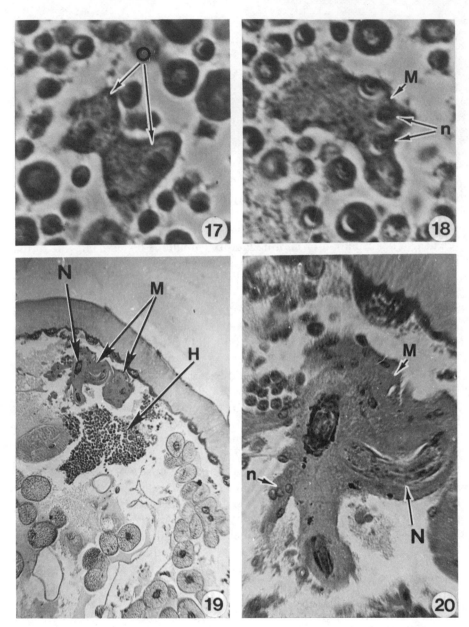

Figures 5.17–5.20. Photomicrographs of the dorsal posterior region of fly larva of *Musca domestica* parasitized by the nematode *H. autumnalis*. One day after infection oenocytoids (O) are seen fusing to form large homogenous masses (M) composed of the cytoplasm and remaining nuclei (n) of the lysed hemocytes. Three days after infection the encapsulated nematodes (N) can be observed surrounded by the cytoplasmic masses and sometimes near large aggregations of hemocytes (H). (From Nappi and Stoffolano, 1971.)

dence using an insect-nematode model awaits development of an in vitro, short-term culture system to demonstrate this early phase of nematode encapsulation.

Only three studies have compared differential hemocyte counts in un-parasitized versus nematode-infected insects. Nappi and Stoffolano (1971) showed that the initial reaction in *M. domestica* larvae to nematode infection was the aggregation and fusion of oenocytoids to form a pigmented layer, or sheath capsule, surrounding the nematode (Figs. 5.17 and 5.18). This reaction was believed to initiate the aggregation and fusion of other hem-ocytes which formed homogenous masses (Figs. 5.19 and 5.20) that com-pletely surrounded the melanized parasite. Nappi and Stoffolano (1971) re-ferred to these masses (Figs. 5.19 and 5.20) as being a syncytium; however, more recent research using transmission electron microscopy has disproved this in other insects (see Section 5.2.3.3). Using another experimental host, Nappi and Stoffolano (1972) demonstrated that the immune reaction was manifested as a change in the hemocyte picture long before any evidence of melanization and encapsulation. Later, Andreadis and Hall (1976), using a different nematode-insect system, provided evidence supporting the work of Nappi and Stoffolano (1971, 1972). Andreadis and Hall (1976) demon-strated a significant decrease in total hemocyte counts in *A. aegypti* larvae infected with *N. carpocapsae*. Based on these three studies, it appears that there is an obvious change in both differential and total hemocyte counts in nematode-infected insects. It should be pointed out, however, that these studies were conducted on larval insects; whether a similar situation exists in adult insects infected with nematodes remains to be shown.

Before we can better understand the encapsulation and melanization re-sponse of insects to nematodes, more studies need to be conducted at the ultrastructural level. Such studies should help in defining more parameters that can be used to separate the humoral response leading to melanization without the participation of blood cells from the cellular response. To date, only three studies are known concerning this topic (Poinar et al., 1968; Poinar and Leutennegger, 1971; and Forton et al., 1985).

In an elegant study of the ultrastructure of the host response of *Aedes trivittatus* adults against inoculated *D. immitis* microfilariae, Forton and as-sociates (1985) have provided new information concerning what many con-sidered only a humoral response. At present there is a dilemma as to what really happens in nematode infection of mosquitoes and, presumably, other Nematocera. Is the host response truly a humoral response without partic-ipation of hemocytes, does it solely involve hemocytes, or is it a combination of both? One must bear in mind that the research of Götz and colleagues has been confined to the larval stage of the host, while the work of Foley (1978) and Forton and associates (1985) has been done on the adult. One question that emerges is whether adult mosquitoes also have low hemocyte counts. The recent review by Hall (1983) focuses mainly on the larval stage. More studies need to focus on the effect of nematode infection on both the

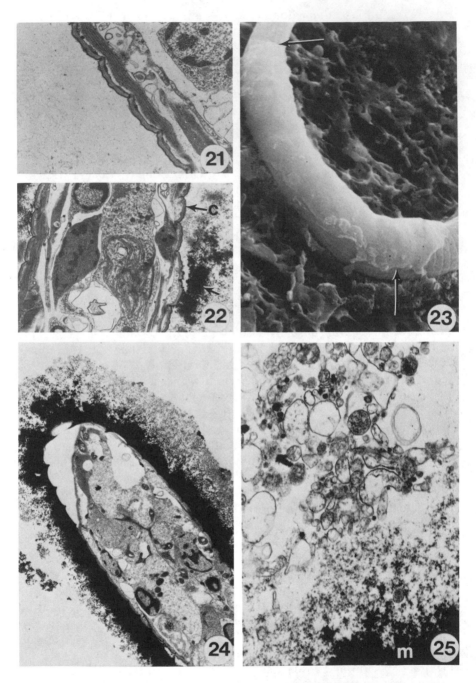

Figure 5.21. Electron micrograph of *Dirofilaria immitis* microfilaria isolated from the blood of an infected dog. (From Forton et al., 1985.)

total and differential hemocyte counts of adult mosquitoes. Following inoculation of microfilariae into adult mosquitoes, Forton and coworkers (1985) dissected adults at 5, 30, and 120 min postinjection. Microfilariae were recovered and prepared for either scanning electron microscopy or transmission electron microscopy. Micrographs showed that control microfilariae had no signs of a host response (Fig. 5.21), while hemocytic lysis and melanization began within 5 min of injection (Figs. 5.22 and 5.23). The melanization response elicited by the host 2–6 hr postinjection represented cellular disruption that produced a cascading of material toward the surface of the parasite, where melanization took place (Figs. 5.24 and 5.25). By 24 hr postinjection, the parasite was completely encased in melanin (Fig. 5.24). Intact hemocytes were seldom seen during the early stages of melanization and never made direct contact with the parasite (Figs. 5.26 and 5.27). There was always a layer of melanin between the hemocyte and the parasite. When present, intact hemocytes showed the formation and secretion of membrane-bound vacuoles (Fig. 5.26). Two days postinjection, there was an increase in remnants of hemocytes around the parasite, and a membranelike structure began to form around the parasite (Fig. 5.28). Initially, this membranelike structure was single, but gradually a distinct second layer began to form (Fig. 5.29). By about 4 days postinjection this structure was double and appeared to completely isolate the parasite from the host's hemolymph and hemocytes (Fig. 5.30). Hemocytes that now contacted the surface of the capsule did not penetrate the capsule, did not release materials onto the capsule, but remained intact and isolated from the nematode by the membranelike structure (Fig. 5.31). This detailed account by Forton and colleagues provides two new insights into the mosquito-nematode model. First, these authors provided evidence that the melanization response, at least in adult mosquitoes, does involve hemocytes. Second, they showed that the termination of the host response, and thus recognition of self, is somehow related to the development of the double, membranelike structure surrounding the parasite. Recognition of self and termination of the host response is discussed later (see Sections 5.2.3.4 and 5.3.3).

Figure 5.22. Within 15 min postinoculation into the hemocoel of an adult *Aedes trivittatus* mosquito, one can see the cascading of melanin (arrow) onto the cuticle (c) of the nematode. (From Forton et al., 1985.)

Figure 5.23. Scanning electron micrograph of the microfilaria of *D. immitis* 15 min postinoculation into an adult *A. trivittatus* mosquito showing the deposits of melanin on the cuticle of the nematode (arrows). (From Forton et al., 1985.)

Figure 5.24. Electron micrograph of a microfilaria 24 hr postinoculation into an adult *A. trivittatus* mosquito. Note the dense inner layer of melanin, the impression of the nematode's cuticular annulations in the melanin, and the more diffuse outer layer. (From Forton et al., 1985.)

Figure 5.25. Electron micrograph showing the dense melanin layer (m) covering the encapsulated microfilaria (*D. immitis*) 2–6 hr postinoculation. At this time large areas containing cellular remnants are evident and indicate that hemocyte lysis has occurred. (From Forton et al., 1985.)

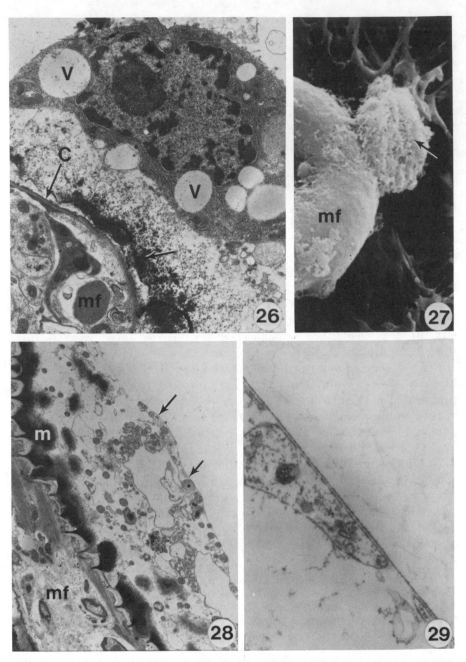

Figure 5.26. An intact hemocyte (plasmatocyte) is seen in close association with a microfilaria (mf) that was injected into an adult *A. trivittatus*. Observe the large vacuoles (V) forming in the hemocyte and their release in the direction of the nematode's cuticle (C). Also note the region of melanin deposits (arrow), which are always seen between intact hemocytes and the parasite's cuticle. (From Forton et al., 1985.)

Two other topics concerning hemocytic encapsulation and melanization that warrant discussion are the ability of certain nematodes to avoid a host response (see Fig. 5.1) and the potential benefit of the host response to the parasite. Poinar (1968) reported that when infective juveniles of *Filipjevimermis leipsandra* penetrate the first- and second-instar larvae of *Diabrotica* sp., the juvenile nematodes are rapidly encapsulated unless the larvae can penetrate the host's ganglion. Once inside the ganglion, they escape the host's reaction, presumably because they are now surrounded by the host's tissues and are consequently recognized as self. Here the nematode grows and eventually returns to the hemocoel as a third-stage larva. Now, for an unknown reason, no host response occurs. If, however, the nematodes enter the third-stage larvae of the host beetles instead of the first- or second-stage larvae, they are rapidly surrounded by blood cells and destroyed before they reach a ganglion (Poinar et al., 1968). This system offers an ideal model for testing the hypothesis that while inside the host's ganglion the cuticular composition of the nematode is somehow altered, possibly through antigen sharing, so that once it reinvades the hemocoel of the host no recognition occurs. This idea of avoiding the host response via antigen sharing is not a new one and is discussed briefly in Section 5.3.3. In Section 5.2.3.4, other examples of how nematodes avoid the host response are presented.

Can the host encapsulation response benefit the nematode parasite in any way? Christensen (1978) asked the interesting question: When does the parasite load limit the ability of a mosquito to transmit the infection? This is an important question, since not only does parasite burden affect the invertebrate host (i.e., adult mosquito), but if the host dies the parasites also die. Christensen (1981) observed that even though *A. trivittatus* was proved to be a natural host of the parasite, the immune response of the adult mosquito was more likely to occur during heavy infections, (i.e., heavy parasite burdens). The average intensity of infection in hosts eliciting an immune response was 21.9 nematodes per host, and in these hosts, 18.8% of the nematodes were melanized. Thus, the host response brought the level down to about 17.8 nematodes per host, a level that is assumed to be even lower in nature. Consequently, the host response reduced the parasite burden to a level below the level that would result in death of the host (Christensen,

---

Figure 5.27. Scanning electron micrograph of an intact hemocyte (plasmatocyte, indicated by the arrow) on the melanized surface of a microfilaria (mf). (From Forton et al., 1985.)

Figure 5.28. Initial formation of a single membrane (arrows) around a microfilaria (mf), which occurs about 2 days postinoculation. Also shown is the layer of melanin (m) that contacts the nematode's cuticle and penetrates the annulations of the parasite. Between the layer of melanin and the single membrane are remains of cellular debris. (From Forton et al., 1985.)

Figure 5.29. The formation of a second membrane on top of the one shown in Figure 5.28, which occurs independently of the formation of the first about 3 days postinoculation. (From Forton et al., 1985.)

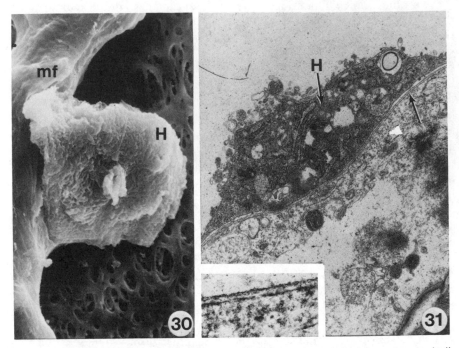

Figure 5.30. Scanning electron micrograph of an intact hemocyte (plasmatocyte, indicated by the H) attached to a melanin-encapsulated microfilaria (mf) that is covered by a double membrane. (From Forton et al., 1985.)

Figure 5.31. A hemocyte (plasmatocyte, indicated by the H) on the surface of a double membrane that completely encloses the melanized microfilaria (5 days postinoculation). Note the complete isolation of the hemocyte from the nematode by the double membrane (arrow), better shown in the inset. (From Forton et al., 1985.)

1978), and, as stated by Christensen (1981), "It is therefore suggested that the immune response might have a survival value for this vector-parasite system by reducing the parasite burdens." Whether the host response to nematodes by other insects benefits the parasite remains to be tested.

5.2.3. Tissue Responses

The least studied of all the nematode-induced host responses is the tissue response. Currently the evasive tactic of entering host cells or tissues, thus avoiding an encapsulation or humoral response, is known to have evolved in seven families of nematodes that use insects as intermediate hosts. Seureau and Quentin (1981) reported that in the primitive heteroxenous nematode (i.e., Subuluridae, Seuratidae, and Maupasinidae), the nematode remains outside the host cells and elicits a hemocytic encapsulation response that does not kill the parasite, while in the more advanced forms (i.e., Rictulariidae, Physalopteridae, Spiruridae, Acuariidae, Gongylonematidae, Diplotriaenidae, and Onchoceridae), the nematodes enter host cells, resulting in the formation of a syncytium. In both the primitive and advanced heteroxenous nematodes studied to date, no melanization is associated with the host structure surrounding the parasite, and the parasite is not destroyed.

According to Seureau and Quentin (1981), the primitive forms elicit a host response that results in the formation by the host's hemocytes of a capsule that eventually surrounds the parasite (Fig. 5.35). Since this response lacks the majority of characteristics associated with the hemocytic encapsulation response described in Section 5.2.2.4, specific differences between the two must keep the parasite from being destroyed. Seureau and Quentin (1981) suggested that the hemocytic response was initiated against degradation products of the parasite that leaked through the gut epithelium and arrived at the hemocoel before the parasite's entry into the hemocoel. It is suggested that since these parasites are slow in penetrating the host's gut, when they finally do cross the basement membrane of the host, they enter a hemocytic capsule that already consists of several layers of host hemocytes. Consequently, the initial direct hemocytic response is targeted at by-products of the host and is not aimed directly at the nematode. This tactic permits the nematode to avoid the usual hemocytic response of the host because it is now surrounded by host cells, or self. More questions can be asked about this host response than there are available documented answers. Because of this novel host response by hemocytes and failure of the hemocytes to destroy the parasite, these nematode-insect systems provide ideal models for approaching some of the historical questions of recognition.

The advanced heteroxenous nematodes, however, enter host cells or tissues, where they induce formation of giant cells (Seureau and Quentin, 1981), thus avoiding a humoral and/or hemocytic encapsulation response. Metalnikov (1924) incorrectly used the term *cellules geantes* to describe the fusion of hemocytes of *Galleria*. Later, Schell (1952) correctly used the term when he stated that "giant cells coalesced to form a syncytium around the parasite." These nematodes penetrate an insect tissue (i.e., epithelial cells of the ileum; see Fig. 5.32), where they destroy adjacent cells, thus forming a syncytium (Fig. 5.33). The tissue penetrated responds by forming a giant cell (Fig. 5.34) that not only houses the nematode parasite but protects the parasite from a humoral and/or encapsulation reaction by the host. The host tissue response to nematodes that results in the formation of giant cells is discussed by Poinar (1969).

5.2.3.1. Initiation. No experimental evidence conclusively identifying the factor or factors inducing the formation of giant cells in insects exists. Hollande (1920) suggested that fat body cells in the host he was studying were induced to hypertrophy as a result of toxins and mechanical irritation caused by the parasites, while Seureau (1973) and Poinar and Hess (1974) suggested that giant cell formation was induced by secretions from the nematode. The

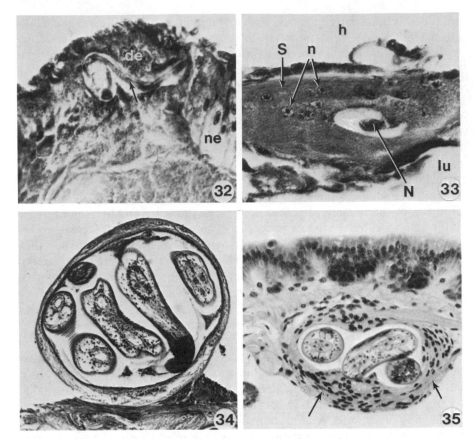

Figure 5.32. Infective-stage larvae of *Turgida turgida* (arrow) 6 hr postinfection in the epithelium of the ileum of *Gryllus pennsylvanicus*. The normal epithelial cells (ne) are shown to the right, while the disturbed cells (de) are evident near the nematode. (From Gray and Anderson, 1983.)

Figure 5.33. Thirty hours postinfection a syncytium (S) has formed around the nematode larvae (N) of *T. turgida*. Remaining nuclei (n) of the disrupted cells are evident, as are the lumen (lu) side and the hemocoel (h) side of the ileum. (From Gray and Anderson, 1983.)

Figure 5.34. Larvae 7 days postinfection of *T. turgida* inside a giant cell attached to and presumably formed by the epithelial cells of the ileum of *Acheta pennsylvanicus*. (From Gray and Anderson, 1982.)

Figure 5.35. Hemocytic capsule (arrow) surrounding the nematode *Echinonema cinctum* 20 days after experimental infection of *Locusta migratoria*. Here the capsule is located just below and attached to the intestine of the experimental host (From Chabaud et al., 1980.)

latter hypothesis was originally proposed for plant parasitic nematodes that cause giant cell formation (Endo, 1971; Bird, 1979); however, Jones (1981a) stated that "the nematodes disrupt or exaggerate normal cell formation rather than directly reprogramming metabolic events." The possibility that the initiation of giant cell formation by nematodes may be similar to the induction of plant neoplasia, or crown gall, (Merlo, 1982) should be tested. In plants, formation of the gall is induced by the virulent strain of *Agrobacterium tumefaciens,* and DNA (Ti plasmid) is transferred from the bacterial cell to the plant cell.

5.2.3.2. Formation. The insect tissues involved in the formaticn of giant cells include the epithelial cells of the digestive tract, body wall, malpighian tubules, and abdominal fat body (Poinar, 1969; Seureau, 1973; Bart ett, 1984; Stoffolano and Yin, 1986). The specificity shown by various nemaode species to a specific host tissue for development is most striking anc must in some way involve recognition of the tissue by the nematode. Geden and Stoffolano (1982) demonstrated that *Thelazia gulosa* larvae are almost exclusively surrounded by giant cells attached to the body wall of experimentally infected *M. autumnalis,* while examination of field-collected flies revealed a tendency toward clumping of the giant cells on the body wall in only 87.4% of the flies multiply parasitized (Geden and Stoffolano, 1981). Since *Thelazia skrjabini* is also reported in Massachusetts (Geden and Stoffolano, 1980), these authors suggested that in multiply parasitized face flies, *T. skrjabini* may represent the 12.6% of flies in giant cells of the fat body (Geden and Stoffolano, 1981). Eye worms of the genus *Thelazia* were never found in field-collected houseflies, *M. domestica,* yet this species is found to overlap with *M. autumnalis* in certain field situations. When *M. domestica* was experimentally fed infective-stage *Thelazia* larvae, the nematodes penetrated the gut and entered the hemoceol but did not enter host tissue. As a consequence, no giant cells were formed, and the nematodes developed slightly but died. Why no host response was elicited should be reexamined, since Poinar (1969) reported that if the nematode fails to invade the correct tisue, a host response results and the parasite dies. The fly genus *Musca* is known to be parasitized by several members of the nematode genus *Thelazia* (Stoffolano, 1970). This insect-nematode group is ideal for studying various aspects of host recognition with respect to the specificity of tissue recognition by the nematode, the initiation and formation of the giant cells, and the specificity of the host defense system by examining the effect of transplanting giant cells formed in *M. autumnalis* into *M. domestica* and vice versa. Another ideal insect-nematode model for studying the specificity of the tissue penetrated by the nematode and the response of the tissue to penetration is that of the locust and the subuluroid and spiruroid nematodes (Seureau, 1973).

5.2.3.3. Structure. Specific differences are obvious between giant cells and hemocytic capsules, yet both are considered to represent a defensive response by the host (Stoffolano and Yin, 1986). Giant cells are 100 times larger than hemocytic capsules (Table 5.1) and entail a tissue response without hemocytic involvement (Seureau and Quentin, 1981). Another major difference is that giant cells have never been reported to form around non-

TABLE 5.1. COMPARISONS BETWEEN GIANT CELLS SURROUNDING *THELAZIA* SP. AND HEMOCYTIC CAPSULES PRODUCED BY INSECTS

Comparison	Giant Cells	Reference[a]	Hemocytic Capsules	Reference[a]
Size	300–600 μm	13, 15, 17, 18	20–50 μm	1, 5
Formation involves	Tissue response	13, 15, 16, 18	Hemocytic response	4, 5, 11
Number of layers	2–4	3, 13, 15, 16	3–4	4, 5, 8, 10
Syncytium	+	13, 15	−	1, 5
Melanin	−	3, 15	+	6, 11
Tracheae	−	15	+	2, 6, 11
Necrosis	−	13, 15	+	4, 5, 7, 12
Will form around a nonbiologic agent	−	—	+	11
Surrounded biologic agent usually dies	−	3, 13, 15, 17	+	11
Hemocytes adhere to outside	+	3, 13, 15	+	11
Hemocytes inside	+	3, 15, 16	+	3, 11
Density differences in layers	−	15	+	5
Lumen around biologic agent	+	15	−	3

Source: Modified from Stoffolano and Yin, 1986.
[a] (1) Baerwald, 1979; (2) Brehélin *et al.*, 1975; (3) Cawthorn, 1980; (4) Ennesser and Nappi, 1984; (5) Grimstone *et al.*, 1967; (6) Nappi, 1975; (7) Nappi, 1977; (8) Norton and Vinson, 1977; (9) Poinar *et al.*, 1968; (10) Rowley and Ratcliffe, 1981; (11) Salt, 1970, (12) Zachary *et al.*, (1975), (13) Poinar and Hess (1974), (14) Stoffolano and Yin (1984), (15) Stoffolano and Yin (1986), (16) Schell, 1952; (17) Poinar and Quentin, 1972; (18) Seureau and Quentin, 1983.

living agents, and, in addition, giant cells do not cause the death of the parasite (Table 5.1). Other differences between the two types of host responses are listed in Table 5.1.

The most detailed study of the ultrastructure of giant cells in insects is that of Stoffolano and Yin (1986). Three types of organelles present in giant cells but lacking in hemocytic capsules are hypertrophied nuclei, annulate lamellae, and smooth-surfaced endoplasmic reticulum (Table 5.2).

A cross section through the giant cell surrounding a second-stage larva (*Thelazia* spp.) reveals four distinct zones, or regions (Fig. 5.36). No signs of necrosis or hemocyte involvement are evident. The organelles present appear normal, and the outer surface of the giant cell is delimited from the host's hemolymph by a connective tissue layer about 0.1–0.3 μm in thickness. The unsuspected complexity of the giant cells formed by *M. autumnalis* against the nematode parasites of the genus *Thelazia* should encourage other investigators to conduct more detailed studies of giant cells in other insect-nematode systems (e.g., *M. domestica* produces giant cells against *Habronema* spp.).

5.2.3.4. Importance to Parasite. Coevolution of nematodes with insects has

TABLE 5.2. COMPARISONS BETWEEN THE ORGANELLES AND OTHER CELLULAR COMPONENTS OF GIANT CELLS SURROUNDING *THELAZIA* SP. AND THOSE OF HEMOCYTIC CAPSULES PRODUCED BY INSECTS

Organelle	Giant Cells	Reference[a]	Hemocytic Capsules	Reference[a]
Hypertrophied nuclei	+	3, 13, 15, 16, 17, 18	−	—
Mitochondria	+	15	+	4, 5
Free ribosomes	+	15	+	2
Smooth endoplasmic reticulum	+	15	−	—
Rough endoplasmic reticulum	+	15	+	2
Annulate lamellae	+	14, 15	−	—
Microtubes	+	13, 15	+	1, 10
Microfilaments	−	15	+	2
Golgi complex	+	15	+	4
Cell junctions	−	15	+	1, 9, 12
Cytolysosomes	−	15	+	4, 5, 7, 10
Outer connective tissue layer	+	15	−	—
Plaques associated with outer layer	+	15	−	—
Disintegrating membranes found inside	+	15	−	—
Basal infolding of outer layer	+	15	−	—

Source: Modified from Stoffolano and Yin, 1986.

[a] (1) Baerwald, 1979; (2) Brehélin *et al.*, 1975; (3) Cawthorn, 1980; (4) Ennesser and Nappi, 1984; (5) Grimstone *et al.*, 1967; (6) Nappi, 1975; (7) Nappi, 1977; (8) Norton and Vinson, 1977; (9) Poinar *et al.*, 1968; (10) Rowley and Ratcliffe, 1981; (11) Salt, 1970, (12) Zachary *et al.*, (1975), (13) Poinar and Hess (1974), (14) Stoffolano and Yin (1984), (15) Stoffolano and Yin (1986), (16) Schell, 1952; (17) Poinar and Quentin, 1972; (18) Seureau and Quentin, 1983.

resulted in selective pressures that have influenced the insect to evolve more effective defensive mechanisms against the parasite while at the same time favoring those parasites that can either avoid the host response or evolve countertactics against the humoral and/or cellular responses. The insect's defensive response to invasion of host cells or tissues has resulted in the formation of giant cells, which are generally accepted by most investigators to protect the parasite by preventing a hemocytic encapsulation and/or melanization response by the host (Stoffolano and Yin, 1986).

Prevention of Hemocytic Encapsulation and Melanization Response. Salt (1970) suggests that nonself, which provokes either a humoral or a cellular defensive response by the host, is equivalent to anything other than intact native basement membrane. This hypothesis is still considered the best explanation of recognition of self (Ratner and Vinson, 1983). Based on it, one

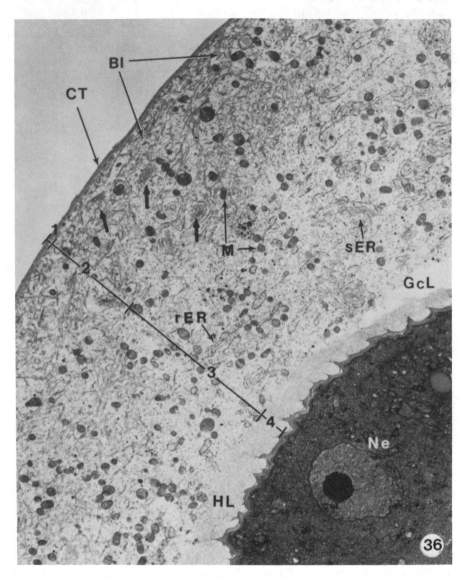

Figure 5.36. Giant cell 6–9 days post infection surrounding a second-stage larval nematode (Ne) of *H. autumnalis* in the host *M. autumnalis*. Note the four regions of the giant cell. The inner layer contacting the nematode is the hyaline or homogenous cytoplasmic layer (HL), which is devoid of organelles. The next layer contains most of the organelles: rough-surfaced endoplasmic reticulum (rER), smooth-surfaced endoplasmic reticulum (sER), mitochondria (M), and Golgi apparatus–like aggregations (arrows). The adjacent layer is a basal infolding (BI) layer and is surrounded by a thin connective tissue layer (CT). (From Stoffolano and Yin, 1986.)

can appreciate that, once the parasite is surrounded by a giant cell, it is also enclosed by host tissue, which is self; consequently, the host does not exhibit a defensive response in the form of hemocytic encapsulation and/or a humoral response. Even though it is generally accepted that the parasite avoids the hemocytic encapsulation response by developing inside the giant cell and that parasites that fail to do so are destroyed (Poinar and Hess, 1974; Seureau, 1977; Cawthorn, 1980; Bartlett, 1984; Stoffolano and Yin, 1986), experimental observations designed to test this hypothesis are lacking.

Exchange of Materials. Only Bartlett (1984) and Stoffolano and Yin (1986) have so far suggested that the giant cell also functions as a metabolic shunt between the nematode and the host. Several investigators have suggested that the giant cell provides the nematodes with nourishment (Schell, 1952; Poinar and Quentin, 1972; Poinar and Hess, 1974; Cawthorn and Anderson, 1977; Gray and Anderson, 1983; Bartlett, 1984; Stoffolano and Yin, 1986). The study of histologic sections revealed the presence of the same cytoplasmic material inside the digestive tract of the nematode and in the giant cell, evidence supporting the proposal that nematodes feed on giant cell material (Poinar and Hess, 1974; Gray and Anderson, 1983; Stoffolano and Yin, 1986). No one, however, has suggested that the nematodes inside the giant cell obtain nutrients from the host's hemolymph by active transport. The absence of a defined cuticle and lack of a completed digestive tract in the first-stage thelazial larva, along with ultrastructural evidence (i.e., basal infolding and associated mitochrondia) of the outer layers of the giant cell (see Fig. 5.36), strongly suggest active uptake of nutrients from the host (Stoffolano and Yin, 1986). In addition, the movement of the nematode's waste products out of the giant cell into the host's hemolymph is also possible. More information is needed on the metabolic functions of giant cells housing nematode parasites.

5.2.3.5. Importance to Host. Even though hemocytic capsules and giant cells differ in several respects (Tables 5.1 and 5.2), both are considered by some investigators to represent a defensive response by the host (Schacher and Khalil, 1968; Spratt, 1972; Ratcliffe, 1982; Stoffolano and Yin, 1986). Bartlett (1984) takes the position that in order to represent a defensive response by the host, the response must involve hemocytes and/or melanin. Even though the parasite is not destroyed within the giant cell, its formation can be considered a defensive response, since the parasite is contained by the host and prevented from further damaging other tissues.

5.3. FUTURE AVENUES FOR RESEARCH

The most important contribution one can make in reviewing any topic is to make predictions concerning the areas that need more concentrated research, to ask questions that need to be answered in order for these areas to advance, and to provide interested readers with the references necessary for entering the literature. This literature is necessary for the synthesis of new ideas and/or for familiarizing readers with new technology.

5.3.1. Effects of Pesticides on Parasitic Encapsulation

A major component of any integrated pest management program using biologic control agents is the compatibility between the pesticides used and the biologic control agents. Croft (1977) reviewed the literature concerning the effects insecticides may have on a parasitoid-host interaction, while Vinson and Iwantsch (1980) reported on the effects biorationals may have on parasitoids. As far as I am aware, no one has studied the effects of pesticides on the development of individuals that represent a pesticide-resistant population that may be either susceptible or refractory to the biologic control agent being used. Two major consequences of pesticide effects on the host's defensive system could occur. Sublethal doses of the pesticide may either enhance or retard the host's defensive system. Also, the development of resistance by the host may favor strains that are either more susceptible or refractory to the parasite in question.

5.3.2. Genetics of Parasite Recognition

Little research has been devoted to the genetic basis of the susceptibility of insect host populations to parasites; yet this topic will become more important if the projections of genetic engineers are realized. Two entirely different strategies may be employed in attempts to alter the host population's susceptibility to a parasite. From a biologic control point of view, what is favored are control agents that either avoid or counter the host's defense response and host strains that fail to destroy the control agent. I am currently unaware of any research on this topic. From an epidemiologic standpoint, what is desired are disease agents that will elicit a host response and host strains that are highly refractory. The ideal system for studying this aspect of the influence of genetics on host responses is the filarial nematode-mosquito model.

Insect biotypes that differ with respect to their susceptibility to encapsulation have already been demonstrated in nature (Salt and Van den Bosch, 1967). An important review by Diehl and Bush (1984) discusses the applied aspects of insect biotypes, but no reference is made to the host defensive system and its importance in control strategies. More information needs to be obtained concerning insect biotypes, their nematode parasites, and their predilection to destroy the parasite. Entomologists now have available new tools to uncover many of the differences that exist in the host as well as the parasite biotypes (Berlocher, 1984). Using these techniques, one can examine the genetic basis of the host's defensive response against nematodes in the field. Again, the mosquito-filarial nematode system would be ideal for such studies.

Some information on the genetic basis of susceptibility of mosquitoes to filarial nematodes already exists (Kartman, 1953; MacDonald, 1962a,b; McGreevy et al., 1974; Baar, 1975). The advantages of using this system are discussed in Section 5.3.3.

5.3.3. Model Systems for Studying the Mechanisms of Recognition

The mechanism whereby the giant cell affords protection to the parasite has not been identified. If the suggestion by Salt (1970) that basement membrane is recognized as self is correct, the connective tissue layer surrounding the giant cell in the study of Stoffolano and Yin (1986) may be similar to the basement membrane of the host and thus may prevent the host from recognizing the parasite within the giant cell. The importance of basement membrane in recognition of self was demonstrated by Rizki and Rizki (1974), who showed that the formation of melanotic tumors in the fat body of the *Drosophila melanogaster* (tumor[w]) mutant occurred as a result of the loss of the basement membrane covering the fat body. The face fly, *M. autumnalis,* and the nematode *H. autumnalis* provide an excellent model for investigating the importance of the covering layer of the giant cell in protecting the parasite from the host's hemocytic encapsulation response. The ability to remove the insect's basal laminae (believed to be comparable to the outer layer of the giant cell) using the enzyme technique developed by Levinson and Bradley (1984) should permit one to remove the outer layer of the giant cell by removing it from a host, using the elastase to remove the covering layer, and then to implant it into another host. Another series of experiments designed to uncover the role and specificity of the outer covering of the giant cell would be to remove the giant cell from *M. autumnalis* (natural host) and to implant it in *M. domestica* (unsuitable host). The ability to manipulate the giant cell should permit the investigator to design experiments that will provide new information concerning the role of the covering layer and its importance in recognition. In order to appreciate the importance of the basement membrane in recognition of self—or in fact the importance of any membranes surrounding parasites or giant cells—more information on their biochemical and biophysical properties must be obtained. Unlike our knowledge of the vertebrate system (Kefalides et al., 1979), our understanding of the basement membrane in insects is scarce.

The success of Franke and Weinstein (1983) in culturing the infective, third-stage larvae of a filarial nematode, even though its arthropod host is a tick, in a cell-free system further enhances the filarial-mosquito system as an ideal model. An in vitro system must now be developed for a mosquito-borne filarial nematode.

The study of Sutherland and colleagues (1984) provided evidence suggesting that the microfilariae of *Brugia pahangi* acquire host antigens on their cuticle during midgut penetration, thus permitting the parasite to avoid recognition. Lafond and associates (1985) also provided evidence suggesting that *B. pahangi* microfilariae avoid the immune recognition response in compatible mosquito hosts by coating themselves with material from the midgut that is obtained during the act of penetrating the host's midgut. Research in the area of naturally occurring surface determinants on the cuticle of nematode parasites and their role in immunity is gaining momentum (Hogarth-Scott, 1968; Simpson and Smithers, 1980; and Ortega-Pierres et al., 1984; and Ouaissi et al., 1984). Research teams must apply new techniques to the nematode-mosquito system in an effort to explain how the parasite either counters or evades the host response. Suggestions for countertactics by parasites against the humoral and/or cellular responses at the level of inactivation of promoting factors (see Fig. 5.1) have been made by Ratner and Vinson (1983) and by Nappi and Silvers (1984) for the next level of interference (see Fig. 5.1).

5.4. CONCLUDING REMARKS

We must be careful in pursuing a single objective, especially one in which blinders permit us only to see what is straight ahead. We must not lose sight of how the information already obtained is related to other disciplines. Briefly, let us consider two areas in which information concerning nematode-induced and other parasite- and pathogen-induced insect responses are important: integrated pest management and vector biology and epidemiology.

5.4.1. Integrated Pest Management

The significance of hemocytic encapsulation by the European larch sawfly, a forest pest, against a hymenopterous parasite was well documented by Turnock and coworkers (1976). These authors discussed how the equilibrium between biotypes of the host that were resistant (refractory) and those that were susceptible to the parasite was influenced by the number of different parasitic species attacking the host. When the parasite in question became the primary parasite, the resistant genotype of the host was selected. The once effective biologic control agent now became ineffective because of the appearance and spread of the refractory biotype of the host, which was capable of destroying the parasite through hemocytic encapsulation. Similar examples of host resistance to potential biologic control agents have been reported for entomophilic nematodes. In a review of the host-specificity literature on entomophilic nematodes, I (Stoffolano, 1973) have noted that physiologic selection in nematodes was influenced by host suitability. Thus, the ability of a nematode to successfully enter the host, avoid host resistance, and exit from the host could influence successful parasitization.

Petersen and colleagues (1968) reported that the biologic control potential of three species of mermithid nematodes against two species of mosquitoes was greatly reduced because of host resistance in the form of encapsulation and melanization. In a brief report on using mermithid parasites as biologic control agents against blackflies, and consequently preventing or reducing onchocerciasis in Africa, one of the areas targeted for investigation was host resistance (Anonymous, 1972; Davies et al., 1984). Rapid melanization, suggested by Poinar (1974) to be a humoral response, against nematode larvae, which had biologic control potential, could reduce their effectiveness in any control program (Petersen et al., 1969; Mitchell et al., 1972). Because of the negative effect of host resistance in any control program, future integrated pest management schemes designed to use nematodes as control agents must include periodic examination of the host population for the presence and/or development of host resistance. Poinar (1979), Petersen (1982), and Nickle (1984) provide reviews concerning the use of nematodes as biologic control agents against insect pests. In fact, the importance of the defensive response of insects against biologic control agents was discussed by Ratcliffe and Rowley (1979); however, the authors gave little attention to nematodes. I (1973) noted that it should be emphasized that "biological control workers are working with a dynamic, evolving host-parasite system and that really they are faced with the same problems (resistance of the host and target specificity of the applied material) originally and still confronting the toxicologist and economic entomologist."

5.4.2. Vector Biology and Epidemiology

The effect of nematodes as causative agents of various diseases of human beings, domestic animals, and wildlife cannot be appreciated until one carefully begins to study the numbers of vertebrate organisms afflicted, the debilitating impact of the disease agent on these populations, and the cost of treatment. Mosquitoes, tabanids, and blackflies are bloodsucking flies that serve as both vector and intermediate hosts for nematodes, which cause the following diseases of vertebrate animals: filariasis, loiasis, onchocerciasis, and elaeophorosis. Geden and Stoffolano (1984) reviewed the literature that focused on nematode parasites that are vectored by flies and cause the following diseases of vertebrate animals: thelaziasis, onchocerciasis, habronemiasis, stephanofilariasis, and parafilariasis.

From an epidemiologic point of view, one important factor in establishing disease foci for insect-vectored nematode diseases is successful parasitization of both the definitive vertebrate host and the intermediate insect host.

Recent studies have shown that the intermediate host or vector often elicits a successful defensive encapsulation and/or melanization reaction. A more comprehensive understanding of both the proportion of naturally occurring refractory versus susceptible insect and vertebrate hosts, as well as the genetic predilection to either of these traits, is vital information currently lacking. In addition, we need to better understand the exact mechanisms behind the vector's refractoriness to infection with nematode parasites.

5.5 SUMMARY

The field of study aimed at understanding parasite-induced host responses should now be expanded to address issues concerning parasite recognition as they are related to other disciplines, such as integrated pest management, vector biology, and epidemiology. Throughout this chapter, an effort has been made to identify nematode-insect models currently available for investigating various problems related to parasite recognition and avoidance.

Currently, investigations of nematode-induced host responses are categorized into humoral, cellular, and tissue responses. These host responses may not be individually called into action by the host but may operate in concert to suppress a particular parasite. Further studies should be designed to examine this aspect of host response.

The emphasis to date has been on the humoral and cellular responses of insects to nematode parasites. More information needs to be obtained concerning the tissue responses of the host. Also, a model has been put forth in this chapter to account for the nematode-insect dynamic system as it is related to parasite survival and suppression. This model should be tested and advanced.

Finally, future avenues for research in this field are numerous and should include studies designed to look at the effects of pesticides on the host response and the genetics of host parasite recognition.

ACKNOWLEDGMENTS

The assistance of B. M. Christensen and A. J. Nappi in reading and making suggestions on the manuscript are greatly appreciated. This chapter is dedicated to A. J. Nappi, who has been an inspiration to me in my studies of host responses by insects to nematodes.

REFERENCES

Anderson, R.S. 1981. Comparative aspects of the structure and function of invertebrate leucocytes, pp. 629–641. In N.A. Ratcliffe and A.F. Rowley (eds.) *Invertebrate Blood Cells,* vol. 2. Academic Press, New York.

Andreadis, T.G. and D.W. Hall. 1976. *Neoaplectana carpocapsae:* Encapsulation in *Aedes aegypti* and changes in host hemocytes and hemolymph proteins. *Exp. Parasitol.* **39**: 252–261.

Anonymous. 1972. *Preventing Onchocerciasis through Blackfly Control: A Proposal for Afro-Canadian Research into the Feasibility of Using Mermithid Parasites as Biological Agents in the Control of Disease-Transmitting Blackflies,* pp. 1–11. International Development Research Centre, publication no. IDRC-006e, Ottawa, Canada.

Ashhurst, D.E. 1982. The structure and development of insect connective tissues, pp. 313–350. In R.C. King and H.Akai (eds.) *Insect Ultrastructure,* vol. 1. Plenum Press, New York.

Ashida, M. and K. Söderhäll. 1984. The prophenoloxidae activating system in crayfish. *Comp. Biochem. Biophysiol.* **77**(B): 21–26.

Baerwald, R.J. 1979. Fine structure of hemocyte membranes and intercellular junctions formed during hemocyte encapsulation, pp. 155–188. In A.P. Gupta (ed.) *Insect Hemocytes.* Cambridge University Press, Cambridge.

Barr, A.R. 1975. Evidence for genetical control of invertebrate immunity and its field significance, pp. 129–135. In K. Maramorosch and R.E. Shope (eds.) *Invertebrate Immunity: Mechanisms of Invertebrate Vector-Parasite Relations.* Academic Press, New York.

Barr, A.R. and R. Shope. 1975. The invertebrate gut as barrier to invading parasites, pp. 113–114. In K. Maramorosch and R.E. Shope (eds.) *Invertebrate Immunity: Mechanisms of Invertebrate Vector-Parasite Relations.* Academic Press, New York.

Bartlett, C.M. 1984. Development of *Dirofilaria scapiceps* (Leidy, 1886) (Nematoda: Filarioidea) in *Aedes* spp. and *Mansonia perturbans* (Walker) and responses of mosquitoes to infection. *Can. J. Zool.* **62:** 112–129.

Berlocher, S.H. 1984. Insect molecular systematics. *Annu. Rev. Entomol.* **29:** 403–433.

Bird, A.F. 1979. Histopathology and physiology of syncytia, pp. 155–171. In F. Lamberti and C.E. Taylor (eds.) *Root-Knot Nematodes* (Meloidogyne *Species*): *Systematics, Biology and Control.* Academic Press, New York.

Brehélin, M., J.A. Hoffmann, G. Matz, and A. Porte. 1975. Encapsulation of implanted foreign bodies by hemocytes in *Locusta migratoria* and *Melolontha melolontha. Cell Tissue Res.* **160:** 283–289.

Bronskill, J.F. 1962. Encapsulation of rhabditoid nematodes in mosquitoes. *Can. J. Zool.* **40:** 1269–1275.

Cawthorn, R.J. 1980. The cellular responses of migratory grasshoppers (*Melanoplus sanguinipes* F.) and African desert locusts (*Schistocerca gregaria* L.) to *Diplotriaena tricuspis* (Nematoda: Diplotriaenoidea). *Can. J. Zool.* **58:** 109–113.

Cawthorn, R.J. and R.C. Anderson. 1977. Cellular reactions of field crickets (*Acheta pennsylvanicus* Burmeister) and German cockroaches (*Blatella germanica* L.) to *Physaloptera maxillaris* Molin (Nematoda: Physalopteroidea). *Can. J. Zool.* **55:** 368–375.

Chabaud, A.-G., C. Seureau, I. Beveridge, O. Bain, and M.-C. Durette-Desset. 1980. Sur les nematodes echinonematinae. *Ann. Parasitol.* (Paris) **55:** 427–443.

Cherbas, L. 1973. The induction of an injury reaction in cultured haemocytes from saturniid pupae. *J. Insect Physiol.* **19:** 2011–2023.

Christensen, B.M. 1978. *Dirofilaria immitis:* Effect on the longevity of *Aedes trivittatus. Exp. Parasitol.* **44:** 116–123.

Christensen, B.M. 1981. Observations on the immune response of *Aedes trivittatus* against *Dirofilaria immitis*. *Trans. R. Soc. Trop. Med. Hyg.* **75:** 439–443.

Christensen, B.M. and D.R. Sutherland. 1984. *Brugia pahangi:* Exsheathment and midgut penetration in *Aedes aegypti*. *Trans. Am. Microsc. Soc.* **103:** 423–433.

Cooper, E.L. (ed.). 1974. *Contemporary Topics in Immunobiology*, vol. 4, *Invertebrate Immunology*. Plenum Press, New York.

Croft, B.A. 1977. Susceptibility surveillance to pesticides among arthropod natural enemies: Modes of uptake and basic responses. *J. Plant Dis. Prot.* **84:** 140–157.

Crossley, A.C. 1975. The cytophysiology of insect blood. *Adv. Insect Physiol.* **11:** 117–221.

Davies, J.B., J.E. McMahon, P. Beech-Garwood, and F. Abdulai. 1984. Does parasitism of *Simulium damnosum* by Mermithidae reduce the transmission of onchocerciasis? *Trans. R. Soc. Trop. Med. Hyg.* **78:** 424–425.

Day, M.F. and M.J. Bennetts. 1953. Healing of gut wounds in the mosquito *Aedes aegypti* (L.) and the leafhopper *Orosiur argentatus* (Ev.). *Aust. J. Biol. Sci.* **6:** 580–585.

Deslongchamps, E. 1824. Encyclopédie méthodique. *Hist. Nat. Zoophyt.* **2:** 396–397.

Dickinson, S. 1959. The behavior of larvae of *Heterodera schachtii* on nitrocellulose membranes. *Nematologica* **4:** 60–66.

Diehl, S.R. and G.L. Bush. 1984. An evolutionary and applied perspective of insect biotypes. *Annu. Rev. Entomol.* **29:** 471–504.

Endo, B.Y. 1971. Nematode-induced syncytia (giant cells): Host-parasite relationships of Heteroderidae, pp. 91–117. In B.M. Zuckerman, W.R. Mai, and R.A. Rohde (eds.) *Plant Parasitic Nematodes,* vol. 2. Academic Press, New York.

Ennesser, C.A. and A.J. Nappi. 1984. Ultrastructural study of the encapsulation response of the American cockroach, *Periplaneta americana. J. Ultrastruct. Res.* **87:** 31–45.

Esslinger, J.H. 1962. Behavior of microfilariae of *Brugia pahangi* in *Anopheles quadrimaculatus. Am. J. Trop. Med. Hyg.* **11:** 749–758.

Foley, D.A. 1978. Innate cellular defense by mosquito hemocytes, pp. 113–144. In L.A. Bulla and T.C. Cheng (eds.) *Invertebrate Models for Biomedical Research,* vol. 4. Plenum Press, New York.

Forton, K.F., B.M. Christensen, and D.R. Sutherland. 1985. Ultrastructure of the melanization response of *Aedes trivittatus* against inoculated *Dirofilaria immitis* microfilariae. *J. Parasitol.* **71:** 331–341.

Franke, E.D. and P.P. Weinstein. 1983. *Dipetalonema viteae* (Nematoda: Filarioidae): Culture of third-stage larvae to young adults in vitro. *Science* (Washington, DC) **221:** 161–163.

Furman, A. and L.R. Ash. 1983. Analysis of *Brugia pahangi* microfilariae surface carbohydrates: Comparison of the binding of a panel of fluorescinated lectins to mature in vitro–derived and immature in utero-derived microfilariae. *Acta Trop.* **40:** 45–51.

Geden, C.J. and J.G. Stoffolano, Jr. 1980. Bovine thelaziasis in Massachusetts. *Cornell Veterinarian* **70:** 344–359.

Geden, C.J. and J.G. Stoffolano, Jr. 1981. Geographic range and temporal patterns of parasitization of *Musca autumnalis* (Diptera: Muscidae) by *Thelazia* sp. (Nematode: Muscidae) as an unsuitable host. *J. Med. Entomol*, **18**: 449–456.

Geden, C.J. and J.G. Stoffolano, Jr. 1982. Development of the bovine eyeworm, *Thelazia gulosa* (Railliet and Henry), in experimentally infected, female *Musca autumnalis* De Geer. *J. Parasitol.* **68**: 287–292.

Geden, C.J. and J.G. Stoffolano, Jr. 1984. Nematode parasites of other dipterans, pp. 849–898. In W.R. Nickle (ed.) *Plant and Insect Nematodes*. Marcel Dekker, New York.

Götz, P. 1964. Der Einfluss unterschiedlicher Befallsbedingungen auf die Mermithogene intersexualitat von *Chironomus* (Dipt.). *Z. Parasitenkd.* (Berlin) **24**: 484–545.

Götz, P. 1969. Die einkapselung von parasiten in der hamolymphe von *Chironomus*-larven (Diptera). *Zool. Anz.* **33**: 610–617, supplement.

Götz, P. 1984. Encapsulation mechanisms in insects, pp. 723. 17th- Int. Congr. Entomol, abstract vol. Hamburg, Germany.

Götz, P., J. Roettgen, and W. Lingg. 1977. Encapsulement humoral en tant que réaction de defense chez les Diptères. *Ann. Parasitol. Hum. Comp.* **52**: 95–97.

Grégorie, C. 1971. Haemolymph coagulation in arthropods, pp. 145–186. In M. Florkin (ed.) *Chemical Zoology,* vol. 6. Academic Press, New York.

Grey, J.B. and R.C. Anderson. 1983. Cellular reactions of the field cricket (*Gryllus pennsylvanicus* [Burmeister]) to *Turgida turgida* (Rudolhi, 1819) (Nematoda: Physalopteroidea). *Can. J. Zool.* **61**: 2143–2146.

Hall, D.W. 1983. Mosquito hemocytes: A review. *Dev. Comp. Immunol.* **7**: 1–12.

Hewitt, C.G. 1914. *The House-Fly*. Cambridge Zoological Society Series, University of Cambridge, Cambridge.

Hoffmann, J.A., D. Zachary, D. Hoffmann, M. Brehélin, and A. Porte. 1979. Postembryonic development and differentiation: Hemopoietic tissues and their functions in some insects, pp. 29–66. In A.P. Gupta (ed.) *Insect Hemocytes*. Cambridge University Press, Cambridge.

Hogarth-Scott, R.S. 1968. Naturally occurring antibodies to the cuticle of nematodes. *Parasitology* **58**: 221–226.

Jansson, H.-B., A. Jeyaprakash, R.A. Damon, Jr., and B.M. Zuckerman. 1984. *Caenorhabditis elegans* and *Panagrellus redivivis:* Enzyme-mediated modification of chemotaxis. *Exp. Parasitol.* **58**: 270–277.

Jeyaprakash, A., H.-B. Jansson, N. Marban-Mendoza, and B.M. Zuckerman. 1985. *Caenorhabditis elegans:* Lectin-mediated modification of chemotaxis. *Exp. Parasitol.* **59**: 90–97.

Jones, M.G.K. 1981. Host cell responses to endoparasitic nematode attack: Structure and function of giant cells and syncytia. *Ann. Appl. Biol.* **97**: 353–372.

Kartman, L. 1953. Factors influencing infection of the mosquito with *Dirofilaria immitis* (Leidy, 1856). *Exp. Parasitol.* **2**: 27—78.

Kefalides, N.A., R. Alper, and C.C. Clark. 1979. Biochemistry and metabolism of basement membranes. *Int. Rev. Cytol.* **61**: 167–228.

Keilin, D. and V.C. Robinson. 1933. The morphology and life history of *Aprocto-*

nema entomophagum Keilin, a nematode parasite in the larva of *Sciara pullula* Winn (Diptera, Nematocera). *Parasitology* **25:** 285–295.

Kirschbaum, J.B. 1985. Potential implication of genetic engineering and other biotechnologies to insect control. *Annu. Rev. Entomol.* **30:** 51–70.

Lackie, A.M. 1981. Humoral mechanisms in the immune response of insects to larvae of *Hymenolepis diminuta* (Cestoda). *Parasit. Immunol.* **3:** 201–208.

Lafond, M.M., B.M. Christensen, and B.A. Lasee. 1985. Defense reactions of mosquitoes to filarial worms: Potential mechanism for avoidance of the response by *Brugia pahangi* microfilariae. *J. Invertebr. Pathol.* **46:** 26–30.

Levinson, G. and T.J. Bradley, 1984. Removal of insect basal laminae using elastase. *Tissue Cell.* **16:** 367–375.

MacDonald, W.W. 1962a. The genetic basis of susceptibility to infection with semiperiodic *Brugia malayi* in *Aedes aegypti. Ann. Trop. Med. Parasitol.* **56:** 373–382.

MacDonald, W.W. 1962b. The selection of a strain of *Aedes aegypti* susceptible to infection with semi-periodic *Brugia malayi. Ann. Trop. Med.* **56:** 368–372.

McGreevy, P.M., G.A.H. McClelland, and M.M.J. Lavoipierre. 1974. Inheritance of susceptibility to *Dirofilaria immitis* infection in *Aedes aegypti. Ann. Trop. Med. Parasitol.* **68:** 97–109.

Merlo, D.J. 1982. Bridging the gap to plants: Bacterial DNA in plant cells, pp. 281–302. In M.S. Mount and G.H. Lacy (eds.) *Phytopathogenic Prokaryotes,* vol. 2. Academic Press, New York.

Metalnikov, S. 1924. Phagocytose et réactions des cellules dans l'immunité. *Ann. Inst. Pasteur* Paris **38:** 787–826.

Mitchell, C.J., P.S. Chen, and H.C. Chapman. 1972. Exploratory trials utilizing a mermithid nematode as a control agent for *Culex* mosquitoes in Taiwan (China). *World Health Organization/Vector Biology and Control* **410:** 1–10.

Nappi, A.J. 1975. Parasite encapsulation in insects, pp. 293–326. In K. Maramorosch and R.E. Shope (eds.) *Invertebrate Immunity.* Academic Press, New York.

Nappi, A.J. 1977. Comparative ultrastructural studies of cellular immune reactions and tumorigenesis in *Drosophila,* pp. 155–188. In L.A. Bulla and T. Cheng (eds.) *Comparative Pathobiology,* vol. 3. Plenum Press, New York.

Nappi, A.J. and M. Silvers. 1984. Cell surface changes associated with cellular immune reactions in *Drosophila. Science* (Washington, DC) **225:** 1166–1168.

Nappi, A.J. and J.G. Stoffolano, Jr. 1971. *Heterotylenchus autumnalis:* Hemocytic reactions and capsule formation in the host, *Musca domestica. Exp. Parasitol.* **29:** 116–125.

Nappi, A.J. and J.G. Stoffolano, Jr. 1972. Distribution of haemocytes in larvae of *Musca domestica* and *Musca autumnalis* and possible chemotaxis during parasitization. *J. Insect Physiol.* **18:** 169–179.

Nickle, W.R. (ed.). 1984. *Plant and Insect Nematodes.* Marcel Dekker, New York.

Nordbring-Hertz, B., E. Friman, and B. Mattiasson. 1982. A recognition mechanism in the adhesion of nematode-trapping fungi, pp. 83–90. In T.C. Bog-Hansen (ed.) *Lectins: Biology, Biochemistry, Clinical Biochemistry,* vol. 2. W. de Gruyter, Berlin.

Norton, W.N. and S.B. Vinson. 1977. Encapsulation of a parasitoid egg within its habitual host: An ultrastructural investigation. *J. Invertebr. Pathol.* **30:** 55–67.

Orihel, T.C. 1975. The peritrophic membrane: Its role as a barrier to infection of the arthropod host, pp. 65–73. In K. Maramorosch and R.E. Shope (eds.) *Invertebrate Immunity: Mechanisms of Invertebrate Vector-Parasite Relations.* Academic Press, New York.

Ortega-Pierres, G., A. Chayen, N.W.T. Clark, and R.M.E. Parkhouse. 1984. The occurrence of antibodies to hidden and exposed determinants of surface antigens of *Trichinella spiralis*. *Parasitology* **88:** 359–369.

Ouaissi, M.A., J. Cornette, and A. Capron. 1984. Occurrence of fibronectin antigenic determinants on *Schistosoma mansoni* lung schistosomula and adult worms. *Parasitology* **88:** 85–96.

Peters, W., H. Kolb, and V. Kolb-Bachofen. 1983. Evidence for a sugar receptor (lectin) in the peritrophic membrane of the blowfly larva, *Calliphora erythrocephala* Mg. (Diptera). *J. Insect Physiol.* **29:** 275–280.

Petersen, J.J. 1982. Current status of nematodes for biological control of insects. *Parasitology* **84:** 177–204.

Petersen, J.J., H.C. Chapman, and O.R. Willis. 1969. Fifteen species of mosquitoes as potential hosts of a mermithid nematode *Romanomermis* sp. *Mosq. News* **29:** 198–201.

Petersen, J.J., H.C. Chapman, and D.B. Woodard. 1968. Bionomics of a mermithid nematode of larval mosquitoes in southwestern Louisiana. *Mosq. News* **28:** 346–352.

Poinar, G.O., Jr. 1968. Parasitic development of *Filipjevimermis leipsandra* Poinar & Welch (Mermithidae) in *Diabrotica u. undecimpunctata* (Chrysomelidae). *Proc. Helm. Soc. Wash.* **35:** 161–169.

Poinar, G.O., Jr. 1969. Arthropod immunity to worms, pp. 173–210. In C.J. Jackson, R. Herman, and I. Singer (eds.) *Immunity to Parasitic Animals,* vol. 1. Appleton-Century-Crofts, New York.

Poinar, G.O., Jr. 1974. Insect immunity to parasitic nematodes, pp. 167–178. In E.L. Cooper (ed.) *Contemporary Topics in Immunobiology: Invertebrate Immunity,* vol. 4. Plenum Press, New York.

Poinar, G.O., Jr. 1979. *Nematodes for Biological Control of Insects.* CRC Press, Boca Raton, FL.

Poinar, G.O., Jr., and C.C. Doncaster. 1965. The penetration of *Tripius sciarae* (Bovien) (Sphaerulariidae: Aphelenchoidae) into its host, *Bradysia paupera* Thom. (Mycetophilidae: Diptera). *Nematologica* **11:** 73–78.

Poinar, G.O., Jr., and R. Hess. 1974. An ultrastructural study of the response of *Blatella germanica* (Orthoptera: Blattidae) to the nematode *Abbreviata caucasica* (Spirurida: Physalopteridae). *Int. J. Parasitol.* **4:** 133–138.

Poinar, G.O., Jr., and J.C. Quentin. 1972. The development of *Abbreviata caucasica* (von Linstow) (Spirurida: Physalopteridae) in an intermediate host. *J. Parasitol.* **58:** 23–28.

Poinar, G.O., Jr., and R. Leutenegger. 1971. Ultrastructural investigation of the melanization process in *Culex pipiens* (Culicidae) in response to a nematode. *J. Ultrastruct. Res.* **36:** 149–158.

Poinar, G.O., Jr., R. Leutenegger, and P. Götz. 1968. Ultrastructure of the formation of a melanotic capsule in *Diabrotica* (Coleoptera) in response to a parasitic nematode (Mermithidae). *J. Ultrastruct. Res.* **25**: 293–306.

Ratcliffe, N.A. 1982. Cellular defense reactions of insects. *Fortschr. Zool.* **27**: 223–244.

Ratcliffe, N.A. and A.F. Rowley. 1979. Role of hemocytes in defense against biological agents, pp. 331–414. In A.P. Gupta (ed.) *Insect Hemocytes.* Cambridge University Press, Cambridge.

Ratcliffe, N.A. and A.F. Rowley. 1984. Opsonic activity of insect hemolymph. *Comp. Pathobiol.* **6**: 187–204.

Ratner, S. and S.B. Vinson. 1983. Phagocytosis and encapsulation: Cellular immune responses in Arthropoda. *Am. Zool.* **23**: 185–194.

Rizki, T.M. and R.M. Rizki. 1974. Basement membrane abnormalities in melanotic tumor formation of *Drosophila. Experientia* **30**: 543–546.

Rodriguez, P.H. and G.B. Craig, Jr. 1973. Susceptibility to *Brugia pahangi* in geographic strains of *Aedes aegypti. Am. J. Trop. Med. Hyg.* **22**: 53–61.

Rowley, A.F. and N.A. Ratcliffe. 1981. Insects, pp. 421–488. In N.A. Ratcliffe and A.F. Rowley (eds.) *Invertebrate Blood Cells,* vol. 2. Academic Press, New York.

Salt, G. 1963. The defense reactions of insects to metazoan parasites. *Parasitology.* **53**: 527–642.

Salt, G. 1970. *The Cellular Defense Reactions of Insects.* Cambridge University Press, Cambridge.

Salt, G. and R. Van den Bosch. 1967. The defense reactions of three species of *Hypera* (Coleoptera: Curculionidae) to an Ichneumon wasp. *J. Invertebr. Pathol.* **9**: 164–177.

Schacher, J.R. and G.M. Khalil. 1968. Development of *Foleyella philistinae* Schacher and Khalil, 1967 (Nematode: Filarioidea), in *Culex pipiens molestus* with notes on pathology in the arthropod. *J. Parasitol.* **5**: 869–878.

Schell, S.C. 1952. Tissue reaction of *Blattella germanica* L. to the developing larva of *Physaloptera hispida* Schell, 1950 (Nematoda: Spiruroidea). *Trans. Am. Microsc. Soc.* **71**: 293–302.

Seureau, C. 1973. Réactions cellulaires provoquées par les nématodes subulures et spirurides chez *Locusta migratoria* (Orthoptère): localisation et structures des capsules. *Z. Parasitenkd.* **41**: 119–138.

Seureau, C. 1977. A cytopathological accumulation of microtubules in the epithelial cells of the gut of an insect parasitized by a heteroxenic nematode. *J. Invertebr. Pathol.* **29**: 240–241.

Seureau, C. and J.-C. Quentin. 1981. Évolution de l'adaptation des Nématodes hétéroxènes à leur hôte intermédiaire: Passage progressif d'un parasitisme extracellulaire à un parasitisme intracellulaire. *C. R. Séances Acad. Sci.* (III) **292D**: 421–425.

Seureau, C. and J.-C. Quentin. 1983. Sur la biologie larvaire de *Cyrnea* (*Cyrnea*) *eurycerca* Seureau, 1914; Nématode *Habronème* parasite du Francolin au Togo. *Ann. Parasitol. Hum. Comp.* **58**: 151–164.

Shapiro, J. 1979. Changes in hemocyte populations, pp. 475–523. In A.P. Gupta (ed.) *Insect Hemocytes.* Cambridge University Press, Cambridge.

Simpson, A.J.G. and S.R. Smithers. 1980. Characterization of the exposed carbohydrates on the surface membrane of adult *Schistosoma mansoni* by analysis of lectin biding. *Parasitology* **81**: 1–15.

Spratt, D.M. 1972. Natural occurrence, histopathology and developmental stages of *Dirofilaria roemeri* in the intermediate host. *Int. J. Parasitol.* **2**: 201–208.

Stoffolano, J.G., Jr. 1970. Nematodes associated with the genus *Musca* (Diptera: Muscidae). *Bull. Entomol. Soc. Am.* **16**: 194–203.

Stoffolano, J.G., Jr. 1973. Host specificity of entomophilic nematodes: A review. *Exp. Parasitol.* **33**: 263–284.

Stoffolano, J.G., Jr., and F.A. Streams. 1971. Host reactions of *Musca domestica, Orthellia caesarion,* and *Ravinia l'herminieri* to the nematode *Heterotylenchus autumnalis*. *Parasitology* **63**: 195–211.

Stoffolano, J.G., Jr., and L.R.S. Yin. 1984. Annulate lamellae in giant cells surrounding nematode parasites (*Thelazia*) of the face fly, *Musca autumnalis. J. Invertebr. Pathol.* **44**: 315–323.

Stoffolano, J.G., Jr., and L.R.S. Yin. 1986. The ultrastructure of the giant cell surrounding larvae of *Thelazia* spp. (Nematoda: Thelazioidae) in the face fly, *Musca autumnalis* De Geer. *Can. J. Zool.* **63**: 2352–2363.

Sutherland, D.R., B.M. Christensen, and K.F. Forton. 1984. Defense reactions of mosquitoes to filarial worms: Role of the microfilarial sheath in the response of mosquitoes to inoculated *Brugia pahangi* microfilariae. *J. Invertebr. Pathol.* **44**: 275–281.

Turnock, W.J., K.L. Taylor, D. Schroder, and D.L. Dahlstein. 1976. Biological control of pests of coniferous forests, pp. 289–311. In C.B. Huffaker and P.S. Messenger (eds.) *Theory and Practice of Biological Control*. Academic Press, New York.

Vinson, S.B. and G.F. Iwantsch. 1980. Host suitability for insect parasitoids. *Annu. Rev. Entomol.* **25**: 397–419.

Welch, H.E. 1963. Nematode infections, pp. 363–392. In E.A. Steinhaus (ed.) *Insect Pathology*. Academic Press, New York.

Yeaton, R.W. 1983. Wound responses in insects. *Am. Zool.* **23**: 195–203.

Zachary, D., M. Brehélin and J.A. Hoffman. 1975. Role of the "thrombocytoids" in capsule formation in the Diptera *Calliphora erythrocephala. Cell Tissue Res.* **162**: 342–348.

Zuckerman, B.M. and H.-B. Jansson. 1984. Nematode chemotaxis and possible mechanisms of host/prey recognition. *Annu. Rev. Phytopathol.* **22**: 95–113.

CHAPTER 6

Surface Changes on Hemocytes during Encapsulation in *Drosophila Melanogaster* Meigen

Tahir M. Rizki
Rose M. Rizki
Division of Biological Sciences
University of Michigan
Ann Arbor, Michigan

6.1. INTRODUCTION

Large foreign objects in the hemocoel of the *Drosophila melanogaster* larva are encapsulated by one type of hemocyte: the lamellocyte. Lamellocytes

157

are large, leaflike cells that adhere to one another to form multilayered capsules. These encapsulating cells differentiate from spherical plasmatocytes and are uniquely adapted for their role in the encapsulation process. Three major questions need to be addressed in analyzing encapsulation by the lamellocytes. First, how does the lamellocyte recognize foreignness? Second, what binds one lamellocyte to another lamellocyte so that capsule walls are formed? Finally, what controls termination of the layering process, or does the outermost lamellocyte layer of the capsule have unique features?

This chapter considers these questions. Although there are no obvious answers, it is worthwhile to reexamine the current status of our knowledge if only to point out deficiencies and difficulties, so that paths for future study can be more sharply defined. We believe that the answers to these questions will emerge as knowledge of lamellocyte surface properties accumulates.

An equally important consideration is the nature of the surfaces of the internal body tissues and the body wall that line the hemocoel. It is these surfaces that the hemocytes recognize as self and to which they remain neutral. Deviations from this self surface property, whether of exogenous or endogenous origin, signifies nonself to the hemocytes.

The *melanotic tumor* (*tu*) mutants of *D. melanogaster* provide excellent material for studying lamellocytes and their reactions. Many lamellocytes are generally present in the hemocoel of a third instar *tu* larva. In these mutants, larval tissues showing signs of aberrant development are encapsulated by lamellocytes, and the resulting encapsulated masses that later melanize are known as melanotic tumors. The encapsulation process in *tu* mutants is the same as that found in the defense reaction against foreign materials. The sole difference appears to be that encapsulation in *tu* mutants is initiated via the action of mutant genes on specific tissues within the body rather than via the entry of a foreign object into the hemocoel. We begin our consideration of the encapsulation reaction with a review of the characteristics of *tu* mutants, examine information on the surface features of body tissues in these mutants and in nontumorous strains, focus on the surface properties of the lamellocytes, and consider cellular interactions during encapsulation.

6.2. MELANOTIC TUMOR MUTATIONS

Although many *tu* mutations have been isolated and described in *D. melanogaster* (Sparrow, 1978), only a few are ideally suited for the study of the encapsulation reaction. Mutant characteristics that are desirable include high penetrance and expressivity. All or most of the mutant larvae should develop melanotic tumors. The pattern of melanotic tumor development should be consistent, so that the presumptive site of melanotic tumor formation can be studied before initiation of encapsulation. We have found two mutations that adequately fulfill these criteria: *tu-W* and *tu-Sz*[ts].

In selected lines of *tu-W*, more than 95% of the larvae develop melanotic masses in the caudalmost region (designated region 5; Rizki, 1978) of the fat body. The masses are generally large and consist of hemocytes surrounding fat body cells (Rizki, 1957a; R. Rizki and Rizki, 1979). The latter, measuring 70–90 μm in the third instar, are considerably larger than the hemocytes, so recognition of cell types is easy, even in freshly dissected, unstained masses observed in the light microscope. Caudal fat body is also the site of melanotic tumors in *tu-Sz*[ts] larvae (Rizki and Rizki, 1980a). In other *tu* strains, however, tissues other than fat body may be the site of encapsulation. For example, in *mt*[A] (allelic to *tu-bw;* Lindsley and Grell, 1976; Sparrow, 1978) and the *tu-l* mutant of *D. yakuba*, lymph glands (also referred to as blood-forming organs or hemopoietic organs) are encapsulated (Rizki, 1957a; Rizki and Rizki, 1984). In some mutants, melanization of body regions or organs may occur in the absence of encapsulation, and this phenotype should not be confused with that of melanotic tumors (Rizki, 1952, 1955).

In the *tu-Sz*[ts] mutant, which is sex linked and temperature sensitive, melanotic masses develop in the fat body when the larvae are raised at 26–27°C but not when they are grown at 18°C. Many circulating lamellocytes are found at both temperatures. With temperature as a tool for controlling melanotic tumor formation in *tu-Sz*[ts] larvae, the sequence of events that culminates in melanotic masses can be manipulated to isolate some of the individual steps at the tumor-forming site. This mutation provides the best evidence that changes in tissue surfaces stimulate the process of encapsulation (see Section 6.2.1). We propose that a mutant gene, such as *tu-W* or *tu-Sz*[ts], alters the biochemical machinery of a specific group of cells and when this genetically modified metabolism affects the tissue surface, the encapsulation process is triggered, because the tissue surface no longer emits the self signal (Rizki and Rizki, 1976; T. Rizki and Rizki, 1980a).

6.2.1. Tissue Surfaces and Basement Membrane

That the site for encapsulation in *tu* larvae results from abnormal tissue development was first reported in the *tu-W* strain, in which individual fat body cells separate from one another and are encapsulated by lamellocytes (Rizki, 1957a). This observation by light microscopy was subsequently confirmed by scanning and transmission electron microscopy (R. Rizki and Rizki, 1974, 1979). Higher magnification also revealed that the basement membrane surrounding the affected fat body cells was disrupted and disintegrating. Plasmatocytes at the tissue surfaces engorge remnants of basement membrane and other materials that leach from the affected fat body cells. Finally, the fat body cells devoid of basement membrane are encapsulated by lamellocytes.

The tissue changes in the caudal fat body of the *tu-Sz*[ts] mutant that precede encapsulation differ from those found in *tu-W*. Some of the caudal fat body cells of the early third instar tu-Sz^{ts} larvae raised at 27°C are hypolipidic (Rizki and Rizki, 1980a), a feature specific to the tu-Sz^{ts} mutant. The remaining caudal fat body cells in tu-Sz^{ts} larvae appear to have a normal content of lipid droplets. In slightly older tu-Sz^{ts} larvae, the basement membrane covering the hypolipidic cells appears swollen, and it is these aberrant tissue surfaces upon which hemocytes accumulate to form melanotic masses. At 18°C, all of the caudal fat body cells of tu-Sz^{ts} larvae have normal lipid contents, no abnormal changes appear in the basement membrane, and melanotic tumors are not formed. The absence of encapsulation at 18°C is not due to lack of lamellocytes, as these larvae have many lamellocytes that are potentially capable of encapsulation (see Section 6.2.2). Since there is no encapsulation reaction at 18°C even though many lamellocytes are present, it appears that the abnormal events in the fat body cells and their atypical basement membrane at 27°C must trigger encapsulation.

6.2.2. Implantation Experiments

There is a direct correlation between basement membrane abnormalities and the development of melanotic capsules in tu-W and tu-Sz^{ts} larvae. If alterations in the basement membrane of the fat body trigger encapsulation in these mutants, then the molecular organization of the basement membrane must possess information for recognition of self. It follows that fat body implants with experimentally modified basement membranes should be encapsulated in these mutant larvae but implants with intact, normal basement membrane should not. That this is so was demonstrated by implanting Ore-R (normal strain) fat body with mechanically damaged surfaces and fat body from which basement membrane had been removed by collagenase into tu-Sz^{ts} larvae growing at 18°C. These implants were encapsulated by lamellocytes even though the caudal fat body of the host was not. Implants with normal, undamaged surfaces were not encapsulated in these experiments (R. Rizki and Rizki, 1980).

Unlike Ore-R fat body implants, fat body of $D.$ $virilis$ larvae is perceived as foreign and encapsulated in tu-Sz^{ts} larvae even though the basement membrane of the implants shows no visible damage (R. Rizki and Rizki, 1980). $D.$ $virilis$ basement membrane differs in its molecular structure from $D.$ $melanogaster$ basement membrane, since it does not bind Drosox 305, a monoclonal antibody specific for $D.$ $melanogaster$ basement membrane (Spragg et al., 1982; R. Rizki et al., 1983). However, the antigen for Drosox 305 cannot be the sole discriminator of self and nonself. When fat bodies from Ore-R donors with everted spiracles—that is, larvae committed to pupariation—and larval fat body from puparia, pupae, and adults were implanted in tu-Sz^{ts} hosts, the implants were encapsulated (R. Rizki et al., 1983). Drosox 305 does not bind to the surfaces of the free fat body cells in pupae and adults. Ultrastructural examination revealed that these cells in pupae and adults lack basement membrane, and according to our hypothesis, we expect them to be encapsulated. However, basement membrane still covers the fat body in larvae with everted spiracles and in puparial stages when the fat body cells have not yet separated from one another. The basement membrane of these stages has Drosox 305 antigen. Since these tissues are encapsulated in tu-Sz^{ts} hosts, we must assume that the tu-Sz^{ts} hemocytes discriminate against the molecular changes in the basement membrane that precede its loss during metamorphosis. The nature of these molecular changes remains to be determined. They are not resolvable at the ultrastructural level used to compare basement membrane from different developmental stages (R. Rizki et al., 1983).

Abnormalities in the basement membrane have also been found in other tu mutants, including mt^A (tu bw) and tu-l of $D.$ $yakuba$ (Rizki and Rizki, 1984, and unpublished). In these mutants, the basement membrane of the lymph glands ruptures, resulting in melanotic capsules. Thus, the general pattern of melanotic tumor formation described for tu-W and tu-Sz^{ts} appears to hold true for other tu mutations as well. Some tu mutants are more variable in either the time or the site of capsule formation, and more than one tissue type may be encapsulated (Chandaman, 1979). It is interesting, however, that fat body and lymph glands are common sites of melanotic tumor formation. In the course of normal development, cells from both the fat body and the lymph glands are freed of basement membrane during the pupal stage. It appears that mutations altering metabolic processes in these tissues may induce precocious disintegration of basement membrane in the larval stage, thus leading to the development of melanotic capsules. During larval life, the fat body cells are polarized with respect to the overlying basement membrane (Rizki and Rizki, 1983a). Either the combined cell–to–basement membrane molecular configuration or the molecular constituents of the basement membrane itself must hold the recognition sites for self. Disruption or removal of these sites at the cell–to–basement membrane interface during larval life evokes the cellular defense response. This response is site specific in tu strains, since it is limited to affected tissue surfaces.

6.3. LAMELLOCYTES AND THEIR ROLE IN ENCAPSULATION

6.3.1. Lamellocyte Surface Features

Sections through fully formed capsules, such as those in tu mutants, show the presence of lamellocytes and no other hemocyte types (Fig. 6.1). The layers of flat cells are tightly apposed. Although plasmatocytes can occasionally be found adhering to the outer surfaces of a capsule wall when intact capsules are examined in the scanning electron microscope, it is not known whether these plasmatocytes will eventually assume the lamellocytic form and be incorporated into the capsule. They may be resting on the capsule surfaces, as plasmatocytes are often found on the body wall and on the surfaces of tissues, on some more than others (Rizki et al., 1976).

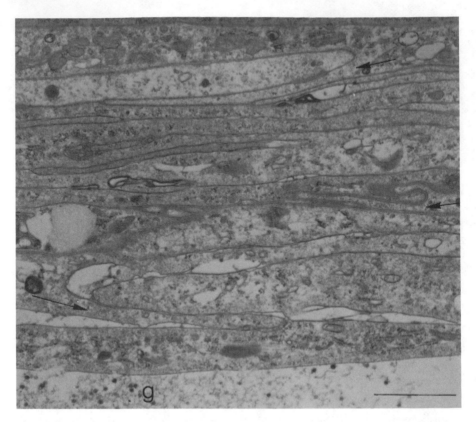

Figure 6.1. Section through a capsule wall showing the layering of lamellocytes. The surfaces of some lamellocytes are folded over the surfaces of other lamellocytes, forming wedges (arrows). Glycogen (g) from the underlying fat body cell is present in the interspace between the first layer of lamellocytes and the fat body cell. (Scale bar = 1 μm.)

    The shape of the lamellocyte is seen best in the scanning electron microscope (Fig. 6.2A; R. Rizki and Rizki, 1979; Rizki and Rizki, 1980a,b). The cell is extremely flattened in one dimension and generally round in the other. With proper mounting, so the cells can be observed from different angles, this feature can also be viewed with phase optics (Rizki, 1962). Undoubtedly, this cell shape is an adaptation for encapsulation. Larvae in which capsule formation is not required have few lamellocytes (T. Rizki, 1957b). When capsule formation is imminent, many lamellocytes are present (Rizki, 1957a).

    Spherical plasmatocytes differentiate into lamellocytes (Rizki, 1956, 1962), and early stages of capsule formation contain lamellocytes in different stages of differentiation (Fig. 6.2C). The cellular adaptation from a spherical to a flattened form increases the surface area available for compact layering of the cells (Fig. 6.2D,F). It is not always evident when spherical plasmatocytes on capsule surfaces have become committed to the lamellocytic state. During differentiation of lamellocytes, materials appear to exude from the cells as they flatten (R. Rizki and Rizki, 1979). There is no firm evidence that this is a necessary feature of the flattening process. A more relevant question is whether these globular materials seen associated with differentiating cells play a role in the encapsulation reaction or whether they are extraneous materials that are lost because they are no longer required by a discoidal cell.

Undulations of the lamellocyte surface have been observed during the differentiation of these cells (Rizki, 1962). This surface characteristic is also seen in fixed lamellocytes in the form of small folds in the surface membrane (Fig. 6.2A). When the dynamic surfaces of two differentiating lamellocytes are in apposition, the folds may wedge, resulting in the interlocking of lamellocyte surfaces. Wedging of lamellocyte surface folds can be seen in sections through capsule walls (Fig. 6.1). The surfaces of fully differentiated lamellocytes also have small knoblike protrusions (Fig. 6.2A,B). These surface projections are apparently sticky (Fig. 6.2E), and it seems reasonable to consider that they may be important either for recognition of foreignness or for adherence between lamellocytes. Thus, observations of lamellocytes by scanning electron microscopy suggest that regions of their surfaces are specialized for the binding reactions required for encapsulation.

The plant lectin wheat germ agglutinin (WGA) conjugated to fluorescein isothiocyanate was used to examine whether there are inhomogeneities in the distribution of lamellocyte surface molecules (Rizki and Rizki, 1983b). For these studies unfixed cells were briefly exposed to the lectin at concentrations that appear to immobilize *Drosophila* cell surfaces (R. Rizki et al., 1975; R. Rizki and Rizki, 1977) so that membrane permeabilization and entry of lectin into the cells might be avoided. WGA binds to β-(1→4)-linked oligomers of N-acetyl-D-glucosamine; N,N'-diacetylchitobiose, which has a high affinity for WGA, is used to show the specificity of WGA binding to cells and tissues (Goldstein et al., 1975).

When lamellocytes of *tu-Sz*[ts] larvae growing at 27°C are treated with WGA and examined in the fluorescence microscope, some regions of the cell surfaces show heavy binding of the lectin, while others do not. This pattern of WGA binding gives a speckled appearance in the fluorescence microscope, and the cells have been designated spk[+] cells (Fig. 6.3A). Most of the lamellocytes in larvae at 18°C are spk[−] (Fig. 6.3B), suggesting that the spk[+] surface characteristic may be associated with capsule formation. Evidence to corroborate this view was obtained by treating hemocytes of two other *tu* mutants—*tu-W* and *tu bw*—with WGA. In both cases a high percentage of the lamellocytes was spk[+]. That the fluorescence observed in these studies was specific to WGA binding was demonstrated by pretreating the WGA with N,N'-diacetylchitobiose. Pretreated WGA did not bind to *Drosophila* hemocyte surfaces (Rizki and Rizki, 1983b).

The correlation between the presence of spk[+] lamellocytes and capsule formation suggested that differentiation of spk[+] lamellocytes may be nec-

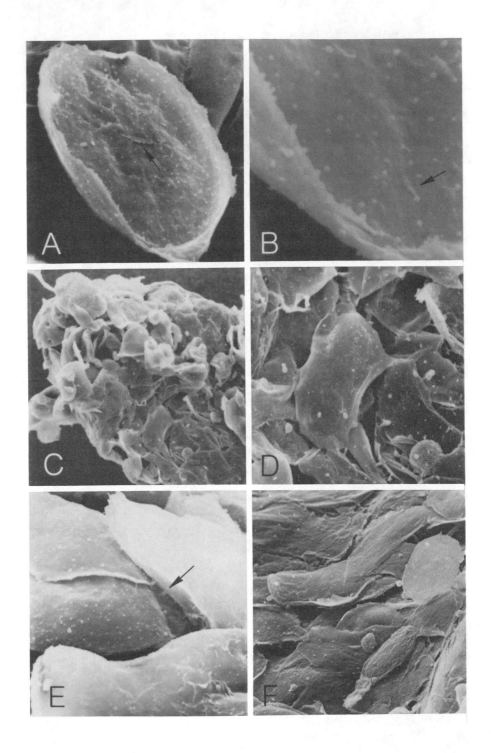

essary for encapsulation. To test this hypothesis, *D. virilis* tissues were implanted into *tu-Sz*[ts] larvae growing at 18°C. There was a marked increase in the percentage of spk[+] lamellocytes in the hosts with heterospecific implants. Since the spk[+] characteristic is stimulated by the presence of foreign bodies in the hemocoel, we concluded that the spk[+] lamellocytes are necessary for encapsulation and are capsule-competent cells (Rizki and Rizki, 1983b; see Chapter 7).

Nappi and Silvers (1984) followed our lead in using WGA to study *Drosophila* hemocytes, but they have not used our terminology (spk[+]) in describing the pattern of lectin binding to the cell surfaces. They refer to the cells with granular fluorescence as immune reactive. This conclusion is based on a correlation between the percentage of cells that binds WGA and the frequency of encapsulated parasitoid wasp eggs (*Leptopilina heterotoma,* formerly *Pseudeucoila bochei*) in host larvae of the *Tum*[1] (sex-linked, temperature-sensitive lethal) strain at two temperatures, 21 and 29°C. Several parameters have been ignored in this study, including the lethality of the hemizygous *Tum*[1] larvae at 29°C (Hanratty and Ryerse, 1981) and the effects of wasp oviposition on the host hemocytes (R. Rizki and Rizki, 1984). It is not clear how Nappi (1970), 1973, 1977), Nappi and Streams (1969), and Nappi and Silvers (1984) overlooked the cellular effects of parasitization, which result in selective destruction of lamellocytes (see Section 6.5). This oversight cannot be attributed to strain differences, since we examined several host and parasite strains in our study (R. Rizki and Rizki, 1984), including the Storrs strain of *L. heterotoma* and the *tu bw* mutant of *D. melanogaster* used by Nappi. Furthermore, Walker (1959) observed spindle-shaped lamellocytes in another strain of *Drosophila* larvae parasitized by *L. heterotoma,* but she did not attach any significance to this variation in cell morphology. In view of the complications that have been overlooked in the study of Nappi and Silvers (1984) and the question of which cell types they are counting in parasitized larvae, their data on WGA binding to *Drosophila* hemocytes are difficult to interpret.

Figure 6.2. Scanning electron micrographs of lamellocytes and capsule formation. (A) Lamellocyte from a *tu-W* larva with minute knobs scattered over its surface and a small fold (arrow). The thinness of the cell can be judged at the upturned edges (×3000). (B) Enlargement of the lower left corner of the lamellocyte in A. The arrow indicates a small microvillus on the cell surface (×7000). (C–F). Stages of lamellocyte layering around the caudal fat body in *tu-Sz*[ts] larvae. (C) An early stage, when lamellocytes in different stages of flattening are loosely packed (×350). (D) Enlargement of the lower right corner of C showing the interdigitation of lamellocyte processes (×1000). (E) A view of two intersurfaces of lamellocytes illustrating the stickiness of the small knoblike processes (arrow) on the lamellocyte surface. Contrast for this photograph was increased by use of a homomorphic processor in the scanning electron microscope (×4200). (F) Late stage of lamellocyte layering, when the cells are fully flattened and tightly packed. A section through such a capsule corresponds to the transmission electron micrograph in Figure 6.1 (×2100).

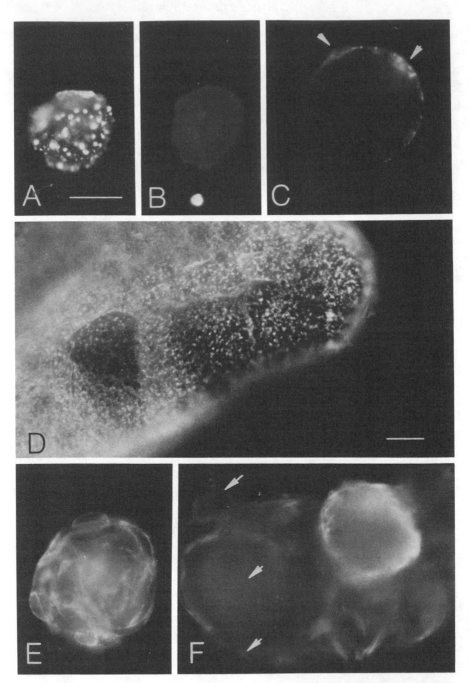

Figure 6.3. Lamellocytes and fat body cells of *tu-Sz*^ts larvae grown at 26°C treated with WGA conjugated to fluorescein isothiocyanate. (A) An spk⁺ lamellocyte. (Scale bar = 25 μm.) (B) An spk⁻ lamellocyte. (C) A fat body cell with two spk⁺ lamellocytes (arrowheads) adhering to its surfaces. (D) Caudal fat body showing the loss of WGA binding sites over three hypolipidic cells. No lamellocytes are present on this fat body. (Scale bar = 25 μm.)

6.3.2. Lamellocyte Interactions during Encapsulation

Cellular capsules in some insects contain alternating layers of flat cells and rounded or slightly flattened cells (Salt, 1970), but capsules in *Drosophila* are composed only of flattened lamellocytes (Fig. 6.1). This finding implies that lamellocytes are capable of at least two types of interactions: first, they bind to foreign surfaces to form the innermost layer of a capsule; second, they bind to one another to form capsule walls. There must be some essential difference or differences between these two types of interactions, since feeding excess glucosamine to *tu-W* larvae interferes with the second and not with the first (Rizki, 1961).

Another, less common behavior of lamellocytes is their ability to wrap around small objects such that surfaces of the same lamellocyte adhere to each other. An instance of this is shown in Figure 6.4, showing a lamellocyte wrapped around a tracheole. Complementary molecules required for lamellocyte-to-lamellocyte surface adhesion must therefore be distributed over a single lamellocyte surface. WGA was used to examine lamellocyte interactions in *tu-Sz*^ts larvae growing at 27°C to follow the pattern of spk^+ and spk^− lamellocyte layering in capsule walls (Rizki and Rizki, 1983b). Two types of interactions were examined: the binding of lamellocytes to fat body cells (i.e., the alienated body surfaces) and the binding between lamellocytes.

6.3.2.1. Early Stages of Capsule Formation. WGA binds to normal fat body cell surfaces and the basement membrane covering these surfaces (Rizki and Rizki, 1983a). There is a loss of WGA binding sites from the surfaces of the hypolipidic cells in *tu-Sz*^ts larvae before the adherence of lamellocytes to form capsules (Fig. 6.3D). This observation confirms that the surfaces of the hypolipidic cells deviate from normal and are therefore targets for lamellocytic activity. However, it is difficult to obtain early stages of encapsulation that are satisfactory for observing WGA bound to the lamellocyte surfaces when the fat body cells are still clustered. The best conditions were found in masses that were loose enough so that individual fat body cells could be examined and photographed. In such masses, we found fat body cells with spk^+ lamellocytes attached to their surfaces (Fig. 6.3C). We also found spk^+ and spk^− lamellocytes attached to fat body cells that showed WGA binding (Rizki and Rizki, 1983b).

The nature and source of the molecules binding WGA on the aberrant fat

---

(E) A single fat body cell that has been encapsulated by lamellocytes. The lateral profiles of the lamellocytes can be seen because of the WGA bound on their surfaces. (F) Encapsulated mass of fat body cells with only one cell having a heavy concentration of spk^+ lamellocytes (right side of the frame). Encapsulation of the remaining fat body cells has been completed, and spk^− lamellocytes (arrows) are mostly visible in these regions. (A, C, D, and F from T. Rizki and Rizki, 1983b.)

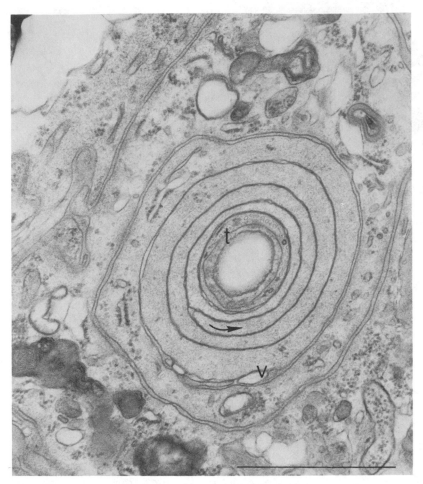

Figure 6.4. A single lamellocyte wrapped around a tracheole (t). Follow the arrow counterclockwise to trace the profile of the cell. Some vesicles (v) are present in the intercellular spaces, whereas cell surfaces are tightly apposed in other regions. (Scale bar = 1 μm.)

body cells of *tu-Sz*^ts larvae are unknown. Ultrastructural studies of developing capsules show a variety of globular or vesicular components on the fat body cell surfaces that are being encapsulated. At present we have no means by which to determine whether these materials (Figs. 6.5 and 6.6) are products of the hemocytes or products of the aberrant fat body cells. We have observed leaching of materials from affected fat body cells (R. Rizki and Rizki, 1979). On the other hand, material from the hemocytes appears to be spewing over the fat body cell surfaces in some cases (R. Rizki and Rizki, 1974; Rizki and Rizki, 1984). Regardless of the source of these materials, they undoubtedly play a role either in attracting or in binding

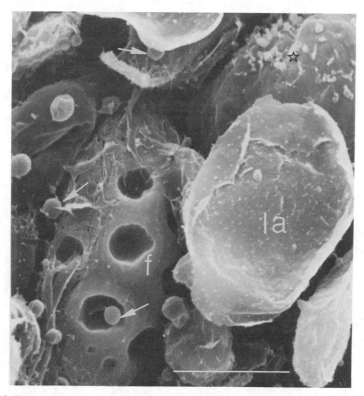

Figure 6.5. A region of a tumor-forming site in a *tu-Sz*ᵗˢ larva showing the type of vesicular material (arrows) and other extracellular materials (∗) often found on the surfaces to be encapsulated. One fat body cell surface (f) is free of basement membrane. Note the presence of a vesicle in the surface cavity of this cell. The fat body cell in the upper right that is coated with extracellular materials is still covered by basement membrane. Note the surface features of the lamellocyte (la). (Scale bar = 10 μm.)

lamellocytes to the aberrant fat body cell surfaces. Since differentiating lamellocytes are often associated with globular extrusions, it is possible that some of the coating materials found on the aberrant fat body cells are produced during the plasmatocyte-lamellocyte transition. Morphologic observations cannot be used to assign the source of the extracellular materials found between fat body cells and lamellocytes. Nor can WGA binding be used to determine the origin of these materials, since intracellular organelles in permeabilized (fixed) fat body cells and hemocytes also bind WGA.

6.3.2.2. Capsule Walls. Lamellocytes adhere tightly to one another, so it is difficult to determine whether capsule walls contain both spk⁺ and spk⁻ lamellocytes. Examination of intact capsules suggests that both lamellocyte types are present (Fig. 6.3E). In such preparations, it is not always possible

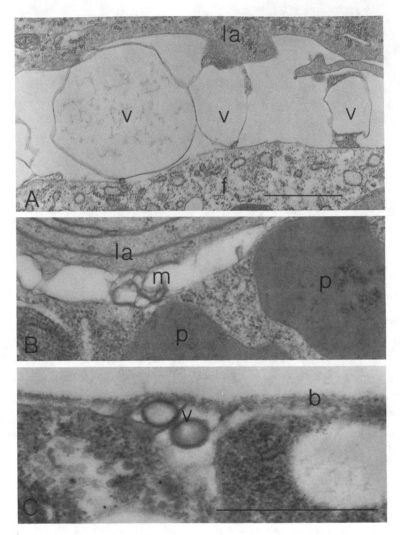

Figure 6.6. Transmission electron micrographs of fat body cell surfaces that are being encapsulated in *tu-Sz*ts larvae. (A,B) Interfaces of lamellocytes (la) and fat body cells (f) show the types of vesicles (v) that are found. The vesicles in A have probably originated from the lamellocytes, since some of their surface structure has the same electron density as the lamellocyte. Only the large vesicle on the left contains electron-dense fibrous material. In B, membranous material (m) between the lamellocyte and the fat body cell (f) contains protein granules (p). Scale bar = 1 μm.) (C) Normal adipose cell surface near the region of encapsulation showing vesicular materials (v) directly underneath the basement membrane (b). In the absence of basement membrane, these vesicles may correspond to those seen in the cavity in Figure 6.5. (Scale bar = 1 μm.)

to determine which cell surface has bound WGA. Also, the absence of WGA binding at some cell surfaces may be due to poor penetration of the lectin between tightly adhering cells.

To examine WGA binding to lamellocytes at a higher magnification, we treated hemocytes and developing capsules of *tu-Sz*^ts larvae with WGA conjugated to ferritin and examined the sectioned material in the transmission electron microscope. The protocol for the lectin treatment followed that used for fluorescein-conjugated WGA. Therefore, the same limitations of lectin penetration between tightly adhering cells of fully formed capsules applies to these preparations.

Hemocytes treated with WGA conjugated to ferritin and examined by transmission electron microscopy confirm the earlier fluorescence study. There are two types of lamellocytes at the tumor-forming site: some have electron-dense ferritin particles bound to their surfaces, and some do not (Fig. 6.7). It is important to note that our method of treating unfixed hemocytes with the lectin results in surface labeling only. There were no ferritin particles bound to intracellular organelles.

Cell surfaces with ferritin particles could be found near cell surfaces that were negative for WGA. In this respect the ultrastructural study confirms the studies by fluorescence microscopy. Another similarity between the two levels of examination concerned the distribution of ferritin particles on the cell surfaces. Some regions of the cell surface have a higher concentration of ferritin particles than do others, and knoblike extensions of the cell membrane often label heavily with ferritin (Fig. 6.7). The knoblike configuration in the cell surface may be partly responsible for the spk⁺ pattern seen in the fluorescence microscope, since a higher concentration of bound WGA on a knoblike structure will "light up" differentially as a speck in the fluorescence microscope. Some of the ferritin particles were bound to vesicles associated with the lamellocyte surfaces (Figs. 6.7B,C).

Alcian blue stains carbohydrate moieties and imparts sufficient electron density to stained cell components to distinguish them in the transmission electron microscope. Therefore, this stain provides a second approach to examining the distribution of carbohydrate moieties in hemocyte surfaces. Figure 6.8 shows the distribution of alcian blue stain in lamellocytes of *tu* larvae. Again, we find patches in the distribution of surface constituents. In a number of cases alcian blue–stained material appears to be concentrated at points where two lamellocyte surfaces meet or are attached to one another, suggesting that these specialized regions of the cell surfaces may be important for initial lamellocyte-to-lamellocyte binding in capsule formation.

6.3.2.3. Termination of Encapsulation. The outer walls of fully formed capsules treated with fluorescein-conjugated WGA do not show bound lectin (Fig. 6.3F). Either the outer wall of the capsule contains only spk⁻ cells, or the lectin-binding sites in the cells of this layer have been modified or lost. It is possible that melanin deposition in the lamellocytes alters their

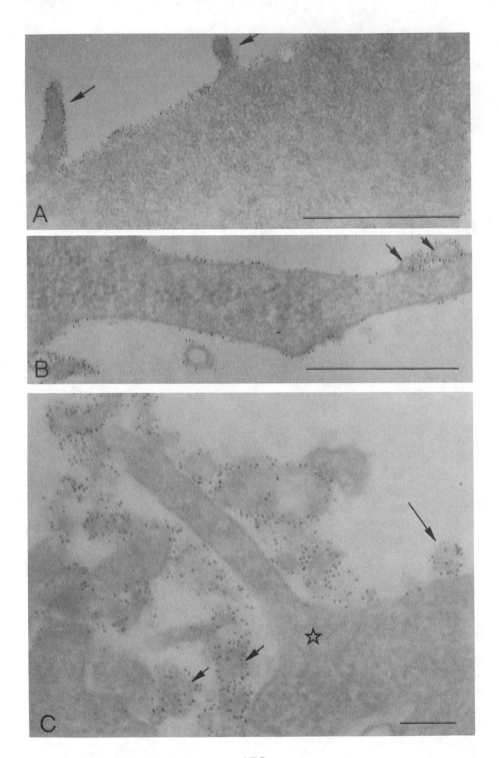

WGA-binding sites and the outermost layer of the capsule wall no longer binds the lectin once the melanization process begins within the capsule. An alternative hypothesis, also open to testing, is based on the relative frequency of spk$^+$ lamellocytes available for capsule formation. This hypothesis (Rizki and Rizki, 1983b) provides a basis for the termination of the layering of lamellocytes to form a capsule. We have suggested that the spk$^+$ lamellocytes are necessary for capsule formation, and only these cells or their products possess the adhesive sites required for the layering process. As the population of these capsule-competent cells is depleted from circulation during the development of a capsule, only spk$^-$ cells will be incorporated in the outermost layer of the capsule wall. This spk$^-$ layer will not have the same adhesive properties as the spk$^+$ layers.

6.4. MELANIZATION OF CELLULAR CAPSULES

Melanization is a consistent feature of cellular capsules in *Drosophila*. The first signs of melanin appear deep within the capsule (Rizki, 1957a), and eventually the entire cellular mass appears black. Two topics for consideration are the source of the melanin that blackens the cellular capsule and the function of melanization.

As noted above, the only cell types that we have been able to distinguish in developing capsules are plasmatocytes and their variant forms, the lamellocytes. This poses a dilemma, for it is quite clear that the only hemocyte type in *Drosophila* that contains phenoloxidases is the crystal cell. Some recent observations shed light on this problem.

6.4.1. Crystal Cells and Phenoloxidases

Crystal cells were so named because they have large paracrystalline inclusions in their cytoplasm (Rizki, 1956). If they contain many inclusions, the spherical cells are larger than the plasmatocytes (Rizki et al., 1976), but some crystal cells are small and contain only one large paracrystalline inclusion. The cells are also unique with respect to their fragility. When hemolymph samples are removed from the larva or the body wall is torn, the

---

Figure 6.7. Transmission electron micrographs of lamellocyte surfaces that have been treated with WGA conjugated to ferritin. The cells were fixed in buffered formaldehyde-glutaraldehyde; sections were stained with vanadatomolybdate. (A) Two knoblike structures (arrows) on the lamellocyte surface have ferritin particles, as do the remaining cell surfaces; the other side of the flattened cell is located at the right end of the scale bar. (Scale bar = 1 μm.) (B) Another lamellocyte surface showing WGA bound to vesicles (arrows) associated with the lamellocyte surface. (Scale bar = 0.5 μm.) (C) Projection (∗) of a lamellocyte surface that is negative for WGA (spk$^-$) and surrounded by vesicles and globular materials that have bound WGA (ferritin particles) on their surfaces (spk$^+$). (Scale bar = 0.1 μm.)

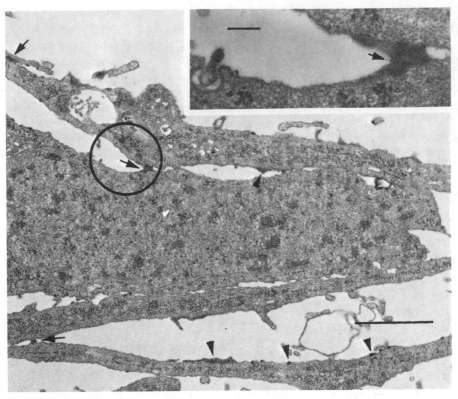

Figure 6.8. Lamellocyte surfaces at the tumor-forming site (lymph gland) in a *tu bw* larva. The dispersion of alcian blue staining material on the cell surface is indicated by arrowheads and at the intercellular adhesion sites by arrows. (Scale bar = 1 μm.) The insert in the upper right is a higher magnification of the circled area. (Scale bar = 0.1 μm.)

crystal cells often swell or rupture, releasing their contents and causing blackening of their immediate environment (Rizki, 1967b; Rizki and Rizki, 1959; Rizki et al., 1985). Crystal cells are homologous to granular cells and oenocytoids of other insects (Gupta, 1979; Ratcliffe et al., 1984). The sensitivity of these cells to changes in the hemolymph must be important to the function they perform in the cellular defense system of the insect.

Insect phenoloxidases exist as inactive precursors that are converted to active enzyme in vitro by a variety of protein denaturants and proteolytic enzymes (Ashida and Dohke, 1980; see Chapters 9, 14, and 15). Experimental studies with crystal cells and studies with mutant strains in which crystal cells are absent or modified clearly demonstrate that only the crystal cells in *D. melanogaster* carry hemolymph prophenoloxidases; thus, melanin in the larval hemolymph must come from the crystal cells (Rizki, 1956; Rizki and Rizki, 1959; Rizki et al., 1985).

To pinpoint the intracellular site of phenoloxidases has proved difficult. In an early study (Rizki and Rizki, 1959), we concluded that the enzymes are present in the extracrystalline cytoplasm of the crystal cells rather than within the paracrystalline inclusions themselves. Recent results refute this conclusion and clearly indicate that the paracrystalline inclusions of the crystal cells contain phenoloxidases (Rizki et al., 1985). This was shown by activating the enzymes within crystal cells that had been previously fixed in paraformaldehyde. Two activators were used: natural activator prepared from pupae according to Mitchell and Weber (1965) and 2-propanol (Batterham and McKechnie, 1980). When fixed activated cells were incubated in dopa substrate, faint pigmentation first appeared in the paracrystalline inclusions, and subsequently the entire cell melanized (Fig. 6.9A). With 2-propanol as the activator, a diffusion gradient of melanization from the crystal cells occurred such that surrounding plasmatocytes, lamellocytes, and hemolymph components melanized (Fig. 6.9C). Such diffusion artifacts are probably responsible for erroneous conclusions about the source of phenoloxidases in the hemocoel and probably account for the results of Shrestha and Gateff (1982). When the prophenoloxidases in fixed crystal cells were activated with natural activator, melanization was confined to the crystal cells (Fig. 6.9B).

Crystal cells can also be blackened by placing living larvae in hot water (70°C; Rizki et al., 1980). This pigmentation is at first limited to the crystal cells and mimics perfectly the distribution of melanized cells in *Bc* larvae in which phenoloxidases are activated in vivo under the influence of the *Bc* mutant allele (Rizki et al., 1980). If larvae treated with hot water are allowed to remain in the liquid as it slowly cools, the pigmentation in the crystal cells diffuses throughout the body.

Additional evidence that the crystal cells are the source of hemolymph phenoloxidases is provided by two nonallelic mutant genes. In *Bc* larvae, there are no hemocytes with paracrystalline inclusions. Instead, these larvae contain melanized hemocytes that are the result of intracellular release of phenoloxidase activity (Rizki et al., 1980). *Bc* larvae do not have detectable phenoloxidase activity, and their hemolymph does not blacken when exposed to the air. The genetic lesion affecting the crystal cells in *lz* mutant larvae differs from that in *Bc* larvae. Some *lz* mutants do not have hemocytes with paracrystalline inclusions, whereas other *lz* alleles have normal crystal cells with paracrystalline inclusions (Rizki and Rizki, 1981). Those *lz* alleles in which crystal cells do not differentiate lack phenoloxidases, and the larvae do not show blackening of the hemolymph upon injury to the body wall. The *lz* alleles with crystal cells have phenoloxidase activity and normal blackening reactions (Rizki et al., 1985). Thus, there is a direct correlation between crystal cells with paracrystalline inclusions and potential phenoloxidase activity.

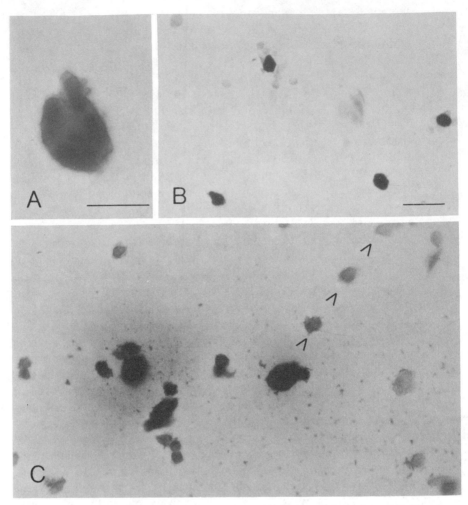

Figure 6.9. Localization of phenoloxidases in crystal cells. (A) A crystal cell that has been fixed in buffered paraformaldehyde, treated with natural activator, and incubated in phosphate buffered dopa shows melanization of the paracrystalline inclusions. (Scale bar = 10 μm.) (B) After 24 hr in dopa the crystal cells are melanized but the surrounding hemocytes are not. (Scale bar = 40 μm.) (C) A preparation similar to that in B except that 2-propanol was used as the activator. Note the diffusion gradient (arrows) of melanin from the melanized crystal cell center to the surrounding plasmatocytes and lamellocytes.

6.4.2. Crystal Cells in Cellular Capsules

Ratcliffe and coworkers (1984) and Ashida and Yoshida (1985) reported that phenoloxidases of granulocytes coat foreign materials in preparatiion for phagocytosis or encapsulation. There is no evidence that crystal cells participate in this function in *Drosophila*. Opsonins are not required for phagocytosis of bacteria by *Drosophila* hemocytes in vitro (Rizki and Rizki, 1982), and the recent report of Dularay and Lackie (1985) does not support the contention that phenoloxidases are opsonins in insects (see Chapter 9).

Mutants that lack phenoloxidases can also be used to study this problem. The lz^s allele does not have detectable phenoloxidase activity, nor do lz^s larvae have cells with paracrystalline inclusions. Their lymph glands contain some cells with abnormal cytoplasmic inclusions, and we presume these are the abortive crystal cells. These abnormal hemocytes are phagocytized by other hemocytes within the lz^s lymph glands (Fig. 6.10A), demonstrating that normal functioning of the crystal cells and phenoloxidases is not required for phagocytosis. Crystal cells are important for wound healing in *Drosophila* larvae, and this role has been confirmed by studies with mutations that block the development of crystal cells (Rizki and Rizki, 1982, 1984).

As noted above, capsule walls contain only flattened cells, and we have never observed a paracrystalline inclusion within these cells. Some plasmatocytes are more electron dense than others (Fig. 6.10B), and plasmatocytes of varying electron densities must differentiate into lamellocytes and become incorporated into capsule walls, since differences in electron density of cells in capsules are apparent. Sections through melanized capsules also show cells that are highly electron dense (possibly melanized) but otherwise resemble the surrounding lamellocyte profiles. The identification of these components is difficult, but it seems reasonable to consider whether they are crystal cells in which phenoloxidase activity has been released.

The single distinguishing feature of the crystal cell is its content of paracrystalline inclusions and the accompanying potential for melanization. In the absence of the paracrystalline inclusions or melanization (either experimentally or genetically induced, both of which cause depletion of the paracrystalline inclusions), the crystal cell cannot be unequivocally identified. This being the case, it is possible that crystal cells are incorporated in capsule walls but that we cannot recognize the cells because they no longer contain materials in paracrystalline form.

To determine whether crystal cells can be incorporated in capsule walls, we used the *Bc* mutant gene. The crystal cells in *Bc* mutant larvae are endogenously labeled by melanization and sclerotization. These black cells are inert and remain unchanged throughout the life of the individual, so their presence in cellular capsules is easy to detect. The *Bc* mutant gene was combined with the *tu-W* gene, and capsule formation was studied in the double mutant (*tu-W Bc/tu-W Bc*) larvae. Double mutant larvae did not form melanotic tumors; instead, their caudal fat body cells were surrounded by layers of lamellocytes with scattered black cells. We refer to these masses as amelanotic tumors.

Figure 6.11 is an amelanotic tumor removed from a *tu-W Bc* larva. Melanized crystal cells are present in the cellular mass, an observation that leaves us with two alternative conclusions: (1) crystal cells do participate in capsule formation and for some reason we have missed their presence in

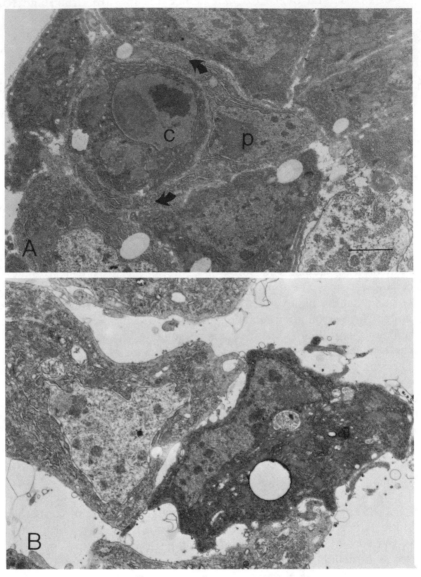

Figure 6.10. (A) An abortive crystal cell (c) in the lymph gland of an *lz*ˢ mutant larva that has been engulfed by a neighboring hemocyte (p). (Scale bar = 1 μm.) (B) Plasmatocytes with various electron densities in the hemocoel of an *Ore-R* larva.

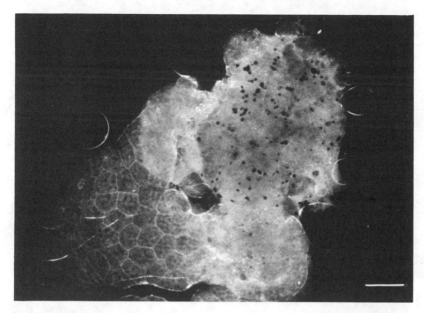

Figure 6.11. Amelanotic tumor mass from a *tu-W Bc* double mutant larva photographed with epi-illumination. The small specks are the black cells in the encapsulating blood cell mass. The lower left of the specimen has normal fat body cells. (Scale bar = 200 μm.)

capsules in *tu* larvae; and (2) the behavior of the melanized crystal cell differs from that of the normal crystal cell, which is not incorporated in the capsule wall. There is some support for the latter suggestion. The black cells in *Bc* larvae are often encapsulated by lamellocytes, and these lamellocytes with entrapped black cells may adhere to developing capsule walls. Lamellocytes have not been seen adhering to normal crystal cells.

Sections through the capsule walls formed in *tu-W Bc* larvae show the black cells. In some regions of the capsule a wide black cell body can be seen (Fig. 6.12A), whereas in other sections of the capsule thin, darkly pigmented cellular extensions can be found (Fig. 6.12B). The latter resemble flattened, electron-dense strands seen in capsules of *tu-W* larvae (Fig. 6.13). There is, however, a difference in the consistency of the capsules in *tu-W* and *tu-W Bc* larvae. Amelanotic capsules with black cells that are formed in *tu-W Bc* larvae are soft. They can be easily distorted and crushed during handling. The melanized capsules in *tu-W* larvae are hard and resilient. Therefore, we can conclude that crystal cell function is required for sclerotization of the proteins of the capsule wall. The unsclerotized, unmelanized capsules of *tu-W Bc* larvae are lysed during metamorphorsis (Rizki and Rizki, 1984), but melanized capsules are retained for the life of the individual (Wilson et al., 1955).

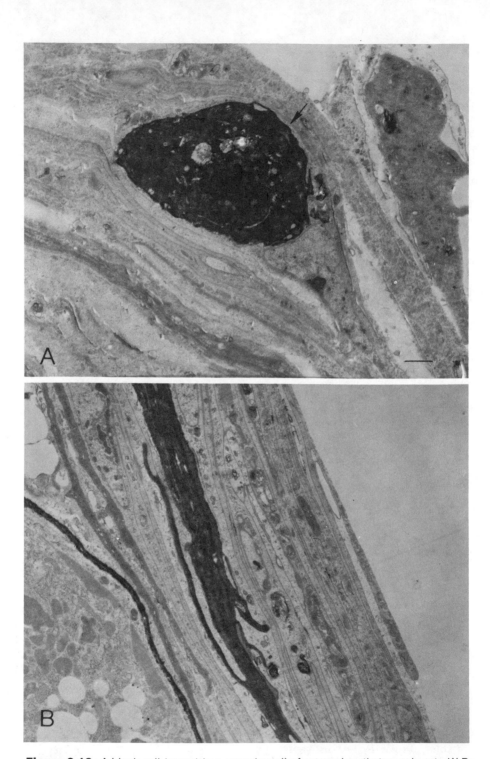

Figure 6.12. A black cell (arrow) in a capsule wall of an amelanotic tumor in a *tu-W Bc* larva. The melanized cell is surrounded by layers of lamellocytes. (Scale bar = 1 μm.) (B) Another region of a capsule from a *tu-W Bc* larva showing melanized cellular layers. Also note the variation in electron density among the layered lamellocytes.

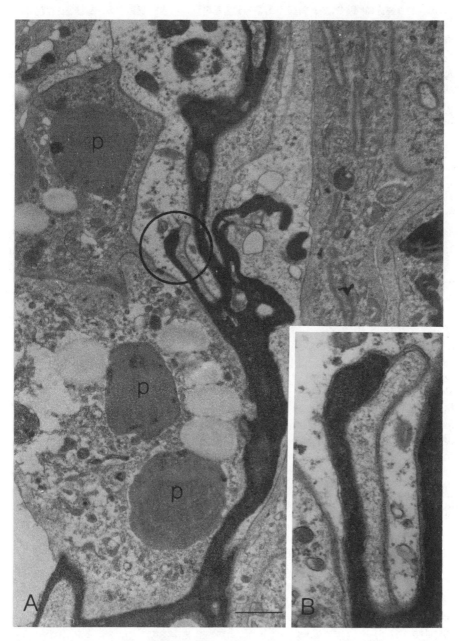

Figure 6.13. (A) Section through a melanized capsule wall from a *tu-W* larva. Note the melanized cellular strands in the capsule wall and the protein granules (p) in the fat body cell to the left. (Scale bar = 1 μm.) (B) Enlargement of the circled region in A showing the plasma membrane of the lamellocyte surrounding the melanized cell strand.

6.5. PARASITISM AND SUPPRESSION OF ENCAPSULATION

The observations of melanotic tumor formation and implanted tissues suggest that any surface within the hemocoel of a tu-Sz^{ts} larva covered with normal larval basement membrane will not elicit an encapsulation reaction, whereas all surfaces not coated with this material will be encapsulated by lamellocytes. How, then, does a parasite avoid the encapsulation reaction and develop to maturity in this hemocoel?

Walker (1959) studied the hemocytes of several *D. melanogaster* strains that had been parasitized by *L. heterotoma*. Her excellent study reported fewer lamellocytes in parasitized *tu* larvae than in unparasitized *tu* larvae, and she suggested that the wasp egg in the hemocoel interferes with the differentiation of lamellocytes. Nappi and Streams (1969) repeated the work of Walker (1959) and supported her conclusion regarding the effect of the parasitoid on the hemocytes of the host. In view of our recent observations on capsule-competent hemocytes (Rizki and Rizki, 1983b), we decided to reexamine the hemocytes of *tu* larvae parasitized by *L. heterotoma* to see whether parasitization affects the capsule-competent lamellocytes.

Some hemocyte samples from parasitized and unparasitized tu-Sz^{ts} larvae growing at 27°C were examined with phase optics (R. Rizki and Rizki, 1984). Other samples from the two groups of larvae were treated with WGA conjugated to fluorescein isothiocyanate and examined in the fluorescence microscope. We found several distinct differences between the lamellocytes in parasitized and unparasitized larvae. Most notable was the difference in cell shape. As described above, lamellocytes are normally discoidal cells. Most of the lamellocytes in parasitized larvae are elongated or bipolar (Fig. 6.14). Filaments often extend from the poles of the elongated cells, and blebs of cellular material can be seen attached to their tips. Similar blebs can be found free in the surrounding hemolymph, suggesting that these materials are being lost from the elongated lamellocytes. That we are viewing the destruction of lamellocytes in parasitized larvae is supported by the fact that hemolymph samples taken 48 hr after parasitization contain only a few lamellocytes (R. Rizki and Rizki, 1984). This observation agrees with that of earlier workers, but it is clear that the decreased lamellocyte frequency results from destruction of these cells.

In addition to the striking difference in shape of the lamellocytes in parasitized and unparasitized larvae, there is a noticeable difference in the adhesive properties of these cells. Discoidal lamellocytes in *tu* larvae are often found adhering in clumps in hemocyte samples. The elongated lamellocytes in parasitized larvae appear to have lost this adhesiveness. The loss of lamellocyte adhesiveness and eventual destruction of these cells must account for the fact that the wasp egg remains unencapsulated. The quality of the egg surface is probably irrelevant.

If the spk$^+$ surface property is related to adhesiveness and the elongated lamellocytes in parasitized hosts are unable to form capsules owing to loss

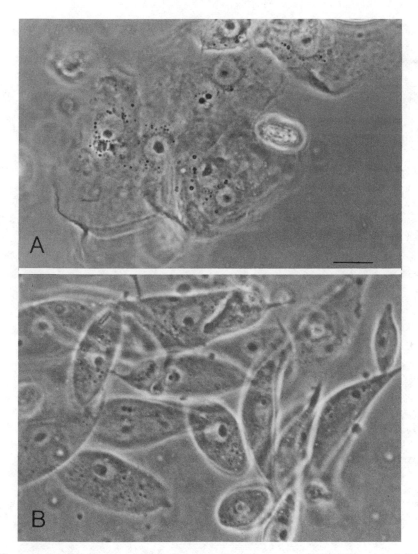

Figure 6.14. (A) Phase micrograph of lamellocytes from an unparasitized *tu-Sz*^{ts} larva. The cells are discoidal and adhere to each other in a clump. (Scale bar = 10 μm.) (B) Elongated lamellocytes from a *tu-Sz*^{ts} larva parasitized by *L. heterotoma*. Since the cell body remains thin, the outlines of underlying cells are visible, as in A.

of their adhesiveness, then these morphologically modified lamellocytes should no longer show the spk$^+$ property. Lamellocytes in hemolymph samples treated with WGA several hours after parasitization were elongated, but the percentage of spk$^+$ cells did not differ from that found in unparasitized larvae (85%). After an additional 4hr, the percentage of spk$^+$ cells in

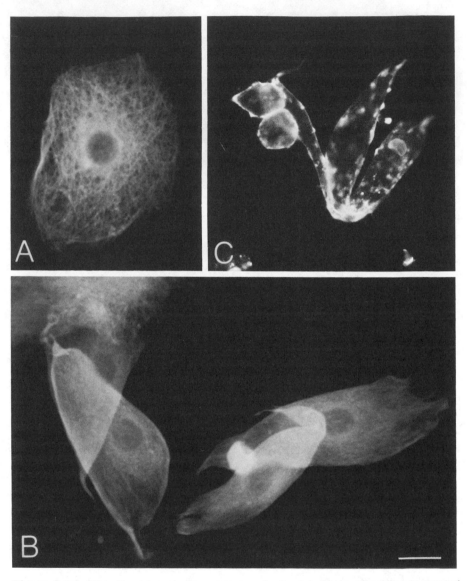

Figure 6.15. (A) Indirect immunofluorescence of microtubules in a lamellocyte reacted with antitubulin. The microtubules radiate from the nuclear region toward the cell periphery. (B) The reorientation of microtubules in lamellocytes from a larva parasitized by *L. heterotoma*. Microtubules are now parallel to the elongating axis of the cells. (Scale bar = 10 μm.) (C) spk⁺ lamellocytes from a parasitized larva treated with WGA conjugated to fluorescein isothiocyanate. Compare this figure with Figure 6.3A.

parasitized larvae decreased to 66%, and many of the cells showed weak speckling (Fig. 6.15C). By 24 hr after infection, the percentage of spk$^+$ lamellocytes in parasitized larvae was reduced to 24% (R. Rizki and Rizki, 1984). The loss of WGA-binding sites on the lamellocyte surfaces must be due to sloughing and shedding of surface molecules along with the blebs seen at the tips of the bipolar cells.

We have considered the discoidal shape of the lamellocyte an important feature for capsule formation. Cell shape is dependent on cytoskeletal elements such as microtubules and microfilaments, and bundles of microtubules can be seen in the lamellocytes within capsule walls (Rizki and Rizki, 1984). Rearrangement of microtubules accompanies or underlies the modification of lamellocyte shape from a disc to a bipolar shape (Fig. 6.15A,B). Modifications in cytoskeletal elements may also disturb the normal arrangements of surface molecules needed for lamellocyte-to-lamellocyte binding and may contribute to the loss of lamellocyte adhesiveness in parasitized larvae.

Since no effects on the morphology and function of the crystal cells and plasmatocytes were detected, the wasp factor responsible for destruction of lamellocytes was named lamellolysin (R. Rizki and Rizki, 1984). Lamellolysin is found in an accessory gland of the female wasp reproductive system and must be injected along with the egg into the host hemocoel. After this factor has been identified and purified (studies in progress), we hope it will be useful for analyzing the adhesive properties of lamellocytes as well as for understanding how a parasitoid foils the defense system of its host. Survival of the parasitoid depends on its ability to counteract those host cells that are potentially harmful to it. Thus, this natural phenomenon of selective destruction of lamellocytes by lamellolysin confirms our experimental analysis that lamellocytes recognize foreign surfaces to be encapsulated, while crystal cell function and plasmatocyte function are not crucial for this cellular defense response.

6.6. APPROACHES FOR FUTURE RESEARCH

We are only now beginning to unravel the complexities of the cellular interactions during encapsulation in *D. melanogaster*. It is clear that the time for descriptive study of encapsulation has peaked. New approaches to examine the surfaces of hemocytes and internal body organs are needed. The use of one lectin, wheat germ agglutinin, has demonstrated differences between hemocytes not previously distinguishable by morphologic criteria. Further study with tools such as this will undoubtedly be useful for establishing details of the encapsulation process.

A far more probing analysis will be possible when we are able to identify specific molecules on the hemocyte surfaces that are involved in recognition and binding phenomena. To this end, we have begun to isolate monoclonal antibodies directed against hemocyte surfaces and hemocyte intracellular

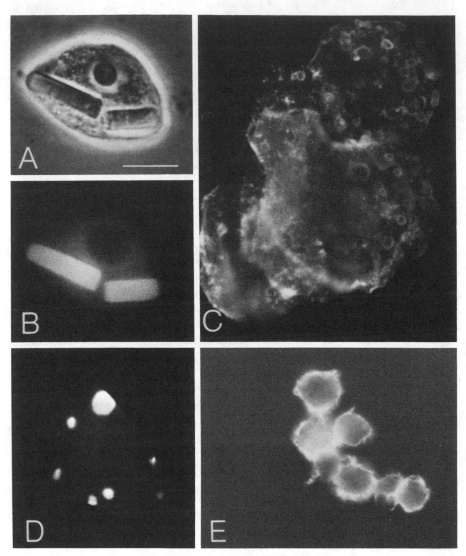

Figure 6.16. Indirect immunofluorescence of *Drosophila* hemocytes reacted with monoclonal antibodies. (A) Phase micrograph of a crystal cell. (Scale bar = 10 μm.) (B) The same crystal cell photographed with fluorescence optics shows the specificity of antibody Drosum 172 for the paracrystalline inclusions. (C) Monoclonal antibody Drosum 16 binds to lamellocyte surfaces. (D) Monoclonal antibody Drosum 23 binds to lamellocytes in a patchy pattern. (E) Monoclonal antibody Drosum 201 lights up plasmatocyte surfaces.

components (Fig. 6.16). Characterization of the antigens for some of these antibodies and establishment of antigen distributions during stages of encapsulation will, we hope, uncover some of the molecular intricacies of this cellular defense reaction. For example, whether components of crystal cells are necessary for opsonization of foreign materials to be phagocytized or encapsulated may be determined by using antibodies specific for paracrystalline inclusions and crystal cells. Also, the cellular origin of the blebs and vesicles seen during encapsulation may be established using antibodies specific to certain cell types.

6.7. SUMMARY

Capsule walls in *D. melanogaster* larvae consist of flattened cells, lamellocytes, which differentiate from spherical plasmatocytes. Lamellocytes are adapted for layering around foreign objects in the hemocoel. There is no experimental evidence that opsonization of foreign objects prepares them for encapsulation by lamellocytes in *Drosophila*. Therefore, lamellocytes or their precursors must recognize foreign objects and adhere to them to form the first layer of a cellular capsule. Lamellocytes also have adhesive surfaces for binding to other lamellocytes to form capsule walls. Small protrusions or knoblike structures on the surfaces of lamellocytes may be important for binding reactions.

Wheat germ agglutinin conjugated to fluorescein isothiocyanate for fluorescence microscopy and the same lectin conjugated to ferritin for electron microscopy was used to label lamellocyte surfaces. These studies distinguish two types of lamellocytes: one to which the lectin binds in a patchy pattern and another that does not exhibit this pattern.

The *Bc* mutant gene, which releases phenoloxidase activity within the crystal cells in vivo, was combined with a melanotic tumor gene to examine the role of crystal cells in capsule formation. The presence of black cells (the mutant form of crystal cells) within the capsules in the double mutant strain suggests that crystal cells may be incorporated within developing capsule walls. Since crystal cells are responsible for cross-linking of proteins, capsule walls are hardened and darkened.

Encapsulation is suppressed in *D. melanogaster* larvae parasitized by the cynipid wasp, *L. heterotoma*. This suppression is mediated by lamellolysin, a factor that the female wasp injects along with its egg into the host hemocoel. Lamellolysin selectively destroys lamellocytes by affecting their shape, adhesivity, surface molecules, and cytoskeletal elements. The molecular target of lamellolysin is unknown, but it is clear that this target must be a crucial component for the structural and functional integrity of the lamellocyte.

ACKNOWLEDGMENT

Preparation of this article was supported in part by NIH Grant AG01945.

REFERENCES

Ashida, M. and K. Dohke. 1980. Activation of prophenoloxidase by the activating enzyme of the silkworm, *Bombyx mori*. *Insect Biochem*. **10**: 37–47.

Ashida, M. and H. Yoshida. 1985. Activation of two zymogens of proteases by peptidoglycan or β-1,3-glucan and the conversion of prophenoloxidase into phenoloxidase in insect plasma. *Dev. Comp. Immunol*. **9**(1): 166.

Batterham, P. and S.W. McKechnie. 1980. A phenoloxidase polymorphism in *Drosophila melanogaster*. *Genetica* **54**: 121–126.

Chandaman, R.K. 1979. Characterization of a second chromosome recessive lethal melanotic tumor mutant $l(2)mt^{kc}$ in *Drosophila melanogaster*. Bachelor's thesis, York University, York, England.

Dularay, B. and A.M. Lackie. 1985. Haemocytic encapsulation and the prophenoloxidase-activation pathway in the locust, *Schistocerca gregaria*. *Dev. Comp. Immunol*. **9**(1): 167.

Goldstein, I.J., S. Hammarstrom, and G. Sundblad. 1975. Precipitation and carbohydrate-binding specificity on wheat germ agglutinin. *Biochim. Biophys. Acta* **405**: 53–61.

Gupta, A.P. 1979. Hemocyte types: Their structures, synonymies, interrelationships, and taxonomic significance, pp. 85–127. In A.P. Gupta (ed.) *Insect Hemocytes*. Cambridge University Press, Cambridge.

Hanratty, W.P. and J.S. Ryerse. 1981. A genetic melanotic neoplasm of *Drosophila melanogaster*. *Dev. Biol*. **83**: 238–249.

Lindsley, D.L. and E.H. Grell. 1976. *Genetic Variations of Drosophila melanogaster*. Publication no. 627, Carnegie Institute, Washington, DC.

Mitchell, H.K. and U.M. Weber. 1965. *Drosophila* phenoloxidases. *Science* (Washington, DC) **148**: 964–965.

Nappi, A.J. 1970. Defense reaction of *Drosophila euronotus* larvae against the hymenopterous parasite *Pseudeucoila bochei*. *J. Invertebr. Pathol*. **16**(3): 408–418.

Nappi, A.J. 1973. Hemocytic changes associated with the encapsulation and melanization of some insect parasites. *Exp. Parasitol*. **33**: 285–302.

Nappi, A.J. 1977. Factors affecting the ability of the wasp parasite *Pseudeucoila bochei* to inhibit tumourigenesis in *Drosophila melanogaster*. *J. Insect Physiol*. **23**: 809–812.

Nappi, A.J. and M. Silvers. 1984. Cell surface changes associated with cellular immune reactions in *Drosophila*. *Science* (Washington, DC) **225**: 1166–1168.

Nappi, A.J. and F.A. Streams. 1969. Haemocytic reactions of *Drosophila melanogaster* to the parasites *Pseudocoila mellipes* and *P. Bochei*. *J. Insect Physiol*. **15**: 1551–1566.

Ratcliffe, N.A., C. Leonard, and A.F. Rowley. 1984. Prophenoloxidase activation: Nonself recognition and cell cooperation in insect immunity. *Science* (Washington, DC) **226**: 557–559.

Rizki, R.M. and T.M. Rizki. 1974. Basement membrane abnormalities in melanotic tumor formation. *Experientia* **30**: 543–546.

Rizki, R.M. and T.M. Rizki. 1977. Modification of *Drosophila* cell surfaces by concanavalin A. *Cell Tissue Res*. **185**: 183–190.

Rizki, R.M. and T.M. Rizki. 1979. Cell interactions in the differentiation of a melanotic tumor in *Drosophila*. *Differentiation* **12:** 167–178.

Rizki, R.M. and T.M. Rizki. 1980. Hemocyte responses to implanted tissues in *Drosophila melanogaster* larvae. *Wilhelm Roux's Arch. Dev. Biol.* **189:** 207–213.

Rizki, R.M. and T.M. Rizki. 1984. Selective destruction of a host blood cell type by a parasitoid wasp. *Proc. Natl. Acad. Sci. U.S.A.* **81:** 6154–6158.

Rizki, R.M., T.M. Rizki, and C.A. Andrews. 1975. *Drosophila* cell fusion induced by wheat germ agglutinin. *J. Cell Sci.* **18:** 113–142.

Rizki, R.M., T.M. Rizki, C.R. Bebbington, and D.B. Roberts. 1983. *Drosophila* larval fat body surfaces: Changes in transplant compatibility during development. *Wilhelm Roux's Arch. Dev. Biol.* **192:** 1–7.

Rizki, T.M. 1952. Pattern of pigmentation associated with *Drosophila* lethals. *Am. Nat.* **86:** 409–412.

Rizki, T.M. 1955. Hereditary melanotic sclerosis in *Drosophila willistoni*. *J. Exp. Zool.* **128:** 591–610.

Rizki, T.M. 1956. Blood cells of *Drosophila* as related to metamorphosis, pp. 91–94. In F.L. Campbell (ed.) *Physiology of Insect Development*. University of Chicago Press, Chicago.

Rizki, T.M. 1957a. Tumor formation in relation to metamorphosis in *Drosophila melanogaster*. *J. Morphol.* **100:** 459–472.

Rizki, T.M. 1957b. Alterations in the hemocyte population of *Drosophila melanogaster*. *J. Morphol.* **100:** 437–458.

Rizki, T.M. 1961. The influence of glucosamine hydrochloride on cellular adhesiveness in *Drosophila melanogaster*. *Exp. Cell Res.* **24:** 111–119.

Rizki, T.M. 1962. Experimental analysis of hemocyte morphology. *Am. Zool.* **2:** 247–256.

Rizki, T.M. 1978. The circulatory system and associated cells and tissues, pp. 397–452. In M. Ashburner and T.R.F. Wright (eds.) *The Genetics and Biology of Drosophila,* vol. 2b. Academic Press, New York.

Rizki, T.M. and R.M. Rizki. 1959. Functional significance of the crystal cells in the larva of *Drosophila melanogaster*. *J. Biophys. Biochem. Cytol.* **5:** 235–240.

Rizki, T.M. and R.M. Rizki. 1976. Cell interactions in hereditary melanotic tumors of *Drosophila melanogaster,* pp. 137–141. *Proceedings First International Colloquium Invertebrate Pathology,* Queen's University, Kingston, Canada.

Rizki, T.M. and R.M. Rizki. 1980a. Developmental analysis of a temperature-sensitive melanotic tumor mutant in *Drosophila melanogaster*. *Wilhelm Roux's Arch. Dev. Biol.* **189:** 197–206.

Rizki, T.M. and R.M. Rizki. 1980b. Properties of the larval hemocytes of *Drosophila melanogaster*. *Experientia* **36:** 1223–1226.

Rizki, T.M. and R.M. Rizki. 1981. Alleles of *lz* as suppressors of the *Bc*-phene in *Drosophila melanogaster*. *Genetics* **97:** s90, abstract.

Rizki, T.M. and R.M. Rizki. 1982. The cellular defense reactions of *Drosophila melanogaster,* pp. 173–176. In H. Akai, R.C. King, and S. Morohoshi (eds.) *The Ultrastructure and Functioning of Insect Cells*. Society for Insect Cells, Japan.

Rizki, T.M. and R.M. Rizki. 1983a. Basement membrane polarizes lectin binding sites of *Drosophila* larval fat body cells. *Nature* (London) **303:** 340–342.

Rizki, T.M. and R.M. Rizki. 1983b. Blood cell surface changes in *Drosophila* mutants with melanotic tumors. *Science* (Washington, DC) **220:** 73–75.

Rizki, T.M. and R.M. Rizki. 1984. The cellular defense system of *Drosophila melanogaster*, pp. 579–604. In R.C. King and H. Akai (eds.) *Insect Ultrastructure*, vol. 2. Plenum Press, New York.

Rizki, T.M., R.M. Rizki, L.F. Allard, and W.C. Bigelow. 1976. Micromanipulation of tissues and cells of the *Drosophila* larva in the SEM, pp. 611–618. In O. Johari (ed.) *Scanning Electron Microscopy: 1976,* vol. 2. Chicago Press, Chicago.

Rizki, T.M., R.M. Rizki, and R.A. Bellotti. 1985. Genetics of a *Drosophila* phenoloxidase. *Mol. Gen. Genet.* **201:** 7–13.

Rizki, T.M., R.M. Rizki, and E.H. Grell. 1980. A mutant affecting the crystal cells in *Drosophila melanogaster. Wilhelm Roux's Arch. Dev. Biol.* **188:** 91–99.

Salt, G. 1970. *The Cellular Defence Reactions of Insects*. Cambridge University Press, Cambridge.

Shrestha, R. and E. Gateff. 1982. Ultrastructure and cytochemistry of the cell types in the larval hematopoietic organs and hemolymph of *Drosophila melanogaster. Dev. Growth Differ.* **24:** 65–82.

Sparrow, J.C. 1978. Melanotic "tumours," pp. 277–313. In M. Ashburner and T.R.F. Wright (eds.) *The Genetics and Biology of Drosophila*. Academic Press, New York.

Spragg, J.H., C.R. Bebbington, and D.B. Roberts. 1982. Monoclonal antibodies recognizing cell surface antigens in *Drosophila melanogaster. Dev. Biol.* **89:** 339–352.

Walker, I. 1959. Die Abwehrreaktion des Wirtes *Drosophila melanogaster* gegen die zoophage Cynipide *Pseudeucoila bochei* Weld. *Rev. Suisse Zool.* **66:** 569–632.

Wilson, L.P., R.C. King, and J.L. Lowry. 1955. Studies on the *tu-W* strain of *Drosophila melanogaster. Growth* **19:** 215–244.

CHAPTER 7

Transplantation
The Limits of Recognition

Ann M. Lackie
Department of Zoology
The University
Glasgow, Scotland

7.1. INTRODUCTION

Studies of the physiology and development of insects have relied heavily on the transplantation of appropriate organs or limbs; such experiments, if performed on mammals, would be impossible unless performed on either immunosuppressed or genetically identical, and thus histocompatible, animals. The lack of response to allografts in insects is not due to the lack of immunocompetence of the recipients, because insects are able to mount very successful cellular and humoral responses against a variety of biotic and abiotic objects, microorganisms, and macromolecules. The question that thus needs to be answered is: What are the limits of immunorecognition? In order to answer this, it is necessary to gain an idea of the range of discrimination between self and nonself and then attempt to understand how the range is delimited. The solution of these problems should allow us to draw conclusions as to why the immune systems of insects do not recognize as foreign certain types of surface, including, most important, those of various habitual parasites.

As experimental animals for investigating immunorecognition, insects have advantages and disadvantages. Experimental implantation of tissues, parasites, and abiotic objects into the body cavity is relatively easy because the hemocoelic space is large and the cellular response, in the form of encapsulation, can be quantified. On the other hand, orthotopic "skin grafts" in the form of pieces of cuticle with the underlying epidermis and connective-tissue layer, while being technically feasible, may present problems owing to the rigidity and opacity of the exoskeleton. Despite this, or perhaps because of this, several interesting assays for immunorecognition have been employed that exploit techniques developed principally by physiologists.

The limits of the recognition process have, accordingly, been tested in a variety of ways, mainly by investigating the response to (1) tissue transplants, (2) abiotic objects, (3) parasites, and (4) soluble molecules. The nature of the response to soluble molecules, whether intact or denatured self (Ashida et al., 1984; Clem et al., 1984), toxins (Rheins et al., 1980) or microbial products (see Chapter 9) will not be considered here. Interpretation of whether the surfaces of parasites are recognized is difficult, because active suppression of the immune response may be involved; despite this, surface phenomena, and thus recognition or its evasion may, however, be important in the *Nemeritis-Ephestia* parasitoid-host system (Salt, 1980) and in the survival of the gregarine *Diplocystis schneideri* within the cockroach *Periplaneta americana* (Rotheram and Lackie, 1975), but discussion of the host-parasite relationship is beyond the scope of this chapter. It is from the first two methods—transplantation immunology and immunity to abiotic objects—that most of the important information on recognition derives, and it is thus upon these two topics that attention is concentrated.

In this chapter, the concept of recognition is discussed from the point of view that intact self is ignored but that nonself is actively recognized and stimulates a response. The alternative hypothesis, that intact self or foreign surfaces treated as such require active recognition and that failure of recognition triggers a response, is discussed elsewhere (Lackie, 1986c).

7.2. TRANSPLANTATION IMMUNOLOGY

In recent years, the immunology of transplantation in vertebrates has become an increasingly fashionable discipline partly because of an urge to understand the phylogenetic origins of that apparent pinnacle of achievement, the histocompatibility system of mammals. In the arthropods, and the insects in particular, transplantation immunology also has great practical importance because, as stated earlier, by knowing the limits of recognition it might be possible to understand why so many of these animals are hosts of parasites or susceptible to pathogens.

Transplantation reactions in insects and in Crustacea have been investigated predominantly by using in vivo assays, including the determination of the fate of implanted tissues and of cuticular grafts from allogeneic (intraspecific) and xenogeneic (interspecific) sources, and the measurement of clearance of foreign particles from the hemocoelic blood space.

7.2.1. Implanted Tissues

7.2.1.1. Alloimplants. Many studies of the physiology and development of insects have relied, sometimes unknowingly, on the fact that alloimplants of various organs are not destroyed by the recipient's hemocytes but continue to function in either production or reception of hormones or nerve signals. Examples of such transplants include the corpora allata in the bug *Rhodnius* (Wigglesworth, 1937) and in larvae of the wax moth *Galleria* (Piepho, 1950); of ovarian tissue in *Drosophila melanogaster* (Ephrussi and Beadle, 1936), in mosquitoes (Larsen and Bodenstein, 1959), in muscid flies (Peterson, 1968), and in cockroaches (Bell, 1972); abdominal ganglia in the cockroach *P. americana* (Guthrie, 1966); and of imaginal discs of eyes (Ephrussi and Beadle, 1936) and of appendages (Hadorn and Buck, 1962; Schubiger, 1971) in *Drosophila*. Some of the early literature has been discussed with reference to the hemocytic defence reaction in an excellent review by Salt (1961).

With the more recent emphasis on comparative immunology, the survival of implanted tissues has been examined from the point of view of the hemocytic response; if the hemocytes do not adhere to and encapsulate the transplanted tissue, then it may be presumed that the surface of the transplant has not been recognized as foreign. Conversely, formation of a multicellular hemocytic capsule around the implant may indicate that immunorecognition has occurred (Fig. 7.1A). There is, of course, a potential problem in interpretation in that hemocytes also respond to and aggregate around wounds; localized encapsulation could thus be indicative of a damaged area of an otherwise unrecognized transplant (Fig. 7.1B). More generalized encapsulation might indicate death of or damage to the transplant due to physiologic causes. If the presence or absence of encapsulation is to be used successfully as an assay, the time interval between implantation and analysis needs to be chosen with care, and relatively large numbers of implants need to be examined. Also, the type of transplant to use will depend on the sex and species of the animal; for example, in some species the ventral nerve cord, although easily recognizable, is unsuitable because it is covered with loose

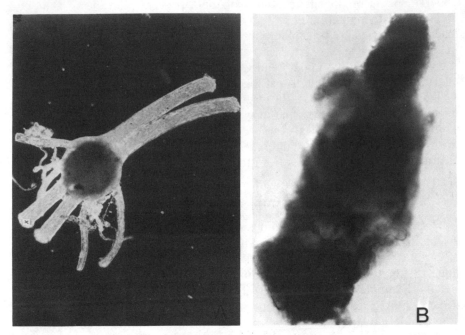

Figure 7.1. (A) Allograft of nerve cord 24 hr postimplantation, its surface completely clear of hemocytes except at the cut ends, indicating that the implant was not recognized as foreign. (B) Xenograft of nerve cord 24 hr postimplantation surrounded by a thick hemocytic capsule.

connective tissue of an indefinite outline. The best types of tissues to use are those that are small and easily manipulable, with a distinctive appearance and a smooth outline.

Some of the earliest studies of the hemocytic encapsulation of implants were carried out by Salt (1960) during his pioneering work on the responses of lepidopteran larvae such as *Diataraxia* and *Ephestia* to parasitoid larvae. In cases in which tissue transplants were not recognized as foreign and remained unencapsulated, hemocytes accumulated only at the cut ends or at sites of wounds on the implant (Salt, 1960). It would seem, therefore, that the transplant was accepted by the recipient's immune system as self, and the hemocytes were responding to the damaged tissue in the same manner in which they would respond to wounded self (Salt, 1961). This phenomenon has been observed by other investigators (Fig. 7.1A; Scott, 1971).

Scott (1971), using the cockroach *P. americana*, found that allografts of ventral nerve cord, whether between individuals of the same or of opposite sex, remained unencapsulated, and this was confirmed in the same species for ovariole (Peterson, 1968; Bell, 1972) and ventral nerve cord (J. Lackie, 1975; Lackie, 1979); the same result is thus obtained irrespective of the type of tissue used. The only cases in which destruction of allogeneic implants has been recorded are in the transplantation of ovarioles between individual milkweed bugs, *Oncopeltus fasciatus*, in which the transplants were not encapsulated but degenerated and became melanized (Peterson, 1968), and in the transplantation of "lymph glands" and tumors between individuals of the fruit fly, *D. melanogaster* (see Chapter 6). Carton's work (1976) on allotransplantation of parasitoid eggs recovered from caterpillar hosts is discussed elsewhere (Lackie, 1983b).

Since most laboratory stocks of insects are highly inbred, individual insects might be more isogeneic than allogeneic. However, transplantation between different stocks of *Periplaneta* gave the same result as transplantation within one stock (Lackie, 1979), and thus, using encapsulation of implanted tissues as an assay, it is apparent that allogeneic recognition is absent in insects.

7.2.1.2. Xenoimplants.

7.2.1.2. Xenoimplants. The fact that xenogeneic recognition of implanted tissues is also absent in some species combinations lends support to the conclusions of the previous section. Much of the early evidence for non-recognition of xenogeneic implants was provided by developmental biologists and physiologists and has been reviewed by Salt (1961). In general, it seems that transplants between related species and genera are more likely to be accepted than are transplants between widely separated genera.

Transplants between different orders, for example, of corpora allata from *Tenebrio* (Coleoptera) or *Carausius* (Orthoptera; Piepho, 1950) or ventral nerve cord of *Schistocerca* (Orthoptera; Schmit and Ratcliffe, 1976) to larvae of *Galleria* (Lepidoptera) are thickly encapsulated, as are pieces of ovariole or nerve cord from *Tenebrio* implanted into *Leucophaea* (Dictyoptera) or two species of locust (Lackie, 1976).

When pieces of testis were transferred from the tomato moth, *Diataraxia*, into larvae of the privet hawk moth, *Sphinx* (different families), or from the flour moth, *Ephestia*, to *Diataraxia* (different families) and retrieved after 24 hr, they were found to have been encapsulated (Salt, 1960). Similarly, ovarioles of the mosquito *Aedes aegypti* implanted into three species of muscid flies degenerated; in this case, no hemocytic encapsulation occurred, but the implants became heavily melanized (Peterson, 1968), a response common in dipterans (see Chapter 15).

In contrast, ovaries transplanted between *Sarcophaga, Phormia, Musca,* and *Calliphora*, all different genera of the same muscid family (Peterson, 1968), and between the mosquitoes *A. aegypti* and *Culex molestus* (Larsen and Bodenstein, 1959) all developed unencapsulated. Kambysellis (1970) used interspecific transplantation of ovaries as a tool to determine phylogenetic affinities between different species of the same genus, *Drosophila*, and found that in many of the species combinations the ovaries grew and matured. In some cases, communication between the donor ovariole and the host reproductive tract was established to such an extent that the donor's eggs were fertilized and sterile F1 hybrids obtained. The presence or absence of a host response was not discussed, but illustrations show that developing implants were neither encapsulated nor melanized. The more recent work of Rizki and Rizki (1980), transplanting pieces of fat body between species of *Drosophila*, confirms the observations that transplants between sibling species are not encapsulated.

7.2.1.3. Cockroaches as Recipients. Cockroaches and other exopterygotes have frequently been used as experimental models for studies of the insect immune response because of the relative ease with which large numbers of individuals of many different species can be reared and manipulated experimentally. Many of these insects have very large numbers of circulating hemocytes (Lackie et al., 1985), and their immune response is manifested in the formation of a hemocytic capsule. The hemocytic response is initiated rapidly, and in the case of *P. americana* is best analyzed at 24 hr postimplantation, when the extent of encapsulation can easily be ascertained and the outline of the enclosed implant can still be seen. The capsule subsequently becomes compressed and its core melanized, making identification of the enclosed object impossible.

Whether interspecifically transplanted ovaries would develop and accumulate host vitellogenins was studied by Bell (1972), using a total of 22 species combinations from several different cockroach families (Fig. 7.2). Ovaries implanted into the thoracic region were examined 10 days to 1 month later for changes in their size, development, and types of yolk proteins, and to determine whether encapsulation of the implant had occurred. A clear pattern emerged from this work to show that lack of encapsulation, followed by ovarian development, was possible with intergeneric and interspecific implants as long as donor and recipient were within the same subfamily.

Immunorecognition of implants by the blattid *Periplaneta* has been investigated by various workers, all of whom have analyzed the hemocytic response within 24 hr of implantation. As might be expected, implants from different orders are thickly encapsulated, exemplified in *Periplaneta* by the response to *Calliphora* (Diptera) supraesophageal ganglia (Scott, 1971) and to nerve cord from two species of Coleoptera and three species of Orthoptera (Figs. 7.1B and 7.3A; Lackie, 1976, 1979). Within the Dictyoptera, recognition and subsequent encapsulation of implants by *Periplaneta* follows approximately the pattern shown by the results of Bell's (1972) longer-term experiments; within the superfamily Blaberoidea, nerve cord taken from the species *Pycnoscelus surinamensis* (Lackie, 1976) and from *Blaberus craniifer* (Lackie, 1983b) is encapsulated. When the oxyhaloine *Leucophaea maderae* is used as recipient for *Periplaneta* tissue—the reciprocal transfer has not been carried out—all of the implants are encapsulated (J. Lackie, 1975). However, although Bell (1972) found that ovarioles transplanted between *Periplaneta* and the oxyhaloine *Nauphoeta* were encapsulated, the results of shorter-term experiments, using nerve cord, are more equivocal;

	Host species [2]																					
Donor species	*P. americana*	*P. australasiae*	*P. fuliginosa*	*P. brunnea*	*B. orientalis*	*E. floridana*	*S. longipalpa*	*B. germanica*	*P. virginica*	*P. pennsylvanicus*	*X. immaculata*	*B. discoidalis*	*B. craniifer*	*B. fumigata*	*E. posticus*	*P. surinamensis*	*P. indicus*	*D. punctata*	*G. portentosa*	*G. brunnea*	*L. maderae*	*N. cinerea*
Superfamily Blattoidea																						
Family Blattidae																						
Subfamily Blattinae																						
Periplaneta americana	●	●	●	●	●							○	○		○				○	○	○	○ ○
Periplaneta australasiae	●	●	●	●	●								○	○							○	
Periplaneta fuliginosa	●	●	●	●	●								○	○							○	
Periplaneta brunnea	●	●	●	●	●								○	○							○	
Blatta orientalis	●	●	●	●	●							○	○		○				○	○	○	○ ○
Subfamily Polyzosteriinae																						
Eurycotis floridana	◐	◐				●																
Superfamily Blaberoidea																						
Family Blattellidae																						
Subfamily Plectopterinae																						
Supella longipalpa	○	○		○																		
Subfamily Blattellinae																						
Blattella germanica	○	○	○	○	○							○	○						○			
Parcoblatta virginica	◐				○			●	●			○	○						○			
Parcoblatta pennsylvanicus	○				○			●	●			○	○						○			
Xestoblatta immaculata	◐				○							○	○						○			
Family Blaberidae																						
Subfamily Blaberinae																						
Blaberus discoidalis	○	○	○	○	○							●	●	●	◐						◐	
Blaberus craniifer	○	○	○	○	○							●	●	●	●							
Byrsotria fumigata	○	○	○	○	○							◐	○	●	●						●	●
Eublaberus posticus	○	○	○	○	○							○	●	●	●						◐	●
Subfamily Pycnoscelinae																						
Pycnoscelus surinamensis	○			○								○		●	●						◐	
Pycnoscelus indicus	○			○								○		●	●						◐	
Subfamily Diplopterinae																						
Diploptera punctata	○			○								○	○					●			○	○
Subfamily Oxyhaloinae																						
Gromphadorhina portentosa	○			○										○	○				●	●	●	●
Gromphadorhina brunnea	○			○								○		○	○				●	●	●	●
Leucophaea maderae	○	○	○	○	○							○		●	◐				●	●	●	●
Nauphoeta cinerea	○			○										○	○				◐	●	●	●

Figure 7.2. Encapsulation of ovarioles transplanted within the Dictyoptera. Closed circles indicate ovarian development without encapsulation; open circles indicate encapsulation. (From Bell, 1972.)

in either direction, a large proportion of the implants remain unencapsulated (Fig. 7.3A), and capsules, when they existed, were thin and did not increase in thickness even after 96 hr (Lackie, 1979). These results are discussed further in Section 7.2.4.

Within the family Blattidae, transplants into *P. americana* from *P. australasiae* (Jackson and Lackie, unpublished) and, more interestingly, from another genus, *Blatta orientalis* (Fig. 7.3A; Lackie, 1979, 1983b) are unrecognized and remain unencapsulated for periods longer than 70 days postimplantation (Lackie, 1983b), allowing plenty of time for rejection to have occurred if it was going to do so.

Recognition of implants from individuals of the same and closely related (Blattinid) species is thus lacking; here both the "histocompatibility" and

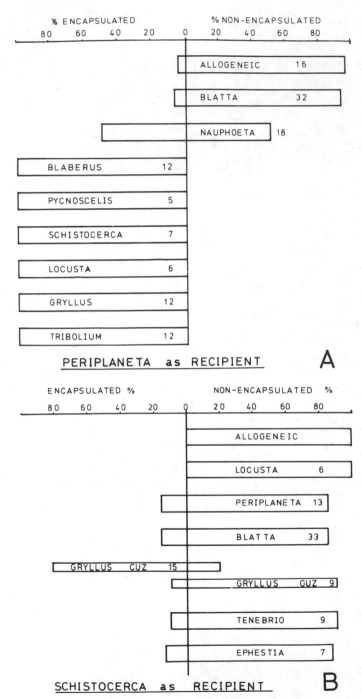

Figure 7.3. Encapsulation of implanted tissues in (A) *Periplaneta* and (B) *Schistocerca*. Numbers within bars indicate number of implants examined. (From Lackie 1976, 1979, 1983b.)

the physiologic compatibility must be sufficiently close so that no immunorecognition occurs; the implant remains healthy and with an intact surface, so that hemocytic involvement in wound-healing or autolysis does not occur.

The importance of testing the degree of responsiveness in both species of a particular xenogeneic combination is obvious when the results of using the locust *Schistocerca gregaria* as recipient are seen (Fig. 7.3B). The hemocytes of *Schistocerca* fail to adhere to and encapsulate tissue implants from a wide variety of species, including those from such different orders as the Dictyoptera and Coleoptera (Lackie, 1976, 1979) and the Lepidoptera (Salt, 1963; Lackie, 1976), as well as from such closely related orthopterans as *Locusta* (Lackie, 1979). This lack of responsiveness is not due to a shortage of potentially reactive hemocytes, for although adult *Schistocerca* do have fewer circulating hemocytes than do adult *Periplaneta* (Lackie et al., 1985), they can mount an impressive cellular response in the form of either nodules or capsules against fungal spores (Gunnarsson and Lackie, 1985), trypanosome protozoa (G. Takle, unpublished; Ingram et al., 1984), and a variety of abiotic objects (see Section 7.3). Nor can it be assumed that the lack of recognition of the xenogeneic implants is due to a close similarity between the surface properties of host and donor tissues for, if *Schistocerca* testis or nerve cord is implanted into caterpillars of *Ephestia* (Salt, 1963) or into adult *Periplaneta* (Fig. 7.3A; Lackie, 1979), the implants are thickly encapsulated within 24 hr, indicating that the recognition is not reciprocated.

From these results two deductions can be made that (1) since xenogeneic implants can survive within *Schistocerca*, there is physiologic compatibility, at least in the short term, between the donor and host species, and (2) the degree of discrimination between self and nonself by the locust immune system is poor (Sections 7.2.5 and 7.3).

7.2.2. Implantation of Cells

Most experiments on the hemocytic recognition of injected foreign cells by insects have been carried out on cockroaches as recipients; studies of the responses to vertebrate erythrocytes (Bettini et al., 1951; Ryan and Nicholas, 1972) and trypanosome protozoa (Ingram et al., 1984; Takle, unpubl.) are not discussed here, but attention is focused on the response to tissue cells and to leucocytes.

Dawe and coworkers (1967) examined the hemocytic response of the cockroach *Leucophaea* to a variety of cultured vertebrate cells, including mammalian carcinoma and sarcoma lines and tissues from three species of freshwater fish. In all cases, hemocytes encapsulated clumps of foreign cells, with the result that necrosis and melanization eventually occurred at the core, not the edges, of the clump.

Rabbit neutrophil leucocytes have been used as a model for investigating the removal of xenogeneic cells from the hemocoel of adult *Periplaneta* (A. Lackie and B. Dularay, unpublished). The neutrophils have the advan-

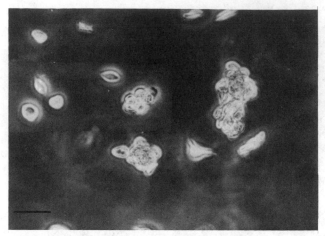

Figure 7.4. Hemocytic aggregates (nodules) and single hemocytes from *Schistocerca*, 3 hr after injection of a nodule-stimulating agent. (Scale bar = 20 μm.) (Courtesy of S. Gunnarsson.)

tages of being easily recognized, surprisingly resistant to damage, and tolerant of immersion in media and physiologic salt solutions designed for work with orthopteran and dictyopteran insects; more important, they are easily obtained as a single-cell suspension. They thus provide a useful xenogeneic cell for transplantation studies. The response of adult *Periplaneta* to injected ^{51}Cr-labeled neutrophils can be followed by measuring sequential changes in the radioactivity of the hemolymph: within an hour, the number of circulating neutrophils plummets. Reduction in their number is partly due to the trapping of the foreign cells within hemocytic aggregates, although by far the majority of the neutrophils are seen, using fluorescently labeled cells, to be trapped within the dorsal diaphragm in dense clusters on each side of the heart. As mentioned previously, insects may respond to the introduction of small particles or soluble molecules into their hemocoels by the formation of aggregates, or nodules (Salt, 1970; Ratcliffe and Gagen, 1976). This nodule response can be quantified (Gunnarsson and Lackie, 1985) and can thus be used as an assay for recognition. Thus, if rabbit neutrophils were injected into *Periplaneta* and the cockroach blood flushed out 2 hr later, large numbers of hemocytic nodules were found (Figs. 7.4 and 7.5). Using fluorescently labeled neutrophils, it was found that more than 90% of the nodules contained neutrophils. By 24 hr postinjection, the number of nodules that could be flushed out had dropped to approximately one-tenth of the value at 2 hr, and small melanized aggregates were found adhering to the fat body and the malphighian tubules and associated with the dorsal diaphragm (A. Lackie and B. Dularay, unpublished).

The nodule assay can, of course, be used to investigate recognition of allogeneic and xenogeneic cells, and the obvious cell type to use is the

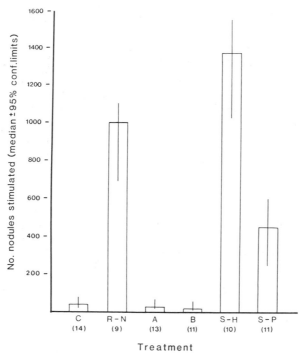

Figure 7.5. Nodule stimulation in adult male *Periplaneta* 3 hr after transfer of xenogeneic cells and whole hemolymph. (C, control injected with 20 μl saline solution; R-N, rabbit neutrophils, 2 × 10⁵ in 20 μl saline solution; A, 20 μl allogeneic hemolymph; B, 20 μl *Blatta* hemolymph; S-H, 20 μl *Schistocerca* hemolymph; S-P, 20 μl *Schistocerca* plasma; numbers in brackets indicate number of recipients injected.) (From Lackie 1986a.)

circulating hemocyte. The main problem with using hemocytes as transplants is that on removal from the donor insect and suspension and washing in medium, the morphology (Lackie et al., 1985) and physiology of the hemocyte changes and the cells might therefore be expected, for a variety of reasons (Lackie, 1986a), to disappear from the recipient's circulation. Consequently, an assay was devised in which a known volume of hemolymph from a chilled donor was transferred immediately to a chilled recipient; as controls, cell-free plasma or washed hemocytes were used. The recipients were returned to their normal maintenance temperature, and after 3 hr their blood was flushed out and the number of nodules recovered was counted. It was apparent that there were large differences in the host response to hemolymph from different donor species (Fig. 7.5). Comparison of these results with Figure 7.3A reveals that, as with tissue transplantation and the encapsulation assay, *Periplaneta* fails to respond to cells from an allogeneic source or to xenogeneic cells from *Blatta*, but does respond to xenogeneic cells or cell-free plasma from *Schistocerca* (Lackie, 1986a).

An insight into the biologic consequences of failing to respond to cells of allogeneic or xenogeneic origin has been provided by the interesting results of Tsang and Brooks (1980, 1982), in which transformed malignant cells, derived from embryonic *Blattella germanica*, were injected into eight species of cockroaches; in all but one species, *B. craniifer*, the malignant cells apparently were not recognized by the recipient's immune system, because rapid mitosis and tumor formation, accompanied by death of the cockroach, occurred. In contrast, and as might be expected, cells derived from an orthopteran *Melanoplus sanguinipes* and from two species of Lepidoptera had no adverse effects on the recipient cockroach; whether the orthopteran cells were entrapped within nodules was not mentioned.

Hemocytes have also been used as transplants, not as single-cell suspensions, but in the form of hemocytic capsules around abiotic material. For example, 4-day-old capsules formed in caterpillars of the tomato moth, *Diataraxia*, and transferred to other caterpillars of the same species were not encapsulated further (Fig. 7.6A). Salt (1960) stated, "All the indications were that the second host had accepted the encapsulated nylon as it would have accepted an intraspecific transplant." In contrast, 3-day-old capsules formed in larvae of *Tenebrio* and transferred to *Diataraxia* were thickly encapsulated by the recipient's hemocytes (Fig. 7.6A). A similar situation is obtained with hemocytic capsules formed around Sepharose beads in *Periplaneta*; 4–6-week-old capsules over whose surface sheets of coating material have been laid (Fig. 7.6B; Lackie et al., 1985) are not encapsulated further after transfer to *Schistocerca* (Fig. 7.6C), a recipient species in which *Periplaneta* tissue is not usually recognized as foreign (Figs. 7.3A, and 7.5).

7.2.3. Cuticular Transplants

The bases of the epidermal cells that secrete the cuticle are covered by a layer of connective tissue containing proteoglycans and collagen (Ashhurst, 1979, 1982) that is in direct contact with the circulating hemolymph. Since the cuticle and epidermis are firmly adherent to each other, except during apolysis just before the moult, a "skin graft" necessarily comprises the exoskeleton plus a layer of living tissue.

When cuticular grafts are performed between mammals, for example, a period of healing in and vascularization usually ensues, during which the cells of the recipient's immune system are able to come in contact with and recognize cells from the donor tissue. The essentially two-dimensional nature of the insect cuticular graft and its immediate proximity to the hemolymph mean that contact between donor and recipient cells and the opportunity for recognition to occur are rapid.

7.2.3.1. Nymphs as Recipients. As with tissue implants, cuticular transplants have been used by developmental biologists and insect physiologists chiefly as methods of investigating the intriguing problems of morphogenesis, pattern-formation, and the hormonal control of molting. Much of the early lit-

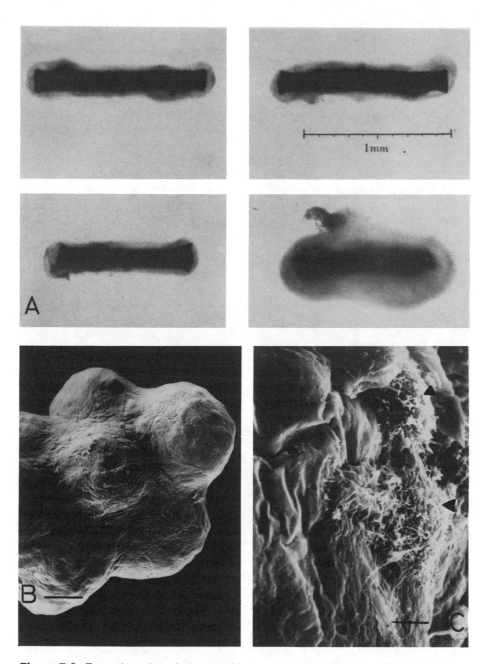

Figure 7.6. Transplantation of completed hemocytic capsules. (A, top left) Four-day capsule around nylon thread from *Diataraxia*. (top right) The same capsule retrieved 24 hr after injection into another caterpillar of *Diataraxia*. (bottom left) Three-day capsule around nylon thread from *Tenebrio*. (bottom right) The same capsule retrieved 24 hr after injection into *Diataraxia*. (From Salt, 1961.) (B) Sheets of coating material over a 6-week capsule around Sepharose beads, from *Periplaneta*. (Scale bar = 75 μm.) (C) Surface of a capsule such as in B 24 hr after transplantation to *Schistocerca*. Although two or three locust cells (arrowhead) adhered, there was no evidence of encapsulation. (Scale bar = 5μm.)

erature has been discussed by Wigglesworth (1954), but some relevant examples of allotransplantation include the work of Piepho (1943) on *Galleria* larvae and of Wigglesworth (1954) and Locke (1974) on *Rhodnius* nymphs. Pieces of limb in the cricket *Acheta (Gryllus,* Sahota and Edwards, 1969), the stick insect *Carausius* (Steinberg, 1959), the cockroach *Leucophaea* (Bohn, 1972), and even posterior abdominal segments of the cockroach *Periplaneta* (Bodenstein, 1953) have been transplanted onto their respective allogeneic nymphs, and the transplanted subcuticular epidermis has fused with that of the recipient, producing new cuticle-covered appendages in synchrony with the molting recipient; the percentage take of these grafts, in terms of the proportion of recipients on which new cuticular structures are produced postmolt, corresponds fairly closely with the complexity of the operation.

Although most of these experiments on the control of morphogenesis and regeneration were probably carried out on inbred laboratory stocks of insects, the apparent compatibility of allograft and host is made more plausible when the dramatic results of xenografting pieces of nymphal limb cuticle are observed. Such experiments have been carried out between nymphs of the hemipterans *Rhodnius* and *Triatoma* (Locke, 1974), of the stick insects *Clitumnus, Carausius,* and *Sipyloidea* (Steinberg, 1959; Bart, 1974), of the cockroaches *Gromphadorhina* and *Leucophaea* (Bohn, 1972), and of the crickets *Teleogryllus* and *Acheta* (French, 1981; Fig. 7.7A). In the latter case, metathoracic legs of nymphal *Teleogryllus* were grafted onto *Acheta* nymphs at the level of the proximal tibia, and, because the donor epidermis was not recognized as foreign, it fused with that of the recipient and produced new cuticle with the characteristic patterning and color of the *Acheta* limb after the molt.

None of these experiments was aimed at determining compatibility in the immunologic rather than the physiologic sense, but the system obviously lends itself equally well to the former type of analysis in that there is a definitive end point: whether the graft has taken can be clearly judged once the recipient has molted.

It has been shown that urodele Amphibia fail to reject alloimplants of organs but reject allogeneic skin grafts in a chronic manner, owing, it is suggested, to weak histocompatibility differences (Cohen and Borysenko, 1970); so the possibility exists that the apparent failure of *Periplaneta*, for example, to reject alloimplants might be due to presentation of the "antigen" in an inappropriate manner. Accordingly, the response to nymphal cuticle transplanted among the cockroaches *Periplaneta, Blatta, Nauphoeta,* and *B. craniifer* was investigated as a comparison with the response to implanted tissue (see Table 7.1). Large nymphs, approximately tenth instar, of *Periplaneta* were used as recipients; their immunocompetence vis-à-vis tissue implants from *Blatta* or *Blaberus* was found to be identical to that of adult cockroaches (Lackie, 1983b). Small squares of cuticle and its epidermis, of side length approximately 2mm, were transplanted onto the dorsal abdominal

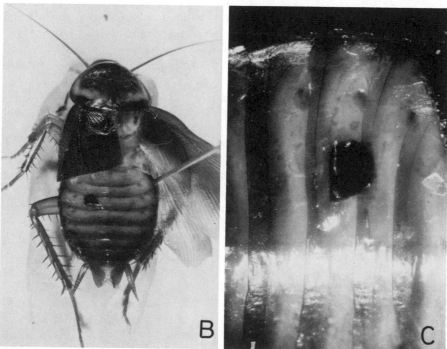

Figure 7.7. (A) Postmolt graft area of Metathoracic leg of *Acheta* (*Gryllus*) grafted at the level of the tibia with *Teleogryllus* (dark cuticle). The donor epidermis has fused with that of the recipient and produced new cuticle in synchrony with the recipient's ecdysis. (Courtesy of V. French.) (B) *Periplaneta* grafted with *Blatta* cuticle at the tenth instar showing dark *Blatta*-like cuticle at the graft site postmolt (arrow). (From Lackie, 1983b.) (C) Enlargement of postmolt graft site from B. (From Lackie, 1983b.)

205

TABLE 7.1. CUTICULAR TRANSPLANTS AMONG
NYMPHAL COCKROACHES

Donor	Recipient	Proportion of Grafts that Survive
Periplaneta	Periplaneta	9/9
Blatta	Periplaneta	148/151
Periplaneta	Blatta	76/85
Nauphoeta	Periplaneta	0/10
Blaberus	Periplaneta	0/10

Source: From Lackie, 1983b, and A. Lackie and B. Dularay,
unpublished.
Note: Proportion of recipients on which donor-type cuticle
reappears at the graft site postmolt.

surface of recipient nymphs; allografts were reversed through 180° so that
the origin of the cuticle at the graft site could be distinguished postmolt by
the orientation of the scale patterns. In all cases, grafts were sealed in by
coagulated hemolymph and usually exhibited a small amount of melanization
at the edges. Whether a graft had been rejected was almost impossible to
determine from examination of gross morphology before the molt. For example, some *Blaberus* grafts as late as 80 days postoperation still appeared
unchanged superficially, retaining their normal pigmentation. However, examination of the graft site postmolt revealed that neither the allografts nor
the xenografts from *Blatta* to *Periplaneta* had been rejected (Figs. 7.7B,C;
Tables 7.1 and 7.2). Donor epidermis had obviously fused with the epidermis
of the recipient (Fig. 7.8B) and had undergone the complex series of cellular
and metabolic events associated with molting in synchrony with the recipient. Analysis of cuticular proteins of the postmolt graft area by gel electrophoresis revealed that the *Blatta*-like graft area contained proteins found
normally in *Blatta* cuticle but not in *Periplaneta* cuticle (K. Lockey,
B. Dularay, and A. Lackie, unpublished). That the hormones associated with
molting are not species specific within arthropods has long been known
(Wigglesworth, 1954), and this *Blatta-Periplaneta* system may be useful in
providing insights into such phenomena as the reported lack of specificity
of the lipophorin that is involved in the transport of cuticular hydrocarbons
from fat body to epidermis (Katase and Chino, 1984).

In contrast, cuticle from *Blaberus* and *Nauphoeta* did not reappear postmolt but was replaced by distorted cuticle of the *Periplaneta* recipient. In
this case, the donor epidermis had been destroyed—a massive accumulation
of hemocytes was observed underneath the graft (Fig. 7.8A)—and recipient
epidermis had grown across the wound, presumably over the accumulated
hemocytic material (Bohn, 1977). The rejection of *Blaberus* and lack of rejection of *Blatta* "skin grafts", even, in the latter case, when multiple simultaneous grafts were made on a recipient (Lackie, 1983b), thus parallels the

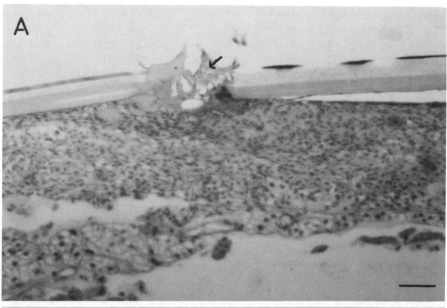

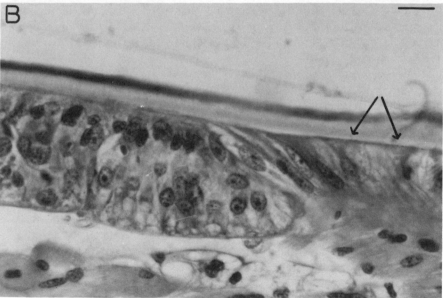

Figure 7.8. (A) Section through *Nauphoeta* cuticle (right) grafted onto *Periplaneta*. There is a small region of coagulated hemolymph at the junction between donor and recipient cuticle (arrow) and a massive accumulation of hemocytes beneath the entire region. (Scale bar = 80 μm.) (B) Postmolt *Blatta* cuticle on *Periplaneta* recipient. The donor epidermis on the left has fused with the recipient epidermis (arrows show region of fusion) and produced new cuticle at ecdysis. No *Periplaneta* hemocytes are seen associated with the *Blatta* epidermis. (Scale bar = 15 μm.)

type of response to tissue implants from these species. However, *Nauphoeta* skin grafts were unequivocally rejected, whereas the tissue implants were not; whether the rejection of skin grafts was due to the difference in mechanical properties (*Nauphoeta* cuticle is more rigid than that of *Periplaneta*) or to the different ways of presenting the "antigen" is discussed in Section 7.2.4.

7.2.3.2. Adults as Recipients. As an assay for immunorecognition, cuticular transplantation onto adult insects poses problems of interpretation, since, as mentioned in Section 7.1, the opacity of the cuticle often makes it impossible to determine whether the underlying epidermis has been destroyed. Thomas and Ratcliffe (1980, 1982) used melanization, cracking of the graft, and, when possible, presence of new recipient cuticle underneath the graft as criteria for assessing rejection. They found that neither autografts nor allografts on the cockroach *Blaberus* and the stick insect *Extatosoma* were rejected, whereas xenografts from a range of donor species were rejected, the mean graft survival times corresponding fairly well with phylogenetic relatedness of the species combination. Thus, *Blaberus* rejected cuticle from *Periplaneta* in approximately 14 days and from the beetle *Tenebrio* in approximately 8 days. Although attempts were made to carry out the reciprocal transfer, using adult *Periplaneta* as recipient, these experiments were unsuccessful because the animals died of septicemia (Thomas and Ratcliffe, 1982).

However, adult *Periplaneta* have been used as recipients by Jones and Bell (1982) and by George and associates (1984) for a different method of assessing graft viability. In both cases, donor integument (3–5 mm side length) was transplanted onto an abdominal tergite and the graft area was excised, fixed, and examined histologically after various intervals; the number of hemocyte nuclei were counted in a 100-μm transect perpendicular to the center of the graft. George and coworkers (1984) compared the response against autografts and against xenografts from *Blaberus giganteus*, both types of graft having been placed on the same recipient, whereas Jones and Bell (1982) quantified the response to autografts, allografts, and a range of xenografts placed on individual recipients.

A remarkable similarity in the number of hemocytes per transect 7–8 days postoperation for autografts (median 54, Jones and Bell; mean 51, George et al.) and for xenografts from *Blaberus* species (median 177, Jones and Bell; mean 172, George et al.) was obtained. A significantly elevated cellular response against *B. giganteus* grafts was maintained for at least 70 days postgrafting, much as would be expected in a typical encapsulation reaction (George et al., 1984). That significantly elevated cellular responses at 8 days postgrafting were found against *Periplaneta brunnei* and *Blatta* (Jones and Bell, 1982) is, in the light of investigations described above (see Sections 7.2.2 and 7.2.3), surprising. Nymphal *Blatta* grafts onto nymphal *Periplaneta americana* were able to molt even at 80 days postgrafting (Lackie, 1983b), and no hemocytes were associated with the epidermis postmolt (Fig. 8A). Large grafts seem to cause greater problems with host survival—as shown by 10% (Jones and Bell, 1982) and 30% mortality (George et al., 1984) of *Periplaneta*—and tend also to have fat body and muscle adhering to them; removal of this "excess tissue" very easily disrupts the integrity of the subepidermal connective tissue layer (A. Lackie and L. Tetley, unpublished) and can lead to wounded epidermis, which acts as a focus for hemocytic aggregation (Wigglesworth, 1979; Cherbas, 1973). Nevertheless, this does not fully explain the difference in *Periplaneta*'s response to *Blatta* found by Lackie (1983b) and Jones and Bell (1982).

7.2.3.3. Cuticle Implantation. A third method of investigating immune responses to cuticular grafts is by heterotopic transplantation: the implantation of pieces of cuticle and epidermis into the abdominal hemocoel of larval insects. This was performed by Piepho (1943) in *Galleria*, followed by Wigglesworth (1954) in *Rhodnius*, and more recently by Riddiford (1976) in *Manduca*, in studies of the influence of molting hormone on cuticular development. This technique produces delightful results, in that when the donor epidermis survives, it grows around the cut edges of the implanted cuticle onto the upper surface of the implant, eventually forming a confluent monolayer completely enclosing the piece of cuticle within an epidermal cyst. At ecdysis the old cuticle is shed into the interior of the cyst and becomes surrounded by newly secreted cuticle.

The fact that the epidermis of the implant can be stimulated to undergo the complex sequence of events associated with ecdysis provides ample evidence that it has not been rejected by the recipient's immune system. Cuticular allografts implanted into tenth-instar *Periplaneta* undergo molting, as do xenografts from *Blatta* (Fig. 7.9B). Implants from *Nauphoeta* and *B. craniifer*, however, are encapsulated by hemocytes and, when the implants are examined histologically after the recipient *Periplaneta* has molted, the epidermal layer is seen to have disappeared (Fig. 7.9A). The disappearance of orthotopic cuticular transplants of *Nauphoeta*, when *Periplaneta* molts, cannot, therefore, be due to mechanical instability of the graft.

7.2.4. The Cockroach Model and What We Can Learn from It

Using *P. americana* and four different methods of assaying for a hemocytic response—tissue implants, cell transfer, and orthotopic and heterotopic cuticular grafting—it has been shown that allogeneic recognition is absent; these results corroborate those of insect physiologists using implantation or cuticular grafting on many other species of insect.

These same four methods indicate that xenografts between *Periplaneta* and *Blatta* are not recognized as foreign; the lack of recognition applies in both directions, and the survival and continued development of transplants must be dependent on both physiologic and immunologic compatibility of

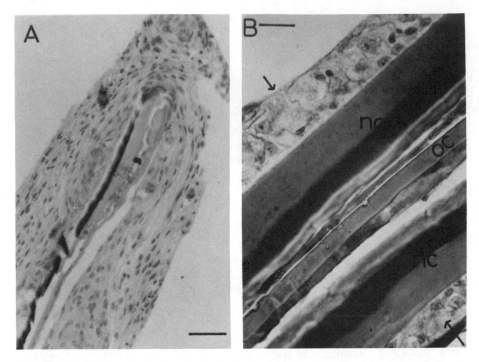

Figure 7.9. Sections through cuticular implants recovered from the recipient *Periplaneta* postmolt. (A) *Nauphoeta* implant. No epidermis is visible, and the remaining piece of cuticle is thickly encapsulated by *Periplaneta* hemocytes. (Scale bar = 50 μm.) (B) Part of a *Blatta* implant that has molted in synchrony with the recipient to form an epidermal cyst. Epidermis (arrow) to which no hemocytes adhere covers the cyst, new cuticle has been formed (nc), and the sheet of old cuticle (oc) has been shed into the center of the cyst. (Scale bar = 25 μm.)

donor and recipient species. Such lack of rejection of transplants from closely related genera or species has been reported for many other insects.

That hemocytes rapidly accumulate around transplants from more distantly related species, such as *Blaberus*, has been shown by several authors using implantation and three different assays for cuticular rejection. However, different assays for recognition of *Nauphoeta* transplants by *Periplaneta* provide different answers; an indeterminate response, either weak or totally lacking, is mounted against implanted tissue, whereas subcuticular epidermis is always destroyed.

It could be argued that *Periplaneta* does not actually reject any tissue transplant but that, because of the alien physiologic environment, foreign cells from more distantly related sources die and hence stimulate the host's hemocytes to carry out their normal "housecleaning" routine. The observed cellular response would thus merely reflect the degree of physiologic unsuitability. Against this argument are two pieces of evidence. First, if the hemocytes responded only to physiologically damaged transplants, then a differential response would be implausible; any weak encapsulation responses, such as those between *Nauphoeta* and *Periplaneta*, would be augmented with time. This is not the case, since the responses to implants remain weak or absent even after 72 hr (Lackie, 1979). Second, the rejection of *Schistocerca* tissue in *Periplaneta* is not reciprocated, and implants from *Periplaneta* and from several other species, including *Tenebrio* and *Ephestia* (see Section 7.2.1.2), can survive unencapsulated within the locust. Indeed, larvae of parasites that are naturally rather host specific, such as *Moniliformis dubius*, which develops naturally within *Periplaneta* (Lackie and Lackie, 1979); of *Hymenolepis diminuta*, which develops naturally within *Tenebrio* (Lethbridge, 1971; Lackie, 1976); and of *Nemeritis*, which develops naturally within *Ephestia* (Salt, 1963), will all grow and develop within *Schistocerca*. The locust might thus, with a few exceptions, be described as the universal recipient; this anomaly is discussed in Section 7.3.

In view of this evidence, and since hemocytes respond rapidly to recognized xenografts (Schmit and Ratcliffe, 1976), it is more likely that we are dealing with a true recognition phenomenon, aimed presumably against the connective tissue (Salt, 1970) that covers all hemocoelic surfaces. Differences in, for example, the proteoglycan composition or the charge (see Section 7.3) of this layer between different species could thus be recognized directly or indirectly by the hemocytes.

It is plausible that the encapsulation of implants is merely an exaggerated wound healing response during which the hemocytes attempt to repair the tissue and make it appear like self (Lackie, 1986c). Whether the acid mucopolysaccharide component within hemocytes (Costin, 1975) is used in the formation of connective tissue is controversial (Ashhurst, 1979, 1982; Wigglesworth, 1979), but the end result of hemocytic encapsulation is that the foreign surface is coated in self material bearing some similarities to connective tissue (Lackie et al., 1985). In the case of those cuticular xenografts against which a response is mounted, the end result is that the recipient's epidermis migrates across the wound, possibly using the hemocytic layer (Williams, 1946; Bodenstein, 1953) or material derived from the hemocytes (Bohn, 1975, 1977) as a guide, and lays down new cuticle of the recipient's type (Williams, 1946; Thomas and Ratcliffe, 1982).

As discussed in Section 7.3, and as has been shown by Jones and Bell (1982) for the hemocytic response to cuticular xenografts and by Lackie (1979; see also Fig. 7.3) for the response to tissue implants, the extent of the hemocytic reaction varies according to the nature of the foreign surface. It can be assumed fairly safely that the weaker the response, the more similar to self—in terms of immunorecognition—the foreign surface is likely to be (Lackie, 1981). It is apparent that, to the immune system, self is not restricted to isogeneic or even allogeneic transplants but is extended to include transplants of closely related xenogeneic origin. In this respect, the range over which self extends is much greater for the immunorecognition system of the locust than for that of the cockroach. The resultant dangers for the individual locust, if not for the species, are obvious. This response is in marked contrast to the xenogeneic transplantation responses in *Periplaneta*, in which the acuity of recognition is much greater, presumably because the thresholds beyond which recognition of foreignness occurs are much closer together, thereby excluding the majority of foreign biotic surfaces.

A range must be delimited at each end by thresholds upon which a xenograft might totter. The lack of conviction of *Periplaneta*'s response to *Nauphoeta* implants might be due to just such an example of weak histoincompatibility (Lackie, 1979, 1981); presentation of foreignness in a different manner, by skin grafting, for example, might be sufficient to bring the incompatibility to the attention of the immune system by pushing it across the threshold toward recognition.

7.2.5. Is Immunization Possible?

Immunization in arthropods could theoretically lead to two different phenomena that need not necessarily be linked. In one case, the immunized organism might produce an enhanced, specific response to a particular antigenic stimulus. In the second case, which applies to a system where histocompatibility apparently exists between a wide range of individuals, it might be possible to teach the recipient to recognize and respond to a previously unrecognized transplant.

There are numerous reports of an enhanced protective response being elicited in insects immunized with bacteria (Chadwick, 1975; Hoffmann, 1980; Boman and Hultmark, 1981; Hultmark et al., 1983) or with soluble microbial products such as endotoxins (Chadwick, 1975), but these have generally resulted from the introduction of a secondary challenge within a matter of hours or a few days of the primary immunizing dose. A protective secondary response against honey bee phospholipase A with kinetics comparable to that of the mammalian immune system can be induced in cockroach hemolymph (Rheins and Karp, 1984; Rheins et al., 1980), suggesting that some form of immunization against soluble molecules is possible. The latter investigation, in which the success of immunization is manifested as either death or survival of the insect, has obvious advantages in comparison with attempts to immunize against tissue transplants, in which measurement of the response is problematical.

The use of hemocytic encapsulation as an assay to search for an enhanced secondary response to implanted tissue is inappropriate because the primary response to a recognized implant occurs so rapidly that any change in rate of encapsulation is unlikely to be measurable. However, it is conceivable that the rate of hemocyte recruitment to the capsule would be accelerated and that a thicker capsule around the implant would result (see Section 7.3). Precisely this type of assay, measuring the rate of accumulation of hemocytes

TABLE 7.2. ATTEMPTED INDUCTION OF RECOGNITION OF
***BLATTA* CUTICULAR GRAFTS BY *PERIPLANETA* USING**
CUTICULAR IMPLANTS AS INDUCER

Experiment Number	Cuticular Implant	Cuticular Graft	Proportion of Grafts that Survive
1	*Periplaneta*	*Blatta*	14/14
	Blatta	*Blatta*	15/15
2	*Periplaneta*	*Blatta*	10/10
	Blatta	*Blatta*	10/10

Source: From A. Lackie and B. Dularay, unpublished.
Note: Proportion of recipient *Periplaneta* on which *Blatta*-type cuticle reappears at the graft-site postmolt.

beneath a primary (George et al., 1984) and a secondary cuticular graft, might indicate whether an accelerated secondary response could occur. Since the degree of incompatibility between donor and recipient must be relatively stable, the postulated enhanced recruitment rate would have to be a function of the number of reactive cells and/or soluble intermediaries, which could be involved in recognition.

The survival time of cuticular grafts, when it can be confidently assigned, provides a useful assay for investigating whether rejection of second-set grafts is enhanced. Using adult insects (see Section 7.2.3.2), Thomas and Ratcliffe (1980) provided evidence of a small but significant decrease in rejection time of *Blaberus* cuticle on *Extatosoma* and of *Schistocerca* cuticle on *Blaberus*. In these experiments, second-set grafts were placed as soon as rejection of the first-set graft was deemed to have occurred, that is, within 1–3 weeks in *Blaberus*. According to George and colleagues (1984), since the hemocytic response to *Blaberus* grafts on adult *Periplaneta* persists for at least 70 day postgrafting, the interpretation of the responses to second-set cuticular grafts poses some problems, although they are surmountable.

Attempts to induce a response to previously unrecognized transplants produce fewer problems in interpretation, since the presence or absence of a hemocytic response need be the only criterion of assessment. Multiple simultaneous grafts of nymphal *Blatta* cuticle onto a nymph of *Periplaneta* fail to stimulate rejection of any of the grafts; new *Blatta*-type cuticle appears at all the graft sites postmolt (Lackie, 1983b). Neither prior implantation of *Blatta* nerve cord at days 9 and 3 pregrafting (Lackie, 1983b) nor the more appropriate implantation of *Blatta* cuticle and epidermis pregrafting is able to prevent the grafted *Blatta* epidermis from producing new cuticle when the recipient *Periplaneta* molts (Table 7.2). The possibility that the immunorecognition system needs to be kicked into activity by an unrelated, and normally recognized, transplant is under investigation.

7.3. THE INFLUENCE OF PHYSICOCHEMICAL PARAMETERS ON IMMUNORECOGNITION

The use of transplanted living tissues to test the limits of immunorecognition has the disadvantage that the properties of their surfaces cannot be easily defined; thus, countless experiments have been carried out in which the hemocytic response to abiotic particles has been investigated. Although many of the objects against which a hemocytic response has been mounted (listed in Salt, 1970) have been chosen for their convenience of handling, some, such as Teflon-coated objects and paraffin wax droplets, have been used to determine whether hemocytes will adhere to nonadhesive or very hydrophobic surfaces (Salt, 1970). With all these particles, irrespective of the insect species tested, hemocyte adhesion and encapsulation has occurred.

Using a more closely defined physicochemical parameter, that of surface charge, Vinson (1974) in caterpillars of *Heliothis* species, Dunphy and Nolan (1980) in caterpillars of *Choristoneura*, and Lackie (1983a) in adult *Periplaneta* and *Schistocerca* have found that differently charged ion-exchange beads are encapsulated to different extents. Negatively charged beads are not encapsulated at all in the caterpillars in vivo, nor do the plasmatocytes of silkworms adhere to them in vitro (Walters and Williams, 1967).

In *Periplaneta* and *Schistocerca*, the system is more complicated, and the magnitude of the in vivo encapsulation response appears to correspond rather well to the adhesive behavior of the hemocytes in vitro. If polystyrene is treated with concentrated sulphuric acid, its wettability and negativity increase (Gingell and Vince, 1982; Lackie, 1983a). This increased negativity is apparently due to the increased density of the hydroxyl groups on the surface, and the possible reasons for the effect of these changes on cell adhesion are discussed elsewhere (Curtis et al., 1983). When polystyrene dishes of a range of negativity and wettability are used as substrata for adhesion by washed hemocytes, it is found that only a low proportion of locust hemocytes adhere, irrespective of the nature of the surface. On the contrary, the proportion of cockroach cells adhering increases markedly as the substratum hydrophobicity decreases and negative charge increases. Within the cockroach hemocoel, polystyrene beads previously acid-treated for 90 sec provoke thicker hemocytic capsules than do beads acid-treated for only 10 sec; in locusts, the majority of beads of either category remains unencapsulated or provokes only patchy capsules. Similarly, negatively charged CM-Sepharose beads remain unencapsulated within *Schistocerca*, even after 9 days. In cockroaches, CM-Sepharose beads are encapsulated more thickly than are neutral Sepharose-6B beads (Fig. 7.10).

There are, thus, interesting differences between the two species, with *Schistocerca* again showing itself to be a poor responder (see Section 7.2.1) and *Periplaneta* showing itself capable of mounting a cellular response to all the differently charged surfaces tested. That *Schistocerca* has not only

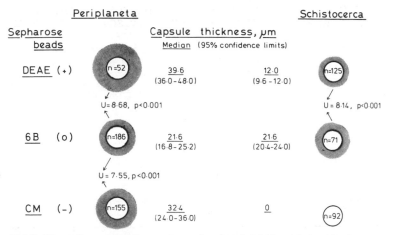

Figure 7.10. The influence of the surface charge of abiotic objects (Sepharose beads) on the extent of encapsulation by hemocytes of *Periplaneta* and *Schistocerca*. Capsule thickness was measured 24 hr postimplantation (From Lackie 1983a.)

a smaller proportion but also a smaller total number of plasmatocytes relative to *Periplaneta* (Lackie et al., 1985) might partly account for the production of thinner capsules than by the cockroach, but this cannot account for the total absence of response to negatively charged surfaces in vivo. These negatively charged surfaces obviously lie below the threshold of recognition and hence are encompassed within the range of foreign surfaces treated as self by the locust's immune system. Since neutral surfaces are encapsulated by *Schistocerca*'s hemocytes, it should be possible, using a range of negatively charged surfaces, to determine where the threshold lies. By injecting agarose beads with various surface densities of carboxyl groups, and thus of different negativities, into locusts and noting the magnitude of the encapsulation response after 24 hr it has been found that the threshold lies remarkably close to neutrality (Lackie, 1986b). Thus, as with the results of transplantation studies (see Section 7.2.1), a low degree of discrimination is found.

The influence of substratum charge on hemocyte adhesion and encapsulation in vivo might be due either to its effect on adsorption of soluble intermediaries whose presence is necessary for cell adhesion to occur or to a direct effect on cell adhesion itself, with cells being attracted or repelled according to their own surface charge (Curtis, 1972).

In support of the latter hypothesis, and on the basis of the differences in their adhesive behavior toward charged substrata, the hemocytes of *Schistocerca* and *Periplaneta* might be expected to have rather different surface charges. This has indeed been found to be the case. Using the technique of cell electrophoresis, it has been shown that hemocytes of *Periplaneta* are less negatively charged than those of *Schistocerca* (Table 7.3). This result has been confirmed by measuring the thickness of the layer of cationized

**TABLE 7.3. ELECTROPHORETIC MOBILITY, ZETA POTENTIAL,
AND THICKNESS OF CATIONIZED FERRITIN LAYER FOR HEMOCYTES
OF *PERIPLANETA* AND *SCHISTOCERCA***

Hemocytes	EPM (TU)[a] (mean ± SD)	Zeta Potential (mV)[b] (mean ± SD)	CF Layer (nm)[c] (mean ± SD)
Live cells			
Periplaneta	7.8 ± 3.6 (n = 470)	12.8 ± 5.8	
Schistocerca	16.3 ± 4.1 (n = 338)	26.6 ± 6.7	
Glutaraldehyde-fixed cells			
Periplaneta	10.0 ± 3.5 (n = 98)	16.30 ± 5.7	22.8 ± 11.8 (n = 464)
Schistocerca	17.6 ± 4.8 (n = 112)	28.6 ± 7.8	47.6 ± 16.3 (n = 414)

Source: Modified from Takle and Lackie, 1985.
Note: Values for *Periplaneta* and *Schistocerca* are significantly different ($P < .001$) in all cases.
[a] Electrophoretic mobility in Tiselius units (TU).
[b] Zeta potential derived from TU by calculation.
[c] Thickness of cationized ferritin (CF) bound by cell surface.

ferritin bound by glutaraldehyde-fixed hemocytes from the two species, the locust cells binding a significantly thicker layer of ferritin than the cockroach cells (Takle and Lackie, 1985). Since most biotic surfaces, whether the connective tissue surrounding tissue transplants or the surfaces of metazoan or protozoan parasites, bear a predominance of negative charges, it might be surmised that the locust hemocytes would be at a disadvantage on the basis of the DLVO theory of electrostatic repulsion and attraction in cell-substratum adhesion (Pethica, 1980). Many of the results discussed in this and the previous section tend to support this view.

One important question that has yet to be answered concerns the assignment of a particular charge to a particular cell type within the hemocyte population; if the coagulocyte is the cell that initiates encapsulation, then the net electrostatic repulsion or attraction between coagulocyte surface and foreign substratum will be the force that plays the most important role in determining whether adhesion and a cellular response subsequently occur.

In spite of the difficulties involved in determining the surface charge of individual hemocyte types by cell electrophoresis, it can be inferred from differential hemocyte counts and from the minimal overlap of electrophoretic mobilities of hemocytes from the two species that the surface charge of coagulocytes from *Periplaneta* and from *Schistocerca* must be very different.

That the relative hydrophobicity of the surfaces of a leucocyte and a foreign particle plays an important role in whether cell-particle interaction occurs, has been shown for mammalian neutrophils and bacteria by several workers (van Oss and Gillman, 1972; Edebo et al., 1980; Magnusson et al.,

TABLE 7.4. ADHESION OF *PERIPLANETA* HEMOCYTES IN VITRO AND IN VIVO TO SUBSTRATA OF DIFFERENT WETTABILITY

	Substratum Wettability (mm)[a]	Hemocyte Adhesion (no./mm^2)	Hemocyte Spreading (μm)	24-Hr Capsule Thickness (μm)[b]
Glass	5.5 ± 0.2 (25)	771 ± 126 (15)	35.6 ± 6.3 (200)	38.6 ± 19.1 (173)
polyHEMA[c]	4.7 ± 0.2 (24)	286 ± 146 (15)	26.2 ± 8.8 (250)	77.3 ± 17.2 (61)
PVA-NaOH[d]	4.0 ± 0.1 (25)	128 ± 78 (16)	21.5 ± 6.4 (200)	75.4 ± 22.5 (46)

Source: From G. Takle, unpublished.

Note: Results expressed as mean ± SD (number of observations).

[a] Diameter of spreading 10-μl water droplet.

[b] Beads injected into *Periplaneta*.

[c] Polyhydroxyethylmethacrylate.

[d] Polyvinylacetate treated with sodium hydroxide for 10 min.

1980). The wettability of a foreign substratum is also an important factor in hemocyte adhesion in vitro (Lackie, 1983a) and in vivo. In the latter case, thick capsules are formed around paraffin wax globules (Salt, 1970), and glass beads coated with hydroxide-treated polyvinyl acetate or poly-HEMA, both substrata presenting very nonwettable surfaces, are more thickly encapsulated than are untreated polyvinyl acetate–coated beads and untreated, uncoated glass beads (Table 7.4).

The relationship between adhesion, electrostatic charge, and hydrophobicity is complex (Gingell and Vince, 1980; Pethica, 1980). Whereas acid-treated polystyrene becomes more hydrophilic and slightly more negatively charged, encouraging greater hemocyte spreading in vitro and encapsulation in vivo, the hydrophobic poly-HEMA or polyvinyl acetate–coated surfaces reduce hemocyte spreading in vitro but also stimulate thicker capsules in vivo. Obviously, then, in the cockroach at least, different factors are involved in recognition and subsequent cellular adhesion. Direct interaction between cell and substratum is one possibility, as explained previously, but it might also be expected that molecules from the plasma will be nonspecifically adsorbed onto an abiotic surface and as a result may undergo steric alteration; denaturation or activation in contact with hydrophobic or charged surfaces may occur. The predominant charge of a foreign surface need not, of course, be completely masked by this interaction, but it is also possible that the circulating hemocytes could respond either nonspecifically to the altered charge of the self molecules or specifically through membrane-bound receptors to newly exposed molecular configurations.

Valuable information about the limits of immunorecognition can thus be gained with the more closely defined physicochemical properties of abiotic surfaces using *Periplaneta* and *Schistocerca* as model systems. As with the cellular responses to tissue implants, dramatic differences are found between the two species in the acuity of recognition and the magnitude of the cellular response. Such differences in magnitude may be related to the number and

different proportions of hemocyte types in the two species, but the differences in acuity may also be related to the interspecific differences in overall
surface charge of the hemocyte populations. With respect to the parameter
of negative surface charge, the thresholds on each side of which immunorecognition occurs in *Schistocerca* are again set so that a comparatively large
range of foreignness is enclosed and hence considered self.

7.4. SUMMARY

The efficacy of the recognition arm of the insect immune response can be
measured in a variety of ways; the results of all the methods used show that
recognition of transplanted, intact, unmodified allogeneic tissue is absent.
Whether intact xenografted tissue is recognized depends to a large extent
on the phylogenetic relatedness of recipient and donor species but also depends on the discriminatory ability of the recipient species itself: reciprocal
transplants between two species may not necessarily provoke the same degree of recognition and response.

The surfaces of tissues and abiotic objects that are not recognized as
foreign must fall within the range of properties that are treated by the recognition system as intact self and that therefore must not provoke a response.
For xenografts near the limits of this range, the way in which the "antigen"
is presented may determine on which side of the limit it will fall, and thus
whether or not it can be recognized. Immunization of a recipient against a
normally unrecognized xenograft cannot be stimulated by prior implantation
of the homologous tissue, but the possibility that recognition can be stimulated by implantation of a heterologous xenograft, which normally provokes
a strong response, requires investigation.

All tissue transplants are covered in a negatively charged layer of connective tissue containing proteoglycans. Recognition of a transplant could,
therefore, be due to specific recognition of molecular components, such as
carbohydrate moieties. Examination of the hemocytic response to abiotic
surfaces of known physicochemical properties suggests that the charge and
wettability of a foreign surface, and perhaps also of the hemocyte surface,
may play important roles in whether recognition and hemocyte adhesion
occur.

ACKNOWLEDGMENTS

I am very grateful to Bubbly Dularay and Garry Takle for permission to use
unpublished data, to Vernon French for Figure 7.7A and for helpful comments, to Stefan Gunnarsson for Figure 7.4, to Sandra Jones for supplying
some extra information, and to William J. Bell and Alan Liss Publishing Co.
for permission to use Figure 7.2. George Salt's published works continue to

be an inspiration, and I am very grateful to him and to the Royal Society for permission to use Figure 7.6A.

Much of my recent work on tissue transplantation has been supported by grants GRC01351 and GRC87959 from the Science and Engineering Research Council.

REFERENCES

Ashhurst, D. 1979. Hemocytes and connective tissue, pp. 319–330. In A.P. Gupta (ed.) *Insect Hemocytes*. Cambridge University Press, Cambridge.

Ashhurst, D.E. 1982. The structure and development of insect connective tissues, pp. 31–50. In R.C. King and H. Akai (eds.) *Insect Ultrastructure*, vol. 1. Plenum Press, New York.

Ashida, M., Y. Ishizaki, and H. Iwahana. 1983. Activation of prophenoloxidase by bacterial cell walls or β-1,3-glucans in plasma of the silkworm, *Bombyx mori*. *Biochem. Biophys. Res. Commun.* **113:** 562–568.

Bart, A., 1974. Sur la détermination de l'épiderme de la patte chez les Phasmes. *C. R. Séances Acad. Sci.* (III) **279**D: 1293–1296.

Bell, W.J. 1972. Yolk formation by transplanted cockroach oocytes. *J. Exp. Zool.* **181:** 41–48.

Bettini, S., D.S. Sarkaria, and R.L. Patton. 1951. Observations on the fate of vertebrate erythrocytes and hemoglobin injected into the blood of the American cockroach, *Periplaneta americana. Science* (Washington, DC) **113:** 9–10.

Bodenstein, D. 1953. Studies on humoral mechanisms in the growth and metamorphosis of the cockroach *Periplaneta americana. J. Exp. Zool.* **123:** 189–232.

Bohn, H. 1972. The origin of the epidermis in supernumerary regenerates of triple legs in cockroaches (Blattaria). *J. Embryol. Exp. Morphol.* **28:** 185–208.

Bohn, H. 1975. Growth-promoting effects of haemocytes on insect epidermis *in vitro. J. Insect Physiol.* **21:** 1283–1293.

Bohn, H. 1977. Enzymatic and immunological characterisation of the conditioning factor for epidermal outgrowth in the cockroach *Leucophaea maderae. J. Insect Physiol.* **23:** 1063–1073.

Boman, H.G. and D. Hultmark. 1981. Cell-free immunity in insects. *Trends Biochem. Sci.* **6:** 306–309.

Carton, Y. 1976. Isogenic, allogenic and xenogenic transplants in an insect species. *Transplantation* **21:** 17–22.

Chadwick, J.S. 1975. Hemolymph changes with infections or induced immunity in insects and ticks, pp. 241–71. In K. Maramorosch and R.E. Shope (eds.) *Invertebrate Immunity*. Academic Press, New York.

Cherbas, L. 1973. The induction of an injury reaction in cultured haemocytes from saturniid pupae. *J. Insect Physiol.* **19:** 2011–2023.

Clem, L.W., K. Clem, and L. McCumber. 1984. Recognition of xenogeneic proteins by the blue crab. *Dev. Comp. Immunol.* **8:** 31–40.

Cohen, N. and M. Borysenko. 1970. Acute and chronic graft rejection and possible phylogeny of transplantation antigens. *Transplant. Proc.* **2:** 333–336.

Costin, N.M. 1975. Histochemical observations of the haemocytes of *Locusta migratoria. Histochem. J.* **7:** 21–43.

Curtis, A.S.G. 1972. Adhesive interactions between organisms, pp. 1–18. In A.E.R. Taylor and R. Muller (eds.) *Functional Aspects of Parasite Surfaces,* BSP Symposium, vol. 10. Blackwell, London.

Curtis, A.S.G., J.V. Forrester, C. McInnes, and F. Lawrie. 1983. Adhesion of cells to foreign surfaces. *J. Cell Biol.* **97:** 1500–1506.

Dawe, C.J., W.D. Morgan, and M.S. Slatick. 1967. Cellular response of a cockroach, *Leucophaea maderae,* to transplants of cell culture lines of vertebrates. *Fed. Proc.* **26:** 1698–1702.

Dunphy, G.B. and R.A. Nolan. 1982. Cellular immune responses of spruce budworm larvae to *Entomophthora egressa* protoplasts and other test particles. *J. Invertebr. Pathol.* **39:** 81–92.

Edebo, L., E. Kihlstrom, K-E. Magnusson, and O. Stendahl. 1980. The hydrophobic effect and charge effects in the adhesion of enterobacteria to animal cell surfaces and the influences of antibodies of different immunoglobulin classes, pp. 65–101. In A.S.G. Curtis and J.D. Pitts (eds.) *Cell Adhesion and Motility,* BSCB Symposium, vol. 3. Cambridge University Press, Cambridge.

Ephrussi, B. and G.W. Beadle. 1936. A technique for transplantation for *Drosophila. Am. Nat.* **70:** 218–225.

French, V. 1981. Pattern regulation and regeneration. *Philos. Trans. R. Soc. Lond.* (Biol.) **295:** 601–617.

George, J.F., R.D. Karp, and L.A. Rheins. 1984. Primary integumentary xenograft reactivity in the American cockroach, *Periplaneta americana. Transplantation* **37:** 478–484.

Gingell, D. and S. Vince. 1980. Long-range forces and adhesion: An analysis of cell-substratum studies, pp. 1–37. In A.S.G. Curtis and J.D. Pitts (eds.) *Cell Adhesion and Motility,* BSCB Symposium, vol. 3. Cambridge University Press, Cambridge.

Gingell, D. and S.A. Vince. 1982. Substratum wettability and charge influence the spreading of *Dictyostelium* amoebae and the formation of ultrathin cytoplasmic lamellae. *J. Cell Sci.* **54:** 255–285.

Gunnarsson, S. and A.M. Lackie. 1985. Haemocytic aggregation in the locust *Schistocerca gregaria* and the cockroach *Periplaneta americana* in response to injected molecules of microbial origin. *J. Invertebr. Pathol.,* **46:** 312–319.

Guthrie, D.M. 1966. Physiological competition between host and implanted ganglia in an insect (*Periplaneta americana*). *Nature* (London) **210:** 312–313.

Hadorn, E. and D. Buck. 1962. Über Entwicklungsleistungen transplantierter Teilstücke von Flügel-Imaginalscheiken von *Drosophila melanogaster. Rev. Suisse Zool.* **69:** 302–310.

Hoffmann, D. 1980. Induction of antibacterial activity in the blood of the migratory locust *Locusta migratoria. J. Insect Physiol.* **26:** 539–549.

Hultmark, D., A. Engstrom, K. Andersson, H. Steiner, H. Bennich, and H.G. Boman. 1983. Insect immunity: Attacins, a family of antibacterial proteins from *Hyalophora cecropia. EMBO J.* **2:** 571–576.

Ingram, G.A.. J. East, and D.H. Molyneux. 1984. Agglutinins of *Trypanosoma, Leishmania* and *Crithidia* in insect haemolymph. *Dev. Comp. Immunol.* **6**: 35–42.

Jones, S.E. and W.J. Bell. 1982. Cell-mediated immune-type response of the American cockroach. *Dev. Comp. Immunol.* **6**: 35–42.

Kambysellis, M.P. 1970. Compatibility in insect tissue transplantations. *J. Exp. Zool.* **175**: 169–180.

Katase, H. and H. Chino. 1984. Transport of hydrocarbons by haemolymph lipophorin in *Locusta migratoria. Insect Biochem.* **14**: 1–6.

Lackie, A.M. 1976. Evasion of the haemocytic defence reaction of insects by larvae of *Hymenolepis diminuta* (Cestoda). *Parasitology* **73**: 97–104.

Lackie, A.M. 1979. Cellular recognition of foreignness in 2 insect species, the American cockroach and the desert locust. *Immunology* **36**: 909–1007.

Lackie, A.M. 1981. Immune recognition in insects. *Dev. Comp. Immunol.* **5**: 191–204.

Lackie, A.M. 1983a. Effect of substratum wettability and charge on adhesion *in vitro* and encapsulation *in vivo* by insect haemocytes. *J. Cell Sci.* **63**: 181–190.

Lackie, A.M. 1983b. Immunological recognition of cuticular transplants in insects. *Dev. Comp. Immunol.* **7**: 41–50.

Lackie, A.M. 1986a. Haemolymph transfer as an assay for immunorecognition in insects. *Transplantation,* **41**: 360–363.

Lackie, A.M. 1986b. The role of substratum surface-charge in adhesion and encapsulation by locust haemocytes *in vivo. J. Invertebr. Pathol.,* **47**: 377–378.

Lackie, A.M. 1986c. Transplantation immunity in arthropods: Is it merely wound-healing? pp. 125–138 In M. Bréhélin, (ed.) *Invertebrate Immune Mechanisms,* Springer-Verlag, New York.

Lackie, A.M. and J.M. Lackie. 1979. Evasion of the insect immune response by *Moniliformis dubius* (Acanthocephala): Further observations on the origin of the envelope. *Parasitology* **79**: 297–303.

Lackie, A.M., G.B. Takle, and L. Tetley. 1985. Haemocytic encapsulation in the locust *Schistocerca gregaria* and the cockroach *Periplaneta americana. Cell Tissue Res.* **240**: 343–351.

Lackie, J.M. 1975. Host-specificity of *Moniliformis dubius,* a parasite of cockroaches. *Int. J. Parasitol.* **5**: 301–309.

Larsen, J.R. and D. Bodenstein. 1959. Humoral control of egg maturation in mosquitoes. *J. Exp. Zool.* **140**: 343–382.

Lethbridge, R.C. 1971. The locust as an intermediate host for *Hymenolepis diminuta. J. Parasitol.* **57**: 445–446.

Locke, M. 1974. Structure and formation of the integument in insects, pp. 129–21. In M. Rockstein (ed.) *Physiology of the Insecta,* vol. 6. Academic Press, New York.

Magnusson, K.E., J. Davies, T. Grundstrom, E. Kihlstrom, and S. Normark. 1980. Surface charge and hydrophobicity of *Salmonella, E. coli* and *Gonococci* in relation to their tendency to associate with animal cells. *Scand. J. Infect. Dis.* **24**: 135–140, supplement.

Peterson, L. 1968. Cellular immune responses of insects to foreign tissue implants. Doctoral dissertation, University of Illinois, Urbana, Illinois.

Pethica, B.A. 1980. Microbial and cell adhesion, pp. 19–45. In R.C.W. Berkeley, J.M. Lynch, J. Melling, P.R. Rutter, and B. Vincent (eds.) *Microbial Adhesion to Surfaces*. Ellis Horwood, London.

Piepho, H. 1943. Wirkstoffe in der Metamorphose von Schmetterlingen und anderen Insekten. *Naturwissenschaften* **31:** 329–335.

Piepho, H. 1950. Über das Ausmass der Arlungspezifizitat von Metamorphoshormone bei Insekten. *Biol. Zentralbl.* **69:** 1–10.

Ratcliffe, N.A. and S.J. Gagen. 1976. Cellular defence reactions of insect haemocytes *in vivo*: Nodule formation and development in *Galleria mellonella* and *Pieris brassicae* larvae. *J. Invertebr. Path.* **28:** 373–383.

Rheins, L.A. and R.D. Karp. 1984. The humoral response in the American cockroach, *Periplaneta americana*: Reactivity to a defined antigen from honeybee venom, phospholipase A$_2$. *Dev. Comp. Immunol.* **8:** 791–802.

Rheins, L.A., R.D. Karp, and A. Butz. 1980. Induction of specific humoral immunity to soluble proteins in the American cockroach, *Periplaneta americana*. 1. Nature of the primary response. *Dev. Comp. Immunol.* **4:** 447–458.

Riddiford, L.M. 1976. Hormonal control of insect epidermal cell commitment *in vitro*. *Nature* (London) **259:** 115–117.

Rizki, R.M. and T.M. Rizki. 1980. Haemocyte responses to implanted tissues in *Drosophila melanogaster* larvae. *Wilhelm Roux's Arch. Dev. Biol.* **189:** 207–213.

Rotheram, S. and J.M. Lackie. 1975. The ultrastructure of the surface of *Diplocystis schneideri* Kunstler (Sporozoa; Eugregarinida), a parasite of *Periplaneta americana*. *Parasitology* **70:** 385–388.

Rowley, A.F. and N.A. Ratcliffe. 1981. Insects, pp. 471–90. In N.A. Ratcliffe and A.F. Rowley (eds.) *Invertebrate Blood Cells*, vol. 2. Academic Press, New York.

Ryan, M. and W.L. Nicholas. 1972. The reaction of the cockroach *Periplaneta americana* to the injection of foreign particulate material. *J. Invertebr. Pathol.* **19:** 299–307.

Sahota, T.S. and J.S. Edwards. 1969. Development of grafted supernumerary legs in the house-cricket *Acheta domesticus*. *J. Insect Physiol.* **15:** 1367–1373.

Salt, G. 1960. Experimental studies in insect parasitism. XI. The haemocytic reaction of a caterpillar under varied conditions. *Proc. R. Soc. Lond.* (Biol.) **151:** 446–457.

Salt, G. 1961. The haemocytic reaction of insects to foreign bodies, pp. 175–192. In J.A. Ramsay and V.B. Wigglesworth (eds.) *The Cell and the Organism*. Cambridge University Press, Cambridge.

Salt, G. 1963. Experimental studies in insect parasitism. XII. The reactions of 6 exopterygote insects to an alien parasite. *J. Insect Physiol.* **9:** 647–669.

Salt, G. 1970. *The Cellular Defence Reactions of Insects*. Cambridge University Press, Cambridge.

Salt, G. 1980. A note on the resistance of 2 parasitoids to the defence reactions of their insect hosts. *Proc. R. Soc. Lond.* (Biol.) **207:** 351–353.

Schmit, A.R. and N.A. Ratcliffe. 1976. The encapsulation of foreign tissue implants in *Galleria mellonella* larvae. *J. Insect Physiol.* **23:** 175–180.

Schubiger, G. 1971. Regeneration, duplication and transdetermination in fragments of the leg disc of *Drosophila melanogaster. Dev. Biol.* **26:** 277–295.

Scott, M. 1971. Recognition of foreignness in invertebrates: Transplantation studies using the American cockroach, *Periplaneta americana. Transplantation* **11:** 78–85.

Steinberg, D.M. 1959. Regeneration of homografted and heterografted limbs in the stick insects (Phasmodea). *Dokl. Akad. Sci. U.S.S.R. Biol. Sci.* **129:** 1001–1003.

Takle, G.B. and A.M. Lackie. 1985. Surface-charge of insect haemocytes measured by cell electrophoresis and cationised ferritin-binding. *J. Cell Sci.* **75:** 207–214.

Thomas, I.G. and N.A. Ratcliffe. 1980. Studies on recognition of foreignness in insects utilising integumental transplants, pp. 105–110. In J.B. Solomon (ed.) *Aspects of Developmental and Comparative Immunology*. Pergamon Press, Oxford.

Thomas, I.G. and N.A. Ratcliffe. 1982. Integumental grafting and immunorecognition in insects. *Dev. Comp. Immunol.* **6:** 643–654.

Tsang, K.R. and M.A. Brooks. 1980. Dose response of insects to malignant transformed cells. *In Vitro* **16:** 469–474.

Tsang, K.R. and M.A. Brooks. 1983. Species specificity of immunity to malignant transformed cultured insect cells. *Dev. Comp. Immunol.* **7:** 241–252.

Van Oss, C.J. and C.F. Gillman. 1972. Phagocytosis as a surface phenomenon. *J. Reticuloendothel. Soc.* **12:** 283–291.

Vinson, S.B. 1974. The role of the foreign surface and female parasitoid secretions on the immune response of an insect. *Parasitology* **79:** 297–306.

Walters, D.R. and C.M. Williams. 1967. Reaggregation of insect cells as studied by a new method of tissue and organ culture. *Science* (Washington, DC) **154:** 516–518.

Wigglesworth, V.B. 1937. The function of the corpus allatum in the growth and reproduction of *Rhodnius prolixus* (Hemiptera). *Q. J. Microsc. Soc.* **79:** 91–121.

Wigglesworth, V.B. 1954. *The Physiology of Insect Metamorphosis*. Cambridge University Press, Cambridge.

Wigglesworth, V.B. 1979. Hemocytes and growth in insects, pp. 303–318. In A.P. Gupta (ed.) *Insect Hemocytes*. Cambridge University Press, Cambridge.

Williams, C.M. 1946. Physiology of insect diapause: The role of the brain in the production and termination of pupal dormancy in the giant silkworm *Platysamia cecropia. Biol. Bull.* (Woods Hole) **90:** 234–243.

PART II
HUMORAL IMMUNITY

CHAPTER 8

Biochemistry of Arthropod Agglutinins

Kenneth D. Hapner
Department of Chemistry
Montana State University
Bozeman, Montana

Mark R. Stebbins
Genetic Systems Corporation
Seattle, Washington

8.1. INTRODUCTION

Agglutinins are naturally occurring proteins or glycoproteins that possess multiple carbohydrate binding sites. The agglutinins bind with various affinity and specificity, and without enzymatic modification, to cells and gly-coconjugates (glycoproteins and glycolipids), causing their respective agglutination or precipitation. The term *agglutinin* as used in this chapter to describe agglutinating activity present in arthropods is synonymous with the term *lectin* as defined by Goldstein and colleagues (1980). Carbohydrate specificity of agglutinins is normally functionally defined by the chemical structure of monosaccharides, oligosaccharides, and glycoproteins that inhibit the interaction of the agglutinin with cells and/or glycoconjugates. An agglutinin may be multispecific and agglutinate numerous types of erythrocytes and other cells, or it may be monospecific and interact with a single type of glycoconjugate. The term *heteroagglutinin* refers to structurally different agglutinins found in the same organism. Thus, *multispecificity* applies to agglutinins of functional heterogeneity, and *heteroagglutinin* implies structural heterogeneity. Heteroagglutinins may exhibit similar or disparate carbohydrate binding properties.

The presence of agglutinins is usually detected by hemagglutination assay. In this assay, vertebrate erythrocytes are added to an experimental sample serially diluted with buffer in a plastic microtiter dish. Subsequent aggregation or hemagglutination of the erythrocytes signifies the presence of agglutinin. The greatest dilution of the original sample without loss of visible hemagglutination is considered the end point. The mathematic reciprocal of the highest dilution causing agglutination is the hemagglutination titer. Hemagglutination titer value is consequently proportional to the amount of functional agglutinin present in the original sample. Hemagglutination titers vary, depending on the carbohydrate binding preference of the agglutinin and on the type of erythrocyte used in the assay. Some erythrocytes are strongly agglutinated, whereas others may not be affected by a particular agglutinin. This phenomenon is interpreted to mean that susceptible erythrocytes contain outer membrane glycoconjugates that are bound and cross-linked by the multivalent agglutinin, resulting in hemagglutination. Several types of erythrocytes are generally used in screening for agglutinin activity in order to ensure its detection. Accordingly, in describing characteristics of an agglutinin, the type of erythrocyte involved in the assay should be noted. Erythrocytes are sometimes modified by chemical (formaldehyde, glutaraldehyde) or enzymatic (trypsin, pronase, neuraminidase) procedures that alter the chemical nature of surface carbohydrate structures and their accessibility. These treatments may change titer values and/or specificity of erythrocyte agglutination and thereby serve to explore the carbohydrate binding properties of the relevant agglutinin.

It should be noted that the standard hemagglutination assay is just a simple and useful laboratory tool for the quick detection of agglutinins. It is not necessarily related to their potential in vivo roles. Hence, carbohydrate binding "specificity" of an agglutinin, as judged with relatively simple saccharides and model cell surfaces of unknown relevance to the organisms' own cells and tissues, must be considered circumstantial until in vivo agglutinin receptors are identified and characterized. Also, substances other than protein agglutinins may cause agglutination of erythrocytes (Tsivion and Sharon, 1981; Ravindranath and Cooper, 1984). Descriptions of agglutinins should contain indications of their protein-lectin nature, such as nondialysability, precipitation by ammonium sulfate, heat and protease lability, inhibition by carbohydrates, and so on. Ideally, the agglutinin should be isolated and purified to molecular homogeneity and its physicochemical and carbohydrate binding characteristics firmly established.

Cell agglutination specificity and agglutination inhibition profiles are dependent on both erythrocyte type and agglutinin purity. Inhibition studies with unpurified agglutinins present in hemolymph can quickly become complex and difficult to interpret in terms of the activity of individual agglutinins. Cross-absorption experiments that attempt to selectively remove certain agglutinins in order to expose the activity of others are minimally successful at clearly distinguishing between heteroagglutinins and a single multispecific agglutinin. To best establish the binding and inhibitory characteristics of an agglutinin, it is necessary to examine the molecule in the purified state. It should be realized, however, that in vivo agglutinin function may be influenced by other hemolymphatic components.

8.2. SOURCE AND OCCURRENCE

8.2.1. Hemolymph and Other Tissues

Agglutinins are ubiquitously present throughout the Arthropoda (Gold and Balding, 1975). Most arthropodan agglutinins that have been described are derived from the hemolymph of adult organisms (Yeaton, 1981b; Vasta and Marchalonis, 1983). The reason for this probably involves the convenience of sample collection, relatively large sample size, and associated facile isolation procedures. A few agglutinins are known from the hemolymph of nonadult developmental stages and from various arthropodan tissues, membranes, and cell types. For example, hemolymphatic agglutinin from *Homarus americanus* is strongly associated with hemocyte extracts, suggesting that hemocytes are the source of the agglutinin (Cornick and Stewart, 1973). Differences in hemagglutination titer among individual lobsters cannot be correlated with changes in population of any one type of circulating hemocyte (Cornick and Stewart, 1978). This indicates that hemocytes may generally act as repositories of agglutinin, some of which escapes or is secreted into the hemolymph. Smaller amounts of agglutinin activity, presumably the same as that found in the hemolymph, are associated with extracts of several lobster tissues, including lymphatic gland, pericardium, testis, ovary, hepatopancreas, and kidney (Hall and Rowlands, 1974a). Because of possible preparative contamination and the nonspecific nature of the hemagglutination assay, however, one must be cautious in concluding that the same agglutinin exists in both the hemolymph and the tissues.

One example exists of a purified membrane-associated agglutinin. The agglutinin present in the epidermal cell membranes of newly apolyzed pupal wings of the large white cabbage butterfly, *Pieris brassicae*, has been described by Mauchamp (1982). It is interesting to note that no hemagglutination activity is detected in the hemolymph or hemocyte extracts from the pharate adults. Cockroaches contain agglutinin (in addition to hemolymphatic agglutinin) activity associated with the membrane fraction of homogenized coxal depressor muscles (Denburg, 1980). Agglutinins are present in larval hemolymph of the beetles *Allomyrina dichotoma* (Umetzu et al., 1984) and *Leptinotarsa decemlineata* (Stynen et al., 1982). The flesh fly, *Sarcophaga peregrina*, contains agglutinin in both larval and pupal hemolymph and fat body, but it is absent in adults (Komano et al., 1980, 1983). Larval and adult hemolymph from the silk moth, *Bombyx mori*, contains agglutinin, whereas the pupal stage lacks activity (Suzuki and Natori, 1983). Nymphal instars of grasshoppers contain hemolymphatic agglutinin activity, as do the adults (Jurenka et al., 1982). An agglutinin is present in the perivitelline fluid (embryonic hemolymph) of the horseshoe crab *Tachypleus gigas* (Shishikura and Sekiguchi, 1984a).

8.2.2. Developmental Aspects

Since most arthropodan agglutinins that have been isolated and studied are derived from a single (adult) metamorphic form, relatively little is known concerning their developmental characteristics. Both sexes contain agglutinins, but the concentration may be dependent on age and metamorphic stage. Examples of agglutinins are present in various developmental stages, including eggs, larvae, pupae, and adults. An organism may contain heteroagglutinins that are associated differently with various developmental forms. For example, the Colorado potato beetle, *L. decemlineata*, has two agglutinins, one present in all developmental stages, including the egg, and the other restricted to the larval and pupal forms (Stynen et al., 1982). A possibly related circumstance is the fact that embryonic agglutinin from *T. gigas* agglutinates human erythrocytes more strongly than those from the horse, whereas the adult agglutinin shows the opposite preference (Shishikura and Sekiguchi, 1984a).

The most thoroughly examined agglutinins from the developmental point of view are those present in the flesh fly, *S. peregrina*, and in the silk moth, *B. mori*. Figure 8.1 shows the relationship between agglutinin activity and metamorphosis in these organisms. Larval hemolymph from *S. peregrina* normally contains very low amounts of active agglutinin, which is induced to higher levels in response to injury of the larval body wall or upon pupation (Fig. 8.1A; Komano et al., 1980). The increased agglutinin activity is sustained throughout the pupal stage but declines to undetectable levels in the newly emerged adult (Komano et al., 1983). At the molecular level, the active agglutinin is generated through proteolytic processing in the fat body of a

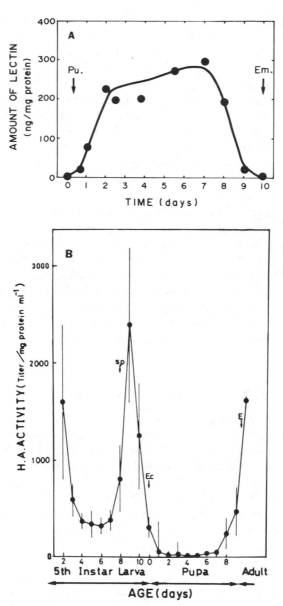

Figure 8.1. Relationship between agglutinin activity and metamorphic stage in two different holometabolous insects. (A) *Sarcophaga peregrina* shows agglutinin activity throughout the pupal (Pu.) stage, which declines upon adult emergence (Em.). (From Komano et al., 1983.) (B) *Bombyx mori* agglutinin is present in larval and adult stages and transiently increases at the onset of pupae formation (Sp). The agglutinin activity declines at larval-pupal ecdysis (Ec) and reappears at emergence of the adult (E). (From Suzuki and Natori, 1983.)

preexisting inactive form of the agglutinin (Komano et al., 1981). After secretion from the fat body, the active molecule is sequestered by the hemocytes, where it may play a role in immune surveillance and stimulate hemocytic phagocytosis of foreign substances and histolytic debris encountered during metamorphosis.

   B. mori also shows a larval agglutinin whose presence is related to pupation (Suzuki and Natori, 1983) but is in a way quite different from that in *S. peregrina*. In *B. mori*, the active agglutinin is normally present in the larval hemolymph (Fig. 8.1B). It declines in the last larval instar and then transiently increases at the onset of puparium formation to again decline to undetectable levels in the pupa. Agglutinin activity then increases in the emerging adult. Thus, in *S. peregrina* agglutinin activity is sustained throughout pupation, whereas in *B. mori* only a transient increase in activity is seen. These agglutinins are perhaps related to pupation in holometabolous insects, but details vary in the two organisms. The *B. mori* agglutinin may be absorbed to histolytic debris or newly developing tissues, which may account for its apparent absence in the pupae. Another difference from *S. peregrina* is that the *B. mori* agglutinin cannot be induced by injury. The two agglutinins are different structurally, developmentally, antigenically, and (presumably) functionally. The suggestion may be made that a humoral agglutinin is essential in metamorphosis of holometabolous insects, but more information is needed to clarify the relationship.

8.2.3. Tissue Localization and Biosynthesis

Several efforts have been made to determine the tissue location and site of biosynthesis of arthropodan agglutinins. As noted in Section 8.2.2., the agglutinin present in *S. peregrina* larvae is synthesized in the fat body and secreted into the hemolymph upon injury to the larval body wall and during pupation (Komano et al., 1983). The agglutinin is then apparently sequestered or recruited by the hemocytes. The hemocytic affinity for the agglutinin increases after injury to the body wall owing to increased availability of binding sites on the hemocyte surface. It is unknown whether certain hemocyte types are involved in sequestration of the agglutinin. It is also uncertain whether the active agglutinin is synthesized de novo by the fat body or whether preexisting inactive agglutinin is withdrawn from the hemolymph, processed to the active form by the fat body, and returned to the hemolymph. Radiolabel incorporation studies with isolated fat body fail to detect newly synthesized hemagglutinin (Takahashi et al., 1984).

   Electron microscopic analysis with colloidal gold-labeled glycoproteins shows the presence of mannose binding agglutinin on the peritrophic membrane of the blowfly, *Calliphora erythrocephala* (Peters et al., 1983). The agglutinin molecules are located exclusively on the lumen side of the membrane, where they presumably participate in adherence interactions with surface mannose residues of mutualistic bacteria.

Two hemocyte types in the cockroach *Leucophaea maderae* contain heteroagglutinins (Amirante, 1976; Amirante and Mazzalai, 1978). One heteroagglutinin is associated with the cytoplasmic contents of the granulocytes and spherulocytes, and the other is localized both on the plasma membrane and generally throughout the two cell types. Disappearance and reappearance of the agglutinins in the respective presence and absence of a protein-synthesis inhibitor indicates that these two types of hemocytes are the synthetic sites of the heteroagglutinins, whose existence in the hemolymph is presumably due to secretion from the hemocytes. A similar approach by specific immunologic labeling procedures with hemocytes from the crustacean *Squilla mantis* has shown that the plasma membrane of a single cell type, the granulocyte, contains agglutinin (Amirate and Basso, 1984). The site of synthesis of the hemolymphatic agglutinin in *S. mantis* was not investigated (see also Chapter 13).

8.3. STRUCTURAL CHARACTERISTICS

8.3.1. Amino Acid Sequence

Little information concerning the primary structure of arthropodan agglutinins is available. The sequence of the amino terminal region of the *Limulus polyphemus* agglutinin, limulin, through approximately 50 amino acid residues is known (Kaplan et al., 1977; Kehoe et al., 1979). The protein has a single, unblocked amino terminal residue (leucine) and shows no heterogeneity through the sequenced region of the molecule. Molecular heterogeneity, often observed during the isolation of limulin, must be due to variable carbohydrate content or undetermined sequence heterogeneity elsewhere in the structure. Isolation and partial sequence analysis of cyanogen bromide fragments show that the single disulfide bond in limulin links half-cystine 38 with a positionally undetermined half-cystine residue in the carboxyl region of the molecule. The structure shows no sequence homology with vertebrate immunoglobulins or with influenza viral hemagglutinin. The general absence of antigenic similarities among arthropodan agglutinins suggests that little or no structural homology is present. The amino terminal tripeptide sequences of two heteroagglutinins from *Tachypleus tridentatus* are different from the limulin sequence (Shimizu et al., 1979). For additional details concerning the primary structure of limulin, see Chapter 12.

8.3.2. Amino Acid Composition

The amino acid compositions of 13 purified agglutinins from 7 arthropodan species are compiled in Tables 8.1 and 8.2. These agglutinins are distributed among the Crustacea (barnacles), Insecta (grasshoppers), and Aquatic Chelicerata (Merostomata, horseshoe crabs). The proteins have in common high

TABLE 8.1. AMINO ACID COMPOSITION[a] OF PURIFIED ARTHROPOD AGGLUTININS

Amino Acid	Balanus[b] balanoides	Megabalanus rosa[c] BRA-1	Megabalanus rosa[c] BRA-2	Megabalanus rosa[c] BRA-3	Malanoplus[d] sanguinipes	Melanoplus[d] differentialis
Aspartic	7.0	14.6	14.8	20.6	12.1	11.6
Threonine	4.0	5.4	5.8	7.6	7.2	6.9
Serine	19.9	8.2	9.2	6.3	6.2	5.8
Glutamic	18.6	14.3	14.9	15.0	12.4	12.2
Proline	1.9	5.1	5.1	3.9	3.3	6.3
Glycine	16.9	4.6	4.1	9.3	6.8	8.8
Alanine	7.7	5.6	9.0	3.0	8.1	8.7
Cystine/2	0	2.0	0.7	6.2	2.3	2.3
Valine	3.7	7.4	6.8	0.7	6.0	5.6
Methionine	0.8	1.4	1.4	0.3	1.6	1.1
Isoleucine	2.1	2.3	2.3	5.8	4.6	4.3
Leucine	3.2	7.0	7.1	7.0	7.4	7.6
Tyrosine	1.7	3.3	3.7	4.9	3.6	3.7
Phenylalanine	1.9	2.6	2.7	2.7	3.7	3.9
Lysine	8.2	3.6	1.4	1.9	4.1	4.0
Histidine	2.2	5.4	5.5	2.9	2.6	2.9
Arginine	0	4.9	5.6	2.0	3.9	4.3
Tryptophan	0.3	ND	ND	ND	ND	ND

Key: BRA-1–3, heteroagglutins; ND, not determined.
[a] Expressed as mole %.
[b] Ogata et al., 1983.
[c] Muramoto et al., 1985.
[d] Stebbins and Hapner, 1985.

amounts of aspartic acid and glutamic acid, which accounts for their acidic ioselectric pH values. Glycine content is high in *Balanus balanoides* and in the merostomes. *B. balanoides* agglutinin is unique in that serine, glutamic acid, and glycine account for 55% of all amino acids and that half-cystine and arginine are absent. The agglutinins generally contain low amounts of half-cystine (with the exception of *M. rosa* BRA-3) and methionine and intermediate amounts of other amino acids. Two of the *Megabalanus rosa* heteroagglutinins, BRA-1 and BRA-2, are related, whereas BRA-3 has a different molecular structure. The two *Melanoplus* (grasshopper) agglutinins have presumably identical structures, and the *T. tridentatus* heteroagglutinins form three structural groups composed of G1 and G2, G3 and G4, and GN3, respectively. There is apparently no shared antigenic similarity among the agglutinins of the different species of horseshoe crabs.

8.3.3. Molecular Weight and Subunit Structure

Selected structural properties of the purified arthropodan agglutinins are compiled in Table 8.3. It is clear that most agglutinins are high molecular weight multimeric structures composed of low molecular weight protein sub-

TABLE 8.2. AMINO ACID COMPOSITION[a] OF PURIFIED ARTHROPOD AGGLUTININS

			Organism				
			Tachypleus tridentatus[d]				
Amino Acid	*Limulus[b] polyphemus*	*Carcinoscorpius[c] rotunda cauda*	G1	G2	G3	G4	GN3
Aspartic	9.7	9.7	16.6	14.4	12.4	11.9	13.6
Threonine	5.9	4.2	5.3	5.6	6.3	6.6	3.7
Serine	6.8	7.6	6.0	6.2	5.9	6.1	8.1
Glutamic	13.1	10.4	9.0	9.4	9.6	10.2	8.3
Proline	4.7	4.9	3.3	3.6	3.7	4.0	5.6
Glycine	9.1	14.6	10.2	10.5	9.0	9.0	11.4
Alanine	5.0	5.6	3.8	3.8	3.8	3.7	3.4
Cystine/2	2.2	ND	2.0	2.0	1.5	1.8	2.2
Valine	7.0	6.9	3.6	3.8	4.7	3.8	10.2
Methionine	1.2	1.4	1.8	1.9	2.5	2.4	1.0
Isoleucine	4.9	5.6	5.5	5.9	7.2	5.3	6.5
Leucine	9.1	8.3	6.4	6.5	6.1	7.4	5.8
Tyrosine	1.6	2.1	7.6	7.7	5.0	5.8	2.5
Phenylalanine	4.1	4.2	3.9	3.9	5.1	4.5	3.6
Lysine	6.6	4.9	7.2	6.2	7.8	8.4	7.0
Histidine	4.9	3.5	3.6	3.8	4.9	5.0	3.0
Arginine	2.0	2.8	4.2	4.8	4.4	4.1	3.9
Tryptophan	2.0	3.5	ND	ND	ND	ND	ND

Key: G1–GN3, heteroagglutins; ND, not determined.
[a] Expressed as mole %.
[b] Kaplan et al., 1977.
[c] Dorai et al., 1981.
[d] Shimizu et al., 1979.

units, as diagramed in Figure 8.2. The quaternary structure of the multimers is generally stabilized by noncovalent forces, but the contributory subunits consist of smaller polypeptide chains that may be stabilized by inter- or intradisulfide bonds. For example, limulin, as indicated in Figure 8.2 and Table 8.3, consists of about 18 subunits that contain no interchain disulfide bonds. The limulin monomer does contain one intradisulfide bond (Kaplan et al., 1977). Carcinoscorpin from *Carcinoscorpius rotunda cauda* and the other horseshoe crab agglutinins have similar noncovalent aggregated structures. *M. rosa* heteroagglutinins are noncovalent aggregates of subunits that contain intradisulfide bonds. *Melanoplus* (grasshopper) agglutinins are similar in this regard.

There are some apparent exceptions to the idea that all arthropodan agglutinins are large multimeric structures. *M. rosa* BRA-3 is a noncovalently associated dimer of 28,000 MW monomers that in turn consist of two 18,000 MW subunits stabilized by intrachain disulfide bonds. The agglutinins from *A. dichotoma* have similar low molecular weights. The smallest arthropodan agglutinin described (43,000 daltons) is that isolated from cell membranes of *P. brassicae*. The agglutinins from *S. mantis* and *Heterometrus bengalensis* may be monomeric structures of relatively high molecular weight.

TABLE 8.3. SELECTED PROPERTIES OF PURIFIED ARTHROPODAN AGGLUTININS

Organism	MW^n	Subunit[n,nr] MW	Glycoprotein	Ca^{2+}	Carbohydrate Inhibitors	Inhibited Eythrocyte
Homarus americanus[a,b,c]						
LAg1	500,000–700,000	$70,000^{r,nr}$	Yes	Yes	NANA, NGNA	Human
LAg2	11S	$35,000^{r,nr}$	Yes	Yes	GalNAc	Mouse
Balanus balanoides[d]	300,000	70,000, 67,000, 26,000	Yes	ND	NANA, GalA, GlcA	Rabbit
Megabalanus rosa[e]						
BRA-1	330,000	$43,000^{nr}, 86,000^{nr}, 22,000^r$	Yes	ND	NANA, GalA, GLcA	Rabbit
BRA-2	140,000	$43,000^{nr}, 22,000^r$	Yes	ND	NANA, GalA, GLcA	Rabbit
BRA-3	64,000	$28,000^{nr}, 18,000^r$	Yes	ND	NANA, GalA, GLcA	Rabbit
Squilla mantis[f]						
Anti A	ND	$193,000^r$	ND	Yes	GalNAc	Human
Anti H	ND	$193,000^r$	ND	Yes	Fucose	Human
Sarcophaga peregrina[i]	190,000	$32,000^{r,nr}$	ND	ND	Galactose, lactose	Sheep
Teleogryllus commodus[g]	>1,000,000	$53,000^r, 31,000^r$	Yes	Yes	NANA, GlcNAc, GalNAc	Human
Melanoplus sanguinipes[h]	500,000–700,000	$70,000^{nr}, 40,000^r, 28,000^r$	Yes	Yes	α,β-Galactosides, α,β-glucosides	Asialo Human
Melanoplus differentialis[h]	500,000–700,000	$70,000^{nr}, 40,000^r, 28,000^r$	Yes	Yes	α,β-Galactosides, α,β-glucosides	Asialo Human
Allomyrina dichotoma[i]						
A-I	65,000	$17,500^r, 20,000^r$	No	No	β-Galactosides	Human
A-II	66,500	$19,000^r, 20,000^r$	No	No	β-Galactosides	Human
Bombyx mori[k]	260,000	ND	ND	ND	GlcA, GalA, heparin	Treated sheep

					Sugar specificity	Erythrocyte source
Pieris brassicae[l]	43,000	23,000[r]	ND	ND	GlcNAc, chitobiose	Treated rabbit
Limulus polyphemus[m,o,p,q,s]	400,000	19,000–29,000[nr]	Yes	Yes	NANA, NGNA, NANG	Horse
Carcinoscorpius rotunda cauda[t,u]	420,000	27,000–28,000[nr]	Yes	Yes	NANA, NGNA, NANG	Rabbit
Tachypleus tridentatus[v]						
GN 2	180,000–480,000[y]	65,000–73,300[nr]	ND	ND	NANA, GalNAc, GlcNAc	Human
GN 3	540,000	32,000–38,000[r] 25,000, 23,000, 21,000[nr,r]	ND	ND	NANA, GalNAc, GlcNAc	Human
Tachypleus gigas[w]	450,000	40,000[r]	ND	Yes	NANA, GalNAc, GlcNAc	Horse
Heterometrus bengalensis[x]	146,000	146,000[r,nr]	No	No	None	Rabbit

Key: ND, not determined; NANA, N-acetylneuraminic acid; NGNA, N-glycoylneuraminic acid; GalNAc, N-acetylgalactosamine; GlcA, glucuronic acid; GlcNAc, N-acetylglucosamine; NANG, N-acetylneuraminyl(2→6)N-acetylgalactosaminitol.
[n] Native molecular weight or sedimentation (S) constant.
[r] Reduced.
[nr] Nonreduced.
[y] Several heteroagglutinins.
[a] Hall and Rowlands, 1974a,b.
[b] VanderWall et al., 1981.
[c] Abel et al., 1984.
[d] Ogata et al., 1983.
[e] Muramoto et al., 1985.
[f] Amirante and Basso, 1984.
[g] Hapner and Jermyn 1981.
[h] Stebbins and Hapner, 1985.
[i] Komano et al., 1980.
[j] Umetsu et al., 1984.
[k] Suzuki and Natori, 1983.
[l] Mauchamp, 1982.
[m] Marchalonis and Edelman, 1968.
[o] Finstad et al., 1974.
[p] Kehoe et al., 1979.
[q] Roche and Monsigny, 1979.
[s] Cohen et al., 1984.
[t] Bishayee and Dorai, 1980.
[u] Dorai et al., 1981.
[v] Shimizu et al., 1979.
[w] Shishikura and Sekiguchi, 1984a.
[x] Basu et al., 1984.

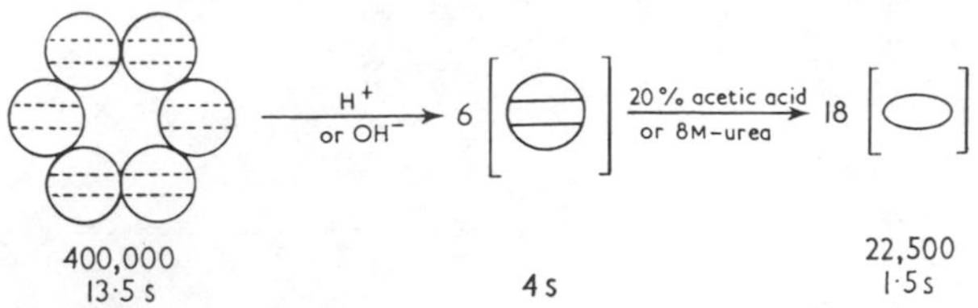

Figure 8.2. Diagrammatic structural model of limulin. The high molecular weight protein aggregate is composed of noncovalently associated smaller subunits. Changes in pH, calcium concentration, or solvent conditions cause the aggregated structure to dissociate. This general scheme may be typical of high molecular weight arthropodan agglutinins. (From Marchalonis and Edelman, 1968.)

Most agglutinins contain Ca^{2+}, which is apparently obligatory for carbohydrate binding and is involved in stabilization of the high molecular weight aggregate (Marchalonis and Edelman, 1968). Circular dichroism measurements show very low content of secondary structure in limulin (Finstad et al., 1972; Roche et al., 1978) and carcinoscorpin (Mohan et al., 1984). In contrast, *M. rosa* heteragglutinins BRA-1 and BRA-2 show large amounts of beta secondary structure (Muramoto et al., 1985). The general lack of effect by Ca^{2+} on the circular dichroic spectra suggests that, although Ca^{2+} is involved in the aggregation (and therefore multivalency) properties of the agglutinins, it is not a major factor in conformational aspects of the protein subunits.

8.3.4. Covalently Bound Carbohydrate

Most arthropodan agglutinins are glycoproteins and contain tightly associated carbohydrate, as listed in Table 8.3. The persistent presence of carbohydrate in the agglutinin molecule, as judged by colorimetric procedures and/or detection of amino sugars in amino acid analysis is generally considered a sufficient indication of glycoprotein character. Since agglutinins presumably interact with glycoconjugates in vivo, the conclusion that the native molecule contains covalently bound carbohydrate must be made with caution. For example, the heteroagglutinins BRA-2 and BRA-3 from *M. rosa* are reported to contain glucose-mannose and glucose, respectively, whereas neither has detectable amino sugars (Muramoto et al., 1985). Both heteroagglutinins are inhibited by glucose and mannose. The possibility exists that the detected carbohydrate may be derived from natural agglutinin binding substances rather than covalently bound carbohydrate. Ideally, pure glycopeptides should be isolated from the purified agglutinins and their chemical nature determined. The possibility that covalently attached carbohydrate performs a modulatory role in the functional character of agglutinins is unexplored.

8.4. FUNCTIONAL CHARACTERISTICS

The function of agglutinins in the arthropods is unclear. It is reasonable to presume that the in vivo roles of agglutinins are related to their capacity to reversibly bind carbohydrates and glycoconjugates. Various proposals have implicated agglutinins in the cellular and tissue transport of carbohydrates, calcium, and glycoproteins. Membrane-associated agglutinin is apparently involved in the resorption of tissue components during molting of insects (Mauchamp and Hubert, 1984). Correlative relationships between agglutinin and pupation have been observed in holometabolous insects (Komano et al., 1983; Suzuki and Natori, 1983). Other possible functional roles include formation of enzyme and glycoprotein aggregates (Sharon and Lis, 1972), antimicrobial agents (Prokop et al., 1968), antibodylike surveillance molecules (Komano et al., 1980), complementlike cytolytic agents (Komano and Natori, 1985), and tissue adherence of beneficial microorganisms (Peters et al., 1983). Most current research emphasizes the possible role of agglutinins as cellular and/or humoral components of the invertebrate immune system. As such, they may be involved in recognition and neutralization of potentially pathogenic and other nonself materials, including histolytic debris and transplanted tissue. The agglutinins may represent a component of an ancient analogue of the vertebrate immune system. Numerous reviews describing agglutinin involvement and immunobiologic significance in invertebrate defense and recognition mechanisms are available (Chorney and Cheng, 1980; Lackie, 1980; Yeaton, 1981a; Renwrantz, 1983; Coombe et al., 1984; Vasta and Marchalonis, 1984). Agglutinins as humoral recognition factors (Vasta and Marchalonis, 1983) and as possible modulating agents of phagocytosis and encapsulation (Ratner and Vinson, 1983) in the Arthropoda have also been described. Among the Insecta, accounts of agglutinin involvement in cellular defense reactions (Ratcliff and Rowley, 1979; Ratcliffe, 1982), immune recognition (Lackie, 1981), and wound response (Yeaton, 1983) are available. A central problem with the role of agglutinins as immunorecognition molecules is the molecular basis for the generation of binding diversity, obligatory for broad nonself recognition. Although a few arthropods have been shown to contain agglutinins of structural (heteroagglutinins) and functional (multispecific agglutinins) heterogeneity, there is still no detailed explanation of wide carbohydrate recognition by a single multispecific agglutinin or by several heteroagglutinins. There is great interest in the biomedical application of agglutinins as molecular probes that can be used to map and decipher cell surfaces and possibly distinguish between normal cells and those altered by disease or developmental modifications (Cohen, 1979). It appears likely that the agglutinins will be shown to have various functional roles, of both immune and nonimmune character, depending on the particular tissue and organism involved. It is also feasible that agglutinins have physiologic and biochemical functions that are only indirectly related to carbohydrate binding and are therefore currently unanticipated.

8.4.1. Nature of the Carbohydrate Binding Site

The specificity of the agglutinin carbohydrate binding site has generally been explored through hemagglutination inhibition studies using various carbohydrates and glycoconjugates. More detailed mapping of the site is possible with natural derivatives and synthetic analogues of the inhibitory carbohydrates. Recently, more quantitative procedures have enabled the determination of equilibrium binding constants of inhibitory substances and Scatchard-type analysis of the carbohydrate binding sites. In all cases, however, the relevance of these measurements to possible agglutinin function in vivo is problematic.

The major carbohydrate inhibitory specificity of the purified arthropodan agglutinins is briefly included in Table 8.3. It is clear that agglutinins isolated from the Merostomata (horseshoe crabs) and most of those from the Crustacea (barnacles) are inhibited by sialic acids and N-acetylhexosamino sugars. Vasta and Marchalonis (1983) have discussed the common sialic acid specificity of the agglutinins from some aquatic and terrestrial chelicerates (horseshoe crabs, scorpions, and spiders). Data in Table 8.3 suggest that inhibition by sialic acid generally extends to the Crustacea as well. The only exceptions are the LAg2 heteroagglutinin from lobster, which has primary specificity directed toward GalNAc, and the heteroaagglutinins from *S. mantis*, which are monospecific for human blood groups A and H and are inhibited by GalNAc and fucose, respectively. In addition to sialic acids, the hexuronic acids show relatively strong inhibition of the crustacean and merostomid agglutinins.

Inhibition by sialic acid is not a prominent feature among agglutinins from the Insecta. The agglutinin from *Teleogryllus commodus* is an exception, and it shows merostomid-crustacean–type carbohydrate inhibition. Other insect agglutinins are generally inhibited by galactosidic and glucosidic carbohydrates. One insect agglutinin, from *B. mori*, is inhibited by hexuronic acids and acidic oligosaccharides. It may be noteworthy that the agglutinins showing narrow carbohydrate specificity are generally those with relatively low molecular weight, which consequently contain fewer subunits. Examples of such agglutinins are the monospecific heteroagglutinins from *S. mantis* that may be monomeric, *A. dichotoma* agglutinins specific for beta-galactosides, and the low molecular weight membrane agglutinin from *P. brassicae*. The lowest molecular weight heteroagglutinin from *M. rosa*, BRA-3, similarly demonstrates restricted carbohydrate inhibition relative to the larger *M. rosa* heteroagglutinins. In most cases, arthropodan agglutinins show broad carbohydrate inhibition properties, including, in the case of merostomes, inhibition by bacterial cell surface components, such as 2-keto-3-deoxyoctonate (KDO) and glycerol phosphate (Dorai et al., 1982a,b; McSweegan and Pistole, 1982; Rostam-Abadi and Pistole, 1982). It is generally observed that treatment of erythrocytes or sialoconjugates with neuraminidase eliminates their agglutination or precipitation by sialic acid binding agglutinins. Although the agglutinins are not strictly specific for sialic acid, this observation shows the relative importance of the sialic acid component for agglutinin binding. The emphasis on sialic acid specificity in the case of many arthropodan agglutinins is curious in light of the relative absence of sialic acid among the arthropods and (potentially pathogenic) bacteria (Schauer, 1982).

The binding site of limulin has broad carbohydrate binding specificity, with principal affinity for sialic acid and sialoconjugates (Cohen et al., 1965; Cohen, 1968). Experiments using natural and synthetic derivatives of sialic acid show that the limulin binding site is characterized by the presence of a hydrophobic region and subsite that binds the negatively charged carboxylate group of NANA (Cohen et al., 1983; Cohen et al., 1984). Incorporation of a triphenylmethyl group or esterification of the carboxyl group results in a large increase or decrease, respectively, of inhibition. Methylglycoside formation, O-acetylation, and fluoridation of NANA have little impact on inhibition, indicating that the limulin site is tolerant of changes in these positions on the inhibitory carbohydrate. In contrast, deacetylation of sialic acid and consequent exposure of the positively charged amino group eliminates inhibition (Roche and Monsigny, 1979). Use of neuraminyl di- and oligosaccharides in inhibition studies shows that α-glycosidic derivatives of NANA and NGNA are better inhibitors than are free NANA and N-glycoylneuraminic acid (NGNA) and the suggestion is made that limulin shows greatest specificity for glycoconjugates that contain the NANA-α-(2 → 3)(6)GalNAc structure (Roche et al., 1975; Roche and Monsigny, 1979).

The nature of the carbohydrate binding site of carcinoscorpin from *C. rotunda cauda* has been investigated by several means. Chemical reagents that modify tryptophan and tyrosine also strongly inhibit hemagglutinating activity, whereas carboxyl and amino group reagents have lesser effect (Bishayee and Dorai, 1980). Acid or base (0.1 M) or 2 M guanidine hydrochloride irreversibly inhibits activity, but the molecule is unaffected by 6 M urea. Alkylating and reducing reagents are not deleterious, indicating that binding site integrity is independent of thiol and disulfide groups. Calcium is essential for hemagglutinating activity. Ligand binding is dependent on two ionizable groups with approximate pK_a of 6.0 and 9.0, respectively. The succinylated and consequently monomerized subunit of carcinoscorpin does not bind to glycoconjugates.

Equilibrium binding studies involving radiolabeled carcinoscorpin and several sialoglycoproteins indicate that at least two different classes of binding sites are present in carcinoscorpin (Mohan et al., 1982). One class contains two high-affinity binding sites, and the other contains six low-affinity sites for fetuin or serotransferrin. There appear to be eight functional sites among the 16 subunits of the molecule. It is uncertain whether two subunits are required for a functional binding site, but the inactivity of succinylated monomers mentioned above is compatible with this idea. It is interesting to note that one-half the sites are unavailable for binding at low temperature, an observation that may be related to aggregation properties of the agglutinin. The carcinoscorpin binding site shows multispecificity toward NANA, NGNA, and other sugar acids, with primary affinity directed toward sialoconjugates. The binding site or sites are responsive not only to terminal sialic acid but also to the configuration, composition, and structure of the attendent saccharide chains. The binding sites show positive cooperativity, a phenomenon that requires molecular conformational alterations that increase binding affinity as ligand occupancy increases. Receptor recognition and extent of binding appear to be influenced by the topography and density of sialic acid residues associated with the glycoconjugates. The extent to which calcium or other endogenous hemolymphatic components may additionally modify the process through effects on the multimeric agglutinin structure is unknown.

Quantitative estimation of ligand association constants determined by protein fluorescence quenching show that O-(N-acetylneuraminyl)-(2 → 6)-N-acetylgalactosaminitol (NANG) has K_a equal to 1.15×10^6 M^{-1} (Mohan et al., 1983). This compound is also a potent inhibitor, as judged by inhibition of fetuin binding and hemagglutination (Mohan et al., 1982). A similar structure inhibits limulin strongly (Roche and Monsigny, 1979). The NANG disaccharide binds carcinoscorpin 100–300 times more strongly than does NANA, NGNA, glucuronic acid, and N-acetylneuraminyl lactose, all of which are potent inhibitors of hemagglutination. The same study shows that two other inhibitors, 2-oxo-3-deoxyoctonate and phosphoryl choline, bind at or near the NANG site and at a site removed, respectively. The NANG disaccharide normalizes all tryptophan residues in carcinoscorpin to quenching by potassium iodide (KI) and renders them more accessible than in the agglutinin in the absence of ligand (Mohan et al., 1983). Phosphoryl choline binds at a different site and does not induce the normalization of tryptophan residues. This is interpreted to mean that occupancy of the carbohydrate sites induces a conformational change in the molecule, a conclusion consistent with the positive cooperativity binding characteristics determined by equilibrium binding studies. Experimental results such as those described here suggest that heterogeneous interactions with receptors may derive from alteration of structural characteristics of the multimeric agglutinins leading perhaps to increased diversity of binding.

8.4.2. Natural Agglutinin Binding Substances

There have been several indications of natural agglutinin binding substances in arthropodan hemolymph. The presence of in vivo inhibitors is interesting because their characterization may provide insight into the physiologic role of the agglutinins. The presence of inhibitors in the hemolymph of horseshoe crabs is suggested by the apparently greater than quantitative yields of agglutinins when they are isolated by affinity chromatographic procedures (Roche and Monsigny, 1979; Shishikura and Sekiguchi, 1983). Muresan and

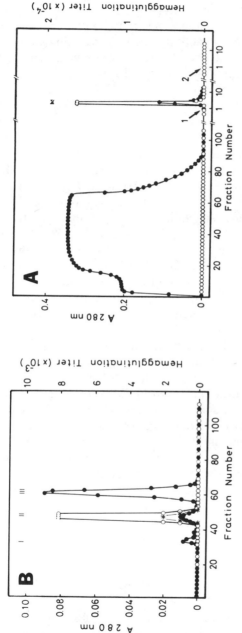

Figure 8.3. (A) Affinity chromatography of perivitelline fluid from *Tachypleus gigas* on a column (1.5 × 5 cm) of BSM-Sepharose 4B. Perivitelline fluid (750 ml) is applied, and the column is washed with pH 7.5 buffer (0.05 M Tris-HCl, 0.5 M NaCl, 0.1 M CaCl$_2$). Bound proteins are eluted by addition of 0.5 M GlcNAc (1) and then 5 M urea (2) to the buffer. All adsorbed proteins are eluted with GlcNAc. Fractions are monitored by absorbancy (●), and hemagglutination activity is determined with human erythrocytes (○). (B) Gel filtration of the affinity-isolated agglutinin in A in the same buffer containing 1 M urea on a column (2.5 × 35 cm) of Fractogel TSK (Toyopearl) HW-60. Fractions are monitored by absorbancy (●), and hemagglutination activity is assayed with human (○) and horse (*) erythrocytes. Peak III contains inhibitory glycoproteins. (From Shishikura and Sekiguchi, 1984a.)

coworkers (1982) removed associated heterogeneous protein substances from limulin by elution of the adsorbed affinity chromatography column with high ionic strength calcium-containing buffers (see Fig. 18.1, in Chapter 18). In this case, the isolated material is noninhibitory to limulin and has neither sialic acid binding nor agglutination activity. Peptide mapping experiments show the limulin-associated substances to be structurally unrelated to limulin, and their significance is unknown. Removal of the associated proteins does, however, result in high specific activity and high yield recovery of pure limulin. It is possible that the substances may represent unknown hemolymphatic components, including other agglutinins, normally associated with the limulin molecule.

Naturally occurring agglutinin inhibitory substances present in the perivitelline fluid of *T. gigas* are comparatively well characterized (Shishikura and Sekiguchi, 1984a,b). Figure 8.3A shows the affinity-chromatographic isolation of agglutinin from *T. gigas*. The inhibitory agglutinin binding substances associated with the single agglutinin peak are subsequently dissociated from the agglutinin by gel filtration in 1 M urea (Fig. 8.3B). Further characterization of the agglutinin binding substances separates them into three homogeneous glycoprotein fractions that have properties different from those of *T. gigas* agglutinin. The glycoproteins are potent inhibitors of *T. gigas* agglutinin, and their binding capacity is inhibited by sugars specific for the agglutinin. This observation suggests that the carbohydrate portion of the binding substances interacts with and inhibits the *T. gigas* molecule at its carbohydrate binding site. In support of this suggestion, the agglutinin binding substances are shown to contain (*N*-acetyl) hexosamines and sialic acid, both carbohydrates that are inhibitory to *T. gigas* agglutinin. Amino acid analysis of the three natural agglutinin binding glycoproteins reveals high amounts of aspartic acid, glutamic acid, and glycine and low content of half-cystine. It is interesting to note that these characteristics are generally seen among the arthropodan agglutinins. It is important to demonstrate that isolated natural agglutinin binding substances represent authentic in vivo receptors for agglutinins rather than possible preparatory contaminants or exogenously derived inhibitory materials. To this end, it appears that the inner egg membrane in the *T. gigas* embryo may contain the agglutinin binding substances and represent the first identification and localization of a natural receptor for an arthropodan agglutinin.

8.5. WHAT IS NEEDED?

What do agglutinins do? Despite much research concerning agglutinins, particularly during the last 10 years, there is still no clear understanding of their physiologic role in arthropods. That they are important and carry out fundamental roles may be deduced from their apparently ubiquitous presence. It is important for sustained support of research in this area that an in vivo function be unequivocally demonstrated for an arthropodan agglutinin. It is now possible to begin to make meaningful comparisons of structural and inhibitory properties among several purified arthropodan agglutinins, as listed in Tables 8.1, 8.2, and 8.3. The systematic purification and characterization of arthropodan agglutinins and putative agglutinin receptors should continue on the premise that more complete knowledge of their molecular properties will suggest, support, or eliminate possible functional roles. When appropriate, experimental procedures should use purified agglutinins in order to minimize possible interference by other hemolymphatic components. Study of the developmental biology of arthropodan agglutinins at all levels, including organismal, tissue, and molecular, is only beginning and should be expanded. Tissue localization of agglutinins and agglutinin receptors by means of specific immunologic probes should provide important clues to agglutinin function. Finally, the supposed working portion of the agglutinin molecule—the carbohydrate binding site—should be investigated further by quantitative means to determine its physicochemical characteristics and its origin within the multimeric agglutinin structure. It is important to confirm or deny the idea that the generation of binding diversity, presumably necessary for broad carbohydrate recognition, arises from conformational and aggregational dynamics of the agglutinin molecule.

8.6. SUMMARY

Little is known concerning the biochemistry of arthropodan agglutinins, and their physiologic function or functions are uncertain. The molecules appear to be ubiquitous among the arthropods; however, they may be undetectable in certain tissues or during certain metamorphic stages. Their biosynthetic origins are essentially unknown, and their common presence in tissues other than hemolymph is tentative. Several agglutinins have been purified to homogeneity, and they generally share glycoprotein character, high molecular weight, noncovalent multimeric structure, Ca^{2+} dependency, little alpha or beta secondary structure, high amounts of aspartic acid and glutamic acid, and low half-cystine content. One agglutinin, limulin, has been subjected to primary structure determination. There appears to be no antigenic similarity among agglutinins of different genera even though they may have similar carbohydrate specificity. Little detailed information is available concerning the carbohydrate binding sites, but initial work shows them to be cooperative and capable of heterogeneous interactions with receptors. Most agglutinins from the Crustacea and Merostomata show binding specificity for sialic acid, but the agglutinins from the Insecta are more diverse in their carbohydrate recognition. Carbohydrate binding specificity is generally broad; however, this must be interpreted in light of possibly irrelevant assay systems. The discovery of natural agglutinin binding substances and their cellular location will be useful in elucidation of agglutinin functions. Additional quantitative characterization of agglutinin molecules is desirable.

ACKNOWLEDGMENT

Some research work described herein and the preparation of this chapter were supported by grants from the National Science Foundation MONTS program and NSF-DCB-8510097, and the Montana Agricultural Experiment Station.

REFERENCES

Abel, C.A., P.A. Campbell, J. VanderWall, and A.L. Hartman. 1984. Studies on the structure and carbohydrate binding properties of lobster agglutinin (LAgl), a sialic acid-binding lectin. *Prog. Clin. Biol. Res.* **157:** 103–114.

Amirante, G.A. 1976. Production heteroagglutinins in haemocytes of *Leucophaea maderae* L. *Experientia* **32:** 526–528.

Amirante, G.A. and V. Basso. 1984. Analytical study of lectins in *Squilla mantis* L. (Crustacea: Stomatopoda) using monoclonal antibodies. *Dev. Comp. Immunol.* **8:** 721–726.

Amirante, G.A. and F.G. Mazzalai. 1978. Synthesis and localization of hemagglutinins in hemocytes of the cockroach *Leucophaea maderae* L. *Dev. Comp. Immunol.* **2:** 735–740.

Basu, P.S., P.K. Datta, P. Agarwal, M.K. Ray, and T.K. Datta. 1984. Purification and partial characterization of an erythroagglutinin from the hemolymph of scorpion, *Heterometrus bengalensis*. *Biochimie* **66:** 487–491.

Bishayee, S. and D.T. Dorai. 1980. Isolation and characterization of a sialic acid–binding lectin (carcinoscorpin) from Indian horseshoe crab, *Carcinoscorpius rotunda cauda*. *Biochim. Biophys. Acta* **623:** 89–97.

Chorney, M.J. and T.C. Cheng. 1980. Discrimination of self and non-self in invertebrates, pp. 37–54. In J.J. Marchalonis and N. Cohen (eds.) *Contemporary Topics in Immunobiology*, vol. 9. Plenum Press, New York.

Cohen, E. 1968. Immunologic observations on the agglutinins of the hemolymph of *Limulus polyphemus* and *Birgus latro*. *Trans. N.Y. Acad. Sci.* **20:** 427–434.

Cohen, E. 1979. Biomedical applications of the horseshoe crab (Limulidae). *Prog. Clin. Biol. Res.* **29:** 1–688.

Cohen, E., A.W. Rowe, and E.C. Wissler. 1965. Heteroagglutinins of the horsehoe crab *Limulus polyphemus*. *Life Sci.* **4:** 2009–2016.

Cohen, E., G.H.V. Ilodi, W. Korytnyk, and M. Sharma. 1983. Inhibition of *Limulus* agglutinins with *N*-acetyl-neuraminic acid and derivatives. *Dev. Comp. Immunol.* **7:** 189–192.

Cohen, E., G.R. Vasta, W. Korytnyk, C.R. Petrie, III, and M. Sharma. 1984. Lectins of the Limulidae and hemagglutination-inhibition by sialic acid analogs and derivatives. *Prog. Clin. Biol. Res.* **157**: 55–69.

Coombe, D.R., P.L. Ey, and C.R. Jenkin. 1984. Self/non-self recognition in invertebrates. *Q. Rev. Biol.* **59**: 231–255.

Cornick, J.W. and J.E. Stewart. 1973. Partial characterization of a natural agglutinin in the hemolymph of the lobster *Homarus americanus. J. Invertebr. Pathol.* **21**: 255–262.

Cornick, J.W. and J.E. Stewart. 1978. Lobster (*Homarus americanus*) hemocytes: Classification, differential counts, and associated agglutinin activity. *J. Invertebr. Pathol.* **31**: 194–203.

Denburg, J.L. 1980. Cockroach muscle hemagglutinins: Candidate recognition macromolecules. *Biochem. Biophys. Res. Commun.* **97**: 33–40.

Dorai, D.T., S. Mohan, S. Srimal, and K. Bachhawat. 1982a. On the multispecificity of carcinoscorpin, the sialic acid binding lectin from the horseshoe crab *Carcinoscorpius rotunda cauda. FEBS Lett.* **148**: 98–102.

Dorai, D.T., B.K. Bachhawat, S. Bishayee, K. Kannan, and D.R. Rao. 1981. Further characterization of the sialic acid–binding lectin from the horseshoe crab *Carcinoscorpius rotunda cauda. Arch. Biochem. Biophys.* **209**: 325–333.

Dorai, D.T., S. Srimal, S. Mohan, B.K. Bachhawat, and T.S. Balganesh. 1982b. Recognition of 2-keto-3-deoxyoctonate in bacterial cells and lipopolysaccharides by the sialic acid binding lectin from the horseshoe crab *Carcinoscorpius rotunda cauda. Biochem. Biophys. Res. Commun.* **104**: 141–147.

Finstad, C.L., R.A. Good, and G.W. Litman. 1974. The erythrocyte agglutinin from *Limulus polyphemus* hemolymph: Molecular structure and biological function. *Ann. N.Y. Acad. Sci.* **234**: 170–180.

Finstad, C.L., G.W. Litman, J. Finstad, and R.A. Good. 1972. The evolution of the immune response. 13. The characterization of purified erythrocyte agglutinins from two invertebrate species. *J. Immunol.* **108**: 1704–1711.

Gold, E.R. and P. Balding. 1975. Receptor-specific proteins: Plant and animal lectins, pp. 251–283. *Excerpta Medica.* American Elsevier, New York.

Goldstein, I.J., R.C. Hughes, M. Monsigny, T. Osawa, and N. Sharon. 1980. What should be called a lectin? *Nature* (London) **285**: 66.

Hall, J.L. and D.T. Rowlands, Jr., 1974a. Heterogeneity of lobster agglutinins. 1. Purification and physiochemical characterization. *Biochemistry* **13**: 821–827.

Hall, J.L. and D.T. Rowlands, Jr. 1974b. Heterogeneity of lobster agglutinins. 2. Specificity of agglutinin-erythrocyte binding. *Biochemistry* **13**: 828–832.

Hapner, K.D. and M.A. Jermyn. 1981. Haemagglutinin activity in the haemolymph of *Teleogryllus commodus* (Walker). *Insect Biochem.* **11**: 287–295.

Jurenka, R., K. Manfredi, and K.D. Hapner. 1982. Haemagglutinin activity in Acrididae (grasshopper) haemolymph. *J. Insect Physiol.* **28**: 177–181.

Kaplan, R., S.S.-L. Li, and J.M. Kehoe. 1977. Molecular characterization of limulin, a sialic acid binding lectin from the hemolymph of the horseshoe crab. *Biochemistry* **16**: 4297–4302.

Kehoe, J.M., R. Kaplan, and S.S.-L. Li. 1979. Functional implications of the covalent structure of limulin: An overview. *Prog. Clin. Biol. Res.* **29**: 617–623.

Komano, H. and S. Natori. 1985. Participation of *Sarcophaga peregrina* lectin in the lysis of sheep red blood cells injected into the abdominal cavity of larvae. *Dev. Comp. Immunol.* **9:** 31–40.

Komano, H., D. Mizuno, and S. Natori. 1980. Purification of lectin induced in the hemolymph of *Sarcophaga peregrina* larvae on injury. *J. Biol. Chem.* **255:** 2919–2924.

Komano, H., D. Mizuno, and S. Natori. 1981. A possible mechanism of induction of insect lectin. *J. Biol. Chem.* **256:** 7087–7089.

Komano, H., R. Nozawa, D. Mizuno, and S. Natori. 1983. Measurement of *Sarcophaga peregrina* lectin under various physiological conditions by radioimmunoassay. *J. Biol. Chem.* **258:** 2143–2147.

Lackie, A.M. 1980. Invertebrate immunity. *Parasitology* **80:** 393–412.

Lackie, A.M. 1981. Immune recognition in insects. *Dev. Comp. Immunol.* **5:** 191–204.

Marchalonis, J.J. and G.M. Edelman. 1968. Isolation and characterization of a hemagglutinin from *Limulus polyphemus. J. Mol. Biol.* **32:** 453–465.

Mauchamp, B. 1982. Purification of an *N*-acetyl-D-glucosamine specific lectin (P.B.A.) from epidermal cell membranes of *Pieris brassicae* L. *Biochimie* **64:** 1001–1008.

Mauchamp, B. and M. Hubert. 1984. Internalization of plasma membrane glycoconjugates and plasma membrane lectin into epidermal cells during pharate adult wing development of *Pieris brassicae* L: Correlation with resorption of molting fluid components. *Biol. Cell.* **50:** 285–294.

McSweegan, E.F. and T.G. Pistole. 1982. Interaction of the lectin limulin with capsular polysaccharides from *Neisseria meningitidis* and *Escherichia coli. Biochem. Biophys. Res. Commun.* **106:** 1390–1397.

Mohan, S., D.T. Dorai, S. Srimal, and B.K. Bachhawat. 1982. Binding studies of a sialic acid–specific lectin from the horseshoe crab *Carcinoscorpius rotunda cauda* with various sialoglycoproteins. *Biochem. J.* **203:** 253–261.

Mohan, S., D.T. Dorai, S. Srimal, B.K. Bachhawat, and M.K. Das. 1983. Fluorescence studies on the interaction of some ligands with carcinoscorpin, the sialic acid specific lectin from the horseshoe crab *Carcinoscorpius rotunda cauda. J. Biosci.* **5:** 155–162.

Mohan, S., D.T. Dorai, S. Srimal, B.K. Bachhawat, and M.K. Das. 1984. Circular dichroism studies on carcinoscorpin, the sialic acid binding lection of horseshoe crab *Carcinoscorpius rotunda cauda. Indian J. Biochem. Biophys.* **21:** 151–154.

Muramoto, K., K. Ogata, and H. Kamiya. 1985. Comparison of the multiple agglutinins of the acorn barnacle *Megabalanus rosa. Agric. Biol. Chem.* **49:** 85–93.

Muresan, V., V. Iwanij, Z.D.J. Smith, and J.D. Jamieson. 1982. Purification and use of limulin: A sialic acid–specific protein. *J. Histochem. Cytochem.* **30:** 938–946.

Ogata, K., K. Muramoto, M. Yamazaki, and H. Kamiya. 1983. Isolation and characterization of *Balanus balanoides* agglutinin. *Bull. Jpn. Soc. Sci. Fish.* **49:** 1371–1375.

Peters, W., H. Kolb, and V. Kolb-Bachofen. 1983. Evidence for a sugar receptor (lectin) in the peritrophic membrane of the blowfly larva *Calliphora erythrocephala* Mg. (Diptera). *J. Insect Physiol.* **29**: 275–280.

Prokop, O., G. Uhlenbruck, and W. Köhler. 1968. Protecktine, eine neue Klasse antikörperähnlicher Verbindungen. *Dtsch. Gesundheitswes.* **23**: 318–325.

Ratcliffe, N.A. 1982. Cellular defense reactions of insects. *Fortschr. Zool.* **27**: 223–244.

Ratcliffe, N.A. and A.F. Rowley. 1979. Role of hemocytes in defense against biological agents, pp. 331–414. In A.P. Gupta (ed.) *Insect Hemocytes.* Cambridge University Press, Cambridge.

Ratner, S. and R.B. Vinson. 1983. Phagocytosis and encapsulation: Cellular immune response in Arthropoda. *Am. Zool.* **23**: 185–194.

Ravindranath, M.H. and E.L. Cooper. 1984. Crab lectins: Receptor specificity and biomedical applications. *Prog. Clin. Biol. Res.* **157**: 83–96.

Renwrantz, L. 1983. Involvement of agglutinins (lectins) in invertebrate defense reactions: The immuno-biological importance of carbohydrate-specific binding molecules. *Dev. Comp. Immunol.* **7**: 603–608.

Roche, A.-C. and M. Monsigny. 1979. Limulin (*Limulus polyphemus* lectin): Isolation, physicochemical properties, sugar specificity and mitogenic activity, pp. 603–616. In E. Cohen (ed.) *Biomedical Applications of the Horseshoe Crab* (Limulidae). Alan R. Liss, New York.

Roche, A.-C., R. Schauer, and M. Monsigny. 1975. Protein-sugar interactions: Purification by affinity chromatography of Limulin, an *N*-acyl-neuraminidyl–binding protein. *FEBS Lett.* **57**: 245–249.

Roche, A.-C., J. Maurizot, and M. Monsigny. 1978. Circular dichroism of Limulin: *Limulus polyphemus* lectin. *FEBS Lett.* **91**: 233–236.

Rostam-Abadi, H. and T.G. Pistole. 1982. Lipopolysaccharide-binding lectin from the horseshoe crab *Limulus polyphemus*, with specificity for 2-keto-3-deoxy-octonate (KDO). *Dev. Comp. Immunol.* **6**: 209–218.

Schauer, R. 1982. Chemistry, metabolism and biological functions of sialic acids. *Adv. Carbohydr. Chem. Biochem.* **40**: 131–234.

Sharon, N. and H. Lis. 1972. Lectins: Cell agglutinating and sugar-specific proteins. *Science* (Washington, DC) **177**: 949–959.

Shimizu, S., M. Ito, N. Takahashi, and M. Niwa. 1979. Purification and properties of lectins from the Japanese horseshoe crab, *Tachypleus tridentatus. Prog. Clin. Biol. Res.* **29**: 625–639.

Shishikura, F. and K. Sekiguchi. 1983. Agglutinins in the horseshoe crab hemolymph: Purification of a potent agglutinin of horse erythrocytes from the hemolymph of *Tachypleus tridentatus*, the Japanese horseshoe crab. *J. Biochem.* **93**: 1539–1546.

Shishikura, F. and K. Sekiguchi. 1984a. Studies on perivitelline fluid of horseshoe crab embryo. 1. Purification and properties of agglutinin from the perivitelline fluid of *Tachypleus gigas* embryo. *J. Biochem.* **96**: 621–628.

Shishikura, F. and K. Sekiguchi. 1984b. Studies on the perivitelline fluid of horseshoe crab embryo. 2. Purification of agglutinin-binding substance from the perivitelline fluid of *Tachypleus gigas* embryo. *J. Biochem.* **96**: 629–636.

Stebbins, M.R. and K.D. Hapner. 1985. Preparation and properties of haemagglutinin from haemolymph of Acrididae (grasshoppers). *Insect Biochem.* **15:** 451–462.

Stynen, D., M. Peferson, and A. Deloof. 1982. Proteins with hemagglutinin activity in larvae of the Colorado potato beetle, *Leptinotarsa decemlineata. J. Insect Physiol.* **28:** 465–470.

Suzuki, T. and S. Natori. 1983. Identification of a protein having hemagglutinating activity in the hemolymph of the silkworm, *Bombyx mori. J. Biochem.* **93:** 583–590.

Takahashi, H., H. Komano, N. Kawaguchi, M. Obinata, and S. Natori. 1984. Activation of the secretion of specific proteins from fat body following injury to the body wall of *Sarcophaga peregrina* larvae. *Insect Biochem.* **14:** 713–717.

Tsivion, Y. and N. Sharon. 1981. Lipid-mediated hemagglutination and its relevance to lectin-mediated agglutination. *Biochim. Biophys. Acta* **642:** 336–344.

Umetsu, K., S. Kosaka, and T. Suzuki. 1984. Purification and characterization of a lectin from the beetle *Allomyrina dichotoma. J. Biol. Chem.* **95:** 239–245.

VanderWall, J., P.A. Campbell, and C.A. Abel. 1981. Isolation of a sialic acid–specific lobster lectin (LAgl) by affinity chromatography on Sepharose-colominic acid beads. *Dev. Comp. Immunol.* **5:** 679–684.

Vasta, G.R. and J.J. Marchalonis. 1983. Humoral recognition factors in the Arthropoda: The specificity of Chelicerata serum lectins. *Am. Zool.* **23:** 157–171.

Vasta, G.R. and J.J. Marchalonis. 1984. Summation: Immunobiological significance of invertebrate lectins. *Prog. Clin. Biol. Res.* **157:** 177–191.

Yeaton, R.W. 1981a. Invertebrate lectins: Diversity of specificity, biological synthesis and function in recognition. *Dev. Comp. Immunol.* **5:** 535–545.

Yeaton, R.W. 1981b. Invertebrate lectins: Occurrence. *Dev. Comp. Immunol.* **5:** 391–402.

Yeaton, R.W. 1983. Wound response in insects. *Am. Zool.* **23:** 195–203.

CHAPTER 9

Prophenoloxidase-Activating Cascade as a Recognition and Defense System in Arthropods

Kenneth Söderhäll
Institute of Physiological Botany
University of Uppsala
Uppsala, Sweden

Valerie J. Smith
University Marine Biological Station
Millport, Scotland

9.1. INTRODUCTION

Despite their lack of immunoglobulins, arthropods, like most invertebrates, display a variety of defense strategies by which they protect themselves against the harmful effects of parasitic and microbial invasion. These host defenses include the possession of mechanical barriers, such as the cuticle, which may itself possess antimicrobial properties (Häll and Söderhäll, 1983); phagocytosis; encapsulation; and the production of agglutinins, lysins, and other antimicrobial factors in the hemolymph. From the numerous investigations of immunity in arthropods, it is clear that within the hemocoel, the recognition of and the response to nonself entities by the hemocytes constitutes the most important step in overcoming systemic infection. However, until recently, very little was known about the biochemical events that initiate, control, and regulate cellular activity in these animals; but evidence is now accumulating that the prophenoloxidase-activating (proPO) system— the enzyme cascade responsible for converting prophenoloxidase into active phenoloxidase during the early stages of melanization—plays a major role in mediating nonself recognition and host defense in arthropods.

That melanization accompanies the host responses to wounding or microbial attack in arthropods (Fig. 9.1) has long been recognized, and many reports describing the phenomenon in various species have been published (for recent reviews see Söderhäll, 1982, and Ratcliffe et al., 1982; see also Chapters 1, 2, 4, 6, and 15). Several workers have proposed that the process is important in host defense in these invertebrates (Taylor, 1969; Salt, 1970; Maier, 1973), but because of the paucity of information about the biochemical events leading to melanin synthesis, the precise involvement of melanization in nonself recognition remains obscure. Certainly, melanization participates in host protection, as shown, for example, by Salt (1956), who found that when melanization in *Carausius morosus* was inhibited by phenylthiourea (PTU), there was better survival of the eggs of the parasite *Nemeritis canescens*. Similarly, Brewer and Vinson (1971) observed decreased encapsulation and reduced melanization in *Heliothis zea* parasitized by *Cardiochiles nigriceps* after injection of PTU or glutathione, while Nappi (1973) reported that PTU-fed *Drosophila algonquin* larvae showed a reduced melanization response as well as reduced encapsulation, and Beresky and Hall (1977) described higher motality in PTU-treated *Aedes aegypti* infected with *Neoplectana carpocapsae* nematodes than in untreated controls. In contrast, other studies of insect and crustacean immunity have revealed that melanin deposition is a relatively late event in the cellular reactions to nonself entities (Salt, 1956; Rizki, 1957; Poinar et al., 1968; Nappi, 1973; Götz and Vey, 1974) and as such cannot be responsible for the recognition of foreign particles. Therefore, to elucidate the physiologic effects that the intermediary products exert in the host, it is necessary to look more closely at the biochemical events that bring about the formation of melanin.

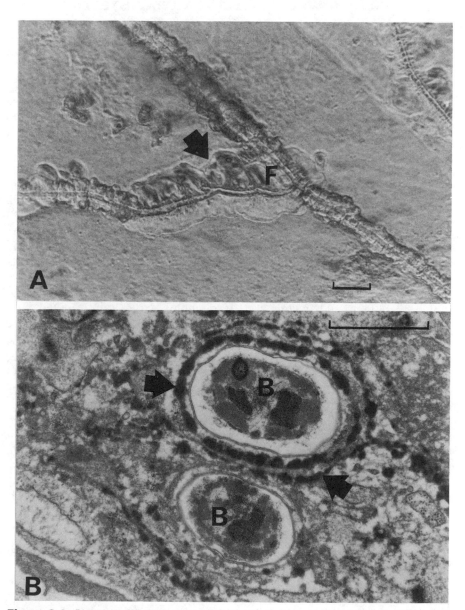

Figure 9.1. Phase contrast micrograph of melanization as a host defense reaction in crustaceans. (A) Melanized hyphae (arrow) of the parasitic fungus (F) *Aphanomyces astaci* in the cuticle of *Pacifastacus leniusculus*. (Scale bar = 100 μm.) (B) Electron micrograph of part of a hemocyte clump in the gills of *Carcinus maenas* enclosing the bacterium *Moraxella* sp. One bacterium is surrounded by a thin ring of electron-dense melanin (arrow). (Scale bar = 0.1 μm.) (From Smith and Ratcliffe, 1980.)

9.2. BIOCHEMISTRY OF MELANIZATION: THE PROPHENOLOXIDASE-ACTIVATING SYSTEM

The key enzyme in melanin synthesis is phenoloxidase (EC 1.14.18.1). Usually in arthropods, the enzyme is the o-diphenol:O_2 oxidoreductase type and not the laccase p-diphenol:O_2 oxidoreductase form, which has a broader specificity and is found more commonly in plants (Mayer and Harel, 1979). Phenoloxidase facilitates melanin formation by oxidizing phenols to quinones, which are converted to melanin through a series of nonenzymic reactions. Because phenoloxidase produces toxic quinones and melanin, which could be deleterious for the animal, it normally occurs in arthropod hemolymph as the inactive proenzyme prophenoloxidase (Bodine et al., 1937; Ohnishi, 1953; Evans, 1967). Conversion of prophenoloxidase to phenoloxidase may be brought about by a variety of agents or treatments, including proteases (Schweiger and Karlson, 1962; Preston and Taylor, 1970; Thomson and Sin, 1970; Ohnishi et al., 1970; Dohke, 1973; Pye, 1974), lipids (Heynemann and Vercauteren, 1968), chloroform (Bodine et al., 1937; Bodine and Allen, 1938), organic solvents (Preston and Taylor, 1979), detergents (Inaba et al., 1963; Thangaraj et al., 1982), products of microbial origin (Pye, 1974; Unestam and Söderhäll, 1977; Ashida et al., 1983), heat (Ashida and Söderhäll, 1984), and aggregation of phenoloxidase itself (Mitchell and Weber, 1965; Ashida and Ohnishi, 1967; Munn and Bufton, 1973). Exactly how each of these exerts its effect on prophenoloxidase is unclear, but in insects and crustaceans the reaction seems to involve proteolytic cleavage, and the naturally occurring proteolytic activator in these animals is a serine protease present in the hemolymph (Ashida, 1981; Söderhäll, 1981). It is interesting to note that the silkworm, *Bombyx mori*, also possesses an activating protease in the cuticle (Dohke, 1973; Ashida and Dohke, 1980), and cleavage of *B. mori* hemolymph prophenoloxidase by this cuticular protease has been shown to release a peptide of MW 5000 (Ashida et al., 1974; Ashida and Dohke, 1980). More important, however, is the finding that the activating proteases of *B. mori* and the freshwater crayfish *Astacus astacus* themselves may require activation by proteolytic or other factors (Söderhäll, 1983; Ashida et al., 1983; Ashida and Söderhäll, 1984), and it is now known that in some arthropods, at least, a whole cascade of enzymes participates in the activation of phenoloxidase (see Figure 9.2 and Sections 9.2.1 and 9.2.2; Söderhäll, 1982).

Early studies of arthropod phenoloxidases have no doubt been confounded by the existence of this activating cascade, and in many of the first reports of the purification of prophenoloxidase, it is likely that it was the active, rather than the inactive, form of the enzyme that was isolated and studied (Inaba et al., 1978; Yamaura et al., 1980). Other factors further contributing to the problem of purifying phenoloxidase include the tendency of the enzyme to aggregate after activation (Mitchell and Weber, 1965; Ashida and Ohnishi, 1967; Munn and Bufton, 1973) and its stickiness, which causes

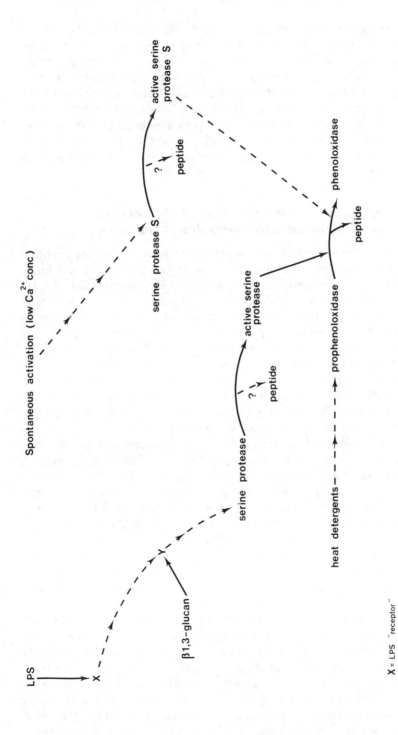

Figure 9.2. Prophenoloxidase-activating system of crustaceans. Dashed lines are inferred pathways. (LPS, lipopolysaccharides.)

it to adhere strongly to adjacent surfaces (Söderhäll et al., 1979; Ashida and Dohke, 1980; Leonard et al., 1985b). In crayfish, four other proteins in the activating sequence also become sticky upon activation (Söderhäll et al., 1984), and it is likely that similar findings will soon be reported for other arthropod species. Nonetheless, despite these complications, a few workers, notably Heynemann (1965), Ashida (1971), and Naqvi and Karlson (1979), have succeeded in obtaining homogeneous prophenoloxidase preparations from insects, and its heterogeneous molecular weight has been shown to be 87,000 in *Calliphora vicina* (Naqvi and Karlson, 1979) and 80,000 in *B. mori* (Ashida, 1971).

9.2.1. Activation by Nonself Molecules, β-1,3-Glucans and Lipopolysaccharides

While in vitro prophenoloxidase may be readily activated by proteases, other treatments, or agents, the question arises as to how activation is brought about in vivo. Since in arthropods melanization is known to nearly always accompany microbial or parasitic invasion, it is reasonable to expect components present on or released from the surface of such invaders to trigger the enzyme system in the host, and in 1974 Pye reported that bacterial or yeast cell walls do indeed enhance the activity of phenoloxidase in *Galleria mellonella*. Later, Unestam and Söderhäll (1977) showed that crayfish prophenoloxidase is activated specifically by β-1,3-glucans (soluble carbohydrates in the cell walls of all fungi except the Mucorales; see Table 9.1) and demonstrated that this activation is accomplished only by β-1,3-linked glycopyranosyl residues with a minimum degree of polymerization of the glucans of five (Söderhäll and Unestam, 1979). To date, prophenoloxidases from several arthropod species have been found to be sensitive to β-1,3-glucans (Ashida, 1981; Smith and Söderhäll, 1983a; Leonard et al., 1985a,b; Dularay and Lackie, 1985; see Table 9.1), as they are to the cell walls of bacteria (Ashida et al., 1983) or their constituent lipopolysaccharides (LPS; Söderhäll and Häll, 1984; see Table 9.1). In crayfish, the prophenoloxidase-activating system is so sensitive to LPS that even concentrations as low as 10^{-10}g/ml are still effective in bringing about activation (Söderhäll and Häll, 1984), although, significantly, concentrations of LPS above 10^{-4}g/ml actually prevent activation of prophenoloxidase (Söderhäll and Häll, 1984). In *B. mori*, while bacterial cell walls readily trigger prophenoloxidase activation, specific activation with LPS has not been achieved (Ashida et al., 1983; Ratcliffe et al., 1984). The reason for this is unknown, but Ashida (personal communication) has recently shown that peptidoglycan extracted from bacterial cell walls can induce proPO activation in *B. mori*.

Regarding the mechanism through which prophenoloxidase is activated, Ashida (1981) has reported that in *B. mori*, while crude prophenoloxidase preparations are readily triggered by β-1,3-glucans, purified prophenoloxidase preparations are not. Söderhäll (1981) has found that ^{14}C-labeled β-1,3-

TABLE 9.1. ACTIVATION OF ARTHROPOD proPO BY NONSELF MOLECULES

Species	Zymosan Yeast Cell Walls	β-1,3-Glucans	Bacterial Cell Walls	LPS	References
Crustacea					
Astacus astacus	Yes	Yes		Yes	Unestam and Söderhäll, 1977 Söderhäll and Unestam, 1979 Söderhäll and Häll, 1984
A. pallipes	Yes				Söderhäll and Vey, unpublished
A. leptodactylus	Yes				Söderhäll and Vey, unpublished
Pacifastacus leniusculus	Yes	Yes		Yes	Johansson and Söderhäll, 1985
Carcinus maenas	Yes	Yes		Yes	Söderhäll et al., 1986
Nephrops norvegicus	Yes	Yes			Smith and Söderhäll, unpublished
Insecta					
Galleria mellonella	Yes	Yes	Yes	No	Pye, 1974 Ratcliffe et al., 1984
Bombyx mori	Yes	Yes	Yes	No	Ashida, 1981, Ashida et al., 1983
Blaberus craniifer	Yes	Yes			Leonard et al, 1985b
Hyalophora cecropia	Yes	Yes			Söderhäll, unpublished

glucans do not bind to activated phenoloxidase, indicating that the glucans do not affect prophenoloxidase directly but trigger instead an intermediary factor that brings about activation of the proenzyme. Since in both crayfish and *B. mori*, soybean trypsin inhibitor (STI), diiosopropyl fluorophosphate (DFP), and *p*-nitrophenyl-*p*-guanidinobenzoate (pNPGB) have all been found to block β-1,3-glucan activation of prophenoloxidase (Ashida, 1981; Söderhäll, 1981; Ashida and Söderhäll, 1984), the intermediate factor is probably a serine protease; confirmation of this proposal has been provided by Söderhäll (1983), Söderhäll and Häll (1984), and Ashida and Söderhäll (1984). Furthermore, this serine protease appears to be highly specific in action and, with the use of synthetic chromogenic peptides, has been shown to hydrolyze only peptides with the structure R-Gly-Arg-pNA (Söderhäll, 1983). In addition, prophenoloxidase can also become activated by low calcium concentrations—that is, without intervention of nonself molecules (see Section 9.2.2; Söderhäll and Häll, 1984; Leonard et al., 1985b)—and this spontaneous activation involves a serine protease with the same specificity as that which is activated by LPS or β-1,3-glucans. The serine protease, which undergoes spontaneous activation at low Ca^{2+} levels, is directly responsible for the conversion of prophenoloxidase to phenoloxidase and is termed the protease S to distinguish it from the other serine protease, which is triggered by β-1,3-glucans or LPS. The activation of prophenoloxidase by nonself molecules thus appears to involve a number of sequential steps in an enzymic cascade (the proPO system; Fig. 9.2), although it is interesting to note that the glucans do not switch on the cascade at the same point as does LPS, since proPO preparations inhibited by high LPS concentrations (see above) are still susceptible to activation by β-1,3-glucans (Söderhäll and Häll, 1984).

9.2.2. Activation of Other Agents or Factors

In crustaceans, the proPO cascade has also been shown to be triggered in other ways, for example, by heat, sodium dodecyl sulphate (SDS), or trypsin (Ashida and Söderhäll, 1984). With heat, activation occurs very rapidly and is usually completed within 3 min (Ashida and Söderhäll, 1984). The optimum temperature is 58°C. Below 45°C, activation does not take place, while at temperatures above 65°C, the proPO-system is inactivated (Ashida and Söderhäll, 1984). However, unlike β-1,3-glucan or LPS stimulation, neither heat nor SDS activation is inhibited by pNPGB (a serine protease inhibitor); so, clearly, activation by these treatments follows a different route than that triggered by nonself molecules (Ashida and Söderhäll, 1984). A similar situation occurs in the insects (Leonard et al., 1985b); and while much remains to be done to resolve the effects of SDS and heat on the proPO cascade in both crustaceans and insects, the evidence available so far from crustaceans indicates that heat, at least, activates only the last step in the cascade, namely, the conversion of prophenoloxidase to phenoloxidase (Ashida and Söderhäll, 1984; Fig. 9.2). Trypsin, in contrast, shows yet another pattern of activation, in that there is always a time lag before phenoloxidase activity can be seen (Ashida and Söderhäll, 1984). Trypsin seems to act by factors (as yet undefined) that are different from the prophenoloxidase-activating enzyme (Ashida and Söderhäll, 1984).

Calcium is always necessary for complete activation of proPO, irrespective of the type of elicitor used (Söderhäll, 1981; Ashida and Söderhäll, 1984), and omission of calcium from the reaction mixtures or treatment of crude proPO fractions with ethylenediaminetetra acetic acid (EDTA) prevents detection of phenoloxidase activity in vitro (Ashida and Söderhäll, 1984). EDTA treatment, in particular, causes irreversible inactivation of the proPO cascade, although prophenoloxidase itself appears to remain intact (Ashida and Söderhäll, 1984). Calcium is probably required for the last step in the activating sequence: the conversion of prophenoloxidase to phenoloxidase (Ashida and Söderhäll, 1984). It is curious, though, that crayfish and *B. mori* seem to differ in this respect, since in *B. mori* prophenoloxidase can still be cleaved to its active form by its activating enzyme in the absence of calcium (Dohke, 1973; Ashida et al., 1983), while in crayfish it cannot be. Ashida and Söderhäll (1984) have suggested that Ca^{2+} ions are necessary for the maintenance of the active conformation of phenoloxidase. The sequence of events that takes place during proPO activation, whether by calcium or other means, is summarized in Figure 9.2.

9.2.3. Regulation of the Activating Pathway

Enzymic cascade reactions generally occur in nature when there is a need for amplification of a chemical response to a small amount of the elicitor. Accordingly, there must also be some way of controlling or regulating the reactions to avoid magnification of the final response or responses when magnification would be inappropriate. Such regulation appears to operate within the proPO system in arthropods, although much remains to be understood about the mechanisms involved. In crayfish and many other arthropods, the proPO cascade is compartmentalized in the circulating blood cells (see Sections 9.3.2 and 9.6). In addition, specific proteolytic inhibitors are present in the hemocytes that may stop or delay excessive activation of prophenoloxidase in the hemocoel (Häll and Söderhäll, 1982), but perhaps of greater importance is the nature of the activation sequence itself. With crayfish, Söderhäll (1983) has shown that the activity of the serine protease in the proPO cascade is always short-lived (approximately 20 min at 37°C) and occurs before any significant phenoloxidase activity is detectable. Moreover, protease activity, but not phenoloxidase activity, is totally lost if hemocyte lysate supernatants (HLS) treated with glucans are left to stand for 4–5 hr at 20°C (Söderhäll, 1983). The biochemical mechanism of this self-inactivation is unknown, but a similar phenomenon is shown by the clotting enzyme of the horseshoe crab *Tachypleus tridentatus* (Nakamura et al., 1982) and could represent a way of avoiding massive in vivo triggering of the system, when the presence of the initiator (presumably infective microorganisms) is not sustained (Söderhäll and Smith, 1984). Recently, an α_2-macroglobulinlike molecule has been detected in the plasma of *Limulus polyphemus* (Quigley and Armstrong, 1983) and crayfish (Hergenhahn and Söderhäll, 1985). In vertebrates, α_2-macroglobulin is known to sequester different proteases, including those of complement, from circulation (Starkey and Barrett, 1977), so it might be expected that an equivalent form in invertebrates serves the same function.

9.3. OCCURRENCE AND DISTRIBUTION OF THE proPO SYSTEM

Prophenoloxidase and possibly the proPO-system occurs widely, if not universally, throughout the Arthropoda. It is also present in the echinoderms (Jacobson and Millott, 1953; Millott, 1953; Smith and Söderhäll, unpublished; Söderhäll and Bertheussen, unpublished), gastropods (de Aragao and

Bacila, 1976), trematodes (Bennett et al., 1978; Seed and Bennett, 1980), and other invertebrates, including some tunicates and bivalves (Smith and Söderhäll, unpublished). It usually predominates in the hemolymph and for a few species has been shown to reside in the hemocytes, particularly the granulocytes (Smith and Söderhäll, unpublished; Söderhäll and Bertheussen, unpublished). For most species, however, the precise location of the key enzymes and their relative abundance in different cell or tissue types still need clarification, as does the sensitivity of these proteins to defined nonself molecules.

9.3.1. Crustaceans

As early as 1929, Pinhey reported the presence of tyrosinase enzymes in *Cancer pagurus* hemocytes. These enzymes were released from the cells by cytolysis and caused blackening of the hemolymph within a short time (Pinhey, 1929). Bhagvat and Richter (1938) subsequently confirmed that this was due to phenoloxidase activity, and Decleir and Vercauteren (1965) established that the enzyme existed in the hemocytes as a proenzyme. Decleir and Vercauteren (1965) also noted that maximum phenoloxidase activity occurred in the hemolymph during stages C_2–C_4 and A_1–B_1 of the molt and resided principally in the granulocytes. Prophenoloxidase has now been identified in the hemocytes of several crustaceans, including *Carcinus maenas*, *C. pagurus*, *Macropipus depurator*, *Eupagurus bernhardus*, *Nephrops norvegicus*, *A. astacus*, and *Pacifastacus leniusculus* (Unestam and Söderhäll, 1977; Söderhäll and Unestam, 1979; Smith and Söderhäll, 1983a; Söderhäll and Smith, 1983 and unpublished). More important, having devised a method of separating the various hemocyte populations of crustaceans, we have been able to show conclusively that prophenoloxidase is strictly confined to the granulocytes and the semigranular cells in these animals (Söderhäll and Smith, 1983; Smith and Söderhäll, 1983b). Greater phenoloxidase activity occurs in the granulocytes than in the semigranular cells and is totally absent from the hyaline cells (or, as they are termed throughout this chapter, to be consistent with other chapters, plasmatocytes; Söderhäll and Smith, 1983; Smith and Söderhäll, 1983b). Phenoloxidase thus represents a convenient marker for purity of the plasmatocytes following their separation on density gradients of Percoll, and a method for the rapid detection of phenoloxidase contamination has been described by Söderhäll and Smith (1983).

9.3.2. Insects

The insects have been subject to considerable investigation, and nearly all the major orders have been shown to display phenoloxidase activity in the hemolymph (for reviews see Messner, 1972; Crossley, 1979; Söderhäll, 1982). Less certain, however, is the distribution of the proPO enzymes within the "blood." Several workers have shown by cytochemical staining with dihydroxyphenylalanine (L-dopa) that phenoloxidase resides in the cells (Vercauteren and Aerts, 1958; Rizki and Rizki, 1959; Decleir et al., 1960; Andreadis and Hall, 1976; Schmit et al., 1977; Crossley, 1979), and recently Leonard and associates (1985b) have provided confirmatory biochemical data for the cellular origin of phenoloxidase activity in *Blaberus craniifer* (see also Chapter 14). Other workers, however, claim that the proPO occur in the plasma (Maier, 1973; Götz and Vey, 1974; Pye, 1974; Vey and Götz, 1975; Ashida, 1981; Ashida et al., 1983). While this may be true, it could also be a reflection of the techniques used to bleed the animals rather than a feature of the taxonomic position of the species tested. Certainly, some insects, for example, the chironomids, have small hemolymph volumes and low cell counts (Rowley and Ratcliffe, 1981), making collection of sufficient hemocytes for study difficult, and in other species, the hemocytes may be so labile (Rowley and Ratcliffe, 1981) that there might be spontaneous release of the proPO proteins during handling or bleeding. With crustacean hemocytes, we have found that such discharge of proPO enzymes is induced by a variety of nonself molecules, including components of the proPO cascade itself (Smith and Söderhäll, 1983b), so it is possible that proPO contamination of the plasma from the cells could happen very easily in insects. Moreover, in view of the potency of the proPO system in eliciting coagulation and as a source of lytic factors (see Section 9.4.5), the presence of enzymes that are highly sensitive to foreign molecules in the plasma is potentially hazardous for the host in case activation of the system occurs accidentally. It is therefore reasonable that arthropods avoid the risk of massive intravascular clotting and its lethal consequences by sequestering the system inside the hemocytes (Söderhäll and Smith, 1984). With *G. mellonella*, Schmit and coworkers (1977) have demonstrated cytochemically that phenoloxidase originates in the granulocytes and oenocytoids. And recently, Iwama and Ashida (1986) showed that proPO is synthesized by the oenocytoids in *B. mori*. Few other analyses of the differential distribution of the proPO system in insect hemocytes have been made, although it would be pertinent to ascertain whether the coagulocytes (which are involved in coagulation) and the plasmatocytes (which execute phagocytosis) enclose any of the proPO enzymes.

9.3.3. Other Species

Only a limited amount of information is available for the onycophorans and myriapods, but phenoloxidase activity has been reported to be present in the hemocytes of spirobolid millipedes (Bowen, 1967), and Krishnan and Ravindranath (1973) have shown that it is localized in the granulocytes. With the chelicerates, despite extensive studies of the granulocytes of the horseshoe crabs (*Limulidae*), prophenoloxidase activity has not been detected in *L. polyphemus* hemolymph (Söderhäll et al., 1985a). These animals do, however, possess a complex clotting system within the granulocytes, which is sensitive to activation by carboxymethylated β-1,3-glucans or LPS (Levin and Bang, 1964; Kakinuma et al., 1981; Morita et al., 1981; Söderhäll et al., 1985a).

9.4. FUNCTIONAL CONSIDERATIONS OF THE proPO SYSTEM

The significance of melanin in arthropod immunity has been the subject of research and speculation for many years, but only recently has attention turned to defining the role of the activating cascade that leads to melanization in the host. So far, these investigations have concentrated on the host defense reactions, and compelling evidence that the proPO cascade plays a crucial part in nonself recognition has now been presented. However, it is possible that components of the proPO system may also have other, peripheral effects on the host, for example, in molting, hemopoiesis, and/or behavior. An evaluation of the part played by the cascade in these processes would be an interesting and potentially important area of investigation.

9.4.1. Phenoloxidase Activity and Host Resistance

With respect to the proPO cascade in *G. mellonella*, immunization has been reported to diminish later melanization in the hemolymph (Stephens, 1962; Ziprin and Hartman, 1971). This loss of melanization capability was considered by Ziprin and Hartman (1971) to confer protection on the host against secondary infection, and Ziprin (1978) suggested that the delay in hemolymph melanization time was in fact a measure of acquired immunity. In a separate study, Pye (1978) reported that phenoloxidase activity was abolished in *G. mellonella* following injection of LPS, but, as LPS is known to activate prophenoloxidase in crustaceans and bacterial cell walls to activate prophenoloxidase in *B. mori* (Ashida et al., 1983), it is likely that the loss of phenoloxidase activity and reduced melanization seen in *G. mellonella* by Pye (1978) and Ziprin (1978) may have been artifactual. It is possible that the LPS caused large-scale lysis and/or degranulation of the blood cells in vivo, thereby releasing the proPO system into the hemolymph, where it may have been either inactivated by an inhibitor or inhibitors present in the plasma or, alternatively, inhibited if the LPS concentration was high (see Section 9.2.2). Certainly, active phenoloxidase is extremely toxic for the host (Zlotkin et al., 1973), probably by producing toxic quinones and melanin (Söderhäll and Ajaxon, 1982); so it is imperative that excess phenoloxidase is removed from the circulation as rapidly as possible. The adhesion of phenoloxidase, a sticky protein (see Section 9.2), to the internal tissues would serve as one way in which the dispersion of the enzyme around the body could be restricted.

A different approach adopted by other workers to assessing the extent of the participation of the proPO system in host defense has been to examine the ways in which certain parasites evade the host immune reactions, for example, by suppressing encapsulation (Nappi and Streams, 1969; Streams and Greenberg, 1969; Kitano, 1974; Vinson, 1976). In such systems, it is reasonable to suppose that the parasites have evolved mechanisms to inhibit the activation and/or activity of phenoloxidase in vivo (Stoltz and Cook, 1983), but, again, evidence in favor of this hypothesis has not always been unequivocal. With *Heliothis virescens*, Sroka and Vinson (1978) found that failure of the host to encapsulate the eggs of hymenopteran wasps was accompanied by a delay or a decrease in phenoloxidase activity of the host hemolymph. Brewer and Vinson (1971) had previously proposed that the degree of phenoloxidase activity correlated with the extent of encapsulation, so Sroka and Vinson (1978) suggested that the parasitoids interfered with the host proPO system in some way. In a more recent report, Stoltz and Cook (1983) have shown conclusively that in *H. virescens* and *Trichoplausia ni*, hymenopteran parasites can suppress phenoloxidase activity in the host hemolymph, and by using pronase to activate *T. ni* hemolymph, these researchers were able to detect phenoloxidase activity in both the parasitized and control larvae, thereby revealing that the enzyme is still present in the hemolymph of infected animals, although activity is not expressed. Stoltz and Cook (1983) also found that phenoloxidase activity could be almost totally abolished in the host by injection of a virus that had been purified from the ovaries of the parasitoid. Exactly how the viruses brought about this effect on the host remains an enigma.

9.4.2. Clotting

Crayfish hemocyte lysate supernatants can be induced to undergo gelation or clotting reaction in vitro by the addition of β-1,3-glucans or LPS (Söderhäll, 1981; Söderhäll and Häll, 1984). The coagulogen responsible for this clotting has not been fully characterized but seems to be converted to coagulin by proteolysis (Durliat and Vranckx, 1981; Söderhäll, 1981). The proteolytic enzyme facilitating this change displays similar hydrolytic activity to that of the protease in the proPO cascade that causes conversion of prophenoloxidase to phenoloxidase (see Section 9.2.2). The clotting reaction seen in crayfish HLS may therefore be related to the proPO system and depends on a serine protease in the cascade (Durliat, 1985). A similar sequence of events occurs in the insects, and as with crustaceans, the clotting protein (coagulogen) is thought to be present in the plasma as well as the hemocytes (Brehélin, 1972, 1979; Barwig and Bohn, 1980; Bohn et al., 1981).

In insects, clotting usually occurs in one of two ways. For the majority, coagulation of the hemolymph is preceded by release of materials from the hemocytes to form a gel, but in a few species there is no gelation of the hemolymph, and instead the hemocytes clump together to form a plug, which prevents blood loss (for reviews see Grégoire, 1970; Rowley and Ratcliffe,

1981; Bohn and Barwig, 1984). In some cases clotting may result in a combination of plasma gelation and hemocyte aggregation, but generally the coagulocytes and granulocytes are the key hemocytes in plasma gelation, and they act by discharging material from their granules, which interacts with the hemolymph to form a coagulum (Rowley and Ratcliffe, 1981). Calcium ions are usually necessary for the rupture of the hemocytes and, in *Leucophaea maderae*, EDTA in high concentration (approximately 5%) prevents cell lysis (Bohn, 1977). After release from the hemocytes, the coagulum associates and cross-links with the plasma coagulogen (Bohn and Barwig, 1984), although the clotting reaction of *L. maderae*, unlike that of *L. polyphemus* (Young et al., 1972; Levin, 1979) and crustaceans (Durliat and Vranckx, 1981; Söderhäll, 1981), does not require a serine protease and lacks transglutaminase activity (Bohn and Barwig, 1984), which in lobsters crosslinks plasma coagulogen (Fuller and Doolittle, 1971). In *G. mellonella*, while the granulocytes are known to contain phendoxidase (Schmit et al., 1977), the presence of a clotting enzyme in the coagulocytes has not been demonstrated. Nonetheless, in the light of work on *L. maderae* (Bohn, 1977; Barwig and Bohn, 1980; Bohn et al., 1981; Bohn and Barwig, 1984), it appears that the clotting process of insects differs from that of other arthropods principally in that a trypsinlike enzyme is not involved in the transformation of coagulogen to coagulin (Bohn and Barwig, 1984).

In the *Limulidae*, there is only one cell type, the granulocyte (or, as it is sometimes termed, the amoebocyte; Armstrong and Levin, 1979), which upon exposure to nonself molecules (LPS) rapidly releases its contents (Levin and Bang, 1964; Levin, 1979; Armstrong and Rickles, 1982; Armstrong, 1985). The biochemical changes that take place during clotting in this group of animals have been well documented and are outlined in Section 9.2.5, where the clotting reaction of the *Limulidae* is compared and contrasted with the proPO cascade of other arthropods.

9.4.3. Opsonins

In the broadest sense, opsonins are nonspecific recognition factors that in vertebrates enhance the response of immunocompetent cells to foreign particles. The term *opsonin* is most often applied to those molecules in the serum or plasma that stimulate phagocytosis. A great effort has gone into the search for opsonins in arthropod hemolymph, although the results of these studies have often been very inconsistent: some arthropod species seem to possess opsonins, while others do not (Tyson et al., 1974; Chorney and Cheng, 1980; Ratcliffe, 1982). Despite this, the proPO cascade in crustaceans has been found to generate a number of sticky or attaching proteins, including phenoloxidase, after activation with nonself molecules (Söderhäll et al., 1979, 1984). It may therefore serve as a source of opsonic factors for the cells. With the crayfish *A. astacus* and the shore crab *C. maenas*, we have recently provided indirect evidence that the proPO cascade does contain factors that serve as opsonins for the hemocytes (Smith and Söderhäll, 1983a). In these experiments, we found that incubation of the hemocyte-bacteria mixtures with β-1,3-glucans in vitro significantly (approximately threefold) raised the percentage of phagocytosis (Smith and Söderhäll, 1983a). Treatment of the cells with other glycans did not evoke the same response, and pretreatment of the hemocytes with β-1,3-glucans before inoculation with the test bacteria slightly depressed the rate of uptake (Smith and Söderhäll, 1983a). As β-1,3-glucans accelerate the rate of degranulation of the hemocytes in vitro (Smith and Söderhäll, 1983a), we suggested that the glucans stimulate phagocytosis by inducing release of proPO components from the cells, which then interact with the foreign particles and facilitate their ingestion by the phagocytes. The phenomenon is time dependent, and if the proPO cascade is completely activated before inclusion of the bacteria in the reaction mixtures, then phagocytosis is imparied, presumably because the opsonins attach to the cells or the nearest adjacent surface and are unavailable for the bacteria. Conclusive evidence that it is the sticky, attaching proteins of the proPO cascade that have opsonic properties has now been obtained with isolated plasmatocytes of *C. maenas* in vitro (Söderhäll et al., 1986). However, phenoloxidase itself is not an opsonin in crabs, as bacteria precoated with HLS activated by heat (in which only phenoloxidase is generated; see Section 9.2.2) are ingested at the same rate as bacteria pretreated with saline solution, whereas bacteria precoated with HLS activated with glucans (in which the entire cascade is stimulated) are phagocytized much more readily (Söderhäll et al., 1986).

Corroborative in vivo findings have also been reported for crayfish by Söderhäll and colleagues (1984). In this study, *A. astacus* hemocytes were found to exhibit stronger in vivo encapsulation reactions toward fungal spores coated with crayfish HLS than to fungal spores coated with plasma or buffer (Söderhäll et al., 1984). Likewise, in crabs, injection of β-1,3-glucans has been shown to specifically induce the formation of numerous small clumps (nodules) in the gills even in the absence of bacteria (Smith et al., 1984). Thus, for crustaceans, the proPO cascade seems to be intimately involved in the recognition and response to foreignness by the cells (Table 9.2).

An opsonic role for the proPO system has now also been implicated for insects (Ratcliffe et al., 1984), in which, in addition to β-1,3-glucans, LPS has also been found to stimulate the cells in vitro. As yet, further studies with other arthropod species have not been forthcoming, and much remains to be learned about the precise nature of the proPO-activating proteins that serve as opsonic factors in both insects and crustaceans (Table 9.2).

9.4.4. Involvement in Cellular Communication

In mammals, efficient immunity against invading microorganisms is achieved through the interaction of the various subpopulations of lymphocytes. While

TABLE 9.2. BIOLOGIC FEATURES OF THE HEMOCYTE POPULATIONS OF *ASTACUS ASTACUS* AND *PACIFASTACUS LENIUSCULUS*

Feature	Plasmatocytes	Semigranular Cells	Granulocytes
Phagocytosis in vitro	Yes	Yes, limited	No
Effect of β-1,3-glucans	NT	Degranulation	No response
Effect of LPS	NT	Degranulation	No response
Effect of proPO proteins	Enhanced phagocytosis	Lysis	Degranulation
Nodule or capsule formation	No	Yes	No
Cytotoxicity	NT	Yes	Yes
Phenoloxidase activity	Absent	Limited	Strong
Protease activity	NT	Limited	Strong

Key: NT, not tested.

equivalent cell cooperation per se may not operate in invertebrates, there does appear to be some communication and interaction among the various hemocyte types (Rowley and Ratcliffe, 1981). A scheme suggesting how insect hemocytes might act in concert in vivo during nodule formation has been provided by Ratcliffe and Gagen (1977). Until recently, the mechanism through which this might be achieved was unknown, but circumstantial evidence that elements of the proPO system are involved in this phenomenon is beginning to emerge. In *G. mellonella*, Ratcliffe and Gagen (1976, 1977) identified the granulocytes as the first cells to respond to the presence of nonself entities in the hemocoel by releasing substances into the hemolymph to form a localized clot around the foreign material. This clot attracts other cells, notably plasmatocytes, to the site of infection and entangles them (Ratcliffe and Gage, 1977), and recently Ratner and Vinson (1983a,b) have shown that an encapsulation-promoting factor is released from the hemocytes of *H. virescens*. This factor may well be one or several of the sticky proteins generated upon activation of the proPO system (Söderhäll et al., 1979; Ashida and Dohke, 1980; Söderhäll et al., 1984; Leonard et al., 1985b). Participation of more than one cell type in capsule or nodule formation also occurs in crayfish and crabs, and in each case it is the granulocytes that have been seen to undergo degranulation or lysis (Unestam and Nylund, 1972; Smith and Ratcliffe, 1980). Since these cells are the major repositories of the proPO-activating proteins (Söderhäll and Smith, 1983; Smith and Söderhäll, 1983b), some of the material released from them must undoubtedly be proPO proteins, and a scheme depicting the probable route for cell communication in arthropods in vivo is shown in Figure 9.3. In vitro, the enhancement of phagocytosis by β-1,3-glucans, reported by Smith and Söderhäll (1983a) and Ratcliffe and associates (1984), must similarly involve cooperation between hemocyte types, particularly as in *C. maenas* phagocytosis is known to be accomplished only by the plasmatocytes (Smith and

**TABLE 9.3. CYTOTOXICITY OF CRAYFISH
HEMOCYTES AGAINST FOUR-TUMOR CELL LINES,
MOUSE FIBROBLASTS, AND SHEEP RED BLOOD
CELLS**

Target Cells	Cytotoxicity (% specific ^{51}Cr release)
P815, a tumor cell line	45
K562, a tumor cell line	20
YAC-1, a tumor cell line	70
RAJI, a tumor cell line	40
Fibroblasts	15
Sheep red blood cells	30

Source: From Söderhäll et al., 1985c.

Ratcliffe, 1978; Söderhäll et al., 1986), whereas key enzymes of the proPO system are strictly confined to the granulocytes and semigranular cells (Söderhäll and Smith, 1983). In *C. maenas*, the elevated levels of uptake observed in the glucan-treated monolayers could only have been due to the activated proPO proteins contributed by the granulocytes and semigranular cells. Data favoring this hypothesis have recently been presented (Söderhäll et al., 1986) where pure populations of (proPO-deficient) plasmatocytes are stimulated to ingest bacteria by the attaching proteins derived from crab HLS by glucan activation.

Cell cooperation may further be inferred for crayfish, in which β-1,3-glucans also enhance the rate of phagocytosis of bacteria in vitro. However, crayfish, unlike crabs, possess large populations of proPO-containing semigranular cells capable of ingesting bacteria (Smith and Söderhäll, 1983b; Table 9.2). Since exposure of these cells to β-1,3-glucans induces profound degranulation of the cytoplasm (Smith and Söderhäll, 1983b) and following exocytosis the semigranular cells loose their capacity to phagocytize (Söderhäll et al., 1986), it is unlikely that the semigranular cells alone account for the elevated levels of bacterial uptake. Since the granulocytes in crayfish are always nonphagocytic (Smith and Söderhäll, 1983b; Söderhäll et al., 1986), the increased uptake of bacteria could have been due to stimulation of the plasmatocytes by the proPO proteins. At present, we have no reason to suppose that the glucans directly affect the plasmatocytes of crayfish or crabs, and research is currently under way to define the nature of the communicating signals of the proPO system in crustaceans (see also Chapters 10 and 11).

9.4.5. Lysis, Exocytosis, and Cytotoxicity

Scattered throughout the literature on arthropod immunity are reports of lysins, lysozymes and other lytic factors in the hemolymph (Cantacuzéne,

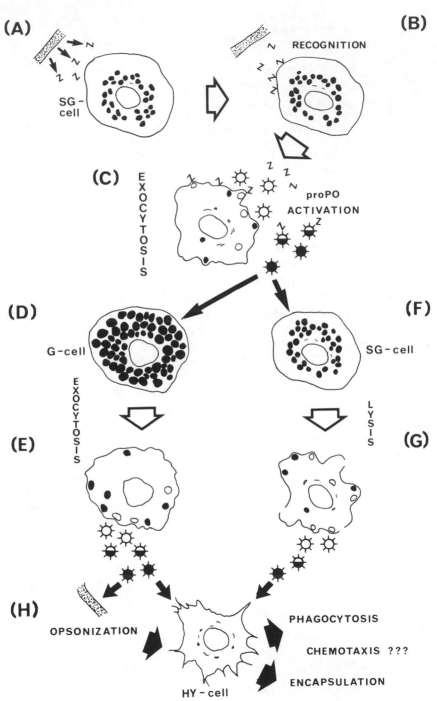

Figure 9.3. Hypothetical scheme of cell cooperation in host cellular defense reactions of crustaceans. (A) Foreign agent releases nonself signals (Z), such as LPS or β-1,3-

1923; Weinheimer et al., 1969). Apart from lysozymes, few characterizations of these factors have been carried out, and there is no information on how the action of such potent molecules is moderated in the hemocoel. Undoubtedly, lytic factors assist in the destruction and elimination of invasive microorganisms from the body, and for the lepidopteran *Hyalophora cecropia* at least, Boman and coworkers have very elegantly shown how one group of such lytic factors is synthesized. These proteins, termed *cecropins*, are induced in *H. cecropia* hemolymph by immunization of the pupae with bacteria (Boman and Steiner, 1981). A wide range of cecropins and related antibacterial proteins have been isolated, purified, and sequenced (Boman, 1980; Steiner et al., 1981), and their effect on a variety of test bacteria has been demonstrated. However, a correlation between phenoloxidase activity and antibacterial immunity in *H. cecropia* has not so far been found, and, in the absence of information about the biochemical events involved in cecropin production, it is reasonable to speculate that peptides liberated during activation of the proPO cascade in *H. cecropia* might be important. Further work on the nature and action of the peptides released from the proPO system in these insects and other arthropods is clearly needed. Melanin and the intermediate compounds generated by phenoloxidase activity are already known to be stable, free radicals (Gan et al., 1976) that possess antimicrobial properties (Kuo and Alexander, 1967; Bull, 1970; Söderhäll and Ajaxon, 1982) or are cytotoxic (Pawelek and Lerner, 1978; see also the review by Menon and Haberman, 1977). Unfortunately, the effects of the proPO-activating compounds have been less well documented, and while HLS from the crab *C. maenas* has been found to kill gram-negative *Moraxella* sp. in vitro (Fig. 9.4; Smith and Söderhäll, unpublished), a role for proPO proteins in bacterial killing has yet to be confirmed. In crayfish, activated prophenoloxidase components lyse isolated crayfish semigranular cells in culture (Smith and Söderhäll, 1983b). The cascade proteins alone are unable to lyse ^{51}Cr-labeled mouse or human tumor cells in vitro (Söderhäll et al., 1985c), but intact crayfish hemocytes are highly cytotoxic toward the target cells, and the killing ability appears to reside inside the semigranular cells and granulocytes (Söderhäll et al., 1985c; Table 9.3). As described in Section 9.4.3, nonself molecules degranulate crayfish and crab hemocytes in vitro (Smith and Söderhäll, 1983b), and more elaborate studies have revealed that,

---

glucans, in the vicinity of a host's semigranular cell. (B) Nonself signals interact with the semigranular cell. (C) Semigranular cell is stimulated to degranulate and release proPO-system (☼) into the surrounding medium. The proPO system is then activated (✸) by the nonself molecules (N) to form the attaching and other active proPO proteins (✹). (D,E) The active proPO proteins interact with and trigger exocytosis of adjacent granulocytes. (F,G) Other semigranular cells (SG) are further stimulated to undergo degranulation, and ultimately lysis, by the active proPO proteins, thereby amplifying the cellular response to the foreign agent. The active proPO proteins then opsonize the foreign material and/or induce reactivity by the plasmatocytes. (SG, semigranular cell; G, granulocytes; HY, plasmatocytes.)

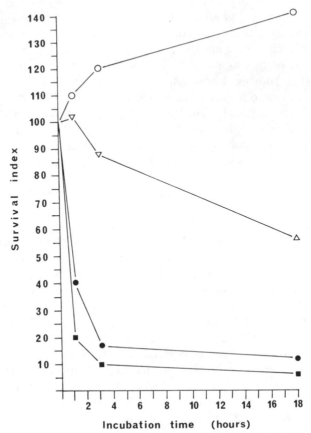

Figure 9.4. In vitro killing of the bacterium *Moraxella* sp. by the hemocytes of *C. maenas*. The HLS, GLS, and PLS were prepared in CS according to the methods described by Smith and Söderhäll (1983a) and Söderhäll and coworkers (1985a). Approximately 1 × 10⁵ bacteria were suspended in 5 ml of HLS, GLS, PLS, or CS at 20°C, and at intervals aliquots were withdrawn, serially diluted, and plated onto Zobell's marine agar (DIFCO 2216E). The survival index was calculated from

$$\frac{\text{colony count at sample time } (t_t)}{\text{colony count at zero time } (t_o)} \times 100$$

(– ▲ –, whole hemocyte lysate supernatant, or HLS; – ● –, granulocyte lysate supernatant, or GLS; – ○ –, plasmatocyte lysate supernatant, or PLS; – △ –, *Carcinus* saline (CS) solution control.)

in crayfish only, semigranular cells are induced to undergo exocytosis by LPS and β-1,3-glucans, the granulocytes being unresponsive (Johansson and Söderhäll, 1985; Table 9.4). In contrast, the Ca^{2+}-ionophore A23187, in the presence of Ca^{2+} ions, causes exocytosis in both the granulocytes and the semigranular cells of crayfish, and the proPO cascade thus released from

TABLE 9.4. IN VITRO EXOCYTOSIS OF CRAYFISH HEMOCYTES AFTER SEPARATION ON GRADIENTS OF PERCOLL

Treatment of Cells[a]	Percent Degranulation of Semigranular Cells	Granulocytes
Laminaran G (1 mg/ml in crayfish saline solution)	40	5
LPS (100 μg/ml in crayfish saline solution)	50	5
A23187 + 10 mM Ca^{2+} in 0.15 M NaCl	80	85
A23187 + 10 mM Ca^{2+} in 1 mM SITS in 0.15 M NaCl	NT	4
A23187 + 10 mM Ca^{2+} + 10 μM calmidazolium in 0.15 M NaCl	NT	4
Laminaran G + 100 μM SITS in crayfish saline solution	6	4
Control (crayfish saline solution)	16	5

Source: From Johansson and Söderhäll, 1985.
Key: NT, not detected.
[a] Monolayers of the hemocytes were incubated with the test reagents for 30–45 min at 20°C.

the cells is stable (i.e., inactive but sensitive to activation by nonself molecules; Johansson and Söderhäll, 1985; Table 9.4). Depletion of the granules from the cells by exocytosis abolishes their cytotoxic action, showing that cytotoxicity in crayfish is not only cell mediated but also seems to depend on proteins of the proPO system (Söderhäll et al., 1985c). Therefore, it is likely that, for crustaceans at least, exocytosis represents a mechanism through which previously cell-bound proPO components are released into

TABLE 9.5. SUMMARY OF THE FUNCTIONS CONNECTED WITH THE proPO SYSTEM IN THE HOST DEFENSE OF ARTHROPODS

Property	Astacus astacus	Carcinus maenas	Bombyx mori	Galleria mellonella	Blaberus craniifer
Opsonic	Yes	Yes	NK	Yes	Yes
Clotting	Yes	Yes	NK	NK	NK
Fungistatic	Yes	NK	NK	NK	NK
Killing of gram-negative bacteria	NK	Yes	NK	NK	NK
Capsule or nodule formation	Yes	Yes	NK	Yes	NK
Chemotaxis	No	No	NK	NK	NK

Source: For references, see text.
Key: NK, not known.

the hemolymph, where they are free to exert their biological function or functions (Table 9.5, Fig. 9.3).

9.5. COMPARISON OF THE proPO SYSTEM TO THE CLOTTING SYSTEM OF *LIMULUS*

In 1956, Bang discovered that clotting occurred when the granulocytes of the horseshoe crab *L. polyphemus* were exposed to gram-negative bacteria. Levin and Bang (1964) later established that LPS similarly induced a clotting reaction in lysates of the granulocytes, a finding that led to the development of the *Limulus* assay for the detection of endotoxin in biologic or diagnostic fluids (Levin and Bang, 1968). Subsequent investigations have revealed that the clotting reaction involves a number of constituent factors: a proclotting enzyme (Sullivan and Watson, 1975; Tai and Liu, 1977; Liu et al., 1979; Nakamura et al., 1982), a coagulogen (Tai et al., 1977), and an endotoxin binding protein (Liang et al., 1981). Endotoxin induces clotting in *Limulus* by stimulating the activation of a proclotting enzyme to form a clotting protein (coagulogen; Tai et al., 1977; Ohki et al., 1980). The coagulogen is then converted to coagulin by limited proteolysis (Young et al., 1972; Tai et al., 1977; Levin, 1979). Like the proPO system of crayfish, the clotting reaction of *Limulus* is calcium dependent (Young et al., 1972) and consists of a cascade of sequentially activated enzymes (Morita et al., 1981). The proclotting enzyme is itself a protease (Solum, 1973; Sullivan and Watson, 1975; Liu et al., 1979) that displays hydrolytic activity toward peptides with the structure R-Gly-Arg-pNA (Liu et al., 1979; Harada et al., 1979) in a manner identical to that shown by the activating protease in crayfish hemocyte lysate (Söderhäll, 1983; Fig. 9.2). The LPS receptor in the clotting cascade has been termed factor B (Ohki et al., 1980) to distinguish it from another factor (G) identified in the horseshoe crab system, which is activated by carboxymethylated β-1,3-glucans (Kakinuma et al., 1981; Morita et al., 1981). An additional factor, which inhibits LPS-mediated clotting, is also present in the *Limulus* system (Tanaka et al., 1982), but unlike factor B and factor G, this so-called anti-LPS factor seems to block activation of the proclotting enzyme in the clotting sequence. Thus, in possessing two separate receptors for LPS and β-1,3-glucans, the *Limulus* clotting system bears a further resemblence to the proPO cascade, although it is interesting to note that in *Limulus* activation with glucans has been achieved successfully only with a carboxymethylated type (Kakinuma et al., 1981; Morita et al., 1981) and not with the type of glucans that trigger proPO in *A. astacus*, *B. mori*, and *B. craniifer* (Söderhäll et al., 1985a). In other respects, the sequence of events during clotting in *Limulus* parallels those during proPO activation in crayfish (Table 9.6); the only major difference between them is the absence of final phenoloxidase activity in the horseshoe crabs (Armstrong, 1985; Söderhäll et al., 1985a).

TABLE 9.6. COMPARISON OF THE CLOTTING SYSTEM OF *LIMULIDAE* WITH THE proPO SYSTEM OF ARTHROPODS

Species Reference	Activation by LPS	Activation by β-1,3-Glucans	Activation Enzyme: Hydrolyzing Activity	Localization of the System	Result of Enzyme Cascade
Limulidae[a,b,c]	Yes	Yes[k]	R-Gly-Arg-pNA	Blood cells	Clotting
Astacus astacus[d,e,f]	Yes	Yes	R-Gly-Arg-pNA	Blood cells	Clotting, proPO Activation
Bombyx mori[g,h]	No	Yes	Not known	Plasma	proPO activation
Galleria mellonella[i]	No	Yes	Not known	Blood cells	proPO activation
Blaberus craniifer[j]	No	Yes	Not known	Blood cells	proPO activation

[a] Levin and Bang, 1964.
[b] Kakinuma et al., 1981.
[c] Morita et al., 1981.
[d] Söderhäll and Unestam, 1979.
[e] Söderhäll and Häll, 1984.
[f] Söderhäll, 1983.
[g] Ashida, 1981.
[h] Ashida et al., 1983.
[i] Ratcliffe et al., 1984.
[j] Leonard et al., 1985b.
[k] Carboxymethylated β-1,3-glucan.

9.6. COMPARISON OF THE proPO SYSTEM TO COMPLEMENT

The clotting system of *Limulus* has often been contrasted with the clotting pathway of mammals, but in the light of present findings on proPO activation, we believe that significant comparisons may further be made between the proPO cascade of arthropods and the complement pathways of mammals (Table 9.7).

Complement is a complex of nine major serum proteins that, together with a number of subunit components, serves to protect vertebrate hosts against microbial infections. The activation and biologic functions of complement have been well documented; for a recent concise review, see Bertheussen (1984). Briefly, the complement pathway may be activated by either IgM or IgG antibody-antigen complexes (the classic pathway) or by the cell envelopes of microorganisms (the alternative pathway) to yield by proteolytic cleavage the key complement C3 factors. Activated C3b (or, more properly, C3bi) binds strongly to nearby foreign particles (usually microorganisms) and facilitates their uptake by phagocytosis. C3b also initiates the assembly of a large lytic complex in the membrane of the foreign cells, and the small peptide fragments released during complement activation cause inflammatory reactions by chemotactically attracting polymorphs and stimulating release of histamines from the host cells.

Since complement is known to be specifically activated through the alternative pathway by β-1,3-glucans or LPS (Diluzio, 1979; Reid and Porter, 1981; Seljelid et al., 1981), in consisting of a complex series of sequentially activated enzymes sensitive to those molecules, the proPO system of arthropods mimics the alternative complement pathway. Further similarities between the proPO system and complement include the possession of activating serine proteases and the release of small peptides by proteolytic cleavage (Table 9.7). The spontaneous activation of the proPO cascade at low calcium concentrations may also be regarded as the arthropod equivalent of the classic pathway of mammalian complement in that those elements of the proPO cascade remaining after activation without nonself molecules are still susceptible to activation by β-1,3-glucans or LPS (see Section 9.2.2). Certainly, the proPO system generates sticky (attaching) proteins that serve as opsonins (Smith and Söderhäll 1983a; Söderhäll et al., 1984, 1986) in a manner akin to that of complement factors C3b and C5b and, like complement, has antimicrobial (Söderhäll and Ajaxon, 1982; Smith and Söderhäll, unpublished) and lytic properties (Smith and Söderhäll, 1983b). At present, it is unknown whether proPO components function as chemoattractants; preliminary experiments in our laboratories have so far failed to show a chemotactic function of hemocyte lysates toward plasmatocytes, semigranular cells, or granulocytes. Moreover, closer analysis of the phylogenetic relationship between complement and the proPO system has revealed that, while receptors for mouse C3bi are present in sea urchins (Bertheussen, 1982, 1983), they are absent from crayfish phagocytes (Söderhäll and Ber-

TABLE 9.7. COMPARISON OF THE proPO SYSTEM OF ARTHROPODS WITH THE COMPLEMENT PATHWAY OF MAMMALS

Property	Complement	proPO System
Structure	Cascade of proteases	A likely cascade of proteases and proPO
Regulation	By proteolytic inhibitors and α_2-macroglobulin	Protease inhibitors, α_2-macroglobulinlike molecules (?), compartmentalization
Localization	In plasma	In blood cells
Activation	By nonself molecules (LPS, β-1,3-glucans) through alternative pathway	By nonself molecules (LPS, β-1,3-glucans)
	By aggregated immunoglobulin through classic pathway	By low calcium
Role in phagocytosis or immune adherence	Provides opsonins	Provides opsonins
Role in cytotoxicity or cell lysis	Membrane damage, hemolysis and cytolysis by the complex C5-9	proPO-bearing hemocytes are cytotoxic, proPO-components lytic toward crayfish semigranular cells
Antimicrobial properties	Lyses gram-negative bacteria (C5-9)	Fungitoxic and probably bactericidal for gram-negative bacteria
Chemotactic agents	Fragment C5a attracts polymorphonuclear leucocytes	Not demonstrated
Anaphylatoxin or histamine release	Achieved by fragments C3a and 5a	Unknown; activated proPO system induces exocytosis of granulocytes

Source: For references, see text.

275

theussen, unpublished). Sea urchins themselves do not display phenoloxidase activity, yet other echinoderms, including sea stars, possess phenoloxidase activity but lack C3bi receptors (Söderhäll and Bertheussen, unpublished). It is very likely, therefore, that the opsonic factors found in crustaceans have a separate phylogenetic origin from those of mammals, and further research in this area is urgently required.

A major difference between the proPO system and complement is that proPO enzymes are located in the blood cells of several arthropods, while complement is located in the plasma. As discussed in Section 9.2.3, such compartmentalization may be necessary in the arthropods to protect the host against the damaging effects that undesirable triggering of the system would incur. Several workers have proposed that invertebrates might possess a complement or complementlike system (Day et al., 1970; Anderson, 1982; Bertheussen, 1982, 1983; Bertheussen and Seljelid, 1982), and so far for arthropods, the prophenoloxidase-activating system constitutes the most likely candidate for the origin of such an "invertebrate complement" (Söderhäll and Smith, 1984).

9.7. SUMMARY

The prophenoloxidase-activating system is a complex cascade of enzymes present in the hemolymph of nearly all arthropods that bring about the conversion of the inactive proenzyme prophenoloxidase to the active enzyme phenoloxidase before melanization. In addition to its role in melanin synthesis, the proPO system plays a major role in nonself recognition by the host and provides the molecular signals that activate the blood cells during the host defense reactions. The biochemical changes that lead to prophenoloxidase activation have been most extensively worked out for crayfish and *B. mori*. Calcium is required for the completion of the reactions (at least in crustaceans), and there appears to be some internal regulation of the cascade through time dependence in the activity of certain components and/or the presence of inhibitors in the hemolymph. The proPO system in vitro may be activated by a variety of agents or treatments but during systemic infection is probably activated either by specific nonself molecules, such as β-1,3-glucans or LPS, or spontaneously by low Ca^{2+} levels. The activating cascade is extremely sensitive to foreign molecules and may still be triggered by LPS or glucan concentrations as low as 10^{-10}g/ml. Functionally, the proPO system serves as a source of opsonins for the hemocytes, participates in coagulation, initiates encapsulation and nodule formation, and mediates microbial killing. Biochemically, many of the enigmas of prophenoloxidase activation still need to be resolved, and further research on other species is required, particularly to define additional factors in the cascade, the function of the released peptides, the nature of the cell communication signals, and the part played by the system in hemopoeisis. For investigations of this type, as with many aspects of invertebrate immunity, the techniques of immunocytochemistry and molecular biology need to be adopted, and from a proper understanding of the proPO system as the basis of nonself recognition and immunity in the arthropods, the phylogenetic relationships of the cascade to mammalian complement can be properly evaluated.

ACKNOWLEDGMENTS

This work was supported by grants from the Swedish Natural Science Research Council and the Royal Society.

REFERENCES

Anderson, R.S. 1982. Comparative aspects of the structure and function of invertebrate and vertebrate leucocytes, pp. 629–641. In N.A. Ratcliffe and A.F. Rowley (eds.) *Invertebrate Blood Cells*, vol. 2. Academic Press, New York.

Andreadis, T.G. and D.W. Hall. 1976. *Neoplectana carpocapsae:* Encapsulation in *Aedes aegypti* and changes in host hemocytes and hemolymph proteins. *Exp. Parasitol.* **39:** 252–261.

Armstrong, P.B. 1985. Adhesion and motility of the blood-cells of *Limulus*, pp. 77–124. In W.D. Cohen (ed.) *Blood Cells of Marine Invertebrates*. Alan R. Liss, New York.

Armstrong, P.B. and J. Levin. 1979. *In vitro* phagocytosis by *Limulus* blood cells. *J. Invertebr. Pathol.* **34:** 145–151.

Armstrong, P.B. and F.R. Rickles. 1982. Endotoxin induced degranulation of the *Limulus* amoebocyte. *Exp. Cell. Res.* **140:** 15–24.

Ashida, M. 1971. Purification and characterization of prephenoloxidase from hemolymph of the silkworm, *Bombyx mori. Arch. Biochem. Biophys.* **144:** 749–762.

Ashida, M. 1981. A cane sugar factor suppressing activation of prophenoloxidase in hemolymph of the silkworm, *Bombyx mori. Insect Biochem.* **11:** 57–65.

Ashida, M. and K. Dohke. 1980. Activation of prophenoloxidase by the activating enzyme of the silkworm, *Bombyx mori. Insect Biochem.* **10:** 34–47.

Ashida, M. and E. Ohnishi. 1967. Activation of prephenoloxidase in the hemolymph of the silkworm, *Bombyx mori. Arch. Biochem. Biophys.* **122:** 411–416.

Ashida, M. and K. Söderhäll, 1984. The prophenoloxidase activating system in crayfish. *Comp. Biochem. Physiol.* **77**B: 21–26.

Ashida, M., K. Dohke, and E. Ohnishi. 1974. Activation of prophenoloxidase. 3. Release of peptide from prophenoloxidase by the activating enzyme. *Biochem. Biophys. Res. Commun.* **57:** 1089–1095.

Ashida, M., Y. Ishizaki, and M. Iwahana. 1983. Activation of prophenoloxidase by bacterial cell walls or by β-1,3-glucans in plasma of the silkworm, *Bombyx mori. Biochem. Biophys. Res. Commun.* **113:** 562–568.

Bang, F.B. 1956. A bacterial disease of *Limulus polyphemus*. *Bull. Johns Hopkins Hosp*. **98**: 325–351.

Barwig, B. and H. Bohn. 1980. Evidence for the presence of two clotting proteins in insects. *Naturwissenschaften* **67**: 47–55.

Bennett, J.L., J.L. Seed, and M. Boff. 1978. Fluorescent histochemical localization of phenoloxidase in female *Schistosoma mansoni*. *J. Parasitol*. **65**: 941–944.

Beresky, M.A. and D.W. Hall, 1977. The influence of phenylthiourea on encapsulation, melanization and survival in larvae of the mosquito *Aedes aegypti*, parasitized by the nematode *Neoplectana carpocapsae*. *J. Invertebr. Pathol*. **29**: 74–80.

Bertheussen, K. 1982. Receptors for complement on echinoid phagocytes. 2. Purified human complement mediates echinoid phagocytosis. *Dev. Comp. Immunol*. **6**: 635–642.

Bertheussen, K. 1983. Complement-like activity in sea urchin coelomic fluid. *Dev. Comp. Immunol*. **7**: 21–31.

Bertheussen, K. 1984. Complement and lysins in invertebrates. *Dev. Comp. Immunol*. **3**: 173–181, supplement.

Bertheussen, K. and R. Seljelid. 1982. Receptors for complement on echinoid phagocytes. 1. The opsonic effect of vertebrate sera on echinoid phagocytes. *Dev. Comp. Immunol*. **6**: 423–432.

Bhagvat, K.D. and D. Richter. 1938. Animal phenolases and adrenaline. *Biochem. J*. **32**: 1397–1406.

Bodine, J.H. and T.H. Allen. 1938. Enzymes in ontogenesis (Orthoptera). 4. Further studies on the activation of the enzyme tyrosinase. *J. Cell. Comp. Physiol*. **12**: 71–84.

Bodine, J.H., T.H. Allen, and E.J. Boell. 1937. Enzymes in ontogenesis (Orthoptera). 3. Activation of naturally occurring enzymes (tyrosinase). *Proc. Soc. Exp. Biol. Med*. **37**: 450–453.

Bohn, H. 1977. Differential adhesion of the haemocytes of *Leucophaea maderae* (Blattaria) to a glass surface. *J. Insect Physiol*. **23**: 185–194.

Bohn, H. and B. Barwig. 1984. Hemolymph clotting in the cockroach *Leucophaea maderae* (Blattaria). *J. Comp. Physiol*. **154**B: 457–467.

Bohn, H., B. Barwig, and B. Bohn. 1981. Immunochemical analysis of hemolymph clotting in the insect *Leucophaea maderae* (Blattaria). *J. Comp. Physiol*. **143**B: 169–184.

Boman, H.G. 1980. A molecular approach to immunity and pathogenicity in an insect-bacterial system, pp. 217–228. In F. Chapeville and A.L. Haenni (eds.) *Molecular Biology, Biochemistry and Biophysics*, vol. 32. Springer-Verlag, Berlin.

Boman, H.G. and H. Steiner. 1981. Humoral immunity in *Cecropia* pupae, pp. 75–91. In W. Henle, P.M. Hofscheider, M. Kaprowski, O. Moaloe, F. Melchers, R. Rott, M.G. Sweiger, and P.K. Vogt (eds.) *Current Topics in Microbiology and Immunology*, vol. 95. Springer-Verlag, Berlin.

Bowen, R.C. 1967. Defence reactions of certain spirobolid millipedes to larval *Macrocanthorhynchus ingens*. *J. Parasitol*. **53**: 1092–1095.

Brehélin, M.M. 1972. Étude du mécanisme de la coagulation de l'hémolymphe d'un acidien, *Locusta migratoria migratorioides* (R. and F.). *Acrida* **1**: 167–175.

Brehélin, M. 1979. Hemolymph coagulation in *Locusta migratoria*: Evidence for a functional equivalent of fibrinogen. *Comp. Biochem. Physol.* **62**B: 329–334.

Brewer, F.D. and S.B. Vinson. 1971. Chemicals affecting the encapsulation of foreign material in an insect. *J. Invertebr. Pathol.* **18**: 287–289.

Bull, A.T. 1970. Inhibition of polysaccharidases by melanin: Enzyme inhibition in relation to mycolysis. *Arch. Biochem. Biophys.* **137**: 345–356.

Cantacuzéne, J. 1923. Le problem de l'immunité chez les invertébrès. *C. R. Soc. Biol.* (Paris) **88**: 48–119, supplement.

Chorney, M.J. and T.C. Cheng. 1980. Discrimination of self and non-self in invertebrates, pp. 37–54. In J.J. Marchalonis and N. Cohen (eds.) *Contemporary Topics in Immunobiology*. Plenum Press, New York.

Crossley, A.C. 1979. Biochemical and ultrastructural aspects of synthesis, storage and secretion in hemocytes, pp. 423–475. In A.P. Gupta (ed.) *Insect Hemocytes*. Cambridge University Press.

Day, N.K.B., H. Gewurz, R. Johanssen, J. Finstad, and R.A. Good. 1970. Complement and complement-like activity in lower vertebrates and invertebrates. *J. Exp. Med.* **132**: 941–950.

de Aragao, G.A. and M. Bacila. 1976. Purification and properties of a polyphenoloxidase from the freshwater snail *Biomphalaria glabrata*. *Comp. Biochem. Physiol.* **53**B: 179–182.

Decleir, W. and R. Vercauteren. 1965. Phenoloxidase activity in crab leucocytes during the intermoult cycle. *Cah. Biol. Mar.* **6**: 163–172.

Decleir, W., F. Aerts, and R. Vercauteren. 1960. The localization of polyphenoloxidase in haemocytes. *Proceedings 11th International Congress Entomology* **3**: 176–179.

Diluzio, N.R. 1979. Lysozyme, glucan activated macrophages and neoplasià. *J. Reticuloendothel. Soc.* **26**: 67–81.

Dohke, K. 1973. Studies on the prophenoloxidase activating enzyme from cuticle of the silkworm, *Bombys mori*. 1. Activation reaction by the enzyme. *Arch. Biochem. Biophys.* **157**: 203–209.

Dularay, B. and A.M. Lackie. 1985. Haemocytic encapsulation and the prophenoloxidase activation pathway in the locust, *Schistocerca gregaria*. *Forsh. Insect Biochem.* **15**: 827–834.

Durliat, M. 1985. Clotting processes in crustacea decapoda *Biol. Rev.* **60**: 473–498.

Durliat, M. and R. Vranckx. R. 1981. Action of various anticoagulants on hemolymphs of lobsters and spiny lobsters. *Biol. Bull.* (Woods Hole) **160**: 55–68.

Evans, J.J.T. 1967. The activation of prophenoloxidase during melanization of the pupal blood of the chinese silkmoth, *Antheraea pernyi*. *J. Insect Physiol.* **13**: 1699–1711.

Fuller, G.M. and R.F. Doolittle. 1971. Studies of invertebrate fibrinogen. 2. Transformation of lobster fibrinogen into fibrin. *Biochemistry* **10**: 1311–1315.

Gan, E.V., H.F. Haberman, and I.A. Menon. 1976. Electron transfer properties of melanin. *Arch. Biochem. Biophys.* **173**: 666–672.

Götz, P. and A. Vey. 1974. Humoral encapsulation in Diptera (Insecta): Defence reactions of *Chironomus* larvae against fungi. *Parasitology* **68**: 193–205.

Grégoire, C. 1970. Haemolymph coagulation in arthropods, pp. 45–74. In R.G. Macfarlane (ed.) *The Haemostatic Mechanisms in Man and Other Animals*. Academic Press, New York.

Häll, L. and K. Söderhäll. 1982. Purification and propertiers of a protease inhibitor from crayfish hemolymph. *J. Invertebr. Pathol.* **39**: 29–37.

Harada, T., T. Morita, S. Iwanaga, S. Nakamura, and M. Niwa. 1979. A new chromogenic substrate method for assay of bacterial endotoxins using *Limulus* hemocyte lysate, pp. 209–220. In E. Cohen and F.B. Bang (eds.) *Biomedical Applications of the Horseshoe Crab* (Limulidae). Alan R. Liss, New York.

Hergenhahn, H.G. and K. Söderhäll. 1985. α_2-macroglobulin-like activity in plasma of the crayfish, *Pacifastacus leniusculus*. *Comp. Biochem. Physiol.* **81**B: 833–835.

Heyneman, R.A. 1965. Final purification of a latent phenolase with mono- and diphenoloxidase activity from *Tenebrio molitor*. *Biochem. Biophys. Res. Commun.* **21**: 162–169.

Heyneman, R.A. and R.E. Vercauteren, 1968. Evidence for lipid activator of prophenoloxidase in *Tenebrio molitor*. *J. Insect Physiol.* **14**: 409–415.

Inaba, T., Y. Suetake, and M. Funatsu. 1963. Studies on tyrosinase in the housefly. 2. Activation of protyrosinase by sodium dodecyl sulphate. *Agric. Biol. Chem.* **27**: 332–339.

Inaba, T., G. Namihira, M. Kawano, and M. Funatsu. 1978. Effect of ionic strength on activation of prophenoloxidase in its crude solution. *Agric. Biol. Chem.* **42**: 2405–2406.

Iwama, R. and M. Ashida. 1986. Biosynthesis of prophenoloxidase in hemocytes of larval hemolymph of the silkworm, *Bombyx mori*. *Insect Biochem.* **16**: 547–555.

Jacobson, F.W. and N. Millott. 1953. Phenolases and melanogenesis in the coelomic fluid of the echinoid *Diadema antilarum*. *Phillipp. Proc. R. Soc.* B **141**: 231–246.

Johansson, M. and K. Söderhäll. 1985. Exocytosis of the prophenoloxidase activating system from crayfish haemocytes. *J. Comp. Physiol.*, B **156**: 175–181.

Kakinuma, A., T. Asano, H. Torii, and Y. Sugino. 1981. Gelation of *Limulus* amoebocyte lysate by an antitumour (1-3)1-β-D-glucan. *Biochem. Biophys. Res. Commun.* **101**: 434–439.

Kitano, H. 1974. Effects of the parasitization of a braconid, *Apanteles*, on the blood of its host. *J. Insect Physiol.* **20**: 315–327.

Krishnan, G. and M.H. Ravindranath. 1973. Blood cell phenoloxidase of millipedes. *J. Insect Physiol.* **19**: 647–653.

Kuo, M.J. and M. Alexander. 1967. Inhibition of the lysis of fungi by melanins. *J. Bacteriol.* **94**: 624–629.

Leonard, C.M., N.A. Ratcliffe, and A.F. Rowley. 1985a. The role of prophenoloxidase activation in non-self recognition and phagocytosis by insect blood cells. *J. Insect. Physiol.*, **31**: 789–799.

Leonard, C.M., K. Söderhäll, and N.A. Ratcliffe. 1985b. Studies on prophenoloxidase and protease activity of *Blaberus craniifer* haemocytes. *Insect Biochem.*, **15**: 803–810.

Levin, J. 1979. The reaction between bacterial endotoxin and amoebocyte lysate, pp. 131–146. In E. Cohen and F.B. Bang (eds.) *Biomedical Applications of the Horseshoe Crab* (Limulidae). Alan R. Liss, New York.

Levin, J. and F.B. Bang. 1964. The role of endotoxin in the extracellular coagulation of *Limulus* blood. *Bull. Johns Hopkins Hosp.* **115**: 265–274.

Levin, J. and F.B. Bang. 1968. Clottable protein in *Limulus*, its localization and kinetics of its coagulation by endotoxin. *Thromb. Haemostasis* **19**: 186–192.

Liang, S.M., C.M. Liang, and T.Y. Liu. 1981. Studies on *Limulus* amoebocyte: Isolation and identification of a membrane bound protein activator of cyclic nucleotide phosphodiesterase from *Limulus* amoebocyte. *J. Biol. Chem.* **256**: 4968–4972.

Liu, T.Y., R.C. Seid, J.P. Tai, S.M. Liang, T.P. Sakmar, and J.B. Robbins. 1979. Studies on *Limulus* lysate coagulation system, pp. 147–158. In E. Cohen and F.B. Bang (eds.) *Biomedial Applications of the Horseshoe Crab* (Limulidae). Alan R. Liss, New York.

Maier, W.A. 1973. Die Phenoloxydase von *Chironomus thummi* und ihre Beeinflussung durch parasitäre Mermithiden. *J. Insect Physiol.* **19**: 85–95.

Mayer, A.M. and E. Harel. 1979. Polyphenoloxidase in plants: A review. *Phytochemistry* **18**: 193–215.

Menon, I.A. and H.F. Haberman. 1977. Mechanisms of action of melanins. *Br. J. Dermatol.* **97**: 109–112.

Messner, B. 1972. Die Rolle des Tyrosinesystems in der immunologischen Abwehr reaktion bei Wirbellosen: Insekten 1. *Zool. Jahrb. Physiol.* **76**: 368–374.

Millott, N. 1953. Observations on the skin pigment and amoebocytes and the occurrence of phenolases in the coelomic fluid of *Holothuria forskali*: Delle chiaje. *J. Mar. Biol. Assn. U.K.* **31**: 529–539.

Mitchell, H.K. and U.M. Weber. 1965. *Drosophila* phenoloxidase. *Science* (Washington, DC) **148**: 964–965.

Morita, T.S., T. Tanaka, T. Nakamura, and S. Iwanaga. 1981. A new (1-3)-β-D-glucan mediated coagulation pathway found in *Limulus* amoebocytes. *FEBS. Lett.* **129**: 318–321.

Munn, E.A. and S.F. Bufton. 1973. Purification and properties of a phenoloxidase from the blowfly *Calliphora erythrocephala*. *Eur. J. Biochem.* **35**: 3–10.

Nakamura, S., T. Morita, T. Harada-Suzuki, S. Iwanaga, K. Takashi, and M. Niwa. 1982. A clotting enzyme associated with the hemolymph coagulation system of horseshoe crab (*Tachypleus tridentatus*): Its purification and characterization. *J. Biochem.* **92**: 781–792.

Nappi, A.J. 1973. The role of melanization in the immune reaction of larvae of *Drosophila algonquin* against *Pseudeucoila bochei*. *Parasitology* **66**: 23–32.

Nappi, A.J. and F. Streams. 1969. Haemocytic reactions of *Drosophila melanogaster* to the parasites *Pseudeucoila mellipes* and *P. bochei*. *J. Insect Physiol.* **15**: 1551–1566.

Naqvi, S.N.H. and P. Karlson. 1979. Purification of prophenoloxidase in the hemolymph of *Calliphora vicina* (R. and D.). *Arch. Int. Physiol. Biochem.* **87**: 687–695.

Ohki, M.T., T. Nakamura, T. Morita, and S. Iwanaga. 1980. A new endotoxin sensitive factor associated with haemolymph coagulation system of horseshoe crab (*Limulidae*). *FEBS. Lett.* **120**: 217–220.

Ohnishi, E. 1953. Tyrosinase activity during puparium formation in *Drosophila melanogaster*. *Jpn. J. Zool.* **11:** 69–74.

Ohnishi, E., K. Dohke, and M. Ashida. 1970. Activation of prophenoloxidase. 2. Activation by chymotrypsin. *Arch. Biochem. Biophys.* **139:** 143–148.

Pawelek, J.M. and A.B. Lerner. 1978. 5,6-Dihydroxyindol is a melanin precursor showing potent cytotoxicity. *Nature* (London) **276:** 627–628.

Pinhey, K.G. 1929. Tyrosinase in crustacean blood. *J. Exp. Biol.* **1:** 19–36.

Poinar, G.O., R. Leutenegger and P. Götz. 1968. Ultrastructure of the formation of a melanotic capsule in *Diabrolica* (Coleoptera) in response to a parasitic nematode (Mermithidae). *J. Ultrastruct. Res.* **25:** 293–306.

Preston, J.W. and R.L. Taylor. 1970. Observations on the phenoloxidase system in the haemolymph of the cockroach, *Leucophaea maderae*. *J. Insect Physiol.* **16:** 1729–1744.

Pye, A.E. 1974. Microbial activation of prophenoloxidase from immune insect larvae. *Nature* (London) **251:** 610–613.

Pye, A.E. 1978. Activation of prophenoloxidase and inhibition of melanization in the hemolymph of immune *Galleria mellonella* larvae. *Insect Biochem.* **8:** 117–123.

Quigley, J.P. and P.B. Armstrong. 1983. An endopeptidase inhibitor, similar to α_2-macroglobulin, present in the plasma of an invertebrate. *Limulus polyphemus. J. Biol. Chem.* **258:** 7903–7906.

Ratcliffe, N.A. 1982. Cellular defence reactions of insects. *Fortschr. Zool.* **27:** (Zbl. Bakt. Suppl. 12): 223–244.

Ratcliffe, N.A. and S.J. Gagen. 1976. Cellular defence reactions of insect hemocytes *in vivo*: Nodule formation and development in *Galleria mellonella* and *Pieris brassicae* larvae. *J. Invertebr. Pathol.* **28:** 373–382.

Ratcliffe, N.A. and S.J. Gagen. 1977. Studies on the *in vivo* cellular reactions of insects: An ultrastructural analysis of nodule formation in *Galleria mellonella*. *Tissue Cell* **9:** 73–85.

Ratcliffe, N.A., C.M. Leonard, and A.F. Rowley. 1984. Prophenoloxidase activation: Non-self recognition and cell co-operation in insect immunity. *Science* (Washington, DC) **226:** 557–559.

Ratcliffe, N.A., K.N. White, A.F. Rowley, and J.B. Walters. 1982. Cellular defence systems of the Arthropoda, pp. 167–256. In N. Cohen and M. Sigel. (eds.) *The Reticulodendothelial System: A Compreshive Treatise*. Plenum Press, New York.

Ratner, S. and S.B. Vinson. 1983a. Encapsulation reactions *in vitro* by haemocytes of *Heliothis virescens*. *J. Insect Physiol.* **29:** 855–863.

Ratner, S. and S.B. Vinson. 1983b. Phagocytosis and encapsulation: Cellular immune responses in Arthropoda. *Am. Zool.* **23:** 185–194.

Reid, K.B.M. and R.R. Porter. 1981. The proteolytic activation systems of complement. *Annu. Rev. Biochem.* **50:** 433–464.

Rizki, M.T.M. 1957. Tumor formation in relation to metamorphosis in *Drosophila melanogaster*. *J. Morphol.* **100:** 459–472.

Rizki, M.T.M. and R.M. Rizki. 1959. Functional significance of the crystal cells in the larvae of *Drosophila melanogaster*. *J. Biophys. Biochem. Cytol.* **5:** 235–240.

Rowley, A.F. and N.A. Ratcliffe. 1981. Insects, pp. 422–488. In N.A. Ratcliffe and A.F. Rowley (eds.) *Invertebrate Blood Cells*, vol. 2. Academic Press, New York.

Salt, G. 1956. Experimental studies in insect parasitism. 9. The reaction of a stick insect to an alien parasite. *Proc. R. Soc. Lond.* (Biol.) **146**: 93–108.

Salt, G. 1970. *The Cellular Defence Reactions of Insects.* Cambridge University Press, Cambridge.

Schmit, A.R., A.F. Rowley, and N.A. Ratcliffe. 1977. The role of *Galleria mellonella* hemocytes in melanin formation. *J. Invertebr. Pathol.* **29**: 232–234.

Schweiger, A. and P. Karlson. 1962. Zum Tyrosinstoffwechsel der Insekten. 10. Die aktivierung der Pröphenoloxydase und Aktivatorenzym. *Hoppe Seyler's Z. Physiol. Chem.* **329**: 210–221.

Seed, J.L. and J.L. Bennett. 1980. *Schistosoma mansoni:* Phenoloxidase's role in eggshell formation. *Exp. Parasitol.* **49**: 430–441.

Seljelid, R., J. Bögwald, and Å. Lundwall. 1981. Glycan stimulation of macrophages *in vitro. Exp. Cell Res.* **131**: 121–129.

Smith, V.J. and N.A. Ratcliffe. 1978. Host defence reactions of the shore crab, *Carcinus maenas* (L) *in vitro. J. Mar. Biol. Assn. U.K.* **58**: 367–379.

Smith, V.J. and N.A. Ratcliffe. 1980. Cellular defense reactions of the shore crab *Carcinus maenas: In vivo* hemocytic and histopathological responses to injected bacteria. *J. Invertebr. Pathol.* **35**: 65–74.

Smith, V.J. and K. Söderhäll. 1983a. β-1,3-glucan activation of crustacean hemocytes *in vitro* and *in vivo. Biol. Bull.* (Woods Hole) **164**: 299–314.

Smith, V.J. and K. Söderhäll. 1983b. Induction of degranulation and lysis of haemocytes in the freshwater crayfish *Astacus astacus* by components of the prophenoloxidase activating system *in vitro. Cell Tissue Res.* **233**: 295–303.

Smith, V.J., K. Söderhäll, and M. Hamilton. 1984. β-1,3-glucan induced cellular defence reactions in the shore crab, *Carcinus maenas. Comp. Biochem. Physiol.* **77**A: 635–639.

Söderhäll, K. 1981. Fungal cell wall β-1,3-glucans induce clotting and phenoloxidase attachment to foreign surfaces of crayfish hemocyte lysate. *Dev. Comp. Immunol.* **5**: 565–573.

Söderhäll, K. 1982. Prophenoloxidase activating system and melanization—a recognition mechanism of arthropods? A review. *Dev. Comp. Immunol.* **6**: 601–611.

Söderhäll, K. 1983. β-1,3-glucan enhancement of protease activity in crayfish hemocyte lysate. *Comp. Biochem. Physiol.* **74**B: 221–224.

Söderhäll, K. and R. Ajaxon. 1982. Effect of quinones and melanin on mycelial growth of *Aphanomyces* spp. and extracellular protease of *Aphanomyces astaci*, a parasite on crayfish. *J. Invertebr. Pathol.* **39**: 105–109.

Söderhäll, K. and L. Häll. 1984. Lipopolysaccharide-induced activation of the prophenoloxidase activating system in crayfish haemocyte lysate. *Biochem. Biophys. Acta* **797**: 99–104.

Söderhäll, K. and V.J. Smith. 1983. Separation of the haemocyte populations of *Carcinus maenas* and other marine decapods and prophenoloxidase distribution. *Dev. Comp. Immunol.* **7**: 229–239.

Söderhäll, K. and V.J. Smith. 1984. The prophenoloxidase activating system: A complement-like pathway in arthropods, pp. 160–167. In J. Aist and D.W. Roberts (eds.) *Infection Processes of Fungi.* Rockefeller Foundation, New York.

Söderhäll, K. and T. Unestam. 1979. Activation of crayfish serum prophenoloxidase: The specificity of cell wall glucan activation and activation by purified fungal glycoprotein. *Can. J. Microbiol.* **25**: 404–416.

Söderhäll, K., A. Vey, and M. Ramstedt. 1984. Hemocyte lysate enhancement of fungal spore encapsulation by crayfish hemocytes. *Dev. Comp. Immunol.* **8**: 23–30.

Söderhäll, K., J. Levin, and P.B. Armstrong. 1985a. The effect of β-1,3-glucans on blood coagulation and amoebocyte release in the horseshoe crab, *Limulus polyphemus. Biol. Bull.* (Woods Hole), **169**: 661–674.

Söderhäll, K., V.J. Smith, and M. Johansson. 1985b. Exocytosis and uptake of bacteria by isolated haemocyte populations of two crustaceans: Evidence for cell cooperation in the host defence reactions of arthropods. *Cell Tissue Res.*, **245**: 43–49.

Söderhäll, K., L. Häll, T. Unestam, and L. Nyhlén. 1979. Attachment of prophenoloxidase to fungal cell walls in arthropod immunity. *J. Invertebr. Pathol.* **34**: 285–294.

Söderhäll, K., A. Wingren, M. Johansson, and K. Bertheussen. 1985c. The cytotoxic reaction of hemocytes from the freshwater crayfish *Astacus astacus. Cell. Immunol.*, **94**: 326–332.

Solum, N.O. 1973. The coagulogen of *Limulus polyphemus* hemocytes: A comparison of the clotted and non-clotted forms of the molecule. *Thromb. Res.* **2**: 55–70.

Sroka, P. and S.B. Vinson. 1978. Phenoloxidase activity in the hemolymph of parasitized and unparasitized *Heliothis virescens. Insect Biochem.* **8**: 399–402.

Starkey, P.M. and A.J. Barrett. 1977. α₂-macroglobulin, a physiological regulator of protease activity, pp. 666–696. In A.J. Barrett. (ed.) *Proteinases in Mammalian Cells and Tissues.* Elsevier North Holland, Amsterdam.

Steiner, H., D. Hultmark, A. Engström, H. Bennich, and H.G. Boman, 1981. Sequence and specificity of two antibacterial proteins involved in insect immunity. *Nature* (London) **292**: 246–248.

Stephens, J.M. 1962. Influence of active immunization on melanization of the blood of wax moth larvae. *Can. J. Microbiol.* **8**: 597–602.

Stolz, D.B. and D.I. Cook. 1983. Inhibition of host phenoloxidase activity by parasitoid Hymenoptera. *Experientia* **39**: 1022–1024.

Streams, F.A. and L. Greenberg. 1969. Inhibition of the defence reaction of *Drosophila melanogaster* parasitized simultaneously by the wasps *Pseudeucoila bochei* and *Pseudeucoila mellipes. J. Invertebr. Pathol.* **13**: 371–372.

Sullivan, J.D., Jr., and S.W. Watson. 1975. Purification and properties of the clotting enzyme from *Limulus* lysate. *Biochem. Biophys. Res. Commun.* **66**: 848–855.

Tai, J.Y. and T.Y. Liu. 1977. Studies on *Limulus* amoebocyte lysate: Isolation of pro-clotting enzyme. *J. Biol. Chem.* **252**: 2178–2181.

Tai, J.Y., R.C. Seid, Jr., R.D. Huhn, and T.Y. Liu. 1977. Studies on *Limulus* amoebocyte lysate. 2. Purification of the coagulogen and the mechanism of clotting. *J. Biol. Chem.* **252**: 4773–4776.

Tanaka, S., T. Nakamura, T. Morita, and S. Iwanaga. 1982. *Limulus* anti-LPS factor: An anticoagulant which inhibits the endotoxin mediated activation of *Limulus* coagulation. *Biochem. Biophys. Res. Commun.* **105**: 717–723.

Taylor, R.L. 1969. A suggested role for the polyphenol-phenoloxidase system in invertebrate immunity. *J. Invertebr. Pathol.* **14**: 427–428.

Thangaraj, T., K. Nellaiappan, and K. Ramalingam. 1982. Activation of prophenoloxidase in the liver fluke, *Fasciola gigantea* Cobboid. *Parasitology* **85**: 577–581.

Thomson, J.A. and Y.T. Sin. 1970. The control of prophenoloxidase activation in larval haemolymph of *Calliphora. J. Insect Physiol.* **16**: 2063–2074.

Tyson, C.J., D. McKay, and C.R. Jenkin. 1974. Recognition of foreignness in the freshwater crayfish *Parachaeraps bicarinatus*, pp.139–166. In E.L. Cooper (ed.) *Contemporary Topics in Immunobiology*, vol. 4. Plenum Press, New York.

Unestam, T. and J.E. Nylund. 1972. Blood reactions *in vitro* in crayfish against a fungal parasite, *Aphanomyces astaci. J. Invertebr. Pathol.* **19**: 94–106.

Unestam, T. and K. Söderhäll. 1977. Soluble fragments from fungal cell walls elicit defence reactions in crayfish. *Nature* (London) **267**: 45–46.

Vercauteren, R.E. and F. Aerts. 1958. On the cytochemistry of the haemocytes of *Galleria mellonella* with special reference to polyphenoloxidase. *Enzymology* **20**: 167–172.

Vey, A. and P. Götz. 1975. Humoral encapsulation in Diptera (Insecta): Comparative studies *in vitro. Parasitology* **70**: 77–86.

Vinson, S.B. 1976. *Microplitis croceipes*: Inhibition of the *Heliothis zea* defense reaction to *Cardiochiles nigriceps. Exp. Parasitol.* **41**: 112–117.

Weinheimer, P.F., E.E. Evans, R.M. Stroud, R.T. Acton, and B. Painter. 1969. Comparative immunology: A natural hemolytic system of the spiny lobster *Panulirus argus. Proc. Soc. Exp. Biol. Med.* **130**: 322–326.

Yamaura, I., M. Yonekura, Y. Katsura, M. Ishiguro, and M. Funatsu. 1980. Purification and some physico-chemical properties of phenoloxidase from the larvae of housefly. *Agric. Biol. Chem.* **44**: 55–59.

Young, N.S., J. Levin, and R.A. Prendergast. 1972. An invertebrate coagulation system activated by endotoxin: Evidence for enzymic mediation. *J. Clin. Invest.* **51**: 1790–1797.

Ziprin, R.L. 1978. Immune responses of the greater wax moth *Galleria mellonella* induced by the marine pseudomonad B-16. *J. Invertebr. Pathol.* **32**: 396–397.

Ziprin, R.L. and P.A. Hartman. 1971. Toxicity of *Pseudomonas aeruginosa* bactericins and cell walls to the greater wax moth *Galleria mellonella. J. Invertebr. Pathol.* **17**: 265–269.

Zlotkin, E., M. Gurevitz, and A. Shulov. 1973. The toxic effects of phenoloxidase from the haemolymph of Tenebrionid beetles. *J. Insect Physiol.* **19**: 1057–1065.

CHAPTER 10

Antibacterial and Antiviral Factors in Arthropod Hemolymph

June S. Chadwick
Department of Microbiology and Immunology
Queen's University
Kingston, Canada

Gary B. Dunphy
Department of Biosciences
Simon Fraser University
Burnaby, Canada

10.1. INTRODUCTION

Naturally occurring substances that exert a detrimental effect on microorganisms have been recognized in the hemolymph of invertebrates for some time. In earlier days, these substances were not specifically termed bactericidal or viricidal. Some investigators attempted to simplify the situation. For example, Mohrig and Messner (1968) stated that lysozyme, a universally occurring substance in the hemolymph of most invertebrates, was likely responsible for all immunity in insects.

Phenoloxidase, another naturally occurring substance in invertebrates has also been considered an efficient defense substance, particularly in the cellular response, in which it is primarily involved in encapsulation and nodule formation (Nappi, 1975, 1978). Its role in antimicrobial defense is unclear.

More recently, the emphasis in research on arthropod immunity has centered around inducible immune substances, and understanding of these has been realized in some depth, especially in insects. In the last decade, reference to antimicrobial agents has been made in a number of comprehensive reviews of arthropod immunity (Maramorosch and Shope, 1975; Chadwick and Aston, 1978; Lackie, 1980; Boman, 1981; Boman and Hultmark, 1981; Ratcliffe, 1985; Cooper, 1985).

This chapter is intended, not as a complete review of antibacterial and antiviral substances, but as an interpretation of some of the more recent reports on these agents in insects and other arthropods. The information on naturally occurring agents is sparse, but an attempt is made to evaluate the contribution of such factors to the overall immune response and to compare their influence to that of induced antimicrobial factors. Though to a lesser degree than with inducible antibacterial substances, it is also in the Insecta that most attention has been given to antiviral factors.

Much of the interpretation depends on whether we consider these antimicrobial factors, particularly antibacterial factors, to function directly or indirectly; for example, they might act indirectly as hemolymph factors aiding phagocytosis (Mohrig et al., 1979a,b) rather than having a directly damaging effect on the microorganism. Since a detailed consideration of phagocytosis is not within the scope of this chapter, phagocystosis per se is discussed only as it is enhanced by antibacterial substances in hemolymph.

The role of complementlike factors as possible antibacterial agents in insects has been discussed by Aston and colleagues (1976), Chadwick and Aston (1976), and Chadwick and Aston (1978), among others.

10.2. ANTIBACTERIAL FACTORS IN INSECT HEMOLYMPH

10.2.1. Naturally Occurring Factors

10.2.1.1. Occurrence and Characteristics. For more than half a century, it has been recognized that insects are resistant to a number of microbial species that affect higher animals (reviewed by Chadwick, 1967; Whitcomb et al., 1974). Moreover, the number of bacterial species classed as insect pathogens represents only a small proportion of existing bacterial species. However, little effort has been made to correlate the resistance to such bacterial species with bactericidal activity in the insect hemolymph. Consequently, the literature contains few references to naturally occurring antibacterial substances in insect hemolymph. Most of the activities reported have not been characterized in terms of a chemical factor.

Stephens (1963) surveyed the hemolymph of a variety of insect species from four orders of insects for bactericidal activity against *Pseudomonas aeruginosa*, *Shigella dysenteriae*, *Salmonella typhi*, *Bacillus cereus*, and *Serratia marcescens*. Results were variable; though the hemolymph of every insect species showed activity against *S. dysenteriae*, none was bactericidal for *S. marcescens*. Human enteric pathogens, such as *S. dysenteriae*, would not be expected to be a natural part of the environment of any of the insect species tested. It is interesting to note that the hemolymph of the housefly and wasp demonstrated limited activity against *P. aeruginosa*, although the hemolymph of *Galleria mellonella* and seven additional Lepidopterous species was inactive against that organism. In that study, no effort was made to determine the active component of the hemolymph.

Chadwick (1975) demonstrated that the hemolymph of untreated *G. mellonella* larvae showed some early antibacterial activity for organisms such as *Escherichia coli*. However, some bacteria survived in the hemolymph and multiplied slowly over a 12-hr period.

Kinoshita and Inoue (1977) showed that significant bactericidal activity against *E. coli* B/SM existed in the pooled cell-free hemolymph from larvae of the silkworm, *Bombyx mori*. They found that two factors were required to kill the organism; one was a lysozymelike enzyme and the other an anionic factor with a smaller molecular weight, referred to as cofactor. These authors believed that the lysozymelike enzyme was similar to that described by Powning and Davidson (1973), but they did not speculate on the nature of the second factor.

Conversely, Boman and associates (1978) found no antibacterial activity against *E. coli* in the hemolymph of untreated larvae and pupae of *Hyalophora cecropia* and/or *Samia cynthia*. Boman (1981) believed that in a large pool of insects used for hemolymph samples a few could be infected, and the so-called naturally occurring immunity, as suggested by Kinoshita and Inoue (1977), might actually be an indication of induced immunity.

A recent report of a naturally occurring substance in insect hemolymph is that of Okai (1985). He isolated an inhibitory factor for DNA synthesis from the hemolymph of *B. mori* as well as from fetal calf serum. The substances from both sources were heat stable and were characterized as 1000 MW peptides. They showed DNA inhibitory effects both on mammalian cells and bacteria. Okai discussed the substance with reference to the cecropins described by Hultmark and coworkers (1982). Cecropins are considered to be induced substances, but as they are 4000 MW peptides, some similarity was suggested. A close relationship with cecropins was not established by the authors. Okai did suggest that these peptides may be responsible for bactericidal activity against foreign microorganisms. It is conceivable that further antibacterial substances, yet to be discovered, exist in insects.

In addition to the examples of antibacterial factors listed above, which act directly, it is possible that indirect antibacterial activity may occur. Rowley and Ratcliffe (1980) suggested that bacterial agglutinins present in insects may aid in clearing the hemocoel of invading microorganisms by promoting rapid agglutination. As a consequence, large masses would be formed, which would be ideal for phagocytosis and nodule formation. This suggestion assigns an important indirect role to agglutinins, which might thus be considered antibacterial factors. However, perhaps agglutinins should not be considered true antibacterial factors.

10.2.1.2. Relationship of Naturally Occurring Factors to Lysozyme. One of the antibacterial factors in the cell-free hemolymph of the silkworm, *B. mori*, described by Kinoshita and Inoue (1977; see Section 10.2.1) was considered to resemble lysozyme. The lysozymelike substance reported by these workers was apparently different from egg-white lysozyme, which would not replace the activity of the substance isolated from silkworms. The substance was tested only against *E. coli*, and it is noteworthy that these investigators did not extend their investigation to other gram-positive bacterial species that might be pathogenic to the silkworm.

Hoffmann and coworkers (1977) suggested that a substance of hemocytic origin occurred in nymphs of *Locusta migratoria*. This substance accounted for nonspecific defense and was believed to be responsible for the fact that a number of bacterial species were considered nonpathogenic in locusts. The substance was suggested to be lysozymelike in nature.

Cheung and colleagues (1978) reported that the injection of *Bacillus thuringiensis* into *Heliothis zea* resulted in the initial clearance of the organism from the hemolymph, followed in a short time by a significant increase in numbers of bacteria. The initial clearance was attributed to a glycopeptide-hydrolyzing enzyme, possibly lysozyme, plus an additional factor present in hemolymph that acted to produce ghost forms of *B. thuringiensis*. The authors mentioned the need for determining whether an attack on the cell membrane had occurred. To our knowledge, this possibility has not been pursued.

10.2.1.3. Role of Hemocytes in Eliciting Factors.
The role of hemocytes, which obviously is very important in cellular defense, is not considered in detail in this chapter. Since many authors report little or no phagocytosis in insects and as there are few reports of naturally occurring antibacterial factors, the question remains as to whether there are nonspecific inducers of antibacterial factors for nonpathogenic bacteria. Horohov and Dunn (1982) suggested that the tremendous increase of numbers of granulocytes and spherulocytes in *Manduca sexta* 1 hr after injection conferred an advantage on the insect in dealing with bacterial infections. These authors did not suggest that antibacterial substances were directly formed.

As mentioned in Section 10.2.1.2, Hoffmann and associates (1977) suggested that a substance of hemocytic origin occurred in nymphs of *Locusta migratoria*. The substance was thought to be lysozymelike, which is a possibility, since *B. thuringiensis* was the susceptible organism.

In a preliminary investigation, Ratcliffe and Walters (1983) monitored the fate in *G. mellonella* of varying doses of two bacterial species *E. coli* and *B. cereus*. They described initial hemocytopenia, stickiness of hemocytes, and nodule formation. These authors reported that beyond a certain dose level effective phagocytosis could not occur unless augmented by nodule formation. In a subsequent investigation, Walters and Ratcliffe (1983) used radioactively labeled *B. cereus* and *E. coli* to determine whether active killing mechanisms occurred within the cell aggregates. The fate of a nonpathogen such as *E. coli* was quite different from that of *B. cereus*: the percent relative viability of *E. coli* was much lower than that of *B. cereus* in nodules. Although no natural antibacterial factors were demonstrated against either organism, the authors discussed the possible interaction of induced antibacterial factors, such as described by Hoffmann and coworkers (1981) with the cellular response. It remains to be demonstrated whether antibacterial substances exist in nodules (see also Chapter 3).

10.2.1.4. Constituents of Normal Hemolymph with Antibacterial Effects.
Certain naturally occurring components of insect hemolymph may have antibacterial effects. Smirnoff and Valero (1980) reported that a disturbance in

essential ions in forest insects may be related to the outcome of infection. In the spruce budworm, *Choristoneura fumiferana*; the tent caterpillar, *Malacasoma disstria*; and a hymenopteran, *Neodiprion swanei*, the level of K^+ ions was significantly increased in both *B. thuringiensis* and *Baculovirus* infections, whereas the level of Ca^{2+} decreased considerably during *B. thuringiensis* infections but tripled during a nuclear polyhedrosis infection. Though there is still insufficient evidence, such results suggest that the normal concentration of essential ions in hemolymph of some insects may be inhibitory for bacteria. Jarosz (1979) reported that *Streptococcus faecalis*, the only bacterium occurring regularly in the gut of *G. mellonella*, has a narrow range of bactericidal activity and releases a lysozymelike enzyme in the presence of proteolytic enzymes. In this manner, many ingested bacteria, including several gram-negative species, which are potential pathogens by injection, were eliminated when ingested. Under stress, *Galleria* larvae die from infections with *S. fecalis*, and moribund insects have high numbers of the organism in the hemolymph as well as the gut (Stephens, 1962b). From time to time, *S. fecalis* has been recovered from the hemolymph of larvae that had been slightly stressed by high temperature, although mortality did not occur (J. Chadwick, unpublished). It is conceivable in the light of this information that bacteremia of this type might result in an antibacterial effect against other bacterial species the insect might encounter.

Boman and colleagues (1981) reported that, on injury, *H. cecropia* pupae produce proteins similar in quantity to the antibacterial cecropins induced by living bacteria. Curiously, though the proteins induced on injury were similar to the cecropins both qualitatively and quantitatively, they had low bactericidal activity and thus were functionally inferior. In the light of these findings, since wounding may occur in a number of situations in nature, it is attractive to speculate that antibacterial substances in hemolymph could be elicited in this manner under natural conditions.

10.2.2. Induced Immunity Resulting in Antibacterial Activity

10.2.2.1. Nature of Induction: Role of Hemocytes and Other Organs. An insect such as *G. mellonella* dies within 1–2 days if injected with any one of a variety of gram-negative species of bacteria that are potential pathogens for the insect. Unlike a higher animal, the insect does not recover after a period of illness by virtue of an immune response consisting of several effector mechanisms. This may be because the immune response is too transient. A typical response is shown in Figure 10.1. The bacterial component, which might induce immunity in a higher animal, might be unable to achieve this in the insect before a degree of bacterial multiplication takes place, which causes fatal septicemia.

On the other hand, it is well established (Chadwick and Aston, 1978) that if the organism is killed or attenuated or if a component such as lipopolysaccharide (LPS) is used, the insect is able to withstand a challenge of the

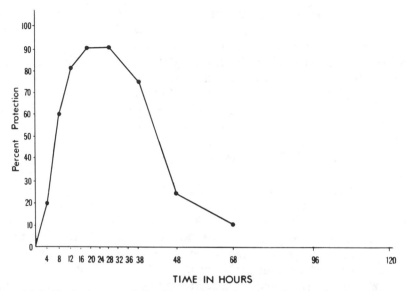

Figure 10.1. Typical curve of development and duration of protective immunity in *Galleria mellonella* larvae against *Pseudomonas aeruginosa* after administration at time zero of homologous heat killed or formalized vaccine.

gram-negative organism provided the time after vaccination is more than 8 hr (Chadwick, 1975, 1977).

Recently, we studied the characteristics of a variety of mutants of *P. aeruginosa* with respect to properties associated with virulence and adherence (Dunphy et al., 1986a,b). It was apparent that rough cell wall mutants had greater LD_{50}'s for *G. mellonella* than the smooth wild type of the organism. In the presence of cell-free insect hemolymph from normal insects, rough-strain *P. aeruginosa* AK 1386 became swollen, suggesting a response possibly due to factors in the serum. Cell wall damage of wild-strain *P. aeruginosa* P11-1 did not occur following exposure to serum from normal *G. mellonella* larvae (Chadwick et al., 1982).

It is attractive to speculate in the above situation that lysozyme may be involved. Lysozyme was shown to damage cell walls of mutants of gram-negative bacteria with defective lipopolysaccharide (Tamaki and Matsuhaski, 1973). It is possible that the lysozyme might have an effect against the mutant strain, though it is not effective enough to kill the bacterium.

Much of the evidence for the presence of antimicrobial factors in insect hemolymph is based on inference from studies of active or passive immunity or adherence to hemocytes, rather than on direct demonstration of the material. For example, Chadwick and coworkers (DeVerno et al., 1983, 1984) stated that hemocytes probably elicit some antibacterial factors, as hemocytes appear to serve as a trigger to the fat body to elicit more active factors. The hemocytic activity probably occurs early in the immune response, perhaps as early as 30 min after the inducing trigger, since DeVerno and associates (1983) showed that hemocytes alone conferred good protection when transferred from 30 min to 4 hr after immunization. These findings confirmed our hypothesis (Chadwick and Aston, 1978) that hemocytes were active in the inductive phase of the immune response. We suggested (DeVerno et al., 1983) that hemocytes could be releasing a substance that stimulated other cells, possibly fat body cells, to synthesize antibacterial factors effective in later steps of the immune response. At that time, it was speculated whether *G. mellonella* might possess a system similar to that described by Faye and Wyatt (1980) for cecropia pupae. However, Abu-Kakimu and Faye (1981) suggested that the granulocytes of cecropia pupae phagocytize bacteria and move to the fat body, where they disintegrate. As Ratcliffe and Rowley (1975) and Rowley and Ratcliffe (1976) showed that granulocytes of *Galleria* are not the main phagocytic cell, we suggested that the mechanism in *Galleria*, if similar to that described by Abu-Hakimu and Faye (1981), would have to occur in the absence of phagocytosis.

To demonstrate the involvement of the fat body of *Galleria* in the generation of antibacterial activity, Chadwick and coworkers devised a totally in vitro system utilizing a mixture of normal hemolymph, LPS, and fat body from nonimmune larvae (DeVerno et al., 1984). In vitro mixtures of fat body and cell-free hemolymph from normal *Galleria* larvae possessed antibacterial activity against *P. aeruginosa*, and this activity was enhanced when LPS was added to the incubation mixture. Unlike Abu-Hakimu and Faye (1981), we did not find that phagocytosis was the first step in the induction of immunity, since antibacterial activity occurred in fat body samples incubated in the absence of hemocytes, although hemocytes did accentuate the activity. Gagen and Ratcliffe (1976) showed that following nodulation, nodules attach to the fat body, and although they do not break down, degranulation of granulocytes is an early step in their formation. The initial steps of nodulation could release factors that result in the production of antibacterial activity by the fat body.

After our report of in vitro generation of antibacterial activity from the fat body of *G. mellonella* (DeVerno et al., 1984), Dunn and colleagues (1985) demonstrated that fat body from *Manduca sexta* previously immunized with a suitable inducing agent synthesized and released lysozyme and cecropin-like peptides. Antibacterial activity was tested only against *E. coli* D-31, so it is not practical to compare the system of Dunn and associates (1985) with that of DeVerno and associates (1984). Dunn and coworkers (1985) stated that the level of synthesis of antibacterial proteins stimulated by LPS was far less than that stimulated by peptidoglycan fragments and also that the *Manduca* system was stimulated by gram-positive organisms, which do not possess LPS. These authors (Dunn et al., 1985) attempted to reconcile the two systems by suggesting that cultures of *M. sexta* fat body may have contained contaminating hemocytes that could be stimulated by peptidoglycan or the samples of *P. aeruginosa* LPS used by DeVerno and colleagues

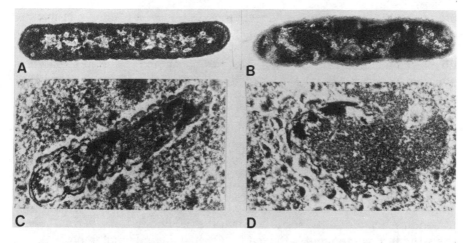

Figure 10.2. Cells of *P. aeruginosa* before and after treatment with normal and immune hemolymph of *G. mellonella*. (A) Untreated control cell ($\times 54{,}600$). (B) Cell exposed to normal hemolymph for 21 min ($\times 71{,}000$). (C) Cell exposed to immune hemolymph up to 7 min. Fragmentation of cell wall is evident at several sites ($\times 60{,}500$). (D) Dispersion of cell debris and cell wall fragments obvious after 7-min exposure to immune hemolymph ($\times 71{,}200$). (From Chadwick et al., 1982.)

(1984) might have contained cell wall material. It is indeed interesting to note similarities in two different insect species, and perhaps, as suggested by Dunn and associates (1985), the future identification of receptors on fat body cells will help resolve the minor differences in the system.

10.2.2.2. Properties of Induced Factors. Chadwick and Aston (1978) discussed several factors that might be involved in the effector phase of the immune response of an insect. Certainly, when dealing with a pathogen, the most obvious demonstration of an effector event is an elevated LD_{50} of the pathogen in a protected insect. Along with this goes demonstration of antibacterial activity, that is, the killing of a pathogen. How the killing action is effected has not often been addressed. We felt it would be most advantageous to determine how the bacterial cell is damaged, for lysis did not occur with *P. aeruginosa*, as reported for some other bacteria. We showed that cells of *P. aeruginosa* rapidly lose structural integrity when exposed to immune hemolymph of *G. mellonella* (Chadwick et al., 1982).

Figure 10.2A shows a normal cell of *P. aeruginosa* similar to that shown by Chadwick and coworkers (1982), while Figure 10.2B shows a cell exposed to normal hemolymph for 21 min. Normal hemolymph had no pronounced effect on *P. aeruginosa*, but it is apparent that the cells in Figures 10.2C and 10.2D, which had been exposed to immune hemolymph for 7 min, suffered very serious damage that caused total breakup of the cell. Chadwick and colleagues (1982) reported that initial damage occurred at the outer membrane of the cell envelope. In view of these findings, it is quite possible that cell wall deficient strains of bacterial test species are much more susceptible to antibacterial factors in hemolymph. It also suggests that reports of antibacterial factors in hemolymph must be interpreted differently, based on whether susceptible organisms are cell wall mutants, wild-type strains, pathogens, or nonpathogens.

Ultrastructural damage reported for *P. aeruginosa* is not an isolated phenomenon. From later work in our laboratory, it became apparent that cells of *Proteus mirabilis* were considerably affected by immune hemolymph of *G. mellonella*, although with that organism, normal hemolymph exerted some effect (Morton et al., 1984). An untreated cell of *P. mirabilis* is shown in Figure 10.3A, while Figures 10.3B–D show cells of *P. mirabilis* at various stages of destruction after 21 min of exposure to immune hemolymph. Morton and associates (1984) suggested that the peptidoglycan layer was one of the structural components affected by *G. mellonella* hemolymph.

These preliminary ultrastructural studies on antibacterial effects of immune hemolymph suggest that the use of more sophisticated techniques in conjunction with the use of test organisms with known chemical deficiencies may yield considerable understanding of the target sites of antibacterial factors.

10.2.3. Cecropins and Cecropinlike Factors

10.2.3.1. Occurrence. Recently, considerable attention has been paid to a group of about 10 inducible proteins reported to occur in most Lepidopterous species (Boman and Hultmark, 1981; Boman and Steiner, 1981). The most important of these proteins are referred to as cecropins. Boman and Steiner (1981) reported that cecropins have been found in two Chinese silk moths (*H. cecropia* and *S. cynthia*), *G. mellonella*, and *M. sexta*. Hoffmann and coworkers (1981) reported cecropinlike activity in three additional Lepidopterous species. Most of the investigations of this group, however, have been focused on *H. cecropia* and to a lesser extent *S. cynthia*.

Hoffmann (1980) demonstrated antibacterial activity in the hemolymph of migratory locusts, *L. migratoria*, after injection of a small dose of *B.thuringiensis, E. coli,* or *P. aeruginosa*. The author considered the similarity of the antibacterial substance to lysozyme to be unlikely in view of differing physicochemical properties. Hoffmann (1980) found some specificity in the antibacterial activity and speculated that it was due to several small molecular weight thermostable substances. This system differed from that described by Pye and Boman (1977), as the multicomponent system of the latter was highly heat sensitive.

10.2.3.2. Properties and Activity. Pupae of *H. cecropia* produce three classes of antibacterial proteins after an injection of live nonpathogenic bacteria or heat-killed pathogenic bacteria, the best response being induced by living

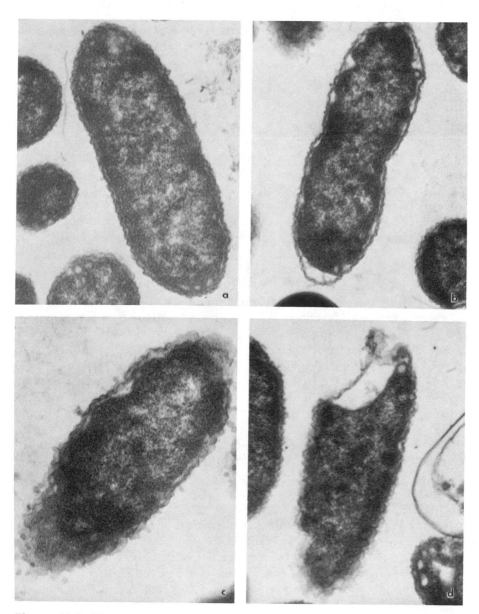

Figure 10.3. Electron micrographs of *Proteus mirabilis* strain 11-1A following (a) no treatment (×36,000) and (B–D) exposure to immune hemolymph of *G. mellonella* for 21 min. (b) Note the spaces between the cell wall and cell membrane (×32,600). (c) The cell wall is discontinuous, and the cytoplasm is uneven and appears to be partially extracted from the cell (×47,000). (d) The cell wall has partially collapsed and withdrawn from the cell membrane (×33,000). (From Morton et al., 1984.)

Enterobacter cloacae (Boman and Steiner, 1981; Engström et al., 1985). The proteins was first classified according to their activity as cecropins, attacins, and lysozyme (Boman and Steiner, 1981).

Lysozyme has been recognized as a possible antibacterial agent in insects by numerous investigators, including Mohrig and Messner (1968), Chadwick (1970), and Powning and Davidson (1973). Cecropia lysozyme was first referred to as protein P7 by Hultmark and coworkers (1980). Engström and colleagues (1985) reported the amino acid sequence of cecropia lysozyme and the isolation and structure of a cDNA clone containing lysozyme information. The lysozyme was described as chicken type.

Cecropins and attacins have in vitro activity against specific strains of several gram-negative bacteria but limited, if any, activity against gram-positive bacteria. Amino acid sequencing and DNA levels have been established for many of the proteins (Steiner et al., 1981; Hultmark et al., 1982; Qu et al., 1982; Engström et al., 1984, 1985).

Cecropins have a molecular weight around 4000 and contain a single basic peptide chain (Boman and Hultmark, 1981). In early investigations, they were referred to as P9 proteins (Hultmark et al., 1980). The most important of these were later described as cecropins A and B (Boman and Steiner, 1981). Boman and Hultmark (1981) considered cecropins a distinct class of antibacterial substances not closely related to other small basic antibacterial proteins. Merryfield and colleagues (1982) and Andreu and colleagues (1983) synthesized cecropin A by a stepwise solid phase method, which resulted in a minimal amount of by products. When the antibacterial activity was tested, the synthetic and synthetic formylated cecropin A showed a twofold reduction in activity against *E. coli*, as compared with the natural cecropin. It is curious that there was no reduction in activity of the synthetic cecropins against *P. aeruginosa*; activity by the natural cecropins against that organism is always minimal and was initially reported as less than one-half of that against *E. coli*.

Attacins have been described as six closely related proteins in the hemolymph of immunized *H. cecropia* (Hultmark et al., 1983). They were considered to be immunologically identical to the P5 protein described by Pye and Boman (1977). Attacins have molecular weights of 2000–2300 and antibacterial activity against *E. coli*, as did the cecropins. Further activity was demonstrated only against *Acinetobacter calcoaceticus* and *Pseudomonas maltophilia*, both bacteria being isolated from the gut of a Chinese oak silkworm. It was curious that only growing cells of *E. coli* were attacked. The authors do not state the range of organisms against which attacins were tested. When P5 proteins were first described, Pye and Boman (1977) stated that the proteins had a synergistic effect when combined with P9 fractions (later cecropins) of hemolymph.

Okada and Natori (1983) described an antibacterial protein induced by injury and referred to as sarcotoxin I from the hemolymph of *Sarcophaga peregrina* larvae. Subsequently, Kubo and associates (1984) showed that antibacterial activity of a similar nature occurred in adults of *S. peregrina*. Sarcotoxin I was a mixture of three proteins with almost identical primary structures. The proteins consisted of 39 amino acids residues and differed in only two to three residues. Antibacterial activity was demonstrated against *E. coli* K12 but was not reported for other bacterial species. The authors state that cecropins from *H. cecropia* have significant similarity to sarcotoxin I.

Lambert and Hoffmann (1985) reported the presence of a 1500-MW antibacterial factor in the blood of *L. migratoria* nymphs. The substance was induced by injection of live *P. aeruginosa* and was said to provide protection against that pathogen. It is difficult to understand how the test insects could withstand the inducing injection if the organism were a true pathogen. The substance may possibly resemble the 1000-MW peptide reported by Okai (1985). Although the molecular weights of these substances and the cecropins are not identical, it would be interesting to know whether they are closely related, particularly as the substance reported by Lambert and Hoffmann (1985) occurs in an insect unrelated to the Lepidoptera.

In all reports of antibacterial factors in insect hemolymph, specificity has been a serious consideration. Often controversial results have been obtained. Boman and coworkers (1985) reported that cecropin B was slightly more potent than cecropin A, while cecropin D had the most narrow antibacterial spectrum. The authors suggested that both cecropins and attacins may have separate target organisms that have not been found. The greatest activity of these substances appeared to be against cell wall deficient mutants of *E. coli*.

Each cecropin and attacin is presumed to have separate targets on the organism on which it acts. According to the authors, this is an attractive theory because it makes it impossible for a bacterium to produce mutants that are resistant to the humoral immunity of the insect. Boman and Steiner (1981) and Steiner and colleagues (1981) tested the activity of whole hemolymph and of isolated cecropin A and cecropin B against several bacterial species. Their comparative results are analyzed in Table 10.1.

Hoffmann and associates (1981) stated that cecropins have high antibacterial activity against both *E. coli* and *P. aeruginosa*. From their results, the activity against *P. aeruginosa* OT97 was only about 25% of that against *E. coli* K12 D31 when whole hemolymph was tested. The analysis in Table 10.1 does not seem to indicate that the antibacterial activity of immune cecropia hemolymph is the same for all bacterial species. The table shows that the activity of purified cecropins A and B against *P. aeruginosa* is only 61% and 78% percent, respectively, to that for *E. coli* K12 D31. It is interesting to note that somewhat better relative activity was obtained when cecropins were purified. However, the inhibition zones of cecropins A and B against *P. aeruginosa* relative to *E. coli* D31, as reported by Boman and Steiner (1981), were slightly different from those reported by Steiner and coworkers (1981). Does this suggest that the activity of cecropins may vary from iso-

**TABLE 10.1. ACTIVITY OF WHOLE HEMOLYMPH AND CECROPINS A AND B FROM
HYALOPHORA CECROPIA ON VARIOUS BACTERIAL SPECIES**

Bacterial Strain	Average Killing Time Whole Hemolymph[a] (min)[b]	Activity Relative to that on *E. coli* K12 D31		
		Whole Hemolymph[a]	Cecropin A[b]	Cecropin B[b]
Escherichia coli K12 D31	0.78	1.0	1.0	1.0
E. coli K12 D21	1.2	0.65	NR[c]	NR
E. coli, wild type	2.0	0.39	NR	NR
Pseudomonas aeruginosa OT97	3.1	0.25	0.61	0.78
Enterobacter cloacae B11	17.0	0.046	NR	NR
Bacillus subtilis BS11	2.8	0.28	0.19	0.42
B. megaterium BM11	0.35	2.23	0.70	0.83

Source: After Boman and Steiner, 1981, and Steiner et al., 1981.
[a] Activity of whole hemolymph originally reported as killing time (min).
[b] Activity of cecropins originally reported as zone of inhibition (mm).
[c] Not reported therein.

lation to isolation, or is the variation due merely to experimental
discrepancy?

10.2.3.3. Bacterial Inhibitors of Cecropins. Boman (1981) has suggested that
cecropinlike substances exist in all Lepidopterous insects, yet there are instances in which the cecropins exhibit no antibacterial activity. Boman (1982)
reported that insect pathogens have developed a counterdefense for dealing
with the cecropins. Sidén and colleagues (1979) and Dalhammer and Steiner
(1984) attribute the inactivity of cecropins against certain bacterial species
to the fact that the bacteria produce inhibitors. Edlund and colleagues (1976)
presented evidence for two immune inhibitors from *B. thuringiensis*, which
interfered with the immune system of saturniid pupae. Sidén and associates
(1979) described an immune inhibitor (InA) of the immune system of saturniid
pupae. InA, a single polypeptide chain with a molecular weight of 78,000,
inhibited in vitro killing of *E. coli* by immune hemolymph but did not affect
the killing of *Bacillus subtilis*.

Flyg and coworkers (1980, 1983) showed that strains of *S. marcescens*
that are pathogenic to insects are resistant to cecropins, whereas some protease-deficient mutants are sensitive to this type of induced immune factor.
The authors believe that in an insect such as *Drosophila melanogaster* passive resistance to immunity is more important than the production of
proteases.

Dalhammar and Steiner (1984) reported that organisms such as *B. thuringiensis* produce exoenzymes that destroy cecropins and attacins. The
substance produced by *B. thuringiensis* was considered analogous to that

described by Sidén and coworkers (1979). Inhibition of antibacterial activity was attributed to a protease that effected proteolytic degradation of the antibacterial proteins. The protease selectively degraded the cecropins and attacins. The investigators were able to identify the susceptible peptide bonds in cecropin A.

These findings add support for the use of *B. thuringiensis* as a microbial control agent. If there is a chance that insect pests might produce cecropins owing to inducement of the substances by some circumstances in their environment, the application of *B. thuringiensis* (in the viable, as opposed to the toxin, state) would still be successful. Moreover, there might be less likelihood of insects' developing resistance to *B. thuringiensis* in nature.

One could speculate further that inducible proteins (i.e., cecropins) are frequently produced but are antibacterial only for certain organisms. The ability of an organism to counteract cecropins might determine its pathogenicity. Such arguments are attractive, but one must keep in mind the questions as to how easily cecropins are induced in nature. Quite likely, virulence factors of pathogens are more complex than this. Certainly, more investigation is necessary in order to substantiate such speculation.

10.2.3.4. *Limitations in Activity*.

According to Boman and Hultmark (1981), cecropins act on both gram-negative and gram-positive bacteria, and in the cases investigated (examples not specified), they caused partial lysis of the bacteria. Using a light microscope, these investigators, observed that cecropins, without contamination by lysozyme, affect the bacterium by producing a ghostlike, rod-shaped form. Moreover, they indicated that in measuring turbidity of the bacterial suspension, complete clearing never occurs unless lysozyme is present. If cecropins are responsible for elimination of bacteria in a natural environment, it is suggested that lysozyme is important as a cleaning agent to rid the host of debris.

Okada and Natori (1983) isolated antibacterial proteins from *Sarcophaga peregrina* larvae and reported that these substances were similar to cecropins. However, the proteins of Okada and Natori had no antibacterial activity against *P. aeruginosa*. Furthermore, among the six organisms inhibited by these proteins, *Proteus vulgaris* was one of the least susceptible. In fact, the minimal inhibitory concentration was 50% higher than for bacterial species, which have been considered nonpathogens by Chadwick (1975). With *P. aeruginosa*, we were able to achieve protective immunity in vivo (Chadwick, 1977) as well as in vitro antibacterial immunity in our fat body assay against *P. aeruginosa* (DeVerno et al., 1984). Moreover, we have demonstrated protective immunity and antibacterial activity against *Proteus mirabilis* and other *Proteus* species (Chadwick, 1967, 1975; Morton et al., 1984). Such diversity in results against specific bacterial species could point to two possibilities: (1) decreased or lost activity on the part of purified hemolymph proteins or (2) the existence of different antibacterial substances in different insect species. In either case, the need for examining the activity of isolated antibacterial substances against more than one bacterial species and for assessments of antibacterial activity done in parallel with in vivo tests for protective immunity seem warranted.

10.2.3.5. Comparison to Other Factors.

Boman and Steiner (1981) reported that immunity in *H. cecropia* resulting in the production of cecropins could be induced by the injection of viable *E. cloacae*. Bacteria of the same species killed by ultraviolet light produced a minimal response, and a heat-killed pathogen such as *P. aeruginosa* also produced a lesser response than did *E. cloacae*. Further, the authors stated that some bacteria, such as *P. aeruginosa* and *Xenorhabdus nematophilus*, may be sensitive to the immune proteins, but in some way they are able to avoid the induced immunity; thus, they start to divide and eventually kill the host. Boman and Steiner also stated that the use of several pathogenic bacteria did not result in any induction, thus theorizing that this may be the reason such organisms are pathogenic. This finding may suggest that workers such as Chadwick (1975), Chadwick and associates (1982), and DeVerno and associates (1983) may have been working with a system different from that of Boman's group. The use of killed immunizing preparations in their hands induced highly protective immunity, the host being able to withstand greatly increased doses of the pathogen. It is not clear from the work of Boman and coworkers whether they analyzed protective immunity in those test insects in which cecropins are formed.

Boman and Steiner (1981) stated that if induction of immunity in cecropia was blocked with actinomycin D, a nonpathogen, such as *E. coli*, would kill its host. These results could be compared to those of Chadwick (1975), who reported that *E. coli* would survive many days in *G. mellonella* hemolymph without apparent effects on the host, the insect being able to withstand doses of about 10^5 bacteria. If the latter situation is comparable to the former, it must be assumed that *E. coli* was inducing cecropins. If this were the case, why did the organisms persist in the hemolymph, and why were they not destroyed, as reported in the in vitro assays of Boman and Hultmark (1981) and Boman and Steiner (1981)? The question remains as to whether in vitro antibacterial assays can be compared to in vivo studies.

Boman (1981) has suggested that cecropinlike substances exist in all Lepidopterous insects, and the possibility exists that such substances may extend to other orders of insects. It can be further inferred from this suggestion that antibacterial factors in all insects are similar and that consequently the immune response is similar. Karp and Rheins (1980; Rheins and Karp, 1982) have not studied or identified antibacterial substances in the humoral immune response of the cockroach *Periplaneta americana*. However, they did state that the inducible immune response in the cockroach was quite different from those described to date in other invertebrates, even suggesting analogy to vertebrate antibody molecules (Rheins and Karp, 1982). In their system, memory was indicated. It is attractive to speculate that if antibacterial factors are demonstrated in the cockroach, they too may be quite distinct from the cecropins of Boman's group.

In most of their assays of antibacterial activity, Boman and coworkers used *E. coli* K12 D31 as the major indicator organisms. Part of the glucose, galactose, and rhamnose was deleted from the core of the lipopolysaccharide of this organism (Boman et al., 1974; Monner et al., 1971). Dunphy and colleagues (1986a,b) found that strains of *P. aeruginosa* deficient in most of the LPS core had lower LD_{50}s for *G. mellonella* than did a smooth strain AK957. All investigations carried out by Chadwick (1975, 1977) and Chadwick and Aston (1976, 1978) involving protective immunity and the demonstration of antibacterial activity have employed a wild-type *P. aeruginosa* with an LD_{50} of the order of 1–10 bacteria per insect. Antibacterial activity against the wild-type *P. aeruginosa* has always been significantly less than against the rough strains (J. Chadwick, unpublished). Although Dunphy and associates (1968a) found little correlation in the degree of roughness and the rate of clearance of the rough strains from the normal hemolymph, there might well have been more variation and correlation if immune hemolymph had been used. Similarly, if cecropins A and B had been tested for antibacterial activity against a wild-type *Pseudomonas*, there may well have been activity considerably different than that reported for *P. aeruginosa* OT97.

10.2.4. Phenoloxidase System and Antibacterial Activity

Söderhäll (1982) has discussed and reviewed the prophenoloxidase-activating system and melanization in arthropods, particularly the role of this system as a recognition mechanism (see Chapter 9). In this chapter, the only reference to the system is a brief consideration of the possibilities of components of the system being antibacterial in nature.

Indirectly, Stephens (1962a), Ziprin and Hartman (1971), and Ziprin (1978) have implied that inhibition of melanization of the hemolymph of actively immunized larvae of *G. mellonella* may be associated with acquired immunity. There has been no direct proof of this. It is attractive to support the theory of Söderhäll (1982) that the inhibition of melanization in insect hemolymph reported by the abovementioned authors and also by Pye (1974) did not necessarily indicate that phenoloxidase itself was excluded from involvement in the defense reactions, since the inhibited phenoloxidase in immunized insects could easily become activated by substances such as proteases in bacteria. Further, Söderhäll reported that, since activated phenoloxidase is toxic to some insects, it is probably rapidly inhibited in order to prevent any damaging effect on the host. According to Söderhäll, phenoloxidase may represent a self-protection mechanism of the host.

Quinones, which are side products of the phenoloxidase system, have been considered antibacterial substances (Pye, 1974; Nappi, 1978). These side products may enhance the immune factors and thus contribute to elimination of certain bacterial species from the insect. Faye and coworkers (1975) found no correlation between phenoloxidase activity and immunity in cecropia pupae. Opinions as to the antibacterial role of components of the polyphenoloxidase system vary, so the importance of their role is as yet undetermined.

10.3. ANTIVIRAL ACTIVITY OF INSECT HEMOLYMPH

Insect resistance to entomopathogenic viruses has only recently been subjected to scrutiny. At the population level, there have been scattered reports of insect viral resistance with prolonged exposure to cytoplasmic polyhedrosis viruses (CPV) (Payne, 1981; Briese and Podgwaite, 1985) and the *Baculoviruses* (Briese, 1981; Briese and Podgwaite, 1985). The antiviral factors have yet to be clearly elucidated and are generally attributed to changes in the physicochemical barrier of the gut. We are not aware of the use of resistant insects to explore possible antiviral factors in the hemolymph or of the generation of resistant insects using attenuated or inactivated virus vaccines.

In nonimmune insects, antiviral activity might be due to the hemolymph or tissues in contact with the hemolymph. However, the general absence of insect antiviral activity may represent artifact due to the use of insect pathogens. Bacterial pathogens are known to either evade or tolerate host defenses (Makela et al., 1980; Horohov and Dunn, 1983; Walters and Ratcliffe, 1983), and insect pathogenic viruses may do likewise. Also, nonpathogenic viruses are readily cleared or inactivated when injected into *P. americana* (Kunis et al., 1978) and beetles (Odier et al., 1974).

10.3.1. Hemocytic Antiviral Activity

The major cellular resistance systems in most insects are the hemocytes (Ratcliffe and Rowley, 1979; Rowley and Ratcliffe, 1981). Kislev and associates (1969) documented phagocytosis of the virions (viropexis) of nuclear polyhedrosis viruses (NPV) by the plasmatocytes of the Egyptian cottonworm, *Spodoptera littoralis*. Although hemocyte-mediated viropexis has been frequently detected in numerous insect species, it is still not known if it is a nonspecific response to nonself particles or a specific response induced by the virus to ensure virus reproduction and/or transmission to secondary tissues.

There are no definitive reports of viral degradation in insect hemocytes. However, Kislev and colleagues (1969) reported the absence of discernible antiviral activity in *S. littoralis* plasmatocytes, based on the development of virogenic stroma. Similarly, *Tipula* iridescens virus, although detected in acid phosphatase–positive lysosomelike structures of *G. mellonella* hemocytes, also produced stroma (Younghusband and Lee, 1970), establishing the inability of the hemocytes to control the virus. It is interesting to note that Kislev and colleagues (1969) reported low levels of virus infection in *S. littoralis* granulocytes and oenocytoids. The authors did not explain this observation, but it may represent (1) differences in virus susceptibility due to the absence of virus receptors and/or concomitant virus-triggered viropexis, (2) a biochemically incompatible cell milieux (passive antiviral activity), or (3) the possession of active antiviral factors within a majority of granulocytes and oenocytoids. A low yield of NPV replication in *Malacosoma disstria* hemocyte cell line IPRI 66, compared with IPRI 108, has been reported (Sohi and Cunningham, 1972). Three possible reasons offered to explain this included (1) the presence of microsporidia in IPRI 66, (2) altered physicochemical membrane properties (IPR 66 cells float, while IPRI 108 cells remain in contact with the substratum), and (3) different numbers of subculturing between the cell lines may have selected for resistant hemocyte lines.

Overall, hemocytic phagocytosis and subsequent activity appear to be ineffective against *Tipula* iridescent virus (Younghusband and Lee, 1969, 1970), *Entomopovirus* (Granados, 1973), and *Baculovirus* (Mazzone, 1985).

10.3.2. Humoral Antiviral Activity

Humoral antiviral activity in insects has not been adequately addressed. In an attempt to induce immunity in *G. mellonella* against cricket paralysis virus, Rohel and colleagues (1980) vaccinated larvae with inactivated virus preparations. No viral protection was found. The hypoproteinemic changes in the electrophoretic patterns of insect hemolymph from virus-infected insects is generally attributed to viral perturbation in fat body metabolism (Watanabe, 1967; Vander Geest and Wassink, 1969; Young and Scott, 1970; Tanada and Watanabe, 1971). Young and Scott (1970) detected two protein species in NPV-infected *Trichoplusia ni* not present in healthy larvae. The nature and function of these proteins remain to be determined.

10.3.3. Additional Antiviral Mechanisms

Additional expressions of antiviral activity in animal cells include interferon production, homologous viral interference, heterologous virus interference (Sherman, 1985), and cell budding (Quiot et al., 1980).

10.3.3.1. Interferon. *Togavirus* infections in mosquito cell lines are generally chronic, producing little cytopathogenic effect (Enzman, 1973). It was suggested that an interferonlike protein or proteins may have created the persistently infected state. This suggestion was based on two facts: (1) enhanced production of Sindbis virus (SV) in infected *Aedes albopictus* cells treated with actinomycin D and (2) medium from persistently infected cells, treated in a manner to isolate interferon, reduced SV production. However, the actinomycin D effect could be mediated by mechanisms other than those inhibiting interferon production (Stellar, 1980). Interferon production has not been detected in mosquito cell lines using Semliki Forest virus (Peleg, 1969) or SV (Murray and Morahan, 1973).

Riedel and Brown (1979) described a low molecular weight virus inhibitor in a culture medium of SV-infected *A. albopictus* cells. The agent appeared in the culture medium 3 days after cell infection. Protease K inactivated the compound, establishing its polypeptide nature. Treating host cells with this factor resulted in a low-yielding persistent SV infection (Scheefer-Borchel et al., 1980). The antiviral agent was cell specific, and unlike interferon, it was virus specific (Riedel and Brown, 1979).

Novokhastaskii and Berezina (1978) failed to reduce arbovirus development in mosquito cells exposed to numerous interferon inducers. This result, in conjunction with those of the aforementioned studies, suggests that interferon does not appear to influence viral infections in insects.

10.3.3.2. Homologous Viral Interference. Homologous viral interference has been detected for *A. albopictus* cells infected with viruses of the *Bunyaviridae*. During initial infection viral proliferation is rapid, giving rise to a persistent infection with low virus titers. Because the cells excluded superinfection with the homologous virus, all or part of the genome was expressed in the cells (Newton et al., 1981). Small plaque variants and temperature-sensitive mutants are known to inhibit wild-type virus multiplication (Youngner et al., 1976). Although these stages were detected in *Bunyavirus* infections, they occurred after the establishment of persistent infections. Newton and Dalgarno (1983) reported that, like Semliki Forest virus infection in *A. albopictus,* persistent *Bunyavirus* infection was associated with a decline in viral protein, RNA, and nucleocapsid synthesis, but these events were not associated with the persistent virus infection state. The low levels of infection may have represented the production of defective particles capable of homologous interference.

Togavirus produces a persistent infection in mammalian cells (Lee and Scholoemer, 1981b). The Banzi virus produces an antiviral factor in mosquito cells. The agent was a low molecular weight compound with specificity activity against the Banzi virus. Anti-Banzi serum, but not anti–mosquito cell serum, inactivated the antiviral factor, suggesting the agent was of viral origin and probably a structural protein (Lee and Scholoemer, 1981a).

10.3.3.3. Heterologous Viral Interference. Heterologous viral interference has been recorded for mixed CPV and NPV infections in *B. mori* (Tanada and Argua, 1967). Because infected cells never contained both types of viruses, viral interference was believed to have occurred at the cellular level. Only live granulosis virus and NPV in *Heliothis armigera* lead to heterologous viral interference as opposed to the use of viral structural proteins (Whitlock, 1977). This implies that interference was due to more than virus competition for a common receptor.

Cells of *Lymantria dispar* infected with *Euxoa scandens* CPV bleb extensively (Quiot et al., 1980). Because of the presence of virus particles enclosed within each bleb, blebbing was believed to represent an antiviral mechanism. Whether the activity is truly antiviral or a mode of virus dispersal remains to be determined.

10.4. ANTIVIRAL ACTIVITY IN NONINSECTAN HEMOLYMPH

Antiviral activity in noninsectan hemolymph has been investigated to a lesser extent than in insects. In general, interferon-mediated antiviral activity has not been detected (Sherman, 1985).

Antiviral activity, expressed as the removal of T-even phage from the hemolymph of the blue crab, *Callinectes sapidus*, has been reported by McCumber and Clem (1977). The T-odd phages—for example, T_3 and T_7— were not removed, and attempts to expedite their removal by "immunizing" the crabs with T-even phage were unsuccessful.

The tissues involved with virus removal from the hemolymph varied with the virus type, with radio-labeled T_2 phage being sequestered in the hepatopancreas and poliovirus in the gills (McCumber and Clem, 1977). The nature of the virus-binding receptors is unknown, and as in the responses of *Cambarus virilis* and *Homarus americanus*, the foreign protein was degraded and externalized (Clem et al., 1984).

Different clearance profiles of different types of protein have been detected in *Procambarus clarkii* (Sloan et al., 1975) and the mud crab *Scylla serrata* (Mullainadhan and Ravindranath, 1984). In the latter animal, hemoglobin and horseradish peroxidase were accumulated in the hepatopancreas and gut, respectively. *Callinectes* is known to clear xenogenic protein from its hemolymph (McCumber and Clem, 1977), but neither humoral nor hemocytic factors are involved (Clem et al., 1984). The similar phage removal patterns in *C. sapidus* implies that virus clearance may be due to the recognition of the nucleocapsid protein as nonself. *Carcinus maenas* gill nephrocytes are known to remove hemocytic and bacterial debris (Smith and Ratcliffe, 1981), but it is not known whether they take up xenogenic proteins (Clem et al., 1984) or virus particles.

McCumber and coworkers (1979) described a T_2 phage–neutralizing factor in *C. sapidus* hemolymph. The 80,000-MW polypeptide did not require cations to neutralize phage activity against the *E. coli* test strain. The factor was inactive against T_3 and T_7 phages.

10.5. ANTIBACTERIAL DEFENSES IN THE HEMOLYMPH OF NONINSECTAN ARTHROPODS

The antibacterial defenses of noninsectan arthropod hemolymph may be divided into cellular defenses and humoral defenses.

The cellular defense systems include the hemocytes, nephrocytes, and phagocytic cells of the fat body. This material has been reviewed by Ravindranath (1981) for the onychophorans and myriapods, Sherman (1981) for the chelicerates, Bauchau (1981) for the crustaceans, and Ratcliffe and coworkers (1982) for the Arthropoda phylum as a whole.

Humoral molecules described in initiating nonself responses in noninsectan arthropod hemolymph include bacterial agglutinins, hemagglutinins (HAs), precipitins, bactericidins, bacteriolysins, hemolysins, opsonins, clotting factors, and lysozymes (Cooper and Lemmi, 1981). However, humoral factors are also known to interact with the cellular defenses, for example, hemagglutinins of the freshwater crayfish *Cherax destructor* (*Parachaeraps bicarinatus*) in opsonin-mediated phagocytosis (McKay et al., 1969) and serum-enhanced bactericidal activity of *Limulus polyphemus* hemocyte lysate (Pistole and Graf, 1984). Thus, although the emphasis of this review is on the humoral components of the hemolymph, their interaction with hemolymph cellular factors is also addressed. The major classes to be considered include Arachnida, Chelicerata, and Crustacea.

10.5.1. Arachnida (Terrestrial Chelicerata)

Hemagglutinins against a diversity of erythrocytes have been detected in the hemolymph of the Saharan scorpion, *Androctonus australis*; the Arizona lethal scorpion, *Centuroides sculpturatus; Hadrurus arizonensis; Mastigoproctus giganteus; Aphonopelina chalcodes; Vaejovis confuscius,* and *Vaejovis spinigerus* (Vasta and Marchalonis, 1983). The molecular properties of the HAs have not been extensively characterized. The HA of *A. australis* had reduced activity after dialysis of the serum (Brahmi and Cooper, 1974). Vasta and Cohen (1982) reported reduced HA titer of *C. sculpturatus* serum with EDTA toward goat and sheep erythrocytes. HA titer was restored with the addition of calcium. Thus, the divalent calcium cation is believed to be important to scorpion lectin activity. Overall, the lectins, like those of *A. australis* (Brahmi and Cooper, 1974), were heat labile, having been denatured at 65°C for 30 min.

The serum of all scorpion species tested by crossed absorption assay has contained heteroagglutinins. The heterogeneity of the serum lectins was also established by the ability of the lectins to differentiate between normal and leukemic lymphocytes (Cohen et al., 1979). The lectin inhibition assay used by Vasta and Cohen (1982) confirmed HA heterogeneity.

The lectins in the hemolymph of seven scorpion species had a main specificity for sialic acids and sialoconjugates with varying degrees of additional specificity for 3-deoxy-D-mannooctulosonic acid (a component of lipopolysaccharide molecules associated with the outer envelope of gram-negative bacteria), chitobiose, and *N*-acetyl-amino sugars (common components of bacterial peptidoglycan layers; Vasta and Marchalonis, 1983). Sialic acids are not part of the arthropod cell membrane. Because many of the lectin-inhibiting sugars are associated with bacterial cell walls and envelopes (Springer, 1970), Vasta and Cohen (1984b) proposed that the lectins may be active against bacterial surfaces and may be part of a moderately nonspecific recognition system (see also Chapters 1 and 19).

Different HA types occur in the serum of *V. confuscius* and *C. sculpturatus* based on the different HA titer responses against selected neuraminidase-treated erythrocytes (Vasta and Cohen, 1984a). Also, unlike the lectins of *V. confuscius*, the lectins of *C. sculpturatus* bind to D-galactosyl residues.

Fetuin, a sialoglycoprotein, and an asialo-orosomucoid were poor inhibitors of *V. confuscius* and *C. sculpturatus* HA, whereas orosomucoid was a good inhibitor. Vasta and Cohen (1984a) suggested that the subterminal monosaccharide and terminal oligosaccharide conformation was important to HA binding. Inhibition studies of *V. spinigerus* and *V. confuscius* HA, using asialobovine submaxillary mucin and fetuin, revealed greater inhibition by the latter glycoprotein than by the former (Vasta and Cohen, 1984b). The lectins bound to terminal oligosaccharides with sialic acid-a-2,6-N-acetylgalactosamine with greater avidity than to sialic acid-a-2,3(6)-galactose.

10.5.2. Chelicerata (Aquatic Chelicerata)

The majority of the antibacterial studies of the chelicerates have revolved around *L. polyphemus* with emphasis on bacterial-induced hemolymph gelation, hemagglutinins, and bacterial agglutinins (Smith and Pistole, 1985; see also Chapter 11).

Specific antibacterial activity has been detected in *L. polyphemus* hemolymph against gram-negative *E. coli* and *Klebsiella pneumoniae* but not against *S. dysenteriae*, *P. aeruginosa*, or gram-positive *Micrococcus luteus* or *Staphylococcus aureus* (Johannsen et al., 1973). Furman and Pistole (1976) found heat-sensitive antimicrobial factors (inactivated at 56°C for 30 min) with maximum activity against microorganisms common to the crab's habitat. The antibacterial titer was influenced less by the environment than by genetic factors (Furman and Pistole, 1976) and varied with the microorganism assayed (Pistole and Furman, 1976). Evidence of humoral participation with hemocyte-killing activity against *E. coli* and *M. luteus* was attributed to increased phagocytosis (Pistole and Britko, 1978).

Broad-spectrum, heat-sensitive, antimicrobial activity has been detected in hemocyte lysate (Nachum, 1979; Nachum et al., 1979). The cations Na^+, K^+, Ca^{2+}, and Mg^{2+} reduced the antimicrobial titer, leading Nachum and colleagues (1980) to speculate that the ions prevented the attachment of the bactericidal factors to the bacteria in a manner analogous to cation abrogation of the antibacterial activity of rabbit granulocyte cation proteins. The antimicrobial factors were localized to the hemocyte granules, which express this activity via the mediation of lectins with affinity to the bacterial surface components, lipopolysaccharides, and teichoic acids (Smith and Pistole, 1985).

The hemolymph and plasma of *L. polyphemus* contains HA against mammalian, avian, reptilian, and amphibian erythrocytes (Cohen et al., 1965). Cross-absorption assays established the existence of heteroagglutinins (see also Chapter 12). Hemocyanin lacked HA activity.

Initial characterization of one HA, limulin, was given by Marchalonis and Edelman (1968). Limulin had a molecular weight of 400,000 and was dissociated into 18 subunits of 22,500 daltons. The subunits were not held together by covalent bonds. The authors predicted that limulin was hexagonal, which was confirmed by Fernández-Moran and associates (1968) using electron microscopy.

Limulin purified by affinity chromatography required Ca^{2+} for activity (Oppenheim et al., 1974). When purified by molecular sieve, ion-exchange, and affinity chromatography, the HA had a molecular weight of 335,000 and an isoelectric point of 5.1. The protein contained high levels of aspartic acid and glutamic acid (thus the low isoelectric point) as well as glycine and alanine. It contained low levels of tryptophan, tyrosine, methionine, and cystine. The most effective HA inhibitors were not monosaccharides but sialoglycoproteins, N,N'-diacetylchitobiose and N,N',N''-triacetylchitotriose (Roche and Munsigny, 1974). Nowak and Barondes (1975) used formalized horse erythrocytes as an affinity adsorbent to isolate limulin and N-acetylneuramic acid as the elutant. The molecular weight by ultracentifugation was 460,000 and the subunits on SDS-PAGE after dissociation in urea and 2-mercaptoethanol appeared as one band with a molecular weight of 22,000.

The sequencing of the first 76 amino acid residues of limulin by Kaplan and colleagues (1977) established that the primary structure was unrelated to that of vertebrate immunoglobulins. Marchalonis and Edelman (1968) stated that vertebrate immunoglobulins were not similar to limulin, based on the absence of covalent linkage between the limulin units and antigenic cross-reactivity with human γ-immunoglobulins (see also Chapter 12).

Four heteroagglutinins active against erythrocytes have been isolated from *Tachypleus tridentatus* hemolymph. The lectins differed in their carbohydrate binding and antigenicity properties (Schimizu et al., 1977). All but one lectin required Ca^{2+}.

A Ca^{2+}-dependent HA lectin has been isolated from *Carcinoscorpius rotundacauda*. The lectin, sarcoscorpin, had a molecular weight of 420,000 and was composed of two subunits differing in molecular weight (Bishayee and Dorai, 1980). HA activity was not altered by alkylating or reducing agents and was relatively insensitive to acetic anhydride (which reacts with free amino groups). Thus, neither carbohydrates, amino groups, nor disulfide bonds were responsible for sarcoscorpin activity. Limulin and sarcoscorpin were antigenically unrelated (see also Chapter 17).

Sialic acid specificity has been detected for lectins of arthropod serum, including *L. polyphemus* (Maget-Dana et al., 1979), the Japanese horseshoe crab, *T. tridentatus*; and the Indian horsehoe crab, *C. rotundicauda* (Vasta and Cohen, 1982). Because the merostomes and arachnids diverged during the Cambrian period and occupied different niches, Vasta and Cohen (1982) proposed that the *N*-acetylneuramic acid binding lectins are a relic trait rather than an adaptive property that appeared simultaneously in both groups.

Pistole (1978, 1979) described the reactions of *L. polyphemus* bacterial agglutinins with both gram-positive and gram-negative bacteria except for *M. luteus* and *S. marcescens*. The titer against gram-negative bacteria, while lowered by adsorption of the serum with gram-negative bacteria, was not affected by adsorption with gram-positive bacteria. The agglutinins were regarded as humoral because the titers were the same for serum and plasma.

The *Re* rough-walled mutant of *Salmonella minnesota* reduced the titer against smooth-walled *S. minnesota*. Sialic acid, a component of the outer membrane of the envelope of gram-negative bacteria, was not the lectin binding component on the Re mutant. The mutant lacked the O side chain and core sugars of the sialic acid–carrying lipopolysaccharide. Only 3-deoxymannooctulosonic acid (KDO) would be accessible to the *Re* mutant agglutinin. Rostam-Abadi and Pistole (1982) detected three isolectins in *L. polyphemus* hemolymph that bound to KDO, one of which was limulin. Binding by limulin was anticipated because of the structural similarities between sialic acids and KDO.

The bacterial agglutinin and toxin-mediated clotting reactions are two distant events for the following reasons: (1) the gelling reaction was not triggered by whole bacterial cells, (2) organic solvents known to block *Limulus* amebocyte (granulocyte) lysate gelation did not inhibit bacterial agglutination, and (3) both plasma and serum were capable of bacterial agglutination, whereas only serum is capable of gelation (Pistole, 1978).

Heteroagglutinins against bacteria occur in *Limulus* plasma. Gilbride and Pistole (1979, 1981) isolated a staphylococcal binding protein that lacked HA activity and agglutinin activity against *S. minnesota*. The lectin was inhibited by D-galactose.

10.5.3. Crustacea

The cultivation of crustaceans is a major commercial industry with substantial potential for the future. In order to exploit the coast and inland water systems, large numbers of crustaceans must be cultivated in small volumes of water. This crowding is conducive to disease outbreaks (Pauley, 1975). Understanding crustacean defense systems is essential to reducing and/or controlling diseases of crustaceans.

In the interest of brevity, the following sections emphasize the American lobster, *Homarus americanus*, and *Panulirus* spp.

10.5.3.1. Homarus americanus. The humoral defenses of *H. americanus* include HAs, bactericidins, bacterial agglutinins, and opsonins (Goldenberg and Greenberg, 1983). Much of the research has been to understand the virulence properties of *Aerococcus viridans homari* and the development of gaffkemia.

The hemolymph of *H. americanus* contains bacterial agglutinins (BAs) against several gram-negative and gram-positive bacteria, including *Pseudomonas perolens* and *Micrococcus sedentarius*. Injecting the bacteria into the lobster resulted in a tenfold decline in BA titer with restoration to the original values by 72 hr postinjection (Cornick and Stewart, 1968). The lectins were heat labile. BA titer was unrelated to lobster sex and varied with the individual. Cornick and Stewart (1978) did not find a correlation between BA titer and hemocyte type; in fact, animals with the same titers had different differential hemocyte counts. Thus, the hemocytes were not considered a source of BA.

The BA did not agglutinate *A. viridans homari*. Lobster serum actually stimulated the growth of this bacterium. The phagocytic cellular defenses of the lobsters (chemocytes, phagocytic cells associated with the heart and hepatopancreas) were also ineffective against *Aerococcus*.

A precipitin active against bovine serum albumin was observed in vitro and in vivo (Stewart and Foley, 1969), but the resulting precipitates were not phagocytized. With the ability of the precipitin to discriminate between self and nonself proteins, it is regrettable that the relationship between the precipitin and other humoral components was not addressed.

Injection of gut bacteria from the West Indian spiny lobster, *Panulirus argus*, or *H. americanus* into the lobster's hemocoel resulted in the induction of bactericidins active against *P. perolens* in the hemolymph and hepatopancreas within 36 hr of injection (Acton et al., 1969; Stewart and Zwicker, 1972). Multiple bactericidins with optimum activity at pH 7.6 (the pH of lobster hemolymph) exhibited synergistic action with lobster hemocytes (Stewart and Zwicker, 1973; Mori and Stewart, 1978a). A decline in total hemocyte counts occurred during the induction of the bactericidins, leading Mori and Stewart (1978b) to suggest that (1) the hemocytes may be required to activate bactericidins by phagocytosis of bactericidin-coated bacteria and/or (2) hemocyte lysis is part of the induction mechanism. The latter is similar to the mechanism inducing immunity in *G. mellonella* (DeVerno et al., 1984) and *M. sexta* (Dunn et al., 1985).

The heat-labile bactericidins did not protect lobsters against *A. viridans homari* (Stewart and Zwicker, 1972), allowing the bacteria to destroy the glucose, glycogen, and ATP reserves of *H. americanus*, culminating in lobster death (Stewart and Cornick, 1972; Stewart and Arie, 1973). The ineffectiveness of the antimicrobial elements to harm *A. viridans* was not due to the absence of these factors adhering to the bacterial cell wall (Mori and Stewart, 1978a). In view of the involvement of bacterial cell walls in the virulence of entomopathogenic bacteria (see Dunphy et al., 1986a, 1986b), it is possible that the tolerance of *A. viridans* to the lobster's humoral and cellular defenses may reside in the bacterial cell wall. An analysis of the cell walls would be an essential, logical step in defining the virulence properties of the bacterium.

The use of virulent *A. viridans* as a vaccine to induce bactericidins in *H. americanus* has resulted in failure. There was also no decline in total hemocyte counts (Mori and Stewart, 1978b). However, avirulent mutants of the bacterium altered the lobster's defenses, resulting in prolongation of the time required by virulent *A. viridans* to kill lobsters (Stewart and Zwicker, 1974). The effect was transient.

Hemagglutinins exist in lobster hemolymph (Cornick and Stewart, 1973; Hall and Rowlands, 1974a). Initially, it was proposed that heat inactivation curves of HA activity indicated the presence of one HA (Cornick and Stewart, 1973). However, the authors also reported that D-glucosamine and *N*-acetylglucosamine could not totally reduce HA activity, which, by itself, contradicts their original contention. Hall and Rowlands (1974a) isolated two HAs: LAg-2, which was active against mouse erythrocytes only; and LAg-1, which was active against mouse and human erythrocytes. Like Cornick and Stewart (1973), they found optimum HA activity between pH 7.5 and 8.0. The lectins had different antigenic determinants. Expanding on the heterogeneous properties of the lectins, Hall and Rowlands (1974b) discovered that sugars inhibiting LAg-2 were not effective against LAg-1. In a puzzling contradiction, they also did not find glucosamine inhibitory to the lectins. Like other arthropodan HAs, lobster HAs were heat labile, but unlike the case with *L. polyphemus* HAs, Ca^{2+} was not required.

Vaccinating lobsters with *P. perolens* did not influence HA titer (Cornick and Stewart, 1973). Thus, the bactericidins and HA were unrelated. However, the addition of hemocytes to plasma substantially boosted the titer, suggesting that the hemocytes are the source of HA. Amirante and colleagues (1976) reported a similar HA-hemocyte association in *L. maderae* (see also Chapters 1 and 13).

Mediation of the phagocytosis of sheep erythrocytes by opsonins was initially described by Paterson and Stewart (1974). Using a Sephadex G-200 column, Goldenberg and Greenberg (1983) isolated a heat-labile opsonin devoid of HA activity from an HA fraction of lobster hemolymph. Further characterization has yet to be done.

Phagocytosis as an antibacterial defense mechanism has been inadequately considered. Lipopolysaccharide from the outer membrane of *P. perolens* is known to enhance the phagocytosis of *A. viridans* (Paterson et al., 1976) and sheep erythrocytes (Goldenberg and Greenberg, 1984). In view of the extremely virulent nature of *A. viridans*, phagocytosis is an ineffective defense.

10.5.3.2. Panulirus spp. and Other Crustaceans. Both *P. argus* and the California spiny lobster, *Panulirus interruptus*, produce inducible bactericidins when vaccinated with formalin-killed, gram-negative bacteria (EMB-1) isolated from the gut of *P. argus* (Evans et al., 1969a,b). Although cidal to *E. coli*, the bactericidin did not harm *P. aeruginosa* or the gram-positive *S. aureus, B. subtilis*, or *Bacillus megaterium*. Neither dialysis nor EDTA inactivated the heat-labile bactericidins. Bactericidins of *P. interruptus* were more heat stable than those of *P. argus*. Although the bactericidins were absorbed by homologous and heterologous bacteria (Evans et al., 1969a), the activity spectrum was very limited (Weinheimer et al., 1969a). The mode by which the cidal factors damaged bacteria was not reported.

Both *Panulirus* spp. exhibited anamnestic secondary responses (Evans et al., 1969a,c), and *P. argus* was capable of a tertiary response.

The bactericidin inducer from EMB-1 was tentatively identified as lipopolysaccharide (Evans et al., 1969c). However, because gram-positive bacteria also induced bactericidin production, the authors rejected their initial proposal. Both LPS and teichoic acid (a major cell wall component of gram-positive bacteria) have similar physiologic properties in vertebrates, for example, complement activation and mitogenicity. Thus, it is possible that Evans and associates (1969c) unjustly rejected their LPS inducer hypothesis.

A potent constitutive hemolysin from the hemolymph of *P. argus* with activity against sheep erythrocytes was first described by Weinheimer and coworkers (1969b). The heat-labile, enzymatic lysin required cations. It readily adhered to sheep erythrocytes, but the mechanism of lysis was not determined. Based on a sigmoidal dose response to lysin with erythrocytes, a multiple lytic step mechanism may be involved.

Neither the hemolysins nor the bactericidins induced bacterial agglutination. Bacterial agglutinins have been reported in hemolymph of *Cancer irroratus* (Pawley, 1974) and *Procambarus clarkii* (Miller et al., 1972). The agglutinins in the latter were active against a diversity of marine bacteria and vertebrate erythrocytes. BA activity was not influenced by dialysis. The lectin (MW 150,000) was not a lipoprotein and was not inhibited by urea, pronase, or trypsin. It was suggested that the BA may assist in removing bacteria from the hemolymph of the decapod. Attempts were not made to identify the BA and HA.

C. destructor hemolymph contains heteroagglutinins (McKay et al., 1969) that are essential for opsonin-mediated phagocytosis (McKay and Jenkin, 1970c). The HA opsonins originated within the hemocytes (Tyson and Jenkin, 1973). Serum factors also facilitate the adhesion of bacteria to hemocytes, often replacing the proteinaceous receptors on the hemocytes (Tyson and Jenkin, 1974).

Yeaton (1981), reviewing the occurrence of invertebrate lectins in general, stated that historically emphasis has been placed on (1) a systematic study of the occurrence of lectins, (2) improved purification, (3) elaboration on the binding specificity of lectins, and (4) analysis of lectin synthesis. Although research has begun to address these issues, the role of crab humoral lectins, lysins, and bactericidins in vivo is subject to speculation. Details of the mode of action of bactericidal factors are also unknown. Ravindranath and Cooper (1984) have proposed that receptor specificity of the lectins may be intimately associated with lectin function. The glycocalyx of many marine bacteria contain o-acetylated sialic acid, which makes them resistant to the hydrolytic enzymes of the phagocytic cells of decapods. O-aryl-sialic acid–specific lectins would mask the resistant sites, making the bacteria more susceptible to the crabs' defenses. Lectins binding to N-acetyl-D-glucosamine could have roles in cuticular resorption, cuticle synthesis, and opsonin-mediated phagocytosis of bacteria. However appealing these postulates may be, it is regrettable that definitive experiments to confirm them are lacking.

10.6. INDUCTION OF IMMUNITY IN NONINSECTAN ARTHROPODS

Despite the observations of bactericidin induction in arthropods, the contribution of these factors to the survival and immunity of the animal has not been considered in any detail. Indeed, the only record of true immunity in noninsectan arthropods is with the freshwater (Yabbie) crayfish, *C. destructor*. The gram-negative bacteria *Pseudomonas fluorescens, Salmonella typhimurium*, and *Pseudomonas* CP and bacterial LPS were effective immunogens. Multiple injections with the immunogens successfully boosted the level of protection in the crayfish (McKay and Jenkin, 1969).

Immunizing the crayfish elevated the total hemocyte counts. The hemocytes phagocytized and degraded the erythrocytes more rapidly than did hemocytes from nonimmune crayfish (McKay and Jenkin, 1970a). The killing rate of bacteria within phagocytes of the hepatopancreas of immune *Cherax* was also accelerated (McKay and Jenkin, 1970b). No bactericidal activity was present in the serum of immune crayfish (McKay and Jenkin, 1970c). Immune crayfish removed *Pseudomonas* CP from the hemolymph more effectively than did nonimmune crayfish (McKay and Jenkin, 1970b). Consequently, the acquisition of the immune state lay in the enhancement of the total phagocytic capabilities of the animal.

10.7. PROBLEMS, CONCLUSIONS, AND FUTURE TRENDS IN RESEARCH

Much of the early work in invertebrate immunity concentrated on the insects, and it is apparent, with respect to antibacterial factors, that the bulk of evidence remains with the insects. Boman stated that his goal and that of the investigators in his group was to describe in chemical terms how the various immune system components of insects interact with their targets and how a lethal effect is exerted on various bacterial species (Boman et al., 1978).

Although it may be too early to establish this finally, the first part of this goal has been met to a large degree by the purification and sequencing of the cecropins, presuming they are the major in vivo immune system substances in insects. As to how these substances exert their lethal effect, much of the interpretation is speculative. Target sites in the cell wall and cell membrane have been indicated by Chadwick and coworkers (1982) and Morton and coworkers (1984), although it has not been established that cecropins were the active agents in either case. Similarly, Dunn and Drake (1983) hypothesized that the killing of *P. aeruginosa* and *E. coli* in immunized larvae of *M. sexta* was probably due to nodule formation at or near the injection site, followed by formation destruction due to phagocytosis and induced humoral bactericidal factors. A wide field of endeavor awaits those who wish to pinpoint the specific sites of damage.

An understanding of what constitutes an efficient inducer of immune factors should be another goal. A living organism such as *E. cloacae* has been reported to be the most efficient inducer of cecropins (Boman and Steiner, 1981). Other living organisms and heat-killed organisms are less effective. The reasons are unknown.

In reports of antibacterial activity in arthropod hemolymph, some discrepancies and contradictory results may have arisen because of difficulty in handling and testing the hemolymph of various species. Melanization is a frequent problem, but the use of antimelanizing agents introduces variables that are often difficult to evaluate in the assessment of antibacterial effects of hemolymph or hemolymph factors.

As test organism in our laboratory, we have always used the bacterial pathogen against which immunity was induced by a heat-killed or formalized vaccine of the homologous organism. In addition, we have always tested for activity against related pathogens. Boman's group (Boman and Steiner, 1981; Steiner et al., 1981) has generally used a plate assay for determining activity of cecropins, measuring the zone of inhibition and placing considerable emphasis on the antibacterial activity against *E. coli* K12 D31 or D21. From time to time they have supplemented their data with tests against other bacterial species both gram-negative and gram-positive.

It seems logical that true antibacterial activity should extend beyond strains of *E. coli* K12. Perhaps investigators are somewhat divided in their purpose. Some wish to characterize induced proteins on a molecular basis, working primarily with a test strain of bacterium that is not necessarily pathogenic (i.e., a cell wall deficient mutant). Others attempt to demonstrate antibacterial activity more specifically, for example, against a pathogenic species that has been used as inducing agent, and, after biologic characterization of the mechanism, proceed to molecular analysis. Most rewarding investigations for the future should combine in vivo and in vitro studies at both the biologic and molecular levels.

In the area of research on noninsectan arthropods, future needs are to (1) continue an accurate biochemical characterization of the hemagglutinins, bacterial agglutinins, and bactericidins; (2) clarify the types of humoral factors in the arthropods to effectively remove the ambiguity of many of the plasma or serum components; (3) determine the mechanisms by which the lysins and bactericidins express their function, and (4) elucidate unequivocally the roles of antibacterial factors in vivo.

10.8. SUMMARY

Antimicrobial substances in arthropod hemolymph may be either naturally occurring or induced. Most attention has been focused on antibacterial factors in insect hemolymph. Naturally occurring antibacterial factors in insect hemolymph have been rarely reported. Among Lepidopterous insects, there are reports of bactericidal activity in hemolymph against organisms, such as *S. dysenteriae*, that are foreign to the insects' environment. Frequently, naturally occurring antibacterial factors in insect hemolymph have been compared to lysozyme. A recent report of a naturally occurring antibacterial factor in *B. mori* suggests a close relationship of the substance with the cecropins. Certain natural components of hemolymph, such as specific ions, have been assumed to be inhibitory to bacterial growth.

Considerable prominence has been given by Swedish investigators to a class of induced antibacterial proteins referred to as cecropins. Cecropins have been demonstrated in several Lepidopterous species, and some workers consider them the major immune defense in all insects. Cecropins have been purified, sequenced, and recently synthesized. Their antibacterial effects vary with the test organism; their greatest demonstrated activity is against strains of *E. coli* K12. It has been suggested that the mode of action of cecropins on bacteria is partially lytic, with final cleanup by lysozyme.

The nature of the induction of antibacterial immunity in insects has been addressed by some workers. The role of hemocytes and the fat body in producing antibacterial factors has been independently demonstrated by researchers in Canada, the United States, and Sweden. An in vitro assay demonstrating the generation of antibacterial activity by fat body in the presence of hemocytes and lipopolysaccharide has been reported in *G. mellonella*. Recently, a similar system has been reported in *M. sexta*.

Some investigators have attempted to demonstrate the target site of antibacterial damage. Immune hemolymph of *G. mellonella* was shown to destroy *P. aeruginosa* in about 7 min. Ultrastructural damage to the cell wall and membrane of *P. mirabilis* has also been demonstrated. More investigations to determine the site of antibacterial damage are recommended.

No clear relationship has been established between the phenoloxidase system in arthropods and antibacterial factors in the hemolymph.

Mechanisms of resistance of arthropods to viruses have generally been assigned a genetic basis. In some instances, control by a single dominant gene has been suggested. Naturally occurring antiviral factors have been reported infrequently, possibly because they have been seldom sought or because they rarely exist.

The humoral defenses of noninsectan arthropods include hemagglutinins, bacterial agglutinins, opsonins, and bactericidins. Many of the antibacterial factors are heat labile and require divalent cations for activity. The overall scarcity of in-depth biochemical characterization of these factors makes comparisons between studies of the same arthropod species or different species difficult. Regrettably, the in vivo functions of these humoral elements remain largely unexplored.

An analysis of the results presented herein points to a need for future investigators to adopt a coordinated approach utilizing both in vivo and in vitro studies.

REFERENCES

Abu-Hakimu, R. and I. Faye. 1981. An ultrastructural and autoradiographic study of the immune response in *Hyalophora cecropia* pupae. *Cell Tissue Res*. **217:** 311–320.

Acton, R.T., P.F. Weinheimer, and E.E. Evans. 1969. A bactericidal system in the lobster *Homarus americanus*. *J. Invertebr. Pathol*. **13:** 463–464.

Amirante, G.A., F.L. Debernardi, and P.C. Magnetti. 1976. Immunochemical studies on heteroagglutinins in the hemolymph of cockroach *Leucophaea maderae* L. (Insecta, Dictyoptera). *Boll. Zoll*. **43:** 63–71.

Andreu, D., R.B. Merryfield, H. Steiner, and H.G. Boman. 1983. Solid-phase synthesis of cecropin A and related peptides. *Proc. Nat. Acad. Sci. U.S.A*. **27:** 6475–6479.

Aston, W.P., J.S. Chadwick, and M.J. Henderson. 1976. Effect of Cobra venom factor on the *in vivo* immune response in *Galleria mellonella* to *Pseudomonas aeruginosa*. *J. Invertebr. Pathol*. **27:** 171–176.

Bauchau, A.G. 1981. Crustaceans, pp. 385–420. In N.A. Ratcliffe and A.F. Rowley (eds.) *Invertebrate Blood Cells*, vol. 2. Academic Press, New York.

Bishayee, S. and D.T. Dorai. 1980. Isolation and characterization of a sialic acid-binding lectin (carcinoscorpion) from the Indian horseshoe crab, *Carcinoscorpius rotundacauda*. *Biochim. Biophys. Acta* **623:** 89–97.

Boman, H.G. 1981. Insect responses to microbial infections, pp. 769–784. In H.D. Burges (ed.) *Microbial Control of Pests and Plant Diseases, 1970–1980*. Academic Press, New York.

Boman, H.G. 1982. Humoral immunity in insects and the counter defence of some pathogens. *Fort. Zool*. **27:** 211–222.

Boman, H.G. and H.G. Hultmark. 1981. Cell free immunity in insects. *Trends Biochem. Sci*. **6:** 306–309.

Boman, H.G. and D.A. Monner. 1975. Characterization of lipopolysaccharides from *Escherichia coli* K-12 mutants. *J. Bacteriol*. **121:** 455–464.

Boman, H.G. and Steiner. H. 1981. Humoral immunity in cecropia pupae. *Curr. Top. Microbiol. Immunol.* **94, 95:** 75–91.

Boman, H.G., A. Boman, and A. Pigon. 1981. Immune and injury responses in cecropia pupae: RNA isolation and comparison of protein synthesis *in vivo* and *in vitro. Insect. Biochem.* **11:** 33–42.

Boman, H.G., I. Nilsson-Faye, K. Paul, and T. Rasmuson. 1974. Insect Immunity. 1. Characteristics of an inducible cell-free antibacterial reaction in hemolymph of *Samia cynthia* pupae. *Infect. Immun.* **10:** 136–145.

Boman, H.G., I. Faye, A. Pye, and T. Rasmuson. 1978. The inducible immunity system of giant silk moths. *Comp. Pathobiol.* **4:** 145–163.

Boman, H.G., I. Faye, P.V. Hofsten, K. Kockum, J.Y. Lee, K.G. Xanthopoulos, H. Bennich, A. Engström, R.B. Merryfield, and D. Andreu. 1985. On the primary structures of lysozyme, cecropins and attacins from *Hyalophora cecropia. Dev. Comp. Immunol.* **9:** 551–558.

Brahmi, Z. and E.L. Cooper. 1974. Characteristics of the agglutinin in the scorpion *Androctonus australis. Contemp. Topics Immunobiol.* **4:** 261–270.

Briese, D.T. 1981. Resistance of insect species to microbial pathogens, pp. 511–545. In E.W. Davidson (ed.) *Pathogenesis of Invertebrate Microbial Diseases.* Allanhead, Osmun, Montclair NJ.

Briese, D.T. and J.D. Podgwaite. 1985. Development of viral resistance in insect populations, pp. 361–398. In K. Maramorosch and K.E. Sherman (eds.) *Viral Insecticides for Biological Control.* Academic Press, New York.

Chadwick, J.M. and W.P. Aston. 1978. An overview of insect immunity, pp. 1–14. In M.E. Gershwin and E.L. Cooper (eds.) *Animal Models of Comparative and Developmental Aspects of Immunity and Disease.* Pergamon Press, Oxford.

Chadwick, J.S. 1967. Serological responses of insects. *Fed. Proc.* **26:** 1675–1679.

Chadwick, J.S. 1970. Relation of lysozyme concentration to acquired immunity against *Pseudomonas aeruginosa* in *Galleria mellonella. J. Invertebr. Pathol.* **15:** 455–456.

Chadwick, J.S. 1975. Hemolymph changes with infection or induced immunity in insects and ticks, pp. 241–271. In K. Maramorosch and R.E. Shope (eds.) *Invertebrate Immunity.* Academic Press, New York.

Chadwick, J.S. 1977. Induction and effector mechanisms of insect immunity. *Comp. Pathobiol.* **3:** 85–102.

Chadwick, J.S. and W.P. Aston. 1976. Effector mechanisms involved in the protective response in *Galleria mellonella* towards bacterial pathogens. *Proc. Int. Colloq. Invertebr. Pathol.* **1:** 204–209.

Chadwick, J.S., P.J. DeVerno, K.L. Chung, and W.P. Aston. 1982. Effects of hemolymph from immune and non-immune larvae of *Galleria mellonella* on the ultrastructure of *Pseudomonas aeruginosa. Dev. Comp. Immunol.* **6:** 433–440.

Cheung, P.Y.K., E.A. Grula, and R.L. Burton. 1978. Hemolymph responses in *Heliothis zea* to inoculation with *Bacillus thuringiensis* or *Micrococcus lysodeikticus. J. Invertebr. Pathol.* **31:** 148–156.

Clem, L.W., K. Clem, and L. McCumber. 1984. Recognition of xenogenic proteins by the blue crab: Dissociation of the clearance and degradation reactions and lack of involvement of circulating hemocytes and humoral factors. *Dev. Comp. Immunol.* **8**: 31–40.

Cohen, E., G.H.U. Illodi, Z. Brahmi, and J. Minowada. 1979. The nature of cellular agglutinins of *Androctonus australis* (Saharan scorpion) serum. *Dev. Comp. Immunol.* **3**: 429–440.

Cohen, E., A.W. Rowe, and F.C. Wissler. 1965. Heteroagglutinins of the horseshoe crab *Limulus polyphemus*. *Life Sci.* **4**: 2009–2016.

Cooper, E.L. 1985. Overview of humoral factors in invertebrates. *Dev. Comp. Immunol.* **9**: 577–583.

Cooper, E.L. and C.A.E. Lemmi. 1981. Invertebrate humoral immunity. *Dev. Comp. Immunol.* **5**: 3–21.

Cornick, J.W. and J.E. Stewart. 1968. Interaction of the pathogen *Gaffkya homari* with the natural defense mechanisms of *Homarus americanus*. *J. Fish. Res. Bd. Can.* **25**: 695–709.

Cornick, J.W. and J.E. Stewart. 1973. Partial characterization of a natural agglutinin in the hemolymph of the lobster *Homarus americanus*. *J. Invertebr. Pathol.* **21**: 255–262.

Cornick, J.W. and J.E. Stewart. 1978. Lobster (*Homarus americanus*) hemocytes: Classification, differential counts and associated agglutinin activity. *J. Invertebr. Pathol.* **31**: 194–203.

Dalhammar, G. and H. Steiner. 1984. Characterization of inhibitor A protease from *Bacillus thuringiensis* which degrades attacins and cecropins, two classes of antibacterial proteins in insects. *Eur. J. Biochem.* **139**: 247–252.

DeVerno, P.J., W.P. Aston, and J.S. Chadwick. 1983. Transfer of immunity against *Pseudomonas aeruginosa* P11-1 in *Galleria mellonella* larvae. *Dev. Comp. Immunol.* **7**: 423–434.

DeVerno, P.J., J.S. Chadwick, W.P. Aston, and G.B. Dunphy. 1984. The *in vitro* generation of an antibacterial activity from the fat body and hemolymph of non-immunized larvae of *Galleria mellonella*. *Dev. Comp. Immunol.* **8**: 537–546.

Dunn, P.E. and D. Drake. 1983. Fate of bacteria injected into naive and immunized larvae of the tobacco hornworm *Manduca sexta*. *J. Invertebr. Pathol.* **41**: 77–85.

Dunn, P.E., W. Dai, M.R. Kanost, and C. Geng. 1985. Soluble peptidoglycan fragments stimulate antibacterial protein synthesis by fat body from larvae of *Manduca sexta*. *Dev. Comp. Immunol.* **9**: 559–568.

Dunphy, G.B., D.B. Morton, and J.M. Chadwick. 1986a. Pathogenicity of lipopolysaccharide mutants of *Pseudomonas aeruginasa* for larvae of *Galleria mellonella*. Hemocyte interaction with the bacteria. *J. Invertebr. Pathol.*, **47**: 56–64.

Dunphy, G.B., D.B. Morton. A. Kropinski, and J.M. Chadwick. 1986b. Pathogenicity of lipopolysaccharide mutants of *Pseudomonus aeruginosa* for larvae of *Galleria mellonella*. Bacterial properties associated with virulence. *J. Invertebr. Pathol.*, **47**: 48–55.

Edlund, T., I. Sidén, and H.G. Boman. 1976. Evidence for two immune inhibitors from *Bacillus thuringiensis* interfering with the humoral defense system of saturniid pupae. *Infect. Immun.* **14**: 934–941.

Engström, A., P. Engström, Z. Tao, A. Carlsson, and H. Bennich. 1984. Insect immunity: The primary structure of the antibacterial protein attacin F and its relation to two native attacins from *Hyalophora cecropia*. *EMBO J*. **3**: 2065–2070.

Engström, Å., K. Xanthopoulos, H.G. Boman, and H. Bennich. 1985. Amino acid and cDNA sequences of lysozyme from *Hyalophora cecropia*. *EMBO J*. **4**: 2119–2122.

Enzman, P.J. 1973. Induction of an interferon-like substance in persistently infected *Aedes albopictus* cells. *Arch. Virus-Forsch*. **40**: 382–389.

Evans, E.E., B. Painter, M.L. Evans, P. Weinheimer, and R.T. Acton. 1969a. An induced bactericidin in the spiny lobster, *Panulirus argus*. *Proc. Soc. Exp. Biol. Med*. **128**: 394–398.

Evans, E.E., P.F. Weinheimer, B. Painter, R.T. Acton, and M.L. Evans. 1969b. Secondary and tertiary responses of the induced bactericidin from the West Indian spiny lobster, *Panulirus argus*. *J. Bacteriol*. **98**: 943–946.

Evans, E.E., J.E. Cushing, S. Sawyers, P.F. Weinheimer, R.T. Acton, and J.L. McNeely. 1969c. Induced bactericidal response in the California spiny lobster, *Panulirus interruptus*. *Proc. Soc. Exp. Biol. Med*. **132**: 111–114.

Faye, I. 1978. Insect immunity: Early fate of bacteria injected in saturniid pupae. *J. Invertebr. Pathol*. **31**: 19–26.

Faye, I. and G.R. Wyatt. 1980. The synthesis and uptake of hemolymph storage proteins by the fat body of the greater wax moth, *Galleria mellonella*. *Experientia* **36**: 1325–1326.

Faye, I., A. Pye, T. Rasmuson, H.G. Boman, and I.A. Boman. 1975. Insect immunity. 2. Simultaneous induction of antibacterial activity and selective synthesis of some hemolymph proteins in diapausing pupae of *Hyalophora cecropia* and *Samia cynthia*. *Infect. Immun*. **12**: 1426–1438.

Fernández-Moran, H., J.J. Marchalonis, and G.M. Edleman. 1968. Electron microscopy of a hemagglutinin from *Limulus polyphemus*. *J. Mol. Biol*. **32**: 453–465.

Flyg, C. and K.G. Xanthopoulos. 1983. Insect pathogenic properties of *Serratia marcescens*: Passive and active resistance to insect immunity studied with protease-deficient and phage-resistant mutants. *J. Gen. Microbiol*. **129**: 453–464.

Flyg, C., K. Kenne, and H.G. Boman. 1980. Insect pathogenic properties of *Serratia marcescens*: Phage-resistant mutants with a decreased resistance to cecropia immunity and a decreased virulence to Drosophila. *J. Gen. Microbiol*. **120**: 173–181.

Furman, R.M. and T.G. Pistole. 1976. Bactericidal activity from the horseshoe crab *Limulus polyphemus*. *J. Invertebr. Pathol*. **28**: 239–234.

Gagen, S.J. and N.A. Ratcliffe. 1976. Studies on the *in vivo* cellular reactions and fate of injected bacteria in *Galleria mellonella* and *Pieris brassicae* larvae. *J. Invertebr. Pathol*. **28**: 17–24.

Gilbride, K.J. and T.G. Pistole. 1979. Isolation and characterization of a bacterial agglutinin in the serum of *Limulus polyphemus*, pp. 525–535. In E. Cohen (ed.) *Biomedical Applications of the Horseshoe Crab* (*Limulidae*). Alan R. Liss, New York.

Gilbride, K.J. and T.G. Pistole. 1981. The presence of copper in a purified lectin from *Limulus polyphemus*: Possible new role for hemocyanin. *Dev. Comp. Immunol*. **5**: 347–352.

Goldenberg, P.Z. and A.H. Greenberg. 1983. Functional heterogeneity of carbohydrate-binding hemolymph proteins: Evidence of a nonagglutinating opsonin in *Homarus americanus*. *J. Invertebr. Pathol*. **42**: 33–41.

Goldenberg, P.Z. and A.H. Greenberg. 1984. Activation of lobster hemocytes for phagocytosis. *J. Invertebr. Pathol*. **43**: 77–78.

Granados, R.R. 1973. Entry of an insect poxvirus by fusion of the virus envelope with the host cell membrane. *Virology* **52**: 305–309.

Hall, J.L. and D.T.J. Rowlands. 1974a. Heterogeneity of lobster agglutinins. 1. Purification and physicochemical characterization. *Biochemistry* **13**: 821–827.

Hall, J.L. and D.T.J. Rowlands. 1974b. Heterogeneity of lobster agglutinins. 2. Specificity of agglutinin-erythrocyte binding. *Biochemistry* **13**: 828–832.

Hoffmann, D. 1980. Induction of antibacterial activity in the blood of the migratory locust *Locusta migratoria*. *J. Insect Physiol*. **26**: 539–549.

Hoffmann, D., M. Brehélin, and J.-A. Hoffmann. 1977. Premiers résultats sur les réactions de défense antibactériennes de larves et d'imagos de *Locusta migratoria*. *Ann. Parasitol. Hum. Comp*. **52**: 87–88.

Hoffmann, D., D. Hultmark, and H.G. Boman. 1981. Insect immunity: *Galleria mellonella* and other Lepidoptera have cecropia-P9–like factors against gram-negative bacteria. *Insect Biochem*. **11**: 537–548.

Horohov, D.W. and P.E. Dunn. 1982. Changes in the circulating hemocyte population of *Manduca sexta* larvae following injection of bacteria. *J. Invertebr. Pathol*. **40**: 327–339.

Horohov, D.W. and P.E. Dunn. 1983. Phagocytosis and nodule formation by hemocytes of *Manduca sexta* larvae following injection of *Pseudomonas aeruginosa*. *J. Invertebr. Pathol*. **41**: 203–213.

Hultmark, D., H. Steiner, T. Rasmuson, and H.G. Boman. 1980. Insect immunity: Purification and properties of three inducible proteins. *Eur. J. Biochem*. **106**: 7–16.

Hultmark, D., Å. Engström, H. Bennich, R. Kapur, and H. Boman. 1982. Insect immunity: Isolation and structure of cecropin D and four minor antibacterial components from cecropia pupae. *Eur. J. Biochem*. **127**: 207–217.

Hultmark, D., Å. Engström, K. Andersson, H. Steiner, H. Bennich, and H. G. Boman. 1983. Insect immunity: Attacins, a family of antibacterial proteins from *Hyalophora cecropia*. *EMBO J*. **2**: 571–576.

Jarosz, J. 1979. Simultaneous induction of protective immunity and selective synthesis of hemolymph lysozyme protein in larvae of *Galleria mellonella*. *Biol. Zentralbl* **98**: 459–471.

Johannsen, R., R.S. Anderson, R.A. Good, and N.K. Day. 1973. A comparative study of the bactericidal activity of the horseshoe crab (*Limulus polyphemus*) hemolymph and vertebrate serum. *J. Invertebr. Pathol*. **22**: 372–376.

Kaplan, R., S.S.-L. Li, and J.M. Kehoe. 1977. Molecular characterization of limulin, a sialic acid binding lectin from the hemolymph of the horseshoe crab *Limulus polyphemus*. *Biochemistry* **16**: 4297–4303.

Karp, R.D. and L.A. Rheins. 1980. Induction of specific humoral immunity to soluble proteins in the American cockroach (*Periplaneta americana*). 2. Nature of the secondary response. *Dev. Comp. Immunol*. **4**: 620–639.

Kinoshita, T. and K. Inoue. 1977. Bactericidal activity of the normal, cell-free hemolymph of silkworms (*Bombyx mori*). *Infect. Immun.* **16**: 32–36.

Kislev, N., I. Harpaz, and A. Zelcer. 1969. Electron microscopic studies on hemocytes of the Egyptian cotton worm, *Spodoptera littoralis* (Boisduval) infected with a nuclear polyhedrosis virus, as compared to noninfected hemocytes. 2. Virus-infected hemocytes. *J. Invertebr. Pathol.* **14**: 245–257.

Kubo, T., H. Komano, N. Okada, and S. Natori. 1984. Identification of hemagglutinating protein and bactericidal activity in the hemolymph of adult *Sarcophaga peregrina* on injury of the body wall. *Dev. Comp. Immunol.* **8**: 283–291.

Kunis, R.M., W.S. Romoser, and C.G. Atkins. 1978. Fate of bacteriophage in *Aedes aegypti, Anopheles quadrimaculatus* (Diptera: Culicidae), and *Periplaneta americana* (Orthoptera: Blattidae). *J. Invertebr. Pathol.* **31**: 27–30.

Lackie, A.M. 1980. Invertebrate immunity. *Parasitol.* **80**: 393–412.

Lambert, J. and D. Hoffmann. 1985. Mise en évidence d'un facteur actif contre des bactéries Gram négatives dans le sang de *Locusta migratoria. C. R. Séances Acad. Sci.* (III) **300**: 425–430.

Lee, C.H. and R.H. Scholoemer. 1981a. Identification of the antiviral factor in culture medium of mosquito cells persistently infected with Banzi virus. *Virology* **110**: 445–454.

Lee, C.H. and R.H. Scholoemer. 1981b. Mosquito cells infected with Banzi virus secrete an antiviral activity which is of viral origin. *Virology* **110**: 402–410.

Maget-Dana, R., A.C. Roche, and M. Monsigny. 1979. Ganglioside-limulin interactions, pp. 567–578. In E. Cohen (ed.) *Biomedical Applications of the Horseshoe Crab* (Limulidae). Alan R. Liss, New York.

Makela, P., R.D. J. Bradley, H. Brandis, M.M. Frank, H. Hahn, W. Henkel, K. Jann, S.A. Morse, J.B. Robbins, D.L. Rusenstreich, H. Smith, K. Timmis, A. Tomasz, M.J. Turner, and D.C. Wiley. 1980. Evasion of host defenses group report, pp. 175–198. In H. Smith, J.J. Skehel, and M. J. Turner (eds.) *The Molecular Basis of Microbial Pathogenicity.* Weinheim Verlag Chemie GmbH, Basel.

Maramorsch, K. and R.E. Shope. 1975. *Invertebrate Immunity.* Academic Press, New York.

Marchalonis, J.J. and G.M. Edleman. 1968. Isolation and characterization of a hemagglutin *Limulus polyphemus. J. Mol. Biol.* **32**: 453–465.

Mazzone, H.M. 1985. Pathology associated with *Baculovirus* infection, pp. 81–120. In K. Maramorosch and K.E. Sherman (eds.) *Viral Insecticides for Biological Control.* Academic Press, New York.

McCumber, L. and L.W. Clem. 1977. Recognition of viruses and xenogenic proteins by the blue crab, *Callinectes sapidus.* 1. Clearance and organ concentration. *Dev. Comp. Immunol.* **1**: 5–14.

McCumber, L., E. Hoffmann, and L.W. Clem. 1979. Recognition of viruses and xenogenic proteins by the blue crab, *Callinectes sapidus:* A humoral receptor for T_2 bacteriophage. *J. Invertebr. Pathol.* **3**: 1–9.

McKay, D. and C.R. Jenkin. 1969. Immunity in the invertebrates: Adaptive immunity in the catfish (*Parachaeraps bicarcinatus*). *Immunology* **17**: 127–137.

McKay, D. and C.R. Jenkin. 1970a. Immunity in the invertebrates: Correlation of the phagocytic activity of haemocytes with resistance to infection in the crayfish (*Parachaeraps bicarcinatus*). *Aust. J. Exp. Biol. Med. Sci.* **48**: 609–617.

McKay, D. and C.R. Jenkin. 1970b. Immunity in the invertebrates: The fate and distribution of bacteria in normal and immunized crayfish (*Parachaeraps bicarcinatus*). *Aust. J. Exp. Biol. Med. Sci.* **48**: 599–607.

McKay, D. and C.R. Jenkin. 1970c. Immunity in the invertebrates: The role of serum factors in phagocytosis of erythrocytes by haemocytes of the freshwater crayfish (*Parachaeraps bicarcinatus*). *Aust. J. Exp. Biol. Med. Sci.* **48**: 139–150.

McKay, D., C.R. Jenkin, and D. Rowley. 1969. Immunity in the invertebrates. 1. Studies on the naturally occurring haemagglutinins in the fluid from invertebrates. *Aust. J. Exp. Biol. Med. Sci.* **47**: 125–134.

Merryfield, R.B., L.D. Vizioli, and H.G. Boman. 1982. Synthesis of the antibacterial peptide cecropin A (1–33) *Biochemistry* **21**: 5020–5031.

Messner, B. 1981. Eine bakterizidie in den Homogenaten von *Locusta migratoria* und *Galleria mellonella* (Insecta). *Zool. Jahrb. Physiol.* **85**: 164–172.

Miller, V.H., R.S. Ballback, G.B. Pauley, and S.M. Krassner. 1972. A preliminary physicochemical characterization of an agglutinin found in the hemolymph of the crayfish *Procambarus clarkii*. *J. Invertebr. Pathol.* **19**: 83–93.

Mohrig, V.W. and B. Messner. 1968. Immunreaktionen bei Insekten. 1. Lysozyme als grundlegender antibakterieller Faktor im humoralen Abwehrmechanismus bei Insekten. *Sonderb. Biol. Zentralbl.* **87**: 439–470.

Mohrig, W., D. Schittek, and R. Hanschke. 1979a. Immunological activation of phagocytic cells in *Galleria mellonella*. *J. Invertebr. Pathol.* **34**: 84–87.

Mohrig, W., D. Schittek, and R. Hanschke. 1979b. Investigations on cellular defense reactions with *Galleria mellonella* against *Bacillus thuringiensis*. *J. Invertebr. Pathol.* **34**: 207–212.

Monner, D.A., D.A.S. Jonsson, and H.G. Boman. 1971. Ampicillin-resistant mutants of *Escherichia coli* K-12 with lipopolysaccharide alterations affecting mating ability and susceptibility to sex-specific bacteriophages. *J. Bacteriol.* **107**: 420–432.

Mori, K. and J.E. Stewart. 1978a. The hemolymph bactericidin of the American lobster (*Homarus americanus*): Adsorption and activation. *J. Fish Res. Bd. Can.* **5**: 1504–1507.

Mori, K. and J.E. Stewart. 1978b. Natural and induced bactericidal activites of the hepatopancreas of the American lobster, *Homarus americanus*. *J. Invertebr. Pathol.* **32**: 171–176.

Morton, D.B., R.I. Barnett, and J.S. Chadwick. 1984. Structural alterations to *Proteus mirabilis* as a result of exposure to hemolymph from the larvae of *Galleria mellonella*. *Microbios* **39**: 177–185.

Mullainadhan, P. and M.H. Ravidranath. 1984. Crustacean defense strategies. 2. Recognition, clearance, accumulation and externalization of soluble foreign proteins by the mud-crab, *Scylla serrata* (Forskal) (Portunidae: Brachyura). *Dev. Comp. Immunol.* **8**: 523–535.

Murray, A.M. and P.A. Morohan. 1973. Studies of interferon production in *Aedes albopictus* mosquito cells. *Proc. Soc. Exp. Biol. Med.* **142**: 11–15.

Nachum, R. 1979. Antimicrobial defense mechanisms in *Limulus polyphemus*, pp.

513–524. In E. Cohen (ed.) *Biomedical Applications of the Horseshoe Crab* (Limulidae). Alan R. Liss, New York.

Nachum, R., S.W. Watson, and S.E. Siegel. 1980. Antimicrobial defense mechanisms in the horseshoe crab *Limulus polyphemus*: Effect of sodium chloride on bactericidal activity. *J. Invertebr. Pathol.* **36**: 382–388.

Nachum, R., S.W. Watson, J.D. Sullivan, and S.E. Siegel. 1979. Antimicrobial defense mechanisms in the horseshoe crab, *Limulus polyphemus:* Preliminary observations with heat-derived extracts of *Limulus* amebocyte lysate. *J. Invertebr. Pathol.* **33**: 290–299.

Nappi, A.J. 1975. Parasite encapsulation in insects, pp. 293–326. In K. Maramorosch and R.E. Shope (eds.) *Invertebrate Immunity.* Academic Press, New York.

Nappi, A.J. 1978. Immune reactions of invertebrates to foreign materials, pp. 15–24. In M.E. Gershwin and E.L. Cooper (eds.) *Animal Models of Comparative and Developmental Aspects of Immunity and Disease.* Pergamon Press, New York.

Newton, S.E. and L. Dalgarno. 1983. Antiviral activity released from *Aedes albopictus* cells persistently infected with Semliki Forest virus. *J. Virol.* **47**: 652–655.

Newton, S.E., N.J. Short, and L. Dalgarno. 1981. Bunyamwera virus replication in cultured *Aedes albopictus* (Mosquito) cells: Establishment of a persistent viral infection. *J. Virol.* **38**: 1015–1024.

Novokhatskii, A.S. and L.H. Berezina. 1978. Arbovirus multiplication in mosquito cells treated with an interferon inducer. *Vopr. Virusol.* **3**: 357–359.

Nowak, T.P. and S.H. Barondes. 1975. Agglutinin from *Limulus polyphemus*: Purification with formalinized horse erythrocytes as the affinity adsorbent. *Biochim. Biophys. Acta* **393**: 115–123.

Odier, F., C. Vago, and M.E. Lamy. 1974. Persistence des virus de vertébrates dans l'hemocoele des insectes: Étude sur le poliovirus chez les larves de coléoptères. *Ann. Soc. Entomol. Fr.* **10**: 449–455.

Okada, M. and S. Natori. 1983. Purification and characterization of an antibacterial protein from haemolymph of *Sarcophaga peregrina* (flesh-fly) larvae. *Biochem. J.* **211**: 727–734.

Okai. Y. 1985. DNA synthesis inhibitory peptides in mammalian serum and insect hemolymph. *Molec. Biol. Rep.* **10**: 123–128.

Oppenheim, J.D., M.S. Nachbar, M.R.J. Salton, and F. Aull. 1974. Purification of a hemagglutinin from *Limulus polyphemus* by affinity chromatography. *Biochem. Biophys. Res. Commun.* **58**: 1127–1134.

Paterson, W.D. and J.E. Stewart. 1974. *In vitro* phagocytosis by hemocytes of the American lobster, *Homarus americanus. J. Fish Res. Bd. Can* **31**: 1051–1056.

Paterson, W.D., J.E. Stewart, and B.M. Zwicker. 1976. Phagocytosis as a cellular immune response mechanism in the American lobster, *Homarus americanus. J. Invertebr. Pathol.* **27**: 95–104.

Pauley, G.B. 1974. Comparison of a natural agglutinin in the hemolymph of the blue crab, *Callinectes sapidus*, with the agglutinins of other invertebrates. *Contemp. Topics Immunobiol.* **4**: 233–239.

Pauley, G.B. 1975. Introductory remarks on disease of crustaceans. *Mar. Fish. Rev.* **37**: 2–3.

Payne, C.C. 1981. Cytoplasmic polyhedrosis viruses, pp. 61–100. In E.W. Davidson (ed.) *Pathogenesis of Invertebrate Microbial Diseases*. Allanhead, Osmun, Montclair, NJ.

Peleg, J. 1969. Inapparent persistent virus infection in continuously grown *Aedes aegypti* mosquito cells. *J. Gen. Virol.* **5**: 463–471.

Pistole, T.G. 1978. Broad spectrum bacterial agglutinating activity in the serum of the horseshoe crab *Limulus polyphemus*. *Dev. Comp. Immunol.* **2**: 65–76.

Pistole, T.G. 1979. Bacterial agglutinins from *Limulus polyphemus*: An overview, pp. 547–553. In E. Cohen (ed.) *Biomedical Applications of the Horseshoe Crab (Limulidae)*. Alan R. Liss, New York.

Pistole, T.G. and J.L. Britko. 1978. Bactericidal activity of amebocytes from the horseshoe crab *Limulus polyphemus*. *J. Invertebr. Pathol.* **31**: 376–382.

Pistole, T.G. and R.M. Furman. 1976. Serum bactericidal activity in the horseshoe crab *Limulus polyphemus*. *Infect. Immun.* **14**: 888–893.

Pistole, T.G. and S.A. Graf. 1984. Bactericidal activity of *Limulus* lectins and amebocytes, pp. 71–81. In E. Cohen (ed.) *Recognition Proteins, Receptors, and Probes: Invertebrates*. Alan R. Liss, New York.

Powning, R.F. and W.J. Davidson. 1973. Studies on insect bacteriolytic enzymes. 1. Lysozyme in hemolymph of *Galleria mellonella* and *Bombyx mori*. *Comp. Biochem. Physiol.* **45**B: 669–686.

Pye, A.E. 1974. Microbial activation of prophenoloxidase from immune insect larvae. *Nature* (London) **251**: 610–613.

Pye, A.E. 1978. Activation of prophenoloxidase and inhibition of melanization in the hemolymph of immune *Galleria mellonella* larvae. *Insect. Biochem.* **8**: 117.

Pye, A.E. and H.G. Boman. 1977. Insect immunity. 3. Purification and partial characterization of immune protein P5 from hemolymph of *Hyalophora cecropia* pupae. *Infect Immun.* **17**: 408–414.

Qu, X., H. Steiner, Å. Engström, H. Bennich, and H. Boman. 1982. Insect immunity: Isolation and structure of cecropins B and D from pupae of the Chinese oak silk moth, *Antheraea pernyi*. *Eur. J. Biochem.* **127**: 219–224.

Quiot, J.M., C. Vago, and S. Bellonick. 1980. Réaction antivirale par bourgeonnement cellulaire: Étude en culture de cellules de lépidoptère infectée par un *Reovirus* de polyedrose cytoplasmique. *C. R. Séances Acad. Sci.* (III) **291**: 481–483.

Ratcliffe, N.A. 1985. Invertebrate immunity: A primer for the non-specialist. *Immunol. Lett.* **10**: 253–270.

Ratcliffe, N.A. and A.F. Rowley. 1975. Cellular defense reactions of insect hemocytes *in vitro*: Phagocytosis in a new suspension culture system. *J. Invertebr. Pathol.* **26**: 225–233.

Ratcliffe, N.A. and A.F. Rowley. 1979. Role of hemocytes in defence against biological agents, pp. 331–415. In A.P. Gupta (ed.) *Insect Hemocytes*. Cambridge University Press, Cambridge.

Ratcliffe, N.A. and J.B. Walters. 1983. Studies on the *in vivo* cellular reactions of insects: Clearance of pathogenic and non-pathogenic bacteria in *Galleria mellonella* larvae. *J. Insect Physiol.* **29**: 407–415.

Ratcliffe, N.A., K.N. White, A.F. Rowley, and J.B. Walters. 1982. Cellular defense systems of the Arthropoda, pp. 167–255. In N. Cohen and M.M. Sigel (eds.) *The*

Reticuloendothelial System: Phylogeny and Ontogeny, vol. 3. Plenum Press, New York.

Ravindranath, M.H. 1981. Onychophorans and myriapods, pp. 327–354. In N.A. Ratcliffe and A.F. Rowley, (eds.) *Invertebrate Blood Cells*, vol. 2. Academic Press, New York.

Ravindranath, M.H. and E.L. Cooper. 1984. Crab lectins: Receptor specificity and biomedical applications, pp. 83–96. In E. Cohen (ed.) *Recognition Proteins, Receptors and Probes: Invertebrates*. Alan R. Liss, New York.

Rheins, L. and R.D. Karp. 1982. An inducible humoral factor in the American cockroach (*Periplaneta americana*): Precipitin activity that is sensitive to a proteolytic enzyme. *J. Invertebr. Pathol.* **40:** 191–196.

Riedel, B. and D.T. Brown. 1979. Novel antiviral activity found in the media of Sinbis virus–persistently infected mosquito (*Aedes albopictus*) cell cultures. *J. Virol.* **29:** 51–60.

Roche, A.-C. and M. Munsigny. 1974. Purification and properties of limulin: A lectin (agglutinin) from hemolymph of *Limulus polyphemus*. *Biochim. Biophysic. Acta* **371:** 242–254.

Rohel, D.Z., J. Chadwick, and P. Faulkner. 1980. Tests with inactivated cricket paralysis virus as a possible immunogen against a virus infection of *Galleria mellonella* larvae. *Int. Virol.* **14:** 61–68.

Rostam-Abadi, H. and T.G. Pistole. 1982. Lipopolysaccharide-binding lectin from the horseshoe crab *Limulus polyphemus* with specificity for 2-keto-3-deoxy-octonate (KDO). *Dev. Comp. Immunol.* **6:** 209–218.

Rowley, A.F. and N.A. Ratcliffe. 1976. The granular cells of *Galleria mellonella* during clotting and phagocytic reactions *in vitro*. *Tissue Cell* **8:** 437–446.

Rowley, A.F. and N.A. Ratcliffe. 1980. Insect erythrocyte agglutinins: *In vitro* opsonization experiments with *Clitumnus extradentatus* and *Periplaneta americana* hemocytes. *Immunology* **40:** 483–492.

Rowley, A.F. and N.A. Ratcliffe. 1981. Insects, pp. 421–488. In N.A. Ratcliffe and A.F. Rowley (eds.) *Invertebrate Blood Cells*, vol. 2. Academic Press, New York.

Scheefers-Borchel, U., B. Riedel, H. Scheefers, and T. Brown. 1980. A novel antiviral activity produced by mosquito (*Aedes albopictus*) cells persistently infected with Sindbis virus. *J. Supramol. Struct.* **4:** 248, supplement.

Sherman, K.E. 1985. Multiple virus interactions, pp. 735–753. In K. Maramorosch and K.E. Sherman (eds.) *Viral Insecticides for Biological Control*. Academic Press, New York.

Sherman, R. 1981. Chelicerates, pp. 355–384. In N.A. Ratcliffe and A.F. Rowley (eds.) *Invertebrate Blood Cells*, vol. 2. Academic Press, New York.

Shimizu, S.M., M. Ito, and M.N. Iwa. 1977. Lectins in the hemolymph of Japanese horseshoe crab *Tachypleus tridentatus*. *Biochem. Biophys. Acta* **500:** 71–79.

Sidén, I., G. Dalhammar, B. Telander, H.G. Boman, and H. Sommerville. 1979. Virulence factors in *Bacillus thuringiensis*: Purification and properties of a protein inhibitor of immunity in insects. *J. Gen. Microbiol.* **114:** 45–52.

Sloan, B., C. Yocum, and L.W. Clem. 1975. Recognition of self from non-self in crustaceans. *Nature* (London) **258:** 521–523.

Smirnoff, W. and J. Valero. 1980. Metabolic exploration in insects: Variations in potassium and calcium levels in insects during various infections. *J. Invertebr. Pathol.* **35:** 311–313.

Smith, R. and T.G. Pistole. 1985. Bactericidal activity of the granules isolated from amebocytes of the horseshoe crab *Limulus polyphemus. J. Invertebr. Pathol.* **45:** 272–275.

Smith, V.J. and N.A. Ratcliffe. 1981. Pathological changes in the nephrocytes of the shore crab, *Carcinus maenas*, following injection of bacteria. *J. Invertebr. Pathol.* **38:** 113–121.

Söderhäll, K. 1982. Prophenoloxidase activating system and melanization—a recognition mechanism of arthropods? A review. *Dev. Comp. Immunol.* **6:** 601–611.

Sohi. S.S. and J.C. Cunningham. 1972. Replication of a nuclear polyhedrosis virus in serially transferred insect hemocyte cultures. *J. Invertebr. Pathol.* **19:** 51–61.

Spinger, G.F. 1970. Importance of blood-group substances in interactions between man and microbes. *Ann. N.Y. Acad. Sci.* **169:** 134–167.

Steiner, H., D. Hultmark, A. Engström, H. Bennich, and H.G. Boman. 1981. Sequence and specificity of two antibacterial proteins involved in insect immunity. *Nature* (London) **292:** 246–248.

Stellar, V. 1980. Togaviruses in cultured arthropod cells, pp. 584–621. In R.W. Schlesinger (ed.) *The Togaviruses: Biology, Structure and Replication.* Academic Press, New York.

Stephens, J.M. 1962a. Bactericidal activity of the blood of actively immunized wax moth larvae. *Can. J. Microbiol.* **8:** 491–499.

Stephens, J.M. 1962b. A strain of *Streptococcus faecalis* Andrewes and Horder producing mortality in larvae of *Galleria mellonella* (Linnaeus). *J. Insect Pathol.* **4:** 267–268.

Stephens, J.M. 1963. Bactericidal activity of hemolymph of some normal insects. *J. Insect Pathol.* **5:** 61–65.

Stewart, J.E. and B. Arre. 1973. Depletion of glycogen and adenosine triphosphate as major factors in the death of lobsters (*Homarus americanus*) infected with *Gaffkya homari. Can. J. Microbiol.* **19:** 1103–1110.

Stewart, J.E. and J.W. Cornick. 1972. Effects of *Gaffkya homari* on glucose, total carbohydrates, and lactic acid of the hemolymph of the lobster (*Homarus americanus*). *Can. J. Microbiol.* **18:** 1511–1513.

Stewart, J.E. and D.M. Foley. 1969. A precipitin-like reaction of the hemolymph of the lobster *Homarus americanus. J. Fish. Res. Bd. Can.* **26:** 1392–1397.

Stewart, J.E. and B.M. Zwicker. 1972. Natural and induced bactericidal activities in the hemolymph of the lobster *Homarus americanus*: Products of hemocyte-plasma interaction. *Can. J. Microbiol.* **18:** 1499–1509.

Stewart, J.E. and B.M. Zwicker. 1974. Induction of internal defense mechanisms in the lobster *Homarus americanus. Contemp. Topics Immunol.* **4:** 233–239.

Tamaki, S. and M. Matsuhaski. 1973. Increase in sensitivity to antibiotic salysozyme on deletion of lipopolysaccharides in *Escherichia coli* strains. *J. Bacteriol.* **114:** 453–454.

Tanada, Y. and H. Argua. 1967. Interference between the midgut nuclear polyhedrosis virus and the cytoplasmic polyhedrosis virus in the silkworm *Bombyx mori. J. Seri. Sci. Jpn.* **36:** 169–176.

Tanada, Y. and H. Watanabe. 1971. Disc electrophoretic patterns of hemolymph proteins of larvae of the armyworm, *Pseudelatia unipunctata*, infected with a nuclear polyhedrosis and granulosis virus. *J. Invertebr. Pathol.* **14**: 419–420.

Tyson, C. and C. Jenkin. 1973. The importance of opsonic factors in the removal of bacteria from the circulation of the crayfish (*Parachaeraps bicarinatus*). *Aust. J. Exp. Biol. Med. Sci.* **51**: 609–615.

Tyson, C.J. and C.R. Jenkin. 1974. Phagocytosis of bacteria *in vitro* by haemocytes from the crayfish (*Parachaeraps bicarinatus*). *Aust. J. Exp. Biol. Med. Sci.* **52**: 341–348.

VanderGeest, L.P.S. and H.J.M. Wassink. 1969. Hemolymph proteins of the cabbage armyworm, *Mamestra brassicae*, after infection with a nucleopolyhedrosis virus. *J. Invertebr. Pathol.* **14**: 419–420.

Vasta, G.R. and E. Cohen. 1982. The specificity of *Centruroides sculpturatus* Ewing (Arizona lethal scorpion) hemolymph agglutinins. *Dev. Comp. Immunol.* **6**: 219–230.

Vasta, G.R. and E. Cohen. 1984a. Humoral lectins in the scorpion *Vaejovis confuscius*: A serological characterization. *J. Invertebr. Pathol.* **43**: 226–233.

Vasta, G.R. and E. Cohen. 1984b. Serum lectins from the scorpion *Vaejovis spinigerus* Wood binds sialic acid. *Experientia* **40**: 485–487.

Vasta, G.R. and J.J. Marchalonis. 1983. Humoral recognition factors in the Arthropoda: The specificity of Chelicerata serum lectins. *Am. Zool.* **23**: 151–157.

Walters, J.B. and N.A. Ratcliffe. 1983. Studies on the *in vivo* cellular reactions of insects: Fate of pathogenic and non-pathogenic bacteria in *Galleria mellonella* nodules. *J. Insect Physiol.* **29**: 417–424.

Watanabe, H. 1967. Electrophoretic separation of the hemolymph proteins in the fall webworm, *Hyphantria cunea*, infected with a nuclear-polyhedrosis virus. *J. Invertebr. Pathol.* **9**: 570–571.

Weinheimer, P.E., R.T. Acton, S. Sawyer, and E.E. Evans. 1969a. Specificity of the induced bactericidin of the West Indian spiny lobster, *Panulirus argus*. *J. Bacteriol.* **98**: 322–326.

Weinheimer, P.E., E.E. Evans, R.M. Stroud, R.T. Acton, and B. Painter. 1969b. Comparative immunology: Natural hemolytic system of the spiny lobster, *Panulirus argus*. *Proc. Soc. Exp. Biol. Med.* **130**: 322–326.

Whitcomb, R.F., M. Shapiro, and R.R. Granados. 1974. Insect defense mechanisms against microorganisms and parasitoids, pp. 447–536. In M. Rockstein (ed.) *Physiology of Insecta*, vol. 5. Academic Press, New York.

Whitlock, V.H. 1977. Simultaneous treatments of *Heliothis armigera* with a nuclear polyhedrosis and a granulosis virus. *J. Invertebr. Pathol.* **29**: 297–303.

Yeaton, R. 1981. Invertebrate lectins. 1. Occurrence. *Dev. Comp. Immunol.* **5**: 391–402.

Young, S.Y., III, and H.A. Scott. 1970. Immunoelectrophoresis of hemolymph of the cabbage looper, *Trichoplusia ni*, during the course of a nuclear polyhedrosis virus infection. *J. Invertebr. Pathol.* **16**: 57–62.

Younghusband, H.B. and P.E. Lee. 1969. Virus-cell studies of *Tipula* iridescent virus in *Galleria mellonella* (L). 1. Electron microscopy of infection and synthesis of *Tipula* iridescent virus in hemocytes. *Virology* **38**: 247–254.

Younghusband, H.B. and P.E. Lee. 1970. Cytochemistry and auto-radiography of *Tipula* iridescent virus in *Galleria mellonella*. *Virology* **40:** 757–760.

Youngner, J.S., E.J. Dubovi, D.O. Quagliana, M. Kelly, and O. T. Preble. 1976. Role of temperature-sensitive mutants in persistent infections initiated with vesicular stomatitis virus. *J. Virol.* **19:** 90–101.

Ziprin, R.L. 1978. Immune response of the greater wax moth, *Galleria mellonella*, induced by the marine *Pseudomonas* B-16. *J. Invertebr. Pathol.* **32:** 396–397.

Ziprin, R.L. and P.A. Hartman. 1971. Toxicity of *Pseudomonas aeruginosa* bacterins and cells to the greater wax moth, *Galleria mellonella*. *J. Invertebr. Pathol.* **17:** 265–269.

CHAPTER 11

Antibacterial Activity in *Limulus*

Thomas G. Pistole
Department of Microbiology
University of New Hampshire
Durham, New Hampshire

Sandra A. Graf
School of Life and Health Sciences
University of Delaware
Newark, Delaware

11.1. INTRODUCTION

Despite the efforts that have been made since the rediscovery of *Limulus* agglutinins independently by Marchalonis (1964) and Cohen and coworkers (1965), we know relatively little about the specific mechanisms by which these arthropods defend themselves from microbial aggression. Furthermore, what does seem confirmed is primarily negative information:

- Host defense mechanisms in the horseshoe crab, and indeed in any invertebrate examined to date, are not mediated by antibody; neither have immunoglobulin-related structures been found in these groups (although Vasta and Marchalonis, 1984, have suggested that tunicate lectin might be included in a large family of recognition molecules that encompasses immunoglobulins).
- Lymphocytes, the cells most directly responsible for specific immune responses in mammals, have not been found in this or any other invertebrate species.
- Inducible immunity has not been described in *Limulus*.

The known habitats of horseshoe crabs, estuary and near-shore waters and the underlying sediment, are well populated with a variety of bacterial species. It is therefore likely that some mechanism exists to regulate the movement of microorganisms into and within the tissues and body fluids of *Limulus*. Studies to date have implicated both serum factors and the circulating amoebocyte, or granulocyte, in host defenses of this animal. This chapter summarizes our current knowledge of the antibacterial system of the horseshoe crab.

11.2. BASIC PROPERTIES OF *LIMULUS* LECTINS

11.2.1. Limulin

The structure of limulin, first reported by Marchalonis and Edelman (1968), has been previously reviewed (see Marchalonis and Waxdal, 1979; Vasta and Marchalonis, 1983). The basic model is a 400,000-dalton, ring-shaped molecule composed of 18 subunits with a molecular weight of 22,500 that are intermediately arranged in six trimers. As with many lectins, the molecule is stabilized by divalent cations. The overall configuration and the amino acid composition provide no evidence for a structural relationship with vertebrate immunoglobulin (see also Chapter 12).

11.2.2. Polyphemin

Much less is known about the physicochemical properties of polyphemin, the second lectin found in *L. polyphemus*, than about limulin. The estimated molecular weight, 200,000 (Gilbride and Pistole, 1979), is distinct from that of limulin. Both lectins, however, exhibit apparent subunit heterogeneity. Purified polyphemin contains high levels of copper, suggesting an association with the major protein in *L. polyphemus* hemolymph, hemocyanin (Gilbride and Pistole, 1981). Competitive assays between the lectin and isolated hemocyanin fractions for binding sites on *Staphylococcus aureus* support this association (E. Brandin, M. Brenowitz, T. Pistole, unpublished data). Whether this finding is due to actual lectin activity of hemocyanin subunits or a strong affinity between hemocyanin and a separate lectin molecule is not known.

11.2.3. Lectins from Other *Limulidae*

Lectins have also been found in other members of the family Limulidae. Serum of the Japanese horseshoe crab, *Tachypleus tridentatus*, has been shown to contain four distinct lectins (Shimizu et al., 1977). The major lectin in this study was subsequently further resolved into at least four subfractions by sequential elution from an *N*-acetyl-D-galactosamine-Sepharose 6B column (Shimizu et al. 1979). The amino acid composition of these four "isolectins" plus that of another lectin (GN3) from this animal differed from that of limulin. Antiserum prepared against crude *T. tridentatus* serum failed to react with *L. polyphemus* serum in immunodiffusion reactions, further supporting the idea that lectins from these two species are structurally distinct (Shimizu et al., 1979).

Bishayee and Dorai (1980) have characterized a lectin (carcinoscorpin) from the Indian horseshoe crab, *Carcinoscorpius rotunda cauda*. The large size (MW 420,000) is comparable to that reported for limulin (Marchalonis and Edelman, 1968), but the size of the subunits is distinct (22,500 for limulin; 27,000 and 28,000 for nonidentical subunits in carcinoscorpin). Immunologic studies suggest no cross-reactivity between these two lectins (Bishayee and Dorai, 1980; see also Chapters 8 and 18).

11.3. BINDING SPECIFICITY

11.3.1. Serum

Limulus serum was first recognized as an erythrocyte-agglutinating protein by Noguchi in 1903 (as cited in Cohen et al., 1965). Among the species of erythrocytes tested, those from human (Cohen et al., 1965; Cohen, 1968), mouse (Cohen, 1968), horse (Marchalonis and Edelman, 1968) and chicken (Rostam-Abadi and Pistole, 1982) are the most active. Human leukemic lymphocytes also react with *L. polyphemus* serum (Cohen et al., 1976, 1979; see also Chapter 19).

A variety of bacteria are agglutinated by *L. polyphemus* serum. Agglutination adsorption studies suggested that the gram-negative bacteria react with a serum component distinct from that which agglutinates the gram-positive organisms (Pistole, 1978). Subsequent studies corroborated these findings, leading to the conclusion that limulin is the serum component reactive with gram-negative bacteria, primarily via its affinity for a core sugar in the lipopolysaccharide molecule (Rostam-Abadi and Pistole, 1982; see Section 11.3.2.1), while the likely source of agglutinating activity for gram-positive cocci is polyphemin (Brandin and Pistole, 1983). Whole *Limulus* serum, along with other lectins, has been used to subgroup certain coagulase-negative staphylococci (Davidson et al., 1982).

11.3.2. Purified Lectins

Although lectins by definition bind to sugars (Goldstein et al., 1981), it has become increasingly clear that the one-lectin–one-sugar model is not sufficient to describe the specificity of lectin binding. Some cautions that need to be realized in ascribing a particular binding specificity to a lectin are itemized here:

- Invertebrates may possess more than one lectin in their body fluids. This is true for *L. polyphemus* (Gilbride and Pistole, 1979; Brandin and Pistole, 1983). Binding studies with unfractionated serum may yield composite results rather than data on specific lectins.
- Although they are easy to use, erythrocytes may not provide the most appropriate information on the binding activity of lectins (Pistole, 1981). Without a firm understanding of the role of invertebrate lectins in situ, we cannot determine the natural receptors for such molecules or even whether such structures exist. Bacteria, for example, contain many sugars not found on mammalian erythrocytes. Data from hemagglutination inhibition studies provide useful markers for lectin activity, but this information should not be viewed as absolute.
- Often the binding site is influenced by adjacent sugars, and, indeed, the site of lectin attachment may be composed of more than one sugar. Simple monosaccharides may exhibit relatively low binding affinity compared with glycoproteins containing the same sugars (Yeaton, 1981; (see also Chapter 20).
- Lectin-binding specificities may be more appropriately expressed in relative terms, as suggested by Goldstein and Hayes (1978). For example, the binding affinity of the well-studied plant lectin concanavalin A is represented by these authors as α-D-mannose > α-D-glucose > α-N-acetyl-D-glucosamine. It should not be assumed that a lectin with a described sugar binding specificity reacts only with that ligand.
- Nonsugar, or aglycon, regions adjacent to the binding site may influence the overall affinity between lectin and receptor (Goldstein and Hayes, 1978).

Despite these cautions, we can make some statements concerning the molecular specificity of lectins from the Limulidae.

11.3.2.1. Limulin. Cohen (1968) showed that neuraminidase-treated erythrocytes failed to agglutinate in the presence of *Limulus* serum. Furthermore, supernatant fluid from these treated cells could inhibit agglutination of untreated erythrocytes. Pardoe and colleagues (1970) reported similar results with human lymphocytes and in addition showed agglutination inhibition by *N*-acetylneuraminic acid (NeuNAc) and *N*-acetyl-neuraminyl–containing glycoproteins. Free *N*-acetyl- and *N*-glycolyl-neuraminic acid inhibited hemagglutination by purified limulin, but only at high concentrations (Roche and Monsigny, 1974). Better inhibition was obtained with glycoproteins containing NeuNAc. Using gangliosides derived from horse erythrocytes (Maget-Dana et al., 1979) or oligosaccharide inhibitors (Roche and Monsigny, 1979), Monsigny's group concluded that limulin preferentially binds *N*-glycolyl-neuraminic acid–containing gangliosides and specifically recognizes glycoproteins containing *N*-acetyl-neuraminyl-α-2-3(6)-*N*-acetyl-galactosamine.

In studies using capsular polysaccharides from strains of the gram-negative bacteria *Escherichia coli* and *Neisseria meningitidis*, McSweegan and Pistole (1982) found that limulin had greatest reactivity with homopolymers of NeuNAc in α(2-9)-ketosidic linkages. *O*-Acetylated homopolymers were more reactive than their *O*-acetyl-negative counterparts.

Limulus serum (Pistole and Rostam-Abadi, 1979) and purified limulin (Rostam-Abadi and Pistole, 1982) also react with the sugar 2-keto-3-deoxyoctonate (KDO), a component of bacterial lipopolysaccharide that is structurally related to NeuNAc (Bhattacharjee et al., 1978). This finding was confirmed with KDO-containing capsular polysaccharides from *N. meningitidis* (McSweegan and Pistole, 1982).

D-Glucuronic acid has been shown to inhibit limulin-mediated hemagglutination, but at concentrations somewhat higher than that of NeuNAc (Nowak and Barondes, 1975). Whole serum from *L. polyphemus* precipitates D-glucuronic acid–containing molecules (Vaith et al., 1979a,b).

11.3.2.2. Polyphemin. The lectin polyphemin binds to *N*-acetyl-D-glucosamine (GlcNAc)–containing molecules in the cell wall of the gram-positive bacterium *Staphylococcus aureus* (Brandin and Pistole, 1983). Affinity gel chromatography, using GlcNAc-associated teichoic acid from staphylococcal cell walls as the specific adsorbent, has been used to recover the lectin from *Limulus* serum. Free GlcNAc was incapable of inhibiting bacterial agglutination by this lectin.

11.3.2.3. Other Activity in Limulus polyphemus. Antigalactan activity has been reported in unfractionated *Limulus* serum (Voigtmann et al., 1971). The proposed binding site for the reactive serum component is β-D-galactose(1→3)D-galactose (Cohen et al., 1975; Uhlenbruck et al., 1976). The relationship of this activity to either limulin or polyphemin is unclear, although Vaith and associates (1979b) have suggested that what was originally viewed as antigalactan activity is in fact anti-D-glucuronyl activity and hence likely to be due to limulin.

11.3.2.4. Lectins from Other Limulidae. Although the relative agglutination titers for erythrocytes from various species differed among the lectin fractions recovered from *T. tridentatus*, all were inhibited by *N*-acetylamino groups and by glycoproteins containing sialic acid (Shimizu et al., 1977, 1979). Thus, *N*-acetyl-D-galactosamine (GalNAc) and GlcNAc exhibited binding inhibitory activity comparable to that of NeuNAc. The authors cite this as a distinct difference from the binding activity of limulin. Nonetheless, it is worth noting that the sugar used in the classic study with *Limulus* serum, which first suggested sialic acid as the binding site for limulin, was GlcNAc (Cohen, 1968).

Carcinoscorpin, the lectin isolated from the Indian horseshoe crab, exhibits a binding specificity comparable to that of limulin. It binds to fetuin, a sialic acid–rich glycoprotein, but not to asialofetuin (Bishayee and Dorai, 1980). Carcinoscorpin immobilized onto Sepharose has been used to resolve isoenzymes of alkaline phosphatase on the basis of their differing sialic acid contents (Dorai et al., 1981). Like limulin, this lectin recognizes KDO, the sugar found in bacterial lipopolysaccharide (Dorai et al., 1982). Agglutination of *E. coli* and *Salmonella minnesota* was inhibited by KDO and by lipopolysaccharide containing this sugar. Lipopolysaccharide devoid of KDO was not inhibitory.

11.4. ANTIBACTERIAL ACTIVITY

11.4.1. Whole Serum

Limulus plasma, the fluid portion of hemolymph from which intact granulocytes have been removed, has no demonstrable bactericidal activity (Furman and Pistole, 1976). In contrast, the fluid obtained after collected hemolymph has been allowed to clot is markedly bactericidal (Furman and Pistole, 1976; Pistole and Furman, 1976). These findings indicate a role for the amoebocyte in antibacterial mechanisms, a point that will be further discussed.

Limulus serum is capable of killing a variety of gram-negative bacteria (Furman and Pistole, 1976). A large degree of individual variation toward different bacteria exists; there is also much fluctuation in bactericidal activity when individual crabs are repeatedly tested. Although it is difficult to generalize from these results, given the inherent variability, it is of interest to note that the greatest bactericidal activity was observed against those gram-negative bacilli found normally in the estuary environment (*Pseudomonas putida* and *Flavobacterium* sp.), while lower activity was observed against those species found normally in warm-blooded animals and present only transiently in seawater (*E. coli, Serratia marcescens,* and *S. minnesota*). The lobster pathogen *Aerococcus viridans* (Pistole and Furman, 1976) and *Micrococcus luteus* and *Salmonella typhimurium* (Furman and Pistole, 1976) were completely unaffected by *Limulus* serum.

11.4.2. Serum Components

The high degree of variability found in these antibacterial studies provided an impetus to examine particular serum components. Our findings in studies on bacterial agglutinins made these lectins likely prospects. Although whole serum from *Limulus* was inherently bactericidal, neither purified limulin nor the serum fraction depleted of limulin exhibited any antibacterial activity (Pistole and Graf, 1984). That this was not due to irreversible loss of biologic activity in the separation process was shown by reconstitution studies. Limulin recombined with adsorbed serum yielded virtually the same bactericidal capabilities as unfractionated serum. Clearly, the killing mechanism is not as simple as was once thought.

11.4.3. Granulocytes

L. polyphemus responds by clot formation following exposure to bacteria (Bang and Frost, 1953) or endotoxin (Bang, 1956). Since bacteria are immobilized in the gelatinous clot (Bang, 1956), it has long been postulated that clotting is an important factor in *Limulus* defenses. Clot formation results in morphologic changes in the circulating blood cells known as granulocytes or amoebocytes (Levin and Bang, 1964a; Shirodkar et al., 1960). The cells aggregate, degranulate, and form long, pseudopodial processes. Electron microscopy has shown that degranulation occurs by fusion of granule and cell membrane, followed by exocytosis of granular contents (Dumont et al., 1966; Hand and Oliver, 1981). These granulocytes are essential to clot formation, as described by Levin and Bang (1964b). Cell-free plasma does not coagulate in the presence of endotoxin until granulocytes are added.

11.4.3.1. Intact Cells. In vitro studies with granulocytes are hampered by the extreme fragility of these cells. At present they can be maintained for a maximum of several hours in a serum-free environment before they begin to degranulate. Thus, studies of antibacterial activity associated with granulocytes are limited to this time period.

Granulocytes incubated with *E. coli, A. viridans, S. marcescens,* or *M. luteus* for 1 hr in the absence of serum exert no bactericidal effect (Pistole and Britko, 1978). When diluted, pooled *Limulus* serum or plasma, which has no inherent killing capacity, is added to these systems, significant killing occurs in the *E. coli* system but not in the remaining three. *Limulus* serum readily agglutinates strains of *E. coli* but not *A. viridans, S. marcescens,* or *M. luteus* (Pistole, 1978). It seems likely that the serum components responsible for bacterial agglutination are also involved in microbial killing.

We have been able to modify in vitro conditions to extend the time these granulocytes can be maintained outside the animal. Under such conditions, significant bactericidal effect of granulocytes alone on *E. coli* is seen after 2 hr (Pistole and Graf, 1984). It is possible that serum components enhance a process that would occur with granulocytes alone, given sufficient time.

The mechanism by which granulocyte-mediated killing occurs is unclear. *Limulus* granulocytes are capable of phagocytizing carbonyl iron particles (Armstrong and Levin, 1979), but to date there is no definitive evidence for uptake of microorganisms by these cells (Armstrong and Levin, 1979; R. Smith, unpublished data). There is also evidence that granulocyte granules are released in the presence of microorganisms (Armstrong and Levin, 1979; Bang, 1956; Mürer et al., 1975; Shirodkar et al., 1960), suggesting a form of extracellular killing. It is known that the cytoplasmic granules of mammalian polymorphonuclear leukocytes are important in mediating the antibacterial effects of these cells (see, e.g., Cohn and Hirsch, 1960). Granulocyte-mediated killing of bacteria may occur in *Limulus* by similar means.

11.4.3.2. Granulocyte Granules. Granulocytes of the horseshoe crab contain granules, and the clotting system associated with these cells (Levin and Bang, 1968; Solum, 1970) is contained within these granules (Mürer et al., 1975). Components of these granules form an extracellular gel in the presence of minute amounts of endotoxin, and this phenomenon is the basis for the *Limulus* amoebocyte lysate test for bacterial endotoxin (Levin and Bang, 1964b).

Using a procedure developed by Mürer and colleagues (1975), we have isolated amoebocyte granules and tested them for bactericidal activity. Intact granules are bactericidal for both *E. coli* and *S. aureus* (Smith and Pistole, 1985). Granule lysates prepared by sequential freezing and thawing are completely devoid of antibacterial activity, although granulocyte lysates prepared by this method retain the ability to clot in the presence of bacterial endotoxin (Mürer et al., 1975). Furthermore, granulocyte lysates are bactericidal for a variety of microorganisms (Nachum, 1979). The reason for these differing results is unclear. Cheng (1983) has reported that granules in molluscan hemocytes act as true lysosomes. Two lysosomal markers, acid phosphatase and β-glucuronidase, have been found in the granulocytes and serum of *L. polyphemus* (R. Smith, unpublished data), suggesting that similar organelles may be found in this animal.

11.4.4. Granulocyte-Serum Interactions

Washed granulocytes have variable in vitro bactericidal activity, depending on the maintenance conditions of the granulocytes and the length of time the granulocytes and bacteria are coincubated (Pistole and Britko, 1978; Pistole and Graf, 1984). Serum from *L. polyphemus* is also known to be capable of killing bacteria (Furman and Pistole, 1976; Pistole and Furman, 1976). Serum is also capable of enhancing the innate killing capacity of granulocytes (Pistole and Britko, 1978), although the mechanism involved is unclear.

Because *Limulus* serum contains lectins that bind to various strains of bacteria, it seems reasonable to assume that these serum components may be responsible for the enhanced killing seen in serum-augmented bactericidal assays with granulocytes. In one such study, we examined the degree of killing of *E. coli* by granulocytes alone and supplemented with whole serum, purified limulin, or serum depleted of this lectin (Pistole and Graf, 1984). The results indicate that our assumption was incorrect. No enhanced killing occurred with the granulocyte-containing system supplemented with either limulin or serum depleted of limulin. As with the studies of killing by serum alone, it appears that the enhancing effect of whole serum on granulocyte-mediated killing is due to more than one component.

11.5. ADDITIONAL COMMENTS

Several points follow from the various studies that have been conducted on host defenses in *L. polyphemus*:

- Invertebrate species such as *L. polyphemus* are not simply prototypes of their vertebrate counterparts. Given the fact that elements of the vertebrate inducible immune system are not present in any invertebrate examined, it is tempting to view these animals as functioning with a primitive defense system. In fact, it is becoming increasingly clear that animals such as *L. polyphemus* have a reasonably complex set of antibacterial defense mechanisms. It is also quite possible that the processes we see in these invertebrates have their counterparts in the vertebrate world. Using mouse and guinea pig systems, Weir and colleagues have shown that various gram-positive and gram-negative bacteria adhere to macrophages in vitro by lectinlike receptors on the surface of these animal cells (Weir, 1980; Glass et al., 1981). This phenomenon, which is distinct from that mediated by cytophilic antibody, may represent a highly conserved process among metozoans.

    As evidence of such conservation, Quigley and Armstrong (1983) have described an endopeptidase inhibitor in *L. polyphemus* with properties very similar to those of vertebrate α_2-macroglobulin. They suggest that such protease inhibitors are "ancient molecules." Likewise, limulin shares both structural and functional properties with the C-reactive protein of mammals (Robey and Liu, 1981).

- The in vivo attachment of antigen to antibody (or lectin to ligand) per se is usually uneventful. The physiologic and pathologic sequelae are mediated by secondary systems activated by the antigen-antibody complexing. The complement system is a classic example of such a system. Although the existence of a complement-mediated hemolytic process has been suggested for *L. polyphemus* (Day et al., 1970), this has not been confirmed (Hall et al., 1972). Still, it is worth noting that in *L. polyphemus*, serum-mediated killing (and likely also serum-enhanced, amoebocyte-mediated killing) requires not only the lectin, or recognition protein, but an additional, as yet uncharacterized, serum factor. It is tempting to note the analogy with complement-mediated phagocytosis in mammals.

- Despite the presence of antibacterial defenses in the horseshoe crab, microorganisms may be found in the hemolymph of this animal (Brandin and Pistole, 1985). The number of microorganisms increases significantly during maintenance of *L. polyphemus* in flowing seawater tanks with no overt distress to the animals. Although virtually all the bacteria isolated from hemolymph samples are gram-negative, there is no detectable decline in the ability of amoebocyte lysate from these animals to detect endotoxin. The mechanism by which gram-negative bacteria and granulocytes coexist in the hemolymph of the horseshoe crab is unknown.

11.6. SUMMARY

L. polyphemus, and probably other members of the family Limulidae, exert a killing effect on a variety of bacterial species. This bactericidal activity is found in serum but not plasma, indicating a probable role for substances found within the granulocytes, since these cells undergo autolysis during routine serum collection. Granulocytes alone also exert a significant, though variable, antibacterial effect. Lectins, found in the serum of these animals, have been shown to agglutinate a wide variety of bacteria and to bind to specific structures on the outer surface of these prokaryotes. These proteins probably function as the recognition units in host defenses, although they exert no bactericidal activity alone. The antibacterial defense system of *L. polyphemus* may represent a primitive and highly conserved mechanism for controlling potential microbial pathogens, which is expressed in vertebrates as a form of innate immunity.

REFERENCES

Armstrong, P.B. and J. Levin. 1979. *In vitro* phagocytosis by *Limulus* blood cells. *J. Invertebr. Pathol.* **34**: 145–151.

Bang, F.B. 1956. A bacterial disease of *Limulus polyphemus*. *Bull. Johns Hopkins Hosp.* **98**: 325–337.

Bang, F.B. and J.L. Frost. 1953. The toxic effect of a marine bacterium on *Limulus* and the formation of blood clots. *Biol. Bull.* (Woods Hole) **105**: 361–362.

Bhattacharjee, A.K., H.J. Jennings, and C.P. Kenny. 1978. Structural elucidation of the 3-deoxy-D-manno-octulosonic acid containing meningococcal 29-e capsular antigen using carbon-13 nuclear magnetic resonance. *Biochemistry* **17**: 645–651.

Bishayee, S. and D.T. Dorai. 1980. Isolation and characterisation of a sialic acid–binding lectin (carcinoscorpin) from Indian horseshoe crab *Carcinoscorpius rotunda cauda*. *Biochim. Biophys. Acta* **623**: 89–97.

Brandin, E.R. and T.G. Pistole. 1983. Polyphemin: A teichoic acid–binding lectin from the horseshoe crab *Limulus polyphemus*. *Biochem. Biophys. Res. Commun.* **113**: 611–617.

Brandin, E.R. and T.G. Pistole. 1985. Presence of microorganisms in hemolymph of the horseshoe crab *Limulus polyphemus*. *Appl. Environ. Microbiol.* **49**: 718–720.

Cheng, T.C. 1983. Internal defense mechanisms of molluscs against invading microorganisms: Personal reminiscences. *Trans. Am. Microsc. Soc.* **102**: 185–193.

Cohen, E. 1968. Immunologic observations of the agglutinins of the hemolymph of *Limulus polyphemus* and *Birgus latro*. *Trans. N.Y. Acad. Sci.* **30**: 427–443.

Cohen, E., B.A. Baldo, and G. Uhlenbruck. 1975. Anti-galactan precipitins in the hemolymph of *Tridacna maxima* and *Limulus polyphemus*, pp. 13–18. In W.H. Hildemann and A.A. Benedict (eds.) *Immunologic Phylogeny*. Plenum Press, New York.

Cohen, E., A.W. Rose, and F.C. Wissler. 1965. Heteroagglutinins of the horseshoe crab *Limulus polyphemus*. *Life Sci.* **4**: 2009–2016.

Cohen, E., L.E. Blumenson, M. Pliss, and J. Minowada. 1979. Differentiation of human leukemic from normal lymphocytes by purified *Limulus* agglutinin, pp. 589–600, In E. Cohen (ed.) *Biomedical Applications of the Horseshoe Crab* (Limulidae). Alan R. Liss, New York.

Cohen, E., J. Minowada, M. Pliss, L. Pliss, and L.E. Blumenson. 1976. Differentiation of human leukemic from normal lymphocytes by *Limulus* serum agglutination. *Vox Sang.* **31**: 117–123.

Cohn, Z.A. and J.G. Hirsch. 1960. The isolation and properties of the specific cytoplasmic granules of rabbit polymorphonuclear leucocytes. *J. Exp. Med.* **112**: 983–1004.

Davidson, S.K., K.F. Keller, and R.J. Doyle. 1982. Differentiation of coagulase-positive and coagulase-negative staphylococcus by lectins and plant agglutinins. *J. Clin. Microbiol.* **15**: 547–553.

Day, N.K.B., H. Gewurz, R. Johannsen, J. Finstad, and R.A. Good. 1970. Complement and complement-like activity in lower vertebrates and invertebrates. *J. Exp. Med.* **132**: 941–950.

Dorai, D.T., B.K. Bachhawat, and S. Bishayee. 1981. Fractionation of sialoglycoproteins on an immobilized sialic acid–binding lectin. *Anal. Biochem.* **115**: 130–137.

Dorai, D.T., S. Srimal, S. Mohan, B.K. Bachhawat, and T.S. Bangalesh. 1982.

Recognition of 2-keto-3-deoxyoctonate in bacteria cells and lipopolysaccharides by the sialic acid–binding lectin from the horseshoe crab *Carcinoscorpius rotunda cauda. Biochem. Biophys. Res. Commun.* **104:** 141–147.

Dumont, J.N., E. Anderson, and G. Winner. 1966. Some cytologic characteristics of the hemocytes of *Limulus* during clotting. *J. Morphol.* **119:** 181–208.

Furman, R.M. and T.G. Pistole. 1976. Bactericidal activity of hemolymph from the horseshoe crab *Limulus polyphemus. J. Invertebr. Pathol.* **28:** 239–244.

Gilbride, K.J. and T.G. Pistole. 1979. Isolation and characterization of a bacterial agglutinin in the serum of *Limulus polyphemus*, pp. 525–535. In E. Cohen (ed.) *Biomedical Applications of the Horseshoe Crab* (Limulidae). Alan R. Liss, New York.

Gilbride, K.J. and T.G. Pistole. 1981. The presence of copper in a purified lectin from *Limulus polyphemus*: Possible new role for hemocyanin. *Dev. Comp. Immunol.* **5:** 347–352.

Glass, E., J. Stewart, and D.M. Weir. 1981. Presence of bacterial binding "lectin-like" receptors on phagocytes. *Immunology* **44:** 529–534.

Goldstein, I.J. and C.E. Hayes. 1978. The lectins: Carbohydrate-binding proteins of plants and animals. *Adv. Carbohydr. Chem. Biochem.* **35:** 127–340.

Goldstein, I.J., R.C. Hughes, M. Monsigny, T. Osawa, and N. Sharon. 1980. What should be called a lectin? *Nature* (London) **285:** 66.

Hall, J.L., D.T. Rowlands, Jr., and U.R. Nilsson. 1972. Complement-unlike hemolytic activity in lobster hemolymph. *J. Immunol.* **109:** 816–823.

Hand, A.R. and C. Oliver. 1981. Comparison of compound with plasmalemmal exocytosis in *Limulus* amoebocytes. *Methods Cell Biol.* **23:** 301–311.

Levin, J. and F.B. Bang. 1964a. A description of cellular coagulation in the *Limulus. Bull. Johns Hopkins Hosp.* **115:** 337–345.

Levin, J. and F.B. Bang. 1964b. The role of endotoxin in the extracellular coagulation of *Limulus* blood. *Bull. Johns Hopkins Hosp.* **115:** 265–274.

Levin, J. and F.B. Bang. 1968. Clottable protein in *Limulus*: Its localization and kinetics of its coagulation by endotoxin. *Thromb. Diath. Haemorrh.* **19:** 186–197.

Maget-Dana, R., A.-C. Roche, and M. Monsigny. 1979. Ganglioside-limulin interactions, pp. 567–578. In E. Cohen (ed.) *Biomedical Applications of the Horseshoe Crab* (Limulidae). Alan R. Liss, New York.

Marchalonis, J.J. 1964. A natural hemagglutinin from *Limulus polyphemus. Fed. Proc.* **23:** 1468.

Marchalonis, J.J. and G.M. Edelman. 1968. Isolation and characterization of a hemagglutinin from *Limulus polyphemus. J. Mol. Biol.* **32:** 453–465.

Marchalonis, J.J. and M.J. Waxdal. 1979. *Limulus* agglutinins: Past, present and future, pp. 665–675. In E. Cohen (ed.) *Biomedical Applications of the Horseshoe Crab* (Limulidae). Alan R. Liss, New York.

McSweegan, E.F. and T.G. Pistole. 1982. Interaction of the lectin limulin with capsular polysaccharides from *Neisseria meningitidis* and *Escherichia coli. Biochem. Biophys. Res. Commun.* **106:** 1390–1397.

Mürer, E.H., J. Levin, and R. Holme. 1975. Isolation and studies of granules of the amoebocytes of *Limulus polyphemus*, the horseshoe crab. *J. Cell. Physiol.* **86:** 533–542.

Nachum, R. 1979. Antimicrobial defense mechanisms in *Limulus polyphemus*, pp. 513–524. In E. Cohen (ed.) *Biomedical Applications of the Horseshoe Crab* (Limulidae). Alan R. Liss, New York.

Nowak, T.P. and S.H. Barondes. 1975. Agglutinin from *Limulus polyphemus*: Purification with formalinized horse erythrocytes as the affinity adsorbent. *Biochim. Biophys. Acta* **393**: 115–123.

Pardoe, G.I., G. Uhlenbruck, and G.W.G. Bird. 1970. Studies on some heterophile receptors of the Burkitt EB2 lymphoma cell. *Immunology* **18**: 73–83.

Pistole, T.G. 1978. Broad-spectrum bacterial agglutinating activity in the serum of the horseshoe crab *Limulus polyphemus*. *Dev. Comp. Immunol.* **2**: 65–76.

Pistole, T.G. 1981. Interaction of bacteria and fungi with lectins and lectin-like substances. *Annu. Rev. Microbiol.* **35**: 85–112.

Pistole, T.G. and J.L. Britko. 1978. Bactericidal activity in the serum of the horseshoe crab *Limulus polyphemus*. *J. Invertebr. Pathol.* **31**: 376–382.

Pistole, T.G. and R.M. Furman. 1976. Serum bactericidal activity in the horseshoe crab *Limulus polyphemus*. *Infect. Immun.* **14**: 888–893.

Pistole, T.G. and S.A. Graf. 1984. Bactericidal activity of *Limulus* lectins and amebocytes, pp. 71–81. In E. Cohen (ed.) *Recognition Proteins, Receptors, and Probes: Invertebrates*. Alan R. Liss, New York.

Pistole, T.G. and H. Rostam-Abadi. 1979. Lectins from the horseshoe crab, *Limulus polyphemus*, reactive with bacterial lipopolysaccharide, pp. 423–426. In H. Peeters (ed.) *Protides of Biological Fluids: Twenty-seventh Colloquium*. Pergamon Press, Oxford.

Quigley, J.P. and P.B. Armstrong. 1983. An endopeptidase inhibitor, similar to mammalian α_2-macroglobulin, detected in the hemolymph of an invertebrate, *Limulus polyphemus*. *J. Biol. Chem.* **258**: 7903–7906.

Robey, F.A. and T.-Y. Liu. 1981. Limulin: A C-reactive protein from *Limulus polyphemus*. *J. Biol. Chem.* **256**: 969–975.

Roche, A.-C. and M. Monsigny. 1974. Purification and properties of limulin: A lectin (agglutinin) from hemolymph of *Limulus polyphemus*. *Biochim. Biophys. Acta* **371**: 242–254.

Roche, A.-C., J.-C. Maurizot, and M. Monsigny. 1978. Circular dichroism of limulin: *Limulus polyphemus* lectin. *FEBS Lett.* **91**: 233–236.

Rostam-Abadi, H. and T.G. Pistole. 1982. Lipopolysaccharide-binding lectin from the horseshoe crab *Limulus polyphemus* with specificity for 2-keto-3-deoxyoctonate (KDO). *Dev. Comp. Immunol.* **6**: 209–218.

Shimizu, S., M. Ito, and M. Niwa. 1977. Lectins in the hemolymph of Japanese horseshoe crab, *Tachypleus tridentatus*. *Biochim. Biophys. Acta* **500**: 71–79.

Shimizu, S., M. Ito, N. Takahashi, and M. Niwa. 1979. Purification and properties of lectins from the Japanese horseshoe crab, *Tachypleus tridentatus*, pp. 625–639. In E. Cohen (ed.) *Biomedical Applications of the Horseshoe Crab* (Limulidae). Alan R. Liss, New York.

Shirodkar, M.V., A. Warwick, and F.B. Bang. 1960. The *in vitro* reaction of *Limulus* amebocytes to bacteria. *Biol. Bull.* (Woods Hole) **118**: 324–337.

Smith, R.H. and T.G. Pistole. 1985. Bactericidal activity of granules isolated from amoebocytes of the horseshoe crab *Limulus polyphemus*. *J. Invertebr. Pathol.* **45:** 272–275.

Solum, N.O. 1970. Some characteristics of the clottable protein of *Limulus polyphemus* blood cells. *Thromb. Diath. Haemorrh.* **23:** 170–181.

Uhlenbruck, G., G. Steinhausen, and B.A. Baldo. 1976. Galactans and antigalactans from invertebrates. *Z. Naturforsch.* **31:** 205–206.

Vaith, P., G. Uhlenbruck, and G. Holz. 1979a. Anti-glucuronyl activity of *Limulus polyphemus* agglutinin, pp. 455–458. In H. Peeters (ed.) *Protides of Biological Fluids: Twenty-seventh Colloquium.* Pergamon Press, Oxford.

Vaith, P., G. Uhlenbruck, W.E.G. Müller, and E. Cohen. 1979b. Reactivity of *Limulus polyphemus* hemolymph with D-glucuronic acid containing glycosubstances, pp. 579–587. In E. Cohen (ed.) *Biomedical Applications of the Horseshoe Crab (Limulidae).* Alan R. Liss, New York.

Vasta, G.R. and J.J. Marchalonis. 1983. Humoral recognition factors in the Arthropoda: The specificity of Chelicerata serum lectins. *Am. Zool.* **23:** 157–171.

Vasta, G.R. and J.J. Marchalonis. 1984. Distribution, specificity and macromolecular properties of tunicate plasma lectins, pp. 125–141. In E. Cohen (ed.) *Recognition Proteins, Receptors, and Probes: Invertebrates.* Alan R. Liss, New York.

Voigtmann, R., B. Salfner, and G. Uhlenbruck. 1971. Studies on broad spectrum agglutinins. 9. Specific and unspecific reactions between *Limulus polyphemus* haemolymph and snail extracts with "anti-A" specificity. *Z. Immunitaetsforsch.* **141:** 488–494.

Weir, D.M. 1980. Surface carbohydrates and lectins in cellular recognition. *Immunol. Today* **2:** 45–51.

Yeaton, R.W. 1981. Invertebrate lectins. 2. Diversity of specificity, biological synthesis, and function in recognition. *Dev. Comp. Immunol.* **5:** 535–545.

CHAPTER 12

Comparative Structural Studies of Limulin

J. Michael Kehoe

Department of Microbiology and Immunology
Northeastern Ohio Universities College of Medicine
Rootstown, Ohio

Rochelle K. Seide

Brumbaugh, Graves, Donohue and Raymond
New York, New York

12.1. INTRODUCTION

The existence of an inducible immune response is now very well established for vertebrates. For invertebrates, a clearly analogous biologic system has not yet been demonstrated to everyone's satisfaction, although clear evi-

345

dence of adaptive response systems has been provided for a number of invertebrate species (Hildemann and Benedict, 1975) and some evidence of induced proteins in invertebrates has been presented as well (Cooper et al., 1984; Ratcliffe, 1985). Clearly, invertebrates are subject to a wide variety of noxious, even lethal, external stimuli that can lead to serious infections (bacterial, viral, etc.). In addition, tumor growths have been observed in at least some vertebrates. Thus, it is not unreasonable to assume that such species would have evolved some defense mechanisms to cope with such insults to their biologic integrity. Such processes might well be somewhat primitive and not possessive of the complete elegance and sophistication of the individual immune systems mediated by lymphocytes in vertebrates. Nonetheless, the systems could be expected to be advanced biochemically (e.g., capable of binding ligands strongly and specifically) and to express itself in the hemolymph, or equivalent fluid component, of the invertebrate animal. Limulin, a hemagglutinating protein found in high concentration in the hemolymph of the horseshoe crab *Limulus polyphemus* is such a candidate protein and, as such, has been subjected to considerable experimental attention (Marchalonis and Edelman, 1968; Kaplan et al., 1977). Current knowledge of the structure and function of this invertebrate protein is the subject of this chapter.

12.2. DISCOVERY AND INITIAL CHARACTERIZATION OF LIMULIN

The story of early work with limulin and the rationale for initial interest in its structure has been well described by Marchalonis and Waxdal (1979). Important initial characterizations of the molecule itself were provided by Marchalonis and Edelman (1968); by Cohen (1968); and, from a primary structure perspective, by Kaplan and coworkers (1977). Characteristics that have made this protein attractive as a model system include its agglutinating capacity for red blood cells, particularly through its sialic acid binding activity (Cohen et al., 1983); the size of its subunits (Marchalonis and Edelman, 1968; Kaplan et al., 1977), which makes the molecule potentially comparable to immunoglobulin subunits of mammals; and, most significantly, its presence in horseshoe crab hemolymph in relatively high concentrations (2–5 mg/ml) such that the isolation of milligram quantities is relatively easily carried out. (See also Chapter 19.)

12.3. ISOLATION OF LIMULIN FROM HEMOLYMPH OF *LIMULUS POLYPHEMUS*

To assure that the limulin used for structural studies represents, as closely as possible, protein in its natural state, direct isolation from freshly drawn *L. polyphemus* hemolymph has proved the best approach. A number of

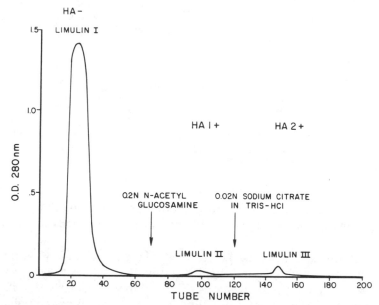

Figure 12.1. Isolation of limulin by an affinity chromatography technique. Bovine submaxillary mucin was attached to CNBr-activated Sepharose. Limulin purified by a combination of gel filtration and ion-exchange methods (Kaplan et al., 1977) was then applied to this column. Elution of bound subfractions was carried out as shown, using 0.2 N *N*-acetylglucosamine and 0.02 N sodium citrate in tris-HCl. The differential agglutinin activity of the limulin subfractions for equine RBCs is indicated.

different methods have been described in the literature and used in our laboratory, including purely chromatographic means (Roche et al., 1975; Kaplan et al., 1977) and affinity chromatography approaches (Fig. 12.1). These methods, in our laboratory as well as those reported in the literature (Oppenheim et al., 1974; Nowak and Barondes, 1975; Roche et al., 1975), suggest that various fractions of limulin can be isolated from horseshoe crab hemolymph by varying the isolation conditions used. However, amino terminal amino acid sequence analysis (Fig. 12.2) of both affinity-purified and non-affinity-purified material showed at least the first 15 amino acid residues to be identical. The possibility of other polymorphic variations in this protein, possibly involving the sialic acid binding site, remains.

In addition to suggestions of some binding heterogeneity on the basis of the abovementioned affinity chromatography results, our laboratory and others have observed some minor heterogeneity on polyacrylamide gel electrophoresis of purified limulin (Kaplan et al., 1977). Without indications of marked primary structure variation (see Section 12.7), such heterogeneity has been attributed to differences in the carbohydrate attached to this lectin, especially since variation in the attachment of carbohydrates has been shown to occur in a number of glycoprotein systems. It is expected that the de-

Affinity	1	5	10
sequence:	Leu–Glu–Glu–Gly–Glu–Ile–Thr–Ser–Lys–Val–Lys–Phe–Pro–		
		15	
	Pro–Ser		

Chromatography	1	5	10
sequence:	Leu–Glu–Glu–Gly–Glu–Ile–Thr–Ser–Lys–Val–Lys–Phe–Pro–		
		15	
	Pro–Ser		

Figure 12.2. The amino terminal amino acid sequence of affinity-purified limulin (corresponding to limulin III in Fig. 12.7) compared with that isolated by conventional chromatography. (From Kaplan et al., 1977.)

velopment and use of monoclonal antibodies (see Section 12.6) will shed some light on the question and nature of possible heterogeneity in the limulin system.

Undoubtedly, other approaches to the isolation of limulin (e.g., affinity chromatography using a bound specific antilimulin antibody) will prove effective in sorting out these questions. This will be particularly true if, once again, monoclonal reagents of this type are used as well as the polyclonal antilimulin sera that have been successfully raised in a number of laboratories. Such an approach would obviously not select molecules on the basis of their ligand binding capacity, as is the case for the affinity methods using a relevant ligand.

12.4. BIOLOGIC ACTIVITY OF LIMULIN

Limulin belongs to that diverse family of proteins referred to as lectins (Lis and Sharon, 1977). These proteins, which have been isolated from both plant and animal sources (see also Chapter 1), have proven to be of significant use as mitogenic agents for lymphocytes, as probes of the cell surface, and as affinity reagents for the specific isolation of either individual molecules (e.g., glycoproteins containing the carbohydrate with which the lectin reacts) or whole cells that display this carbohydrate on their cell surface.

With respect to limulin, it is now clear that more than one ligand can be significantly bound when this protein exerts its lectin activity. As detailed by Cohen and colleagues (1983), as well as others, a prominant ligand for limulin is N-acetyl-neuraminic acid (NANA) and its various derivatives. However, binding has also been clearly demonstrated for N-acetyl-D-glucosamine and for N-acetyl-D-mannosamine, a precursor of NANA. Our laboratory has focused on sialic acid, however, because of the potential importance of this compound in neoplastic conditions in mammalian species, including human being. A number of lectins other than limulin have been shown to display a differential reactivity with neoplastic cells, as compared with their normal counterparts. For example, a differential agglutinability between normal and transformed cells was shown by Burger and Goldberg (1967) for the N-acetyl-D-glucosamine binding lectin wheat germ agglutinin (WGA).

Among the large array of molecular entities—glycoproteins, glycolipids, proteins, and lipids—that are known to be or could be involved in these neoplasia-associated membrane changes, special attention has been focused on sialic acid as a significant surface component (Weiss, 1973). Some tumors have shown absolute increases in sialic acid content (Mabry and Carubelli, 1972). Other neoplasms have shown a decrease in total sialic acid concomitant with an increase in surface sialic acid as measured by accessibility to treatment with neuraminidase (Bosman et al., 1974). Even more striking, a relationship has been discerned between the amount of sialic acid present on the tumor cell surface and the capacity of these cells to metastasize (Yogeeswaran et al., 1978). Thus, the biologic activity of limulin, as expressed in its sialic acid binding capacity, could make this lectin useful as a marker protein or probe for especially invasive tumor cells of various sorts.

12.5. SUBUNIT STRUCTURE OF LIMULIN

A number of invertebrate lectins have been shown to be composed of subunits arranged and maintained in a surprisingly uniform manner (for a summary, see Litman and Kehoe, 1978). With respect to limulin, Marchalonis and Edelman (1968) clearly and convincingly showed that the 400,000-dalton intact unit could be broken down, initially to a hexamer of basic subunits and then, under more stringent conditions, to the ultimate 22,500-dalton subunit, of which there are 18 per intact limulin molecule. As noted below, both amino acid compositional and sequential analyses have detected no differences in structure among these ultimate 1.5 S subunits. The subunit bonding (subunit to subunit) is noncovalent and dependent on Ca^{2+} as a divalent cation.

12.6. SEROLOGIC REACTIVITY OF LIMULIN

Limulin is a highly immunogenic molecule when injected into animals commonly used for the production of antisera (e.g., rabbits and mice). Thus, it has been relatively easy to generate potent antisera against this lectin (Marchalonis and Edelman, 1968; Kaplan et al., 1977). The precipitating nature of the resulting antilimulin antibodies has allowed the use of Ouchterlony analysis (double diffusion in gels) in studying the antigenic nature of the intact limulin molecule and its subunits. As noted by Kaplan and coworkers (1977), distinctions can be made between antigenic determinants associated with the intact Ca^{2+}-bound molecule of 18 subunits and other determinants possessed by the individual subunits themselves (Fig. 12.3). This fact can

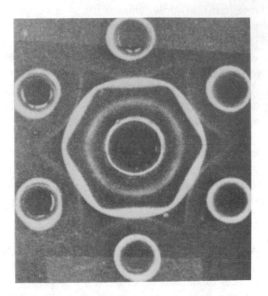

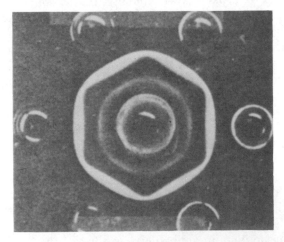

Figure 12.3. Ouchterlony analysis, using a polyclonal rabbit antilimulin antiserum placed in the center well. (a) Intact polymeric limulin in a Ca^{2+}-containing buffer was placed in the peripheral well. (b) The limulin was treated with a chelating buffer (tris-EDTA) to bind the Ca^{2+} ions, allowing the limulin to be broken down into its constituent subunits. Note the additional precipitin line in A, characteristic of the intact lectin. (Reproduced with permission from Kaplan et al., 1977.)

be demonstrated by the treatment of intact limulin with chelating agents, such as ethylenediaminetetraacetic acid (EDTA), which ablate the precipitin ring closest to the antigen (limulin) well (Fig. 12.3B). Such chelating agents are known to cause the intact lectin to be disassembled to its constituent subunits by removing the relevant divalent cation (Ca^{2+} in this case) upon which the structure of the intact molecule depends. Current studies are attempting to refine and expand these observations by using monoclonal antibodies (mAbs) prepared in mice by the method of Kohler and Milstein (1975). The immediate goal is to develop individual mAbs whose specificity is directed against the various antigenic determinants characteristic of both the intact lectin and its ultimate subunit.

Another invertebrate lectin that binds sialic acid has been isolated from the lobster *Homarus americanus* (Hartman et al., 1978). Because of the shared ligand binding activity for sialic acid between this lectin and limulin, polyclonal antisera were raised in rabbits against each protein and cross-reactions searched for by reciprocal Ouchterlony analysis. No evidence of serologic cross-reactions between these two invertebrate lectins was detected using this approach (J. Kehoe and R. Seide, unpublished results).

12.7. LIMULIN COVALENT STRUCTURE

12.7.1. General Principles

The amino acid sequence of a protein, be it derived directly by protein chemistry or by refernce to the nucleotide sequence of the appropriate gene, offers unique information about that protein. Such information includes possible relationships between the protein in question and others, even from widely diverse sources; how a particular function of a protein (e.g., ligand binding) may be related to its primary structure; and at what point in evolutionary time a given protein may have diverged from its more or less homologous counterparts. With respect to limulin, a significant question regarding its possible role in any form of immune response in the horseshoe crab would be whether limulin as a protein bears, in any respect, a structural relationship to the well-characterized immunoglobulin family of vertebrates. A most direct way of examining this is to ask whether any part of the covalent structure of limulin, even a minor constituent peptide, bears any relationship to any vertebrate immunoglobulin whose covalent structure is known. Alternatively, one can ask whether the horseshoe crab lectin limulin bears any relationship to any of the other lectin molecules for which amino acid sequence information is known. In such comparative analyses, one has extensive sequence information pertaining to immunoglobulins (Kabat et al., 1983) but, as yet, relatively little for alternative lectin molecules. In addition, one can search for possible sequence relationships between limulin and any other protein for which information is currently available, as discussed in more detail in Section 12.8.

TABLE 12.1. AMINO ACID COMPOSITION OF AN INTACT SUBUNIT OF LIMULIN

Amino Acid	Residues per Mole of Subunit
Lysine	11.0 (11)
Histidine	8.2 (8)
Arginine	3.3 (3)
CM-cysteine	—
Aspartic acid	16.2 (16)
Threonine	9.8 (10)
Serine	11.4 (12)
Glutamic acid	21.9 (22)
Proline	7.8 (8)
Glycine	75.3 (15)
Alanine	8.4 (8)
Half-cystine	3.7 (3–4)
Valine	11.7 (12)
Methionine	2.1 (2)
Isoleucine	8.2 (8)
Leucine	15.3 (15)
Tyrosine	2.6 (3)
Phenylalanine	6.9 (7)
Tryptophan	(3–4)
Homoserine (lactone)	—

Note: Values are expressed residues per mole rounded off to integer units, as shown in parentheses. Corrections have beem made in certain amino acids to account for destruction during hydrolysis. Other considerations are as described by Kaplan et al., 1977.

12.7.2. Cyanogen Bromide Fragments of Limulin

The amino acid composition of intact limulin is shown in Table 12.1. (Kaplan et al., 1977). This analysis showed that two moles of methionine were present per mole of limulin subunit. Since the chemical cleaving reagent cyanogen bromide is known to cleave proteins at methionine residues with essentially 100% efficiency (Gross, 1967), three fragments were expected from the intact subunit. As reported by Kaplan and associates (1977), three such fragments were detected and isolated. Evidence that these three fragments constituted the total subunit was provided by comparing the summed amino acid compositions of the individual fragments (Table 12.2) with that composition obtained for the intact subunit. Within the reasonable experimental error associated with amino acid composition analysis, the two totals did correspond (Tables 12.1 and 12.2).

12.7.3. Enzymatic Fragmentation of Limulin

The intact limulin subunit has also been fragmented by proteolytic enzymes. Most attention has been directed to the use of trypsin, an enzyme that cleaves on the carboxyl terminal side of lysine and arginine residues. Inspection of the amino acid composition of the intact subunit (Table 12.1) suggests that

TABLE 12.2. AMINO ACID COMPOSITIONS OF INDIVIDUAL CYANOGEN BROMIDE FRAGMENTS OF LIMULIN

Amino Acid	Residues per Mole of Subunit			
	CB I	CB II	CB III	Integer Sum: CB I, II, III
Lysine	2.6 (2)	2.0 (2)	8.0 (8)	(12)
Histidine	0.6 (0)	1.1 (1)	6.9 (7)	(8)
Arginine	1.9 (1)	1.3 (1)	2.0 (2)	(4)
CM-cysteine	—	0.8 (1)	2.1 (2)	(3)
Aspartic acid	1.5 (0)	2.6 (3)	12.1 (12)	(15)
Threonine	2.2 (1)	2.6 (3)	5.8 (6)	(10)
Serine	5.0 (5)	0.9 (1)	6.2 (6)	(12)
Glutamic acid	5.5 (3)	2.9 (3)	14.0 (14)	(20)
Proline	4.4 (4)	1.0 (1)	2.9 (3)	(8)
Glycine	2.1 (1)	2.5 (3)	10.1 (10)	(14)
Alanine	0.4 (0)	0.2 (0)	5.5 (6)	(6)
Half-cystine	—	—	—	—
Valine	3.7 (2)	2.8 (3)	6.3 (7)	(12)
Methionine	—	—	—	—
Isoleucine	1.4 (1)	1.1 (1)	4.1 (5)	(7)
Leucine	3.5 (2)	5.6 (6)	8.6 (9)	(17)
Tyrosine	0.4 (0)	1.2 (1)	1.3 (1)	(2)
Phenylalanine	2.7 (2)	1.3 (1)	3.3 (3)	(6)
Tryptophan	—	(1)	(2–3)	(3–4)
Homoserine (lactone)	(1)	(1)	—	(2)

Note: Since CB I has been only incompletely purified due to technical limitations, its integer values have been derived from the results of amino acid sequence analysis.

a total of approximately 14 such residues (11 lysines and 3 arginines) are present. Thus, it would be expected that about 15 separate tryptic peptides would be generated by a total tryptic cleavage of the subunit. The actual number depends on the precise placement of these lysine and arginine residues. Early attempts to purify the individual tryptic fragments by paper chromatography were only partially successful. However, more recent separations of the constituents of a tryptic digest using high-pressure liquid chromatography (HPLC) have been more satisfying. Figure 12.4 illustrates a peptide map of such a digest, prepared using the HPLC approach. As can be seen, a large number of individual peptides are distinguishable when the column effluent is assayed at 220 nm. Current studies involve the preparation of comparable digests of the limulin subunit by using other enzymes (e.g., chymotrypsin) and the isolation and characterization of the individual peptides.

12.7.4. Limulin Primary Structure and Active Site Localization

Determination of the covalent structure of limulin has been approached by classic protein chemistry procedures, primarily involving automated Edman

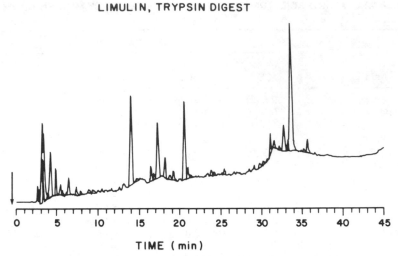

Figure 12.4. A peptide map of a trypsin digest of limulin prepared by reverse-phase HPLC. An aliquot of the trypsin digest was applied to a C_{18} HPLC column and eluted with a linear gradient of acetonitrile in water-trifluoroacetic acid. Peptides were detected at 220 μm.

degradation methods. Initial analyses (Kaplan et al., 1977) primarily involved isolation of the three cyanogen bromide fragments expected on the basis of the subunit composition. The sequence so determined, which also allowed the proper alignment of the cyanogen bromide fragments, is shown in Figure 12.5. The gap in fragment III is currently being filled in using the enzymatically generated (by trypsin and chymotrypsin) peptides that have been isolated by HPLC (e.g., Figure 12.4). Assurance that the full complement of peptides from this region was isolated was difficult to obtain before the application of HPLC technology.

Evidence has also been obtained that at least part of the ligand binding capacity of the intact lectin is retained following fragmentation by trypsin. This evidence is based on inhibition studies wherein the proteolytic digest can be shown to inhibit the capacity of intact limulin to agglutinate horse red blood cells. Current extensions of these results are obviously directed toward determining which of the individual peptides possesses the inhibitory capacity.

12.8. GENERAL RELATIONSHIP OF LIMULIN TO OTHER PROTEINS

It has been concluded on the basis of both physicochemical (Marchalonis and Edelman, 1968) and amino acid sequence studies (Kaplan et al., 1977)

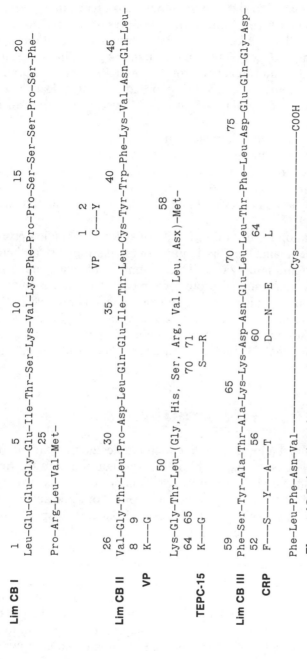

```
Lim CB I
         1                   5                  10                  15                  20
         Leu-Glu-Glu-Gly-Glu-Ile-Thr-Ser-Lys-Val-Lys-Phe-Pro-Pro-Ser-Ser-Ser-Pro-Ser-Phe-
                                         25
         Pro-Arg-Leu-Val-Met-

Lim CB II
         26                  30                  35                  40                  45
         Val-Gly-Thr-Leu-Pro-Asp-Leu-Gln-Glu-Ile-Thr-Leu-Cys-Tyr-Trp-Phe-Lys-Val-Asn-Gln-Leu-
         8  9                                            1  2
     VP  K---G                                       VP  C---Y

TEPC-15
                                                 58
         Lys-Gly-Thr-Leu-(Gly, His, Ser, Arg, Val, Leu, Asx)-Met-
         64  65                                  70  71
         K---G                                    S---R

Lim CB III
         59                  65                  70                  75
         Phe-Ser-Tyr-Ala-Thr-Ala-Lys-Lys-Asp-Asn-Glu-Leu-Leu-Thr-Phe-Leu-Asp-Glu-Gln-Gly-Asp-
         52          56                  60          64
     CRP F---S---Y---A---T              D---N---E    L

         Phe-Leu-Phe-Asn-Val------------------Cys-----------------COOH
```

Figure 12.5. Location of suggested sequence homologies between limulin and certain other proteins. (Marchalonis et al., 1984; Stewart and Channabasavaiah, 1979). No matter how weak, no other limulin homologies have been identified to date. The limulin sequence is written in a three-letter code, while the homology stretches for the other proteins are indicated by the corresponding single-letter amino acid code. Compared proteins are lysine vasopressin (VP), immunoglobulin heavy chain (TEPC-15), and C-reactive protein (CRP). The limulin cyanogen bromide fragments are individualized.

that limulin is not related to the immunoglobulin protein family of vertebrates. However, some recent, interesting analyses (Marchalonis et al.,
1984) have detected what are believed to be alternative indications of relationships at a "minigene" level involving an immunoglobulin (V_H segment
of a TEPC-15 antibody, and C-reactive protein, or CRP). While it is arguable
that such modest relationships should be accorded great significance at this
time, the observations are certainly of sufficient interest to warrant further
scrutiny.

A short stretch of sequence homology has been found between limulin
and the mammalian hormone lysine vasopressin (Stewart and Channabasavaiah, 1979). Once again, the possible functional or evolutionary significance of such a relationship is not completely clear at this time. The various
amino acid sequence homologies with limulin detected to date are illustrated
in Figure 12.5.

Sequence information for one other family of sialic acid binding proteins
has recently become available. This family comprises the surface hemagglutinins of the various influenza viruses, a number of which have been
sequenced either by conventional protein chemistry techniques or by reference to the nucleotide sequence of the relevant gene (Ward, 1981). It has
not yet been possible to find amino acid sequence homologies between limulin and segments of any of the available influenza hemagglutinins. It is,
of course, possible that additional sequence data will reveal such
homologies.

12.9. SUMMARY

One clear example of a biologically active lectin molecule that can be found
in invertebrates is limulin, a sialic acid binding protein found in the hemolymph of the horseshoe crab *L. polyphemus*. This protein binds strongly to
and agglutinates cells that contain significant amounts of sialic acid on their
surface. Examples of such cells are erythrocytes from certain species (e.g.,
the horse) and certain tumor cell variants that carry higher than normal
amounts of sialic acid on their surface. Limulin is a highly antigenic protein
that induces both polyclonal and monoclonal antibody responses in the appropriate systems. There are clear distinctions between the antigenic characteristics of the large intact polymeric lectin and its constituent subunits.
No structural evidence for heterogeneity of the individual subunits has been
established to date. Comparison of amino acid sequence stretches of limulin
with other proteins, including other lectins and immunoglobulins, shows only
very minimal homologies, highlighting the unique character of this invertebrate protein. No serologic evidence could be obtained for cross-reactivity
between limulin and another invertebrate sialic acid binding lectin isolated
from the lobster *H. americanus*.

ACKNOWLEDGMENTS

We express our appreciation to Ms. Ming Ming Fu, Ms. Judy Maw, and Ms. Evelyn Than for their help in various aspects of these studies.

REFERENCES

Bosmann, H.B., K.R. Case, and H.R. Morgan. 1974. Surface biochemical changes accompanying primary infection with Rous sarcoma virus. 1. Electrokinetic properties of cells and cell surface glycoprotein: Glycosyl transfer activities. *Exp. Cell Res.* **83:** 15–24.

Burger, M.M. and A.R. Goldberg. 1967. Identification of a tumor-specific determinant on neoplastic cell surfaces. *Proc. Nat. Acad. Sci. U.S.A.* **57:** 359–366.

Cohen, E. 1968. Immunologic observations of the agglutinins of the hemolymph of *Limulus polyphemus* and *Birgus latro*. *Trans. N.Y. Acad. Sci.* **30:** 427–443.

Cohen, E., G.H.U. Ilodi, W. Korytnyk, and M. Sharma. 1983. Inhibition of Limulus agglutinins with *N*-acetyl-neuraminic acid and derivatives. *Dev. Comp. Immunol.* **7:** 189–192.

Cooper, E.L., E.A. Stein, and A. Wojdani. 1984. Recognition receptors in annelids, pp. 43–54. In E. Cohen (ed.) *Recognition Proteins, Receptors, and Probes: Invertebrates*. Alan R. Liss, New York.

Gross, E. 1967. The cyanogen bromide reaction. *Methods Enzymol.* **11:** 238–255.

Hartman, A.L., P.A. Campbell, and C.A. Abel 1978. An improved method for the isolation of lobster lectins. *Dev. Comp. Immunol.* **2:** 617–625.

Hildemann, W.H. and A.A. Benedict 1975. *Immunologic Phylogeny*. Plenum Press, New York.

Kabat, E.A., T.T. Wu, H. Bilofsky, M. Reid-Miller, and H. Perry. 1983. *Sequences of Proteins of Immunological Interest*. U.S. Department of Health and Human Services, Public Health Service, National Institutes of Health, Bethesda, MD.

Kaplan, R., S.S-L. Li, and J.M. Kehoe. 1977. Molecular characterization of limulin, a sialic acid binding lectin from the hemolymph of the horseshoe carb *Limulus polyphemus*. *Biochemistry* **16:** 4297–4303.

Kohler, G. and C. Milstein. 1975. Continuous cultures of fused cells secreting antibody of predefined specificity. *Nature* (London) **256:** 495–497.

Lis, H. and N. Sharon. 1977. Lectins: Their chemistry and application to immunology, pp. 429–529. In M. Sela (ed.) *The Antigens* vol. 4. Academic Press, New York.

Litman, G.W. and J.M. Kehoe. 1978. The phylogenetic origins of immunoglobulin structure, pp. 205–228. In R.A. Good and S.B. Day (eds.) *Comprehensive Immunology*. Plenum Press, New York.

Mabry, E.W. and R. Carubelli. 1972. Sialic acid in human cancer. *Experientia* **28:** 182–183.

Marchalonis, J.J. and G.M. Edelman. 1968. Isolation and characterization of a natural hemagglutinin from *Limulus polyphemus*. *J. Mol. Biol.* **32:** 453–465.

Marchalonis, J.J. and M.J. Waxdal. 1979. Limulus agglutinins: Past, present and future. In E. Cohen (ed.) *Biomedical Applications of the Horseshoe Crab* (Limulidae). Alan R. Liss, New York.

Marchalonis, J.J., G.R. Vasta, G.W. Warr, and W.C. Barker. 1984. Probing the boundaries of the extended immunoglobulin family of recognition molecules: Jumping domains, convergence and minigenes. *Immunol. Today* **5**: 133–142.

Nowak, T.P. and S.H. Barondes. 1975. Agglutinin from limulus polyphemus purification with formalinized horse erythrocytes as the affinity adsorbent. *Biochem. Biophys. Acta* **393**: 115–123.

Oppenheim, J.D., M.S. Nachbar, M.J.R. Salton, and F. Aull. 1974. Purification of a hemagglutinin from *Limulus polyphemus* by affinity chromatography. *Biochem. Biophys. Res. Commun.* **58**: 1127–1134.

Ratcliffe, N.A. 1985. Invertebrate immunity: A primer for the non-specialist. *Immunol. Lett.* **10**: 253–270.

Roche, A-C., R. Schauer, and M. Monsigny. 1975. Protein-sugar interactions purification by affinity chromatography of limulin: An *N*-acyl-neuraminidyl binding protein. *FEBS Lett.* **57**: 245–249.

Stewart, J.M. and K. Channabasavaiah. 1979. Evolutionary aspects of some neuropeptides. *Fed. Proc.* **38**: 2302–2308.

Ward, C.W. 1981. Structure of the influenza virus hemagglutinin. *Curr. Top. Microbiol. Immunol.* **94**: 1–74.

Weiss, L. 1973. Neuraminidase, sialic acid and interactions. *J. Nat. Cancer Inst.* **50**: 3–19.

Yogeeswaran, G., B.S. Stein, and H. Sebastian. 1978. Altered cell surface organization of gangliosides and sialylglycoproteins of mouse metastatic melanoma variant lines selected in vivo for enhanced lung implantation. *Cancer Res.* **38**: 1336–1344.

CHAPTER 13

Agglutinins and Lectins of Crustacea
Their Composition, Synthesis, and Functions

Gianni A. Amirante
Department of Biology
University of Trieste
Trieste, Italy

13.1. INTRODUCTION

The morphofunctional and biochemical aspects of the immune system of crustaceans have not yet been as thoroughly studied as have those of in insects, although the crustacean circulatory system (Bauchau, 1981) and the morphology, ultrastructure, and cytochemistry of crustacean hemocytes are well known (Bauchau, 1975; Bauchau and Mengeot, 1978; Gupta, 1979). The arthropod hemocytes are morphologically and functionally comparable to vertebrate leucocytes; crustaceans do not possess cells comparable to vertebrate erythrocytes. The crustacean hemocytes were described for the first time by Carus (1824). However, more accurate studies of these cells date back only to 1965. These studies enabled some authors to classify crustacean hemocytes into three or four classes on the basis of their characteristic morphology; cytochemical and immunochemical reactions; behavior in wound repair, coagulation, and defense mechanisms; and their role in the synthesis of some hemolymph components, such as hemocyanin, vitellogenins, agglutinins, and lectins (Bauchau and De Brouwer, 1972, 1974; Johnston et al., 1973; Amirante and Basso, 1984; Amirante et al., 1984).

As sources of agglutinins and lectins, as well as for their roles as membrane receptors, hemocytes are of primary importance in both cellular and humoral immunity in Crustacea.

This chapter is a brief review of the roles of hemolymph and hemocytes in the immune system of the crustaceans known to date; a great deal is still unknown.

13.2. AGGLUTININS

Several of the proteins or glycoproteins known as agglutinins have been reported in Crustacea. Agglutinins, heteroagglutinins, hemagglutinins, lysins, and opsonins are some of the terms that have been used since the work of some of the earliest authors in this field (Cantacuzene, 1912, 1920, 1922; Damboviceanu, 1928), who detected in several groups of invertebrates humoral reactions similar to the immune reactions in vertebrates, including mammals. Although these terms are commonly employed even today by comparative immunologists, they are not indicative of the biochemical, immunologic, and physiologic characteristics of agglutinins, except to indicate that these substances agglutinate red blood cells, regardless of whether these agglutinins also agglutinate other cells (spermatozoa, bacteria, cells in vitro, and so on) and whether they precipitate soluble proteins. This practice perpetuates the errors and inaccuracies that were initially made in the study and classification of vertebrate immunoglobulins.

13.2.1. Nature and Characteristics of Agglutinins

Agglutinins, heteroagglutinins, hemagglutinins, lysins, opsonins, and precipitins do not exist as different classes of proteins. Rather, these terms suggest various stages of a single protein that must have two specific characteristics: (1) they must be able to recognize and react with a specific and characteristic group, present on the cell membrane or along the polypeptidic chain of a protein, and (2) they must be at least bivalent (i.e., the molecule must have at least two specific binding sites).

Researchers in the field of crustacean immunology often use *agglutinin* as a synonyms for *lectin*. The term *agglutinin* should not be used unless it can be shown or specified that the determined agglutinin reacts against an antigenic determinant not characterized by a specific group of sugars or even by one specific sugar (see Section 13.3). Only under those conditions is the distinction between *lectin* and *heteroagglutinin* meaningful; as a matter of fact, the term *heteroagglutinin* should be replaced with a more appropriate one. At any rate, this section includes all those immunoproteins for which the characteristics of the antigenic determinant against which they react is not specified.

Often, the agglutinins studied in various crustaceans are proteins composed of subunits and with high molecular weights. The complex structure of the heteroagglutinin isolated and studied by Fernandez-Moran and co-workers (1968) in the aquatic chelicerate *Limulus polyphemus* could be taken as a model. It has a molecular weight of about 400,000, and it is made up of ringlike structures with a diameter of about 10 nm. Each one has a dense and well-defined central core with a diameter of 2–4 nm and hexagonal shape. This core is surrounded by doughnut-shaped, lighter shell, which confers on the particle its typical polygonal or hexagonal shape (see also Chapters 11, 12). Shishikura and Sekiguchi (1983) also isolated and purified from another aquatic chelicerate, *Tachypleus tridentatus,* four hemagglutinins specific to mammalian eruthrocytes and made up of eight subunits with MW from 22,000 to 45,000. The specificity of these four hemagglutinins varies greatly among the red blood cells of various vertebrates (human, horses, etc.).

13.2.2. Composition and Synthesis of Crustacean Agglutinins

Acton and colleagues (1973) pointed out that the crustacean agglutinins are made up of subunits with a sedimentation constant varying from 13 to 33 S and MW of 20,000–23,000; these subunits tend to associate and dissociate easily both in vivo and in vitro and are stabilized by Ca^{2+} ions.

Ghidalia and associates (1973, 1975) studied *Macropipus puber* hemagglutinins, which react with human red blood cells of A, B, and O groups; this indicates that they are not group specific. They are present in high concentration (agglutination titer about 1,256), and they display an electrophoretic γ-globulin mobility and have a MW of about 300,000. The only report in which no heteroagglutinins were mentioned was by Cushing (1967), who did not find heteroagglutinins in the hemolymph of *Paguristes*.

It is known that in certain developmental stages of some crustaceans, the agglutinins disappear. For example, Bang (1967) demonstrated that in the hemolymph of *Maia squinado,* which normally has agglutinins reactive against a ciliate protozoan, these proteins are absent during the molting period. Table 13.1 shows agglutinins isolated from various crustacean species and the relative cell antigens. The very interesting data provided by Faglioni and coworkers (1971) point out that the hemagglutinating power of the hemolymph of many crustaceans disappears, if, instead of the usual

TABLE 13.1. HETEROAGGLUTININS DETECTED IN CRUSTACEAN SPECIES

Species	Antigens	References
Austropotamobius pallipes	Mouse RBC	Faglioni et al., 1971
Birgus latro	Human group A, B—M, N sialic acid	Cohen et al., 1974
Callinectes guanhumi	ND	Smith and Goldstein, 1971
C. sapidus	Chicken and rabbit	Pauley, 1973; McCumber and Clem, 1977
Cancer irroratus	ND	Cornick and Stewart, 1968
Cardiosoma guanhumi	Sea urchin sperm	Smith and Goldstein, 1971
Corystes cassivelaunus	Hamster RBC	Faglioni et al., 1971
Crangon crangon	Mouse RBC	Faglioni et al., 1971
Dardanus arrosor	Mouse RBC	Faglioni, et al., 1971
Eriphia verrucosa	Mouse RBC	Faglioni, et al., 1971
Eupagurus bernhardus	ND	Cantacuzéne, 1920
E. prideauxxi	ND	Cantacuzéne, 1912
Homarus americanus	Sialic acid, GalNAc	Noguchi, 1903; Acton et al., 1969; Hall and Rowlands, 1974a,b
H. gammarus	Mouse RBC	Faglioni et al., 1971
Macropipus depurator	Mouse RBC	Faglioni et al., 1971
M. puber	HRBC	Ghidalia et al., 1975
Maia squinado	Ciliate protozoan	Bang, 1967
Paguristes ulreyi	ND	Cushing, 1967
Palaemon elegans	Mouse RBC	Faglioni et al., 1971
Palaemonetes varians	Mouse RBC	Faglioni et al., 1971
Panulirus argus	Sheep RBC	Acton et al., 1973
P. interruptus	Teleostean, amphibian, reptilian, avian, and mammalian RBC	Tyler and Metz, 1945
P. vulgaris	Hamster RBC	Faglioni et al., 1971
Paromola cuvieri	Hamster RBC	Faglioni et al., 1971
Parthenope angulifrons	Mouse RBC	Faglioni et al., 1971
Scyllarus arctus	Mouse RBC	Faglioni et al., 1971
Squilla mantis	Human and mouse RBC	Faglioni et al., 1971; Amirante and Basso, 1984
Xantho poressa	Mouse RBC	Faglioni et al., 1971

Key: ND, not determined.

bleeding, a complete extract of the animal is prepared. This indicates that the activity of the agglutinins present in the hemolymph is inhibited by some component of the animal with which the hemolymph does not normally come in contact.

In contrast to the abundant data on hemagglutinins, little is known about agglutinins that agglutinate cells other than the red blood cells. Tyler and

Metz (1945) indicated that the hemolymph of *Panulirus interruptus* agglutinates spermatozoa of various invertebrate and vertebrate animals. Through specific absorptions, it is shown that the serum of this animal contains several heteroagglutinins, eight of which are group specific and two of which react with sperm of more than one group. It is necessary to emphasize that these heteroagglutinins agglutinate the red blood cells of the tested vertebrates in addition to their sperm. This indicates that they react with one or more antigenic determinants common to the red blood cells and the sperm.

More recently, Smith and Goldstein (1971), resuming the studies of Tyler, reported the presence of a native heteroagglutinin in *Cardiosoma granhui* that reacts specifically against the sperm of a single echinoderm, *Echinometre lucunter*. It is still unknown whether invertebrates in general, and crustaceans in particular, are capable of a secondary response or whether they are capable of synthesizing heteroagglutinins ex novo (through induction by a specific antigen).

Taylor and colleagues (1964) injected in the hemocoel of *Carcinus maenas* the bacteriophage Tl and after 30 min extracted samples of the hemolymph. This sampling was repeated after a second injection of the antigen and so on up to a 20-week period. They did not detect any increase in the antiphagic activity of the hemolymph, even though the phagus in the hemolymph disappeared increasingly after the second injection. They ascribed this to a cellular activity; in the light of our current knowledge of the immunology of cells and of membrane receptors, this fact is very interesting.

Pauley (1973) detected in *Callinectes sapidus* an increased concentration of hemagglutinins following injections of rabbit and chicken red blood cells. This increase, however, diminishes very rapidly and is not specific, since it was detected in specimens treated with nonspecific antigens, and even with saline solution.

Much earlier, Damboviceanu (1928) and Drilhon (1936) showed that *C. maenas* infected with *Sacculina carcini* displayed a greater protein concentration in the hemolymph. Also, Manwell and Baker (1963) noted that specimens of *Carcinus sapidus,* infected with *Loxythylacus texanus* displayed a higher level of hemolymph protein, with the highest levels displayed by two hemocyanins, a "fast" and a "slow" one. Uglow (1969 a,b,c), on the other hand, linked the significant variations in protein concentration, not to antigenic stimulations, but to various stages of molt cycle and the nutritional state of the animal. Shapiro and associates (1974) indicated an active defense mechanism in *P. interruptus*. Indeed, these authors infected specimens of the spiny lobster with *Pediococcus homari,* a pathogenic bacterium that is also pathogenic for *Homarus americanus* and *H. vulgaris*. They calculated the mean time of death (MTD) and the LD_{50} after injecting 8×10^5 bacteria/ml of hemolymph. If the animals had been previously immunized with non-virulent bacteria (attenuated *P. homari*), the LD_{50} increased 100 times. This result is of enormous interest because, apart from having an obvious com-

TABLE 13.2. CARBOHYDRATE COMPOSITION OF
HEMAGGLUTININ SUBUNIT OF *PANULIRUS* AND OF HUMAN
IMMUNOGLOBULINS

Subunit	Percent Lobster	IgG	IgA	IgM
Fucose	0.1	0.3	0.2	0.7
Hexose	3.5	1.2	4.8	6.2
GlcNAc[a]	0.8	1.1	3.8	3.3
Sialic acid[b]	0.2	0.2	1.7	2.0
Total CHO[c]	4.6	2.8	10.5	12.2

Source: From Acton et al., 1973.
[a] As free base.
[b] As *N*-acetylneuraminic acid.
[c] Total carbohydrate as sum of monosaccharides.

mercial and practical interest, it demonstrates the presence of an active immunodefense mechanism in this animal.

Finally, comparative studies were performed by various authors in an attempt to show a similarity or structural analogy between the heteroagglutinin and the light or heavy chain of immunoglobulins in vertebrates. Such studies were done with the intent of pointing out a single phyletic line shared by these two classes of proteins, thus indicating some sort of commonality between the defense mechanisms of invertebrates and vertebrates. For example, Smith and Taylor (1975), by means of immunodiffusions of hemolymph of various invertebrates against the antibodies antimammal IgA, IgM, and IgG showed that some precipitation lines were formed; this also happens with the hemolymph of the only studied species of crustacean: the pagurid hermit crab (species unknown). I believe that this information is not very significant because the presence of some antigenic determinants in common between the immunoglobulins and the hemolymph proteins does not necessarily indicate a genic or functional affinity. The information provided by Acton and coworkers (1973) is interesting in this regard. After isolating and purifying heteroagglutinins in *P. argus,* they determined the monosaccharides present by gas chromatography. When compared with data obtained from human immunoglobulins, the data so obtained enabled them to assume that those heteroagglutinins could have been the ancestral subunits from which the light chains of immunoglobulins derived (see Table 13.2).

13.3. LECTINS

The term *lectin,* first used by Boyd and Shapleigh (1954), is derived from Latin, meaning "to choose, to select." These authors assumed that many of these proteins were able to selectively recognize various groups of the human red blood cells. This, however, proved to be only partly correct, because only few lectins have this capacity; nevertheless, the term *lectin* has persisted. A certain confusion among *lectin, hemagglutinin,* and *agglutinin* in general exists.

13.3.1. Differences Between Lectins and Agglutinins

The main difference between lectins and agglutinins is that lectins, by definition, interreact with the oligosaccharides present on the cell membrane or on the surfaces of glycoproteins, which characterizes their antigenic property. On the other hand, the agglutinating property of lectins is strongly inhibited or even destroyed if they are pretreated with particular sugars. I strongly believe that more accurate studies aimed at distinguishing lectins from many "hemagglutinins" found in many crustaceans would prove the latter to be lectins. Lectins are a very hetrogeneous group of complex proteins that are extremely old phylogenetically and that probably share (at least in their primary role) only the physiologic characteristic of linking themselves in a more or less specific way with some groups of sugars.

13.3.2. General Distribution and Nonimmunologic and Immunologic Functions of Lectins

Lectins are found in algae, vegetables, and legumes (broad beans, beans, Conavalia, Genistae), and that of the castor oil plant is very specific (Stillmark, 1888; Hellin, 1891; Elfstrand, 1898). They probably transport synthesized sugars from the site of photosynthesis to the storage tissues (they are particularly abundant in seeds). Other assumed functions are defense against insects that are sensitive to particular lectins and defense against pathogenic fungi that cannot grow in the presence of lectins because they block chitin synthesis (Lis and Sharon, 1973; Reeke et al., 1974; Liener, 1976; Yeaton, 1981 a,b).

Lectins are also present in almost every animal group, including mammals (Yeaton, 1981b). They are supposed to transport sugars or to recognize the latter when they are present on cell membranes. It is also assumed that lectins are important in reproduction of some teleosteans, which possess sugars and specific lectins on their spermatozoa and eggs (Tyler, 1946; Hunt et al., 1977). Lectins also play an important role in ensuring the symbiosis between some algae (rich in sugars) and a mollusk, *Tridachna,* which possesses considerable amounts of specific lectins (Uhlenbruck and Steinhausen, 1977), and in sponge aggregation (Kuhns et al., 1974; Vaith et al., 1979) Because of the role of lectins in defense, their study is becoming more important in comparative immunology (Cohen, 1970; Burnet, 1974). Furthermore, as more lectins are isolated and studied, they are becoming more important because of their high specificity toward certain sugars on the one hand and because of the simplicity in characterizing, purifying, concentrating, and determining proteins or cells characterized by the same sugars on the other. Consequently, numerous highly specific lectins are available commercially, either in pure form or marked with fluorochroms, enzymes, or other markers (fluorescein, rhodamine, peroxidase, biotin, etc.). Among the most common and the oldest are the phytohemagglutinin (PHA), with mythogenic power; concanavalin A (conA); and ricin.

It should be mentioned that several groups of proteins display specific affinities toward membrane oligosaccharides or glycoproteins, such as lectins, vertebrate antibodies, and cartain enzymes (Franz and Ziska, 1981). Of these, the lectins, unlike other groups of oligosaccharide-specific proteins, are glycoproteins with at least two active physiologic sites; they are thermolabile, calcium dependent, precipitable in the absence of monovalent ions (e.g., by dialysis against distilled water; Amirante and Basso, 1984), and not generally as active (unlike the lipohemagglutinins) against glutaraldehyde-fixed erythrocytes as against fresh erythrocytes (Tsivion and Sharon, 1981). Although there are abundant data on plant lectins, those on animal lectins, including of Crustacea, are scarce. Probably, the most likely explanation of this is that, with few exceptions, vegetable lectins are more specific and easier to extract and purify economically in large quantities more than animal lectins, (Uhlenbruck et al., 1968; Baldo and Uhlenbruck, 1974; Pemberton, 1974; Amirante et al., 1984).

13.3.3. Lectins of Crustacea: Their Functions and Binding Sites

Relatively few lectins have so far been separated from Crustacea, and it is likely that many of the agglutinins listed in Table 13.1 are lectins (at least those that react with human erythrocytes). At any rate, as early as in 1903, Noguchi detected hemagglutinins in some crustaceans. Others were described by Cantacuzène (1912, 1918). In a review, Vasta and Marchalonis (1983) showed in a clear and unequivocal way the presence of lectins in various species of decapods. Cornick and Stewart (1973) studied a lectin present in the serum of *H. americanus* that agglutinates red blood cells of monkey, sheep, rabbit, and horse. Agglutination was inhibited by D-galactosamine, with the exception of that of the horse erythrocytes; and because of this fact, they deduced that *Homarus* has in its serum both a lectin and a heteroagglutinin that is nonspecific for oligosaccharides. Other lectins, specific for *N*-acetylgalactosamine (GalNAc) or for sialic acid, were pointed out by Hall and Rowland (1974a,b) in lobster. Cohen and associates (1974) isolated a humoral lectin from *Birgus latro*. Ravindranath and Cooper (1984) indicated that some species of crabs (*Cancer antennarius, C. productus,* and *C. anthonyi*) have lectins that agglutinate monkey, sheep, goat, human, guinea pig, and dog erythrocytes. These lectins are particularly sensitive to the *O*-acetyl group of sialic acid. The same lectins link to *Escherichia coli* if there are *O*-acetyl groups on the capsid; the lectins do not react when *O*-acetyl groups are not present. Vanderwall and coworkers (1981) isolated from lobster a sialic-acid–specific lectin, while Hartman and coworkers (1978) isolated from *H. americanus* two lectins specific to sialic acid and GalNAc, respectively.

The physiologic function of lectins in crustaceans is the same as in invertebrates: the transport of sugars and the recognition of particular mono- or oligosaccharides. They could also perform a specific role during the molting period. Cohen and colleagues (1974) pointed out that the young *Birgus* do not have lectins that are found in the adults. They assumed that lectins could be involved in the transport of sugars or in the formation of the exoskeleton, including chitin.

The only nondecapod studied to date is the stomatopod *Squilla mantis,* from which we (Amirante and Basso, 1984) isolated and purified two highly specific lectins against human red blood cells of groups O and A, respectively. These lectins, called anti-H and anti-A, respectively, are specific against GalNAc and fucose: the sugars that characterize the antigenic determinant H and A. In the presence of strong concentration of specific sugars, they lose their affinity and specificity to O and A determinants. Both are complex proteins, with MW of 192,000, and are composed of three subunits (Figs. 13.1 and 13.6). The assembling of subunits and probably the efficiency of the active sites are ensured by the presence of Ca^{2+} and Mg^{2+} ions. And, indeed, when treated with EDTA, they lose their agglutinating power. They are thermolabile and lose their binding power at nonphysiologic pH. We also studied their biophysical characteristics using monoclonal antibodies and demonstrated that at least one subunit is involved in the physiologic site. Furthermore, the active sites are, from an antigenic point of view, different in the two lectins. Their characteristics are summarized in Table 13.3. We prepared a specific monoclonal antibody for the active site of anti-H lectin and found that, after treating the lectin with the antibody, its agglutinating property decreased but did not disappear. This result supports the assumption that this lectin, like the wheat germ agglutinin (WGA; and possibly lectins in general), has a binding site composed of various binding subsites that react with an oligosaccharide present on the erythrocyte membrane and formed by monosaccharides assembled in various ways according to the type of binding with various amino acids in different species (Brogren and Bisati, 1981). Moreover the *Squilla* lectin appears to be more active against this complex saccharidic structure, given the structure of its binding site, than against the isolated monosaccharide (Fig. 13.3). Nevertheless, it is proved that frequently invertebrate lectins have a stronger affinity for glycoprotein sugars than for monosaccharides (Shimizu et al., 1977; Yeaton, 1980). In fact, studies by Bretting and associates (1978), Hardy and associates (1977), and Baldo and associates (1978) show that the saccharide sequence proximal to the terminal sugar is also important in determining the binding affinity. This could explain the multiplicity of reactions that one lectin presents against various cells and protein components. Indeed, it is necessary to emphasize that there are few sugars that enter in cell glyco-

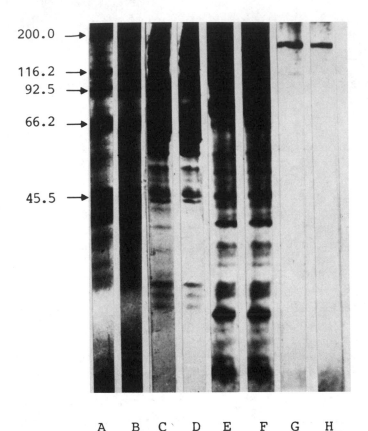

200.0 →
116.2 →
92.5 →
66.2 →
45.5 →

A   B   C   D   E   F   G   H

Figure 13.1. Electrophoresis in SDS polyachrilamide. (A,B) Standard molecular weights ($\times 10^3$). (C,D) Native hemolymph (10 and 5μl), (E,F) Precipitate obtained by dialysis of hemolymph (20 μl). As a result of this process, proteins present at high concentration in the native hemolymph (as well as hemocyanin and vitellogenins) are eliminated and other proteins usually present at low concentrations thus become evident. (G) Purified anti-H lectin. (H) Purified anti-A lectin.

sylation or that have an important role in the characterization of membrane antigenic determinants. The sugars that most frequently inhibit crustacean lectins are GalNAc, GlcNAc, and neuraminic acid (Cornick and Stewart, 1973; Hall and Rowlands, 1974a,b; Shimizu et al., 1977, 1979). As Vasta and Cohen (1984) indicated in *Birgus,* there are lectins that are specific against sialic-acid, which is largely diffused and present on membranes of numerous heterologous cells and bacteria; it is also present on many other proteins in *Birgus* (proteins of the digestive gland) and probably on the membranes of cells in general. We (Amirante et al., 1984) supported this fact in *Squilla mantis.* Indeed, in the mantis shrimp, two lectins, specific to fucose and GalNAc, are found. Furthermore, these two sugars of human erythrocytes

TABLE 13.3. CHARACTERISTICS OF ANTI-A AND ANTI-H PURIFIED LECTINS AND OF NATIVE HEMOLYMPH OF *SQUILLA MANTIS*

Purified and Hemolymph Lectins	Agglutinating Titers[a]	
	HRBC (group O)[b]	HRBC (group A)
Native hemolymph	128	128
Hemolymph + O HRBC	0	64
Hemolymph + A HRBC	64	0
Hemolymph + GalNAc	64	0
Hemolymph + fucose	0	64
Hemolymph at 60°C	0	0
Hemolymph + EDTA	0	0
Supernatant after dialysis[c]	4	4
Precipitate by dialysis[c]	64	64
Lectin anti-H	64	0
Lectin anti-A	0	64

[a] The titer is expressed as the reciprocal of the highest dilution showing positive agglutination.
[b] HRBC, human red blood cell.
[c] Supernatant and precipitate of hemolymph after dialysis against distilled water

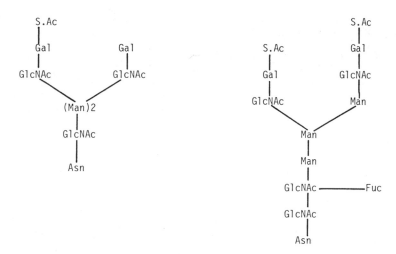

ERYTHROCYTE   PORCINE THYROGLOBULIN

Figure 13.2. Two different configurations of the same sugars in two different antigenic determinants. (From Ochoa et al., 1981.)

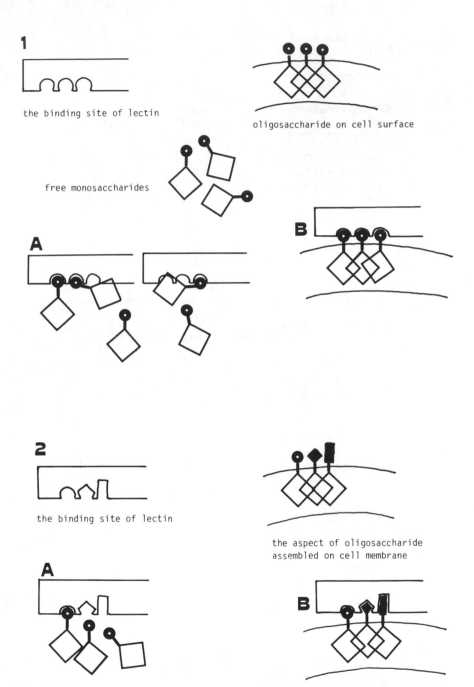

1

the binding site of lectin

oligosaccharide on cell surface

free monosaccharides

A

B

2

the binding site of lectin

the aspect of oligosaccharide
assembled on cell membrane

A

B

Figure 13.3. Two models to explain the lower affinity of lectins to free sugars (monosaccharides) than to assembled oligosaccharides on cell membrane or glycoprotein surface. In 1 the subsites are identical; in 2 they are different. (A) Some subsites of the lectin binding site are inhibited by linked molecules of sugar or are incompatible with the sugar molecules. (B) The three subsites of lectin link the oligosaccharide on the cell membrane.

(and probably also of other erythrocytes) are also present on the cell membranes of many bacteria, as well as on the proteins of the crustacean itself. The multiplicity of these reactions and the various modes of recognition could be explained only by taking into account the fact that, on one hand, the binding site is complex and made up of several interdependent subsites and, on the other, these sugars acquire different configurations on different cellular surfaces or on different proteins. This suggests a close link between the structure of the binding site of lectins and the mechanisms of cell glycosylation. When one considers the defensive role of certain types of hemocytes in the immune system of arthropods, a close link between lectins (or heteragglutinins) and hemocytes becomes apparent. Because of their opsonic property, lectins (Tyson and Jenklin, 1973; Renwrantz and Mohr, 1978) participate in phagocytosis by hemocytes once linked to the antigen; on the other hand, the hemocytes recognize corpuscles, cells, or heterologous tissue fragments and form around them nodules or capsules that are eventually destroyed, eliminated, or isolated.

13.3.4. Synthesis of Lectins

There is very little information on cells responsible for the synthesis of lectins. Amirante (1976) showed for the first time that in a group of insects plasmatocytes (PLs) are responsible for the synthesis of some lectins. Later, he (Amirante, 1982) showed in the stomatopod *S. mantis* the primary role of hemocytes in the synthesis of the two lectins that he isolated and purified (see Section 13.3.3). The anti-H and anti-A lectins were isolated by immunoprecipitation with specific monoclonal antibodies and protein A; and Amirante pointed out the presence of these lectins in the cytoplasm of hyaline cells (PLs; see Fig. 13.7A and on surface of granulocytes (GRs) by means of immunofluorescence or immunoperoxidase, again with monoclonal antibodies (Fig. 13.4). Smears of human erythrocytes of O and A groups were treated with a medium of hemocytes grown in vitro in the presence of ^{3}H methionine. Autoradiographs of these smears gave positive results (Fig. 13.5); autoradiographs of the same smears, but previously treated with fucose and GalNAc, respectively, gave negative results.

This evidence proves that the hyaline cells (PLs) and/or GRs synthesize and secrete proteins that link specifically with determinants of human O and A groups. The presence of these tritiated lectins were also confirmed by means of fluorography of the medium of hemocytes after purification by monoclonal antibodies and protein A and migration in SDS polyacrylamide gel electrophoresis, both under reducing and nonreducing conditions. The fluorography yielded the same results as obtained on the lectins isolated from native hemolymph (Fig. 13.6). All this leads one to assume that the three subunits of *Squilla* lectins are synthesized and assembled in the form of physiologically active protein by hemocytes before being secreted into the hemolymph (see also Chapter 1).

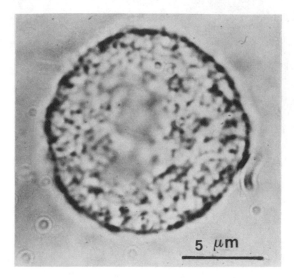

Figure 13.4. Positive immunoperoxidase reaction on the membrane of a *Squilla* granulocyte, showing the presence of lectins on the cell surface.

13.3.5. Lectins as Membrane Receptors and Their Immunocytoadherence

The ability of lectins to recognize and characterize various membrane receptors in invertebrate and vertebrate cells is well known (Ravindranath and Cooper, 1984) and is supported by many encouraging results. Lectins, however, are also found on the membrane of certain cells; these lectins could

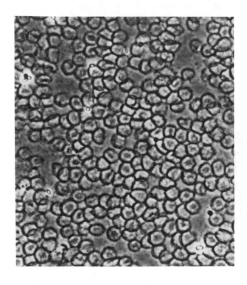

Figure 13.5. Autoradiograph of human red blood cells (group O) treated with tritiated lectins synthesized by hyaline cells (PLs).

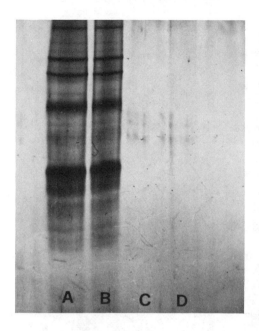

Figure 13.6. Fluorograph of urea SDS-PAGE of (A,B) tritiated proteins synthesized and secreted into a culture medium by hyaline cells (PLs) and (C,D) the three subunits of lectins after immunoprecipitation with monoclonal antibodies.

have an essential role in some cellular immunologic reactions or even act as membrane receptors. In fact, we (Amirante et al., 1984) have demonstrated that on the membrane of the GRs of *Squilla,* lectins specific for fucose and GalNAc are present (Fig. 13.4). These lectins were stressed with monoclonal antibodies and immunoperoxidase. These same lectins are linked to the membrane in a fluid manner. Indeed, when GRs fixed with glutaraldehyde are treated with monoclonal antilectin and subsequently with antimouse Ig labeled with FITC, a diffuse fluorescence is observed all over the membrane (Fig. 13.7B). If these marked antibodies are made to react with fresh GRs, the fluorescence displays the typical aspect of capping (Fig. 13.7C,D), with an appearance very similar to that on mammalian lymphocytes.

Finally, *Squilla* hemocyte cultures treated with human erythrocytes of O and A groups form specific rosettes, demonstrating the property of the GRs immunocytoadherence reaction (Fig. 13.8). The formation of rosettes is inhibited if the GR cultures have been previously treated with fucose or GalNAc. Furthermore, the sugars, present in high concentration, detach the red blood cells from preformed rosettes.

13.4. SUMMARY

Crustaceans, as well as all other arthropods, have an open circulatory system that strictly correlated with immunologic mechanisms. Their blood (hemo-

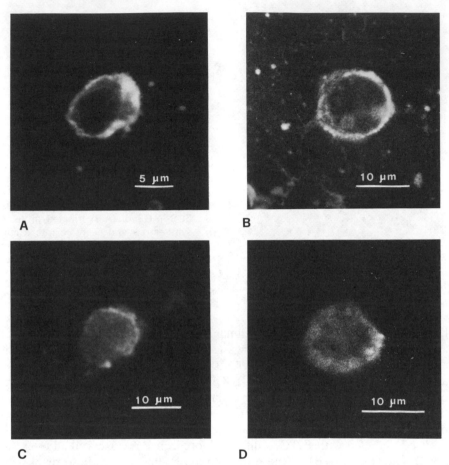

Figure 13.7. Binding of fluorescein-coupled monoclonal antibody antilectin to (A) a hyaline cell (PL), (B) a glutaraldehyde-fixed granulocyte, and (C,D) fresh granulocytes (GRs). Note the capping formation in C and D.

lymph) contains three or four main classes of hemocytes, which are involved in many physiologic activities, including synthesis of some hemolymph components, repair, coagulation, and defense mechanisms. The hemolymph also contains some proteins called agglutinins. There is, however, some controversy about their classification. In fact, they are called agglutinins, heteroagglutinins, hemagglutinins, lysins, precipitins, and opsonins. These various terms actually suggest different stages of a single protein that must recognize and react with an antigenic determinant on the cell membrane or on the polypeptidic chain of a protein and must have at least two specific binding sites. Lectins are a particular class of agglutinins. Lectins are glycoproteins that are more or less specific to some groups of sugars. They are

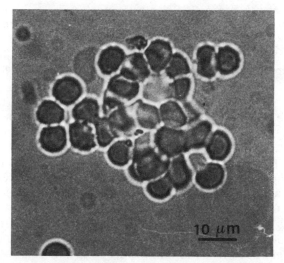

Figure 13.8. Immunocytoadherence between *Squilla* granulocytes (GRs) and fresh HRBC.

present both in vegetables and animals; in Crustacea, some lectins have been studied, isolated, and purified, and it is very likely that many crustacean agglutinins are lectins (at least those that react with HRBC). From the stomatopod. *S. mantis,* two lectins have been isolated and purified; they are specific for O and A human erythrocytes. Using monoclonal antibodies, research on *Squilla* lectins has demonstrated that their binding sites are very complex and probably made up of several interdependent subsites. The *Squilla* lectins are synthesized by hemocytes (plasmatocytes and/or granulocytes) and are also present on the granulocyte membrane. In fact, *Squilla* hemocytes treated with HRBC form rosettes, demonstrating the GR's immunocytoadherence reaction. All these facts lead one to suppose a strict correlation among hemocytes, agglutinins, lectins, and humoral and cellular immunologic mechanisms in crustaceans.

REFERENCES

Acton, R.T., P.F. Weinheimer, and W. Niedermeier. 1973. The carbohydrate composition of invertebrate hemagglutinin subunits isolated from the lobster *Panulirus argus* and the oyster *Crassostrea virginica. Comp. Biochem. Physiol.* **44:** 185–189.

Amirante, G.A. 1976. Production of heteroagglutinins in haemocytes of *Leucophaea maderae. Experientia* **32:** 526–528.

Amirante, G.A. 1982. Caratterizzazione di due lectine in *Squilla mantis* L. (Crustacea Stomatopoda) mediante anticorpi monoclonali. *Boll. Zool.* **49:** 4.

Amirante, G.A. and V. Basso. 1984. Synthesis of lectins in *Squilla mantis* L. (Crustacea Stomatopoda) using monoclonal antibodies. *Dev. Comp. Immunol.* **8:** 721–726.

Amirante, G.A. and F. G Mazzalai. 1978. Synthesis and localization of hemagglutinins in haemocytes of the cockroach *Leucophaea maderae*. *Dev. Comp. Immunol.* **2:** 735–740.

Amirante, G.A., G. Valle, and V. Basso. 1984. Synthesis of lectins in *Squilla mantis* L. haemocytes and the possible role of cell glycosylation in the recognition of "self" and "nonself," p. 759. *Seventeenth International Congress on Entomology,* Hamburg, West Germany.

Anderson, R.S. 1975. Phagocitosis by invertebrate cells in vitro: Biochemical events and other characteristics compared with vertebrate phagocytic systems, pp. 153–180. In K. Maramorosh and R.E. Shope (eds.) *Invertebrate Immunity*. Academic Press, New York.

Baldo, B.A., W.H. Sawier, R.V. Stick, and G. Uhlenbruck. 1978. Purification and characterization of a galactan-reactive agglutinin from the clam *Tridacna maxima* and a study of its combining site. *Biochem. J.* **175:** 467–478.

Baldo, B.A. and G. Uhlenbruck. 1974. Studies on the agglutinin specificities and blood group O(H)-like activities in extracts from the molluscs *Pomacea paludosa* and *Pomacea urceus*. *Vox Sang.* **27:** 67–80.

Bang, F.B. 1967. Serological responses among invertebrates other than insects. *Fed. Proc.* **26:** 1680–1684.

Bauchau, A.G. 1981. Crustaceans, pp. 385–420. In N.A. Ratcliffe and A.F. Rowley (eds.) *Invertebrate Blood Cells*. vol. 2. Academic Press, New York.

Bauchau, A.G. and B. De Brouwer. 1972. Ultrastructure des hémocytes d'*Eriocheir sinensis,* Crustacé Décapode Bracyoure. *J. Microsc.* (Paris) **15:** 171–180.

Bauchau, A.G. and M. De Brouwer. 1974. Étude ultrastructurale de la coagulation de l'hemolymphe chez Crustacés. *J. Microsc.* (Paris) **19:** 37–46.

Bauchau, A.G., M. De Brouwer, E. Passelacq-Guerrin, and J.C. Mengeot. 1975. Étude cytochimique des hémocytes des crutacés décapodes bracyoures. *Histochemistry* **45:** 101–113.

Bauchau, A.G. and J.C. Mengeot. 1978. Structure et function des hémocytes chez les Crustacés. *Arch. Zool. Exp. Gen.* **119:** 227–248.

Boyd, W.C. and E. Shapleigh. 1954. Specific precipitating action of plant agglutinins (lectins). *Science* (Washington, DC) **119:** 479.

Bretting, H., H. Kalthoff, and S. Fehr. 1978. Studies on the relationships between lectins from *Axinella polyploides* agglutinating bacteria and human erythrocytes. *J. Invertebr. Pathol.* **32:** 151–158.

Brogren, C.H. and S. Bisati. 1981. Lectin binding properties of chicken major histocompatibility complex coded lymphocyte plasma membrane glycoproteins, pp. 375–385. In T.C. Bøg-Hansen (ed.) *Lectins: Biology, Biochemistry, Clinical Biochemistry*. Walter de Gruyter, Berlin.

Burnet, F.M. 1974. Invertebrate precursor to immune responses. *Contemp. Top. Immunobiol.* **4:** 13–23.

Cantacuzène, J. 1912. Sur certain anticorps naturales observés chez *Eupagurus prideauxii*. *C. R. Soc. Biol.* (Paris) **73:** 663–664.

Cantacuzène, J. 1920. Formation d'hémolysin dans le serum de *Maia squinado* inoculés avec des hématies de mammiféres existence dans ce serum d'une substance antagoniste qui empedre ou retarde hémolyse. *C. R. Soc. Biol.* (Paris) **83:** 1512–1514.

Cantacuzène, J. 1922. Sur la role agglutinant des urnes chez *Sipunculus nudus. C. R. Soc. Biol.* (Paris) **87:** 259.

Carus, D.C.G. 1824. *Von den äuszeren Lebensnedingungen der heisz und kaltblutigen.* Thiere, Leipzig, East Germany.

Clarke, A.E. and R.B. Kox. 1979. Plants and immunity. *Dev. Comp. Immunol.* **3:** 571–589.

Cohen, E. 1970. A review of the nature and significance of hemagglutinins of selected invertebrates, p. 87. In C.P. Bianchi and R. Hilf (eds.) *Protein Metabolism and Biological Function.* Rutgers University Press, New Brunswick, NJ.

Cohen, E., M. Rozenberg, and E.J. Massaro. 1974. Agglutinins of *Limulus polyphemus* (horseshoe crab) and *Birgus latro* (coconut crab). *Ann. N.Y. Acad. Sci.* **234:** 28–33.

Cornick, J.W. and J.E. Stewart. 1973. Partial characterization of a natural agglutinin in the hemolymph of the lobster *Homarus americanus. J. Invertebr. Pathol.* **21:** 255–262.

Cushing, J.E. 1967. Invertebrates: Immunology and evolution. *Fed. Proc.* **26:** 1666–1670.

Damboviceanu, A. 1928. Variations des substances protéiques coagulables par le chaleur dans le plasma des *Carcinus maenas* normaux et sacculinés. *C. R. Soc. Biol.* (Paris) **98:** 1633–1635.

Drilhon, A. 1936. Quelques constantes chimiques et physiochimiques du milieu intérieur du crabe sacculiné (*Carcinus maenas*) *C. R. Séance Acad. Sci.* (III) **202:** 981–982.

Drilhon, A. and E.A. Pora. 1936. Sacculiné et crabe:Étude chimique et physiochimique. *Trav. Stn. Biol. Roscoft.* **14:** 111–120.

Elfstrand, M. 1898. *Über blutkorperchenagglutinierende Eiweisse Görbersdorfer Veröffentlichungen.* R. Kobert und F. Enke, Stuttgart, West Germany.

Faglioni, M., V. Parisi, and G. Salsi. 1971. Some experimental observations on the heteragglutinins of decapod crustaceans. *Boll. Zool.* **38:** 159–163.

Fernàndez-Moran, H., J.J. Marchalonis, and G.M. Edelman. 1968. Electron microscopy of a hemagglutinin from *Limulus polyphemus. J. Mol. Biol.* **32:** 467–469.

Fontaine, C.T. and D.V. Lightner. 1974. Observations on the phagocytosis and elimination of carmine particles injected into the abdominal musculature of white shrimp *Penaeus setiferus. J. Invertebr. Pathol.* **24:** 141–148.

Franz, H. and P. Ziska. 1981. Affinitins: Combining sites containing proteins, pp. 17–21. In T.C. Bøg-Hansen (ed.) *Lectins: Biology, Biochemistry, Clinical Biochemistry.* Walter de Gruyter, Berlin.

Ghidalia, W., P. Lambin, and J.M. Fine. 1975. Electrophoretic and immunologic studies of a hemagglutinin in the hemolymph of the decapod *Macropipus puber. J. Invertebr. Pathol.* **25:** 151–157.

Ghidalia, W., M. Vicompte, and K. Bien Tan. 1973. Immunoelectrophoretic analysis of *Macropipus puber* male serum. *Comp. Biochem. Physiol.* **49**: 715–724.

Gupta, A.P. 1979. Arthropod hemocytes and phylogeny, pp. 669–735. In A.P. Gupta (ed.) *Arthropod Phylogeny*. Van Nostrand Reinhold, New York.

Hall, J.L. and D.T. Rowlands. 1974a. Heterogeneity of lobster agglutinins. 1. Purification and physicochemical characterization. *Biochemistry* **13**: 821–827.

Hall, J.L. and D.T. Rowlands. 1974b. Heterogeneity of lobster agglutinins. 2. Specificity of agglutinin-erythrocyte binding. *Biochemistry* **13**: 828–832.

Hardy, S.W., T.C. Fletcher, and L.R. Gerrie. 1976. Factors in haemolymph of the mussel *Mytilus edulis* L. of possible significance as defense mechanisms. *Trans. Biochem. Soc.* **4**: 473–80.

Hardy, S.W., P.T. Grant, and T.C. Fletcher. 1977. A hemagglutinin in the tissue fluid of the pacific oyster *Crassostrea gigas,* with specificity for sialic acid residues in glycoproteins. *Experientia* **33**: 767–768.

Hartman, A.L., P.A. Campbell, and C.A. Abel. 1978. An improved method for the isolation of lobster lectins. *Dev. Comp. Immunol.* **2**: 617–625.

Hellin, H. 1891. Der giftige Eiweisskörper Abrin und dessen wirtung das Blut. Doctoral Dissertation, Doprat.

Hunt, R.C., P.J. Letts, L. Pinteric, and H. Schachter. 1977. Lectin-mediated agglutination properties of spermatocytes, spermatids and spermatozoa isolated from rat testis. *Fed. Proc.* **36**: 709.

Johnston, M.A., H.Y. Elder, and P.S. Davies. 1973. Cytology of *Carcinus* hemocytes and their functions in carbohydrate metabolism. *Comp. Biochem. Physiol.* **46**: 569–581.

Komano, H. and S. Natori. 1985. Participation of *Sarcophaga peregrina* humoral lectin in the lysis of sheep red blood cells injected into the abdominal cavity of larvae. *Dev. Comp. Immunol.* **9**: 31–40.

Kuhns, W.J., G. Weinbraun, R. Turner, and M.H. Barger. 1974. Sponge aggregation: A model for studies on cell-cell interactions. *Ann. N.Y. Acad. Sci.* **234**: 58–64.

Liener, I.E. 1976. Phytohemagglutinins (phytolectins). *Annu. Rev. Plant Physiol.* **27**: 291–319.

Lis, H. and N. Sharon. 1973. The biochemistry of plant lectins. *Annu. Rev. Biochem.* **42**: 541–574.

Manwell, C. and C.M.A. Baker. 1963. Starch gel electrophoresis of sera from some marine arthropods: Studies on the heterogeneity of hemocyanin and on a "ceruloplasmin-like protein." *Comp. Biochem. Physiol.* **8**: 193–208.

McCumber, L.J. and L.W. Clem. 1977. Recognition of virus and xenogeneic proteins by the blue crab, *Callinectes sapidus. Dev. Comp. Immunol.* **1**: 5–14.

Mullainadhan, P., M.H. Ravindranath, R.K. Wright, and E.L. Cooper. 1984. Crustacean defense strategies. 1. Molecular weight dependent clearance of dyes in the mud crab *Scylla serrata. Dev. Comp. Immunol.* **8**: 41–50.

Noguchi, H. 1903. On the multiplicity of the serum haemagglutinins of cold-blooded animals. *Z. Bakteriol. Parasitenk. Abt.* **34**: 286–288.

Ochoa, J.L., A. Sierra, and F. Cordoba. 1981. On the specificity and hydrophobicity of lectins, pp. 73–80. In T.C. Bøg-Hansen (ed.) *Lectins: Biology, Biochemistry, Clinical Biochemistry*. Walter de Gruyter, Berlin.

Pauley, G.B. 1973. An attempt to immunize the blue crab, *Callinectes sapidus,* with vertebrate red blood cells. *Experientia* **29:** 210–211.

Pemberton, R.T. 1974. Anti A and anti B of gastropod origin. *Ann. N.Y. Acad. Sci.* **234:** 95–121.

Prowse, R.H. and N.N. Tait. 1969. *In vitro* phagocytosis by amoebocytes from the haemolymph of *Helix aspersa:* Evidence for opsonic factor(s) in serum. *Immunology* **17:** 437–443.

Ravindranath, M.H. and E.L. Cooper. 1984. Crab lectins: Receptor specificity and biomedical applications, pp. 83–96. In E. Cohen (ed.) *Recognition Proteins, Receptors and Probes: Invertebrates.* Alan R. Liss, New York.

Reeke, G.N., J.W. Becker, B.A. Cunningham, G.R. Gunther, J.L. Weng, and G.M. Edelman. 1974. Relationships between the structure and activities of conA. *Ann. N.Y. Acad. Sci.* **234:** 369.

Renwrantz, L. and W. Mohr. 1978. Opsonizing effect of serum and albumin gland extracts on the elimination of human erythrocytes from the circulation of *Helix pomatia. J. Invertebr. Pathol.* **31:** 164–170.

Salt, G. 1970. *The Cellular Defence Reactions of Insects.* Cambridge University Press, Cambridge.

Shapiro, H.C., J.H. Mathewson, J.F. Steenbergen, S. Kellog, G. Nierengarten, C. Ingram, and H. Rabin. 1974. Gaffkemia in the California spiny lobster, *Panulirus interruptus:* Infection and immunization. *Aquaculture* **3:** 403–408.

Shimizu, S., M. Ito, and M. Niwa. 1977. Lectins in the haemolymph of Japanese horseshoe crab *Tachypleus tridentatus. Biochim. Biophys. Acta* **500:** 71–79.

Shimizu, S., M. Ito, and M. Niwa. 1979. Purification and properties of lectins from Japanese horseshoe crab *Tachypleus tridentatus. Prog. Clin. Biol. Res.* **29:** 625–639.

Shishikura, F. and K. Sekiguchi. 1983. Agglutinins in the horseshoe crab hemolymph: Purification of a potent agglutinin of horse erythrocytes from the hemolymph of *Tachypleus tridentatus,* the Japanese horseshoe crab. *J. Biochem.* **93:** 1539–1546.

Sloan, B., C. Yocura, and L.W. Clem. 1975. Recognition of self from non self in Crustaceans. *Nature* (London) **258:** 521.

Smith, A.C. and R.A. Goldstein. 1971. "Natural" agglutinins against sea urchin sperm in the hemolymph of the crab *Cardiosoma guanhumi. Mar. Biol.* **8:** 6.

Smith, A.C. and R.L. Taylor. 1975. Immunoglobulins or similar substances in invertebrate body fluids. A preliminary study. *Aquaculture* **6:** 295–299.

Stillmark, H. 1888. Über Ricin, ein giftiges Ferment aus den Sauren von *Ricinus communis* L. und einigen anderen Euphorbiaceen. Doctoral dissertation, Doprat.

Taylor, A.E., G. Taylor, and P. Collard. 1964. Secondary immune response to bacteriophage Tl in the shore crab, *Carcinus maenas. Nature* (London) **203:** 775.

Tsivion, Y. and N. Sharon. 1981. Lipid-mediated hemagglutination *Biochim. Biophys. Acta* **642:** 336–344.

Tyler, A. 1946. Natural heteroagglutinins in the body-fluids and seminal fluids of various invertebrates. *Biol. Bull.* (Woods Hole) **90:** 213–219.

Tyler, A. and C.B. Metz. 1945. Natural heteroagglutinins in the serum of the spiny lobster, *Panulirus interruptus. J. Exp. Zool.* **100:** 387–406.

Tyson, C.J. and C.R. Jenkin. 1973. The importance of opsonic factors in the removal of bacteria from the circulation of the crayfish (*Parachnap bicarinata*). *Aust. J. Exp. Biol. Med. Sci.* **51:** 609–688.

Uglow, R.F. 1969a. Haemolymph protein concentration in portunid crabs. 1. Studies on adult *Carcinus maenas. Comp. Biochem. Physiol.* **30:** 1083–1090.

Uglow, R.F. 1969b. Haemolymph protein concentration in portunid crabs. 2. The effects of imposed fasting on *Carcinus maenas. Comp. Biochem. Physiol.* **31:** 959–967.

Uglow, R.F. 1969c. Haemolymph protein concentration in portunid crabs. 3. The effect of Sacculina. *Comp. Biochem. Physiol.* **31:** 969–973.

Uhlenbruck, G., H. Otten, U. Rehfeldt, U. Reifenberg, and O. Prokop. 1968. Enzymatischen Abbau der Erythrozytenmembran: Topochemie verschiedner A_{hel}-Rezeptoren sowie Nachweis "inkompletter" Antikorper durch Subtilisin A-Behandlung. *Z. Immunol. Forsch.* **134:** 476–491.

Uhlenbruck, G. and G. Steinhausen. 1977. Tridacnins: Symbiosis or defense-purpose? *Dev. Comp. Immunol.* **1:** 183–192.

Vaith, P., G. Uhlenbruck, W.E.G. Muller, and G. Holz. 1979. Sponge aggregation factor and sponge hemagglutinin: Possible relationships between two different molecules. *Dev. Comp. Immunol.* **3:** 399–416.

VanderWall, J., P.A. Campbell, and C.A. Abel. 1981. Isolation of a sialic acid–specific lectin (LAgl) by affinity chromatography on sepharose colominic acid beads. *Dev. Comp. Immunol.* **5:** 679–684.

Vasta, G.R. and E. Cohen. 1984. Carbohydrate specificities of *Birgus latro* serum lectins. *Dev. Comp. Immunol.* **8:** 197–202.

Vasta, G.R. and J.J. Marchalonis. 1983. Humoral recognition factors in the Arthropoda: The specificity of Chelicerata serum lectins. *Am. Zool.* **25:** 157–171.

Yeaton, R.W. 1980. Lectins of a North American silkmoth (*Hyalophora cecropia*): Their molecular characterization and developmental biology. Doctoral dissertation, University of Pennsylvania, Philadelphia.

Yeaton, R.W. 1981a. Invertebrate lectins. 1. Occurrence. *Dev. Comp. Immunol.* **5:** 391–402.

Yeaton, R.W. 1981b. Invertebrate lectins. 2. Diversity of specificity, biological synthesis and functions in recognition. *Dev. Comp. Immunol.* **5:** 535–545.

CHAPTER 14

Humoral Recognition Factors in Insects, with Particular Reference to Agglutinins and the Prophenoloxidase System

Andrew F. Rowley
Norman A. Ratcliffe
Catherine M. Leonard
Elaine H. Richards
Biomedical and Physiological Research Group
School of Biological Sciences
University College of Swansea
Swansea, Wales

Lothar Renwrantz
Zoological Institute and Zoological Museum
University of Hamburg
Hamburg, West Germany

14.1. INTRODUCTION

Insects represent one of the most successful groups of animals. This success is often attributed to such factors as a high reproductive potential and a tough, impermeable exoskeleton, but their resistance to parasite invasion and infection has no doubt been of paramount importance too.

Central to the immune system of all organisms is the ability to distinguish self from nonself; without this, no group could survive and reproduce before succumbing to an array of invading parasites and pathogens. Hence, evolutionarily successful groups, such as insects, would be expected to have well-developed immunosurveillance and immunorecognition mechanisms capable of triggering both cellular and humoral defenses. Although we now understand some of the processes involved in these defenses (see reviews by Whitcomb et al., 1974; Ratcliffe and Rowley, 1979; Rowley and Ratcliffe, 1981; Götz and Boman, 1985; Ratcliffe and Rowley, 1986), the controlling mechanisms are still unclear.

Recognition processes in "professional" phagocytes, such as vertebrate macrophages and granulocytes, are aided by immunoglobulins and complement fragments that recognize and coat foreign material, leading to more efficient ingestion and killing. This type of phagocytosis has been termed *immunological phagocytosis* (Rabinovitch, 1968; Rabinovitch and De-Stefano, 1970) and can be compared with *nonimmunologic* or *nonspecific phagocytosis,* in which direct interaction occurs between receptors on the phagocyte and the nonself material (Weir and Ögmundsdóttir, 1977; Sharon, 1984). A specialized form of nonspecific phagocytosis has recently been shown to involve carbohydrate-lectin interactions, with the lectins present on the phagocyte and the foreign particle and/or free in the body fluids, where they act as opsonins (Sharon, 1984). This method of recognition is believed to be more primitive than that involving immunoglobulins, as it has been reported in organisms from protozoans to mammals (Tripp, 1966; Tripp and Kent, 1967; Brown et al., 1975; Anderson and Good, 1976; Weir, 1980; Renwrantz, 1983; Renwrantz and Stahmer, 1983). For example, Renwrantz and Stahmer (1983) showed in the bivalve *Mytilus edulis* that an agglutinin (lectinlike molecules that bind the carbohydrate residues found on the outer walls or membranes of foreign material) naturally present in the hemolymph acted as an opsonin by facilitating the attachment of the test particles to the phagocytic hemocytes. Similar molecules have also been reported in the coelomic fluid and hemolymph of many other invertebrates.

More recently, other substances that are responsible for nonself recognition have been found in invertebrates, including components of the prophenoloxidase-activating system, which is responsible for melanin formation, parasite destruction, coagulation, and opsonization in crustaceans (Söderhäll, 1981, 1982; Söderhäll and Ajaxon, 1982; Smith and Söderhäll, 1983a; Smith et al., 1984) and probably also in insects (Ratcliffe et al., 1984; Leonard et al., 1985a,b; see also Chapters 8 and 9).

This chapter reviews some of the research carried out on these two recognition systems in insects and suggests possible avenues for future work aimed at producing a better understanding of immune recognition in this important group of animals.

14.2. AGGLUTININS: SYNTHESIS, STIMULATION, DISTRIBUTION, AND PHYSICOCHEMICAL PROPERTIES

Agglutinins are widely distributed in the body fluids of both invertebrates and vertebrates (Gold and Balding, 1975; Ey and Jenkin, 1982) and react with a variety of foreign particles, such as bacteria, protozoans, spermatozoa, vertebrate lymphocytes, and erythrocytes (Ratcliffe et al., 1985). Despite a wealth of information on crustacean, molluscan, and tunicate agglutinins, surprisingly little is known about these molecules in insects. The following account gives a brief summary of the work to date and stresses the areas that still need further consideration.

14.2.1. Synthesis of Agglutinins

In insects, agglutinins have been reported to be synthesized in both hemocytes and fat body cells (Amirante, 1976; Amirante and Mazzalai, 1978; Yeaton, 1980, 1981a,b; Stein and Cooper, 1981; Komano et al., 1983). In *Hyalophora cecropia,* Yeaton (1980, 1983) used polyvalent fluorescein-labeled antisera to hemagglutinins to demonstrate the presence of these molecules in the cytoplasm of the granular hemocytes (granulocytes) and on the surface of the plasmatocytes. Hemocyte cultures of *H. cecropia* were also shown to synthesize and release agglutinins into the culture medium, whereas fat body cultures failed to do so (Yeaton, 1980). In *Sarcophaga peregrina,* however, Komano and coworkers (1983) suggested that although the agglutinins could be found on the surfaces of some hemocytes, they were synthesized by the fat body and released into the hemolymph (Figure 14.1). These workers deduced that, since the agglutinin was released into the supernatant following incubation of hemocytes with galactose or lactose (sugars involved in binding with the agglutinin) without appreciable cell lysis, the lectin is located on the cell surface and not intracellularly. The release of agglutinin might, however, have occurred by exocytosis from the hemocytes without any appreciable cell lysis. One difference that could account for these conflicting reports on the site of agglutinin synthesis is that *S. peregrina,* like many other dipterans, may lack cells equivalent to the granular cells and coagulocytes of other insect orders (Zachary and Hoffmann, 1973; Price and Ratcliffe, 1974), and consequently the role of these cells in agglutinin production would be taken over by the fat body cells (for further details see Section 14.3 and Chapters 1 and 8).

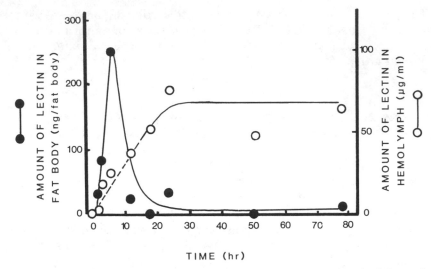

Figure 14.1. Changes in the amount of lectin in fat body and hemolymph of *Sarcophaga peregrina* as determined by radioimmunoassay following body wall injury. (Redrawn with permission from Komano et al., 1983.)

14.2.2. Stimulation of Agglutinin Activity

Most studies of insect agglutinins have observed the levels of these substances in naive animals. As the properties of agglutinins are often compared with those of immunoglobulins, some workers have attempted to induce the production of new agglutinins or stimulate the level of existing molecules. The results of these studies are often equivocal due to variation in the response from one animal to another or one immunogen to another (Table 14.1). For example, in a series of reports, Komano and colleagues (Komano et al., 1980, 1981, 1983; Kubo et al., 1984) demonstrated that an agglutinin was formed in *S. peregrina* following cuticular wounding, while Yeaton (1980, 1983) failed to stimulate agglutinin levels by a similar procedure in *H. cecropia*. According to the former study, the agglutinin formed following injury is composed of α and β subunits. In uninjured larvae, only the α subunits are present, while following injury the conversion of some α to β subunits occurs to form the active agglutinin $\alpha_2\beta_4$. These results partially explain how nonantigenic stimulation, such as wounding, can lead to agglutinin production. Further evidence of the importance of the nature of the immunogen and the choice of experimental animal employed has recently been produced by Pendland and Boucias (1985). Injection of viable hyphal bodies of *Nomuraea rileyi* into the velvet bean caterpillar, *Anticarsia gemmatalis*, resulted in a massive increase in agglutinin titers to both rabbit and human erythrocytes, while similar experiments with heat-killed hyphal bodies produced little or no stimulation. Furthermore, similar experiments with

TABLE 14.1. EXPERIMENTS ON THE STIMULATION OF AGGLUTININ LEVELS IN INSECTS

Insect	Immunization Agent	Time Used for Immunization	Particles Used for Agglutinin Tests	Stimulation of Agglutinin Levels	Specificity of Stimulation	References
Anticarsa gemmatalis	Hyphal bodies of *Nomurea rileyi*	1 Day	Rabbit, sheep, and human erythrocytes	+	—	Pendland and Boucias, 1985
Apis mellifera	*Bacillus larvae*	1 Day	*Bacillus subtilis, B. larvae, Salmonella thompson*	+	+	Gilliam and Jeter, 1970
Clitumnus extradentatus	Sheep erythrocytes	1–28 Days	Sheep erythrocytes	—	—	Ratcliffe and Rowley, 1984
Extatosoma tiaratum	Sheep erythrocytes	1–28 Days	Sheep erythrocytes	+/−	ND	Ratcliffe and Rowley, 1984
Hyalophora cecropia	Wounding only	7 Days	Human erythrocytes	—		Yeaton, 1980, 1983
Periplaneta americana	Sheep erythrocytes	?	Sheep erythrocytes	—		Scott, 1971a
P. americana	Sheep erythrocytes	1–28 Days	Sheep erythrocytes	—		Ratcliffe and Rowley, 1984
Sarcophaga peregrina	Wounding only	2 Days	Sheep erythrocytes	+	ND	Komano et al., 1980, 1981, 1983; Kubo et al., 1984
Schistocerca gregaria	*Trypanosoma brucei*	1–7 Days	*T. brucei, Leishmania hertigi, Crithidia fasciculata*	—		Ingram et al., 1983
S. gregaria	*Leishmania hertigi*	1–7 Days	*T. brucei, Leishmania hertigi, Crithidia fasciculata*	+	—	Ingram et al., 1983

Key: ND, not determined.

TABLE 14.2. DISTRIBUTION OF NATURALLY OCCURRING HEMAGGLUTININS IN INSECT HEMOLYMPH

Order	Species	Hemagglutinin Activity	Mean of Titer[a,b]	Range
Odonata	*Anax imperator* (L)	+	16 (1)	—
	Brachytron pratense (L)	+	32 (45)[c]	—
Orthoptera	*Locusta migratoria* (A)	+	7.2 (15)	0–16
	Schistocerca gregaria (A)	+	4.7 (26)	0–16
Phasmida	*Clitumnus extradentatus* (A)	+	947.4 (54)	512–2048
	Carausius morosus (A)	+	144.8 (28)	4–1024
	Extatosoma tiaratum (A)	+	47.3 (39)	4–256
Dictyoptera	*Blaberus craniifer* (A)	+	46.9 (13)	20–160
	Leucophaea maderae (A)	+	8.8 (24)	0–80
	Periplaneta americana (A)	+	35.1 (63)	8–160
	Pycnoscelus surinamensis (A)	+	64 (15)[d]	—
	Sphodromantis sp. (A)	+	14.9 (10)	4–32
Dermaptera	*Forficula auricularia* (A)	−	− (20)[c]	—
Isoptera	*Zootermopsis nevadensis* (A)	−	− (25)[c]	—
Hemiptera	*Notonecta glauca* (A)	+	128 (45)[c]	—
Lepidoptera	*Bombyx mori* (L)	−	− (15)[c]	—
	Galleria mellonella (L)	−	− (20)[c]	—
	Pieris brassicae (L)	−	− (20)[c]	—
	Manduca sexta	+	120 (10)	16–512
Diptera	*Calliphora erythrocephala* (L)	−	−	—
	Drosophila melanogaster (A)	−	− (50)[d]	—
	Tipula sp. (L)	+/−	1.7 (8)	0–2
Hymenoptera	*Apis mellifera* (L)	−	− (15)[c]	—
	Apis mellifera (P)	+/−	1.4 (20)[c]	—
Coleoptera	*Tenebrio molitor* (L)	+	8 (20)[c]	—
	Dytiscus sp. (A)	+/−	5.7 (10)	0–128

Source: Modified from Ratcliffe and Rowley, 1983.
Key: L, larvae; P, pupae; A, adults.
[a] Agglutinating activity was determined by twofold dilutions of hemolymph against saline solution mixed with an equal volume of 2% sheep erythrocytes.
[b] Mean of reciprocal of end point; figures in brackets represent the number of insects used.
[c] Value based on pooled hemolymph.
[d] In *Drosophila* a whole-body extract from 50 insects was used.

the cabbage looper, *Trichoplusia ni,* and the fall armyworm, *Spodoptera frugiperda,* failed to elicit any significant rise in agglutinin titers.

One further question regarding the stimulation of agglutinins is that of specificity: Does injection of one immunogen lead to an increase in agglutinin activity to this antigen alone? In *Apis mellifera,* Gilliam and Jeter (1970) found that, following immunization with the bacterium *Bacillus larvae,* the agglutinin formed was active in agglutination assays against this microorganism and the closely related *Bacillus subtilis,* while no reactivity was recorded against *Salmonella thompson.* Such results appear to suggest a high level of specificity. In contrast, however, in *Schistocerca gregaria,* following immunization with the protozoan *Leishmania hertigi,* the agglu-

**TABLE 14.3. AGGLUTINATION OF
ERYTHROCYTES FROM DIFFERENT SPECIES
BY SERUM FROM *EXTATOSOMA TIARATUM***

Erythrocyte Type	Titer[a]
Sheep[b,c]	16
Chicken	4
Goat	4
Horse	64
Calf	32
Rabbit	65, 536
Human	
O	32
B	32
AB	16
O	32

Source: From E. Richards, N. Ratcliffe, and L. Renwrantz, unpublished.
[a] Results are expressed as the reciprocal of the lowest dilution in which agglutination occurs.
[b] Two percent erythrocyte in 0.9% saline solution containing 0.01 M $CaCl_2$ solution used in all cases.
[c] For method of microtitration used, see Ratcliffe and Rowley, 1983.

tinin formed was active against a number of unrelated protozoan parasites (Ingram et al., 1983, 1984).

In summary, no clear picture emerges from the studies to date on agglutinin induction or stimulation and the specificity of the reaction. At first glance, the production of specific agglutinins seems improbable in animals such as insects, without lymphocytes or their equivalent cells, but the immunization studies of Karp and colleagues with the cockroach *Periplaneta americana* have detected antigen-specific secondary responses of the type seen in vertebrates (Karp and Rheins, 1980a,b; Rheins et al., 1980; Rheins and Karp, 1982). Thus, careful additional studies on the induction and specificity of all humoral factors in insects are required.

14.2.3. Distribution of Agglutinins

With over 1 million species of insects already identified, it is not surprising that few workers have attempted to determine the distribution of agglutinins even with a reasonable range of these animals. Bernheimer (1952) examined the distribution of hemagglutinating activity in 46 species of lepidopterans and Jurenka and colleagues (1982) in 19 species of grasshoppers. With the lepidopterans, it was suggested that parasitized insects were more likely to have high agglutinating activity than were healthy individuals, although this was not subsequently tested. In a more comprehensive study with 26 species

TABLE 14.4. PHYSICOCHEMICAL PROPERTIES OF SOME INSECT AGGLUTININS

Insect	Purified and Isolated Agglutinin and Nomenclature	Chemical Nature	Molecular Weight	Heat Sensitivity	Requirement for Divalent Cations	Binding Specificity[a]	References
Allomyrina dichotoma	+, allo A-I[b]	Proteinaceous	65,000 (subunits 17,500 and 20,000)	ND[c]	−	Lactose, β-linked D-galactose	Umetsu et al., 1984
	+, allo A-II[b]	Proteinaceous	66,500 (subunits 19,000 nd 20,000)	ND	−	Lactose, β-linked D-galactose	
Drosophila melanogaster	−	ND	ND	ND	ND	Heparin, N-acetyl-D-glucosamine	Ceri, 1984
Extatosoma tiaratum	+[d]	Glycoprotein	Approximately 80,000	ND	+	Lactose, D-galactose-containing molecules (see text)	E. Richards, N. Ratcliffe, and L. Renwrantz, unpublished
Leptinatarsa decemlineata	+, larval-pupal	ND	ND	ND	ND	Heparin, mucin	Stynen et al., 1982
	+, chromoprotein$_2$	Lipoprotein	356,000 (subunits 95,500 and 90,000)	ND	ND	Heparin, mucin	Peferoen et al., 1982

Species						Reference
Melanoplus sanguinipes	−	ND	+ (65°C)[e]	+	Many complex and simple sugars except mannose	Jurenka et al., 1982
Periplaneta americana	−	Protein containing	+ (56°C)	+	ND	Scott, 1971a, 1972
P. americana	−	ND	ND	ND	− (?)	Lackie, 1981
P. americana	−	Protein or glycoprotein	+ (65°C)	−	ND	Ingram et al., 1983, 1984
Sarcophaga peregrina	+	Protein 190,000	ND	ND	Lactose, D-galactose	Komano et al., 1980, 1981, 1983
Schistocerca gregaria	−	ND	ND	ND	Sucrose, fetuin	Lackie, 1981
Teleogryllus commodus	−	Aggregated metalloprotein >10⁶ (subunits of 31,000 and 53,000)	+ (56°C)	+	N-acetyl-D-galactosamine, N-acetyl-D-glucosamine	Hapner and Jermyn, 1981

[a] Binding specificity is determined by sugar and glycoprotein inhibition tests. Only those substances giving the greatest inhibition are listed.
[b] Allo A-I and allo A-II are probably isolectins.
[c] ND, not determined.
[d] See the text for details.
[e] Temperature at which agglutinating activity is lost.

from a range of insect orders (Ratcliffe and Rowley, 1983, 1984), high agglutinin titers to sheep erythrocytes were found in the dictyopterans, phasmids, and hemipterans tested, with some activity in most other orders (Table 14.2). This preliminary work, although of great use in selecting suitable insects for further examination, has many limitations. For example, agglutinating activity in hemolymph varies dramatically, depending on the type of erythrocytes used. Our recent results with hemagglutinins from the giant stick insect, *Extatosoma tiaratum,* illustrate this point, as high titers were recorded against rabbit erythrocytes but far lower activity occurred with sheep, chicken, and goat red blood cells (E. Richards, N. Ratcliffe, and L. Renwrantz, unpublished; Table 14.3). Such variations should be borne in mind when choosing suitable insects for further experimentation.

14.2.4. Physicochemical Properties of Agglutinins

Work aimed at determining the physicochemical characteristics and purification of agglutinins in insects has lagged behind that in other invertebrates. Some of the main results to date are summarized in Table 14.4. They show that agglutinins, though apparently all proteinaceous, vary greatly in a range of properties, such as molecular weight and binding specificity as determined by sugar inhibition studies.

Few conclusions can be drawn on the physicochemical properties of agglutinins from the majority of studies undertaken, as whole hemolymph rather than purified agglutinins were used (see also Chapter 8). This practice has hindered investigations of the nature and functions of agglutinins, as whole hemolymph may contain several such substances, each with a unique structure and sugar specificity (Jurenka et al., 1982). This situation is particularly well illustrated by work on hemolymph agglutinins from noninsectans, such as the tunicate *Botrylloides leachii.* Here, two agglutinins, designated HA-1 and HA-2, have been identified. HA-1 is specific for guinea pig erythrocytes, has a molecular weight of 200,000, requires Ca^{2+} ions, and binds lactose and galactose, while HA-2 agglutinates a wide range of erythrocytes, has a molecular weight of 63,000, is independent of divalent cations, and binds lactose (Schluter et al., 1981; Ey and Jenkin, 1982). More important, only HA-2 has opsonic activity (Coombe et al., 1982). This elegant work reinforces the importance of using purified agglutinins whenever possible. Recently, we (N. Ratcliffe, L. Renwrantz, and E. Richards, unpublished) have been working on the isolation and characterization of agglutinins from a number of insect species, including *Clitumnus extradentatus, E. tiaratum,* and *Manduca sexta.* Most of the following preliminary results have been obtained with the giant stick insect, *E. tiaratum,* because of its large hemolymph volume. The sugar specificity was determined before purification of the agglutinin from whole hemolymph using a range of monosaccharides, oligosaccharides, *N*-acetylated sugars, and glycoproteins. A high level of inhibition occurs with lactose (galactose β-1,4 glucose) but is not so

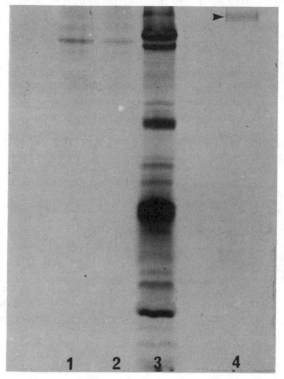

Figure 14.2. Polyacrylamide gel electrophoresis of *Extatosoma tiaratum* whole hemo-lymph (lane 3), lactose (lane 4), and MgCl$_2$-eluted material (lanes 1 and 2) from a Sepharose 4B column. The purity of the material from lactose elution is evidenced by a single band (unlabeled arrow).

marked with either of its constituent monosaccharides. Other saccharides with D-galactose configurations and desialyzed fetuin, in which the D-gal-actose groups of the glycoprotein are exposed, provide strong inhibition. Hence, the agglutinin or agglutinins are probably specific for D-galactose–containing molecules. Similar results have also been obtained with other insects (Table 14.4), in contrast to results with some other arthropods, in which the agglutinins are inhibited by sialic acid (Vasta and Marchalonis, 1983; see also Chapter 8). Isolation of the *E. tiaratum* agglutinin has been achieved by affinity chromatography with Sepharose 4B. Incubation of the column with 0.2 M lactose for 60–90 min, followed by elution with lactose, produces a distinct protein peak, while further washing of the column with 3 M MgCl$_2$ gives a second peak. Both of these peaks have agglutinating activity, which, following concentration and electrophoresis, result in the banding pattern seen in Figure 14.2. The material from the lactose peak gives a single band in the acrylamide gels that stains with PAS and Coomassie brilliant blue, indicating its purity and chemical nature, respectively. On the other hand, the MgCl$_2$-eluted peak has several bands following electrophoresis. Preliminary measurements of the molecular weight of the agglutinin from the lactose peak, using gel filtration, indicate a value for this glycoprotein of approximately 80,000. This agglutinin, when further purified and characterized, will be used in experiments on the synthesis and function of these molecules in insects.

14.3. AGGLUTININS: POSSIBLE ROLE IN IMMUNE RECOGNITION AND DEFENSE REACTIONS

Invertebrate hemolymph often contains opsoninlike molecules that aid in the phagocytosis of nonself material. Since the majority of the studies upon which this statement is based used whole hemolymph, the role of agglutinins as opsonic principles is mainly inferred. Indeed, nonagglutinating opsonins may well exist in invertebrates (Goldenberg and Greenberg, 1983; Ratcliffe and Rowley, 1984).

Definite proof of the role of agglutinins as recognition molecules in invertebrates has come from work with purified molecules and hemocyte monolayer cultures (Hardy et al., 1977; Renwrantz and Stahmer, 1983). Renwrantz and Stahmer (1983) used purified agglutinin, diluted whole hemolymph, or saline solution to presensitize yeast cells before placing them on washed monolayers of hemocytes from the bivalve mollusk *Mytilus edulis*. They found that yeast treated with either whole hemolymph or purified agglutinin resulted in the ingestion of approximately 55% of the hemocytes by these particles, while saline-incubation alone produced a value of only 5% phagocytosis. Moreover, immunocytochemical staining of the agglutinin showed it to be associated with the surface of phagocytic hemocytes, where it may act as a surface-bound receptor.

Similar studies with insects have, unfortunately, not been carried out with purified agglutinins, but experiments designed to demonstrate opsonic activity using whole hemolymph have failed (Scott, 1971b; Anderson et al., 1973; Rowley and Ratcliffe, 1980; Ratcliffe and Rowley, 1984). For example, Anderson and associates (1973) reported that bacteria sensitized in concentrated hemolymph were not internalized and killed by the hemocytes of the cockroach *Blaberus craniifer* more efficiently than were untreated bacteria. However, since these workers did not test the hemolymph for agglutinins to any of the bacteria, this lack of opsonic activity may have been due to the absence of these molecules. Scott (1971b), using monolayers of cockroach (*P. americana*) hemocytes, failed to demonstrate any opsonic activity in the serum for sheep and chicken erythrocytes, even though agglutinins to these red cells were present. In our own investigations with *C. extradentatus* and *P. americana* (Rowley and Ratcliffe, 1980; Ratcliffe and Rowley, 1984), we found that hemolymph pretreatment of formalized sheep

TABLE 14.5. EFFECT OF SERUM PRETREATMENT ON THE PHAGOCYTOSIS OF FORMALIZED ERYTHROCYTES BY THE HEMOCYTES OF *PERIPLANETA AMERICANA*

Particle Treatment	Percent Phagocytosis	
	Half above End Point	Half below End Point
Serum incubated	$5.8 \pm 3.7^{a,b}$	7.9 ± 3.8^{c}
Saline incubated	10.1 ± 6.9	8.7 ± 4.4

Source: From Rowley and Ratcliffe, 1980.
Note: End point calculated by titration of serum against a 2% solution of formalized erythrocytes.
[a] Mean value $\pm$ SD (n = 10).
[b] $P <$ .02 compared with saline-incubated control.
[c] $P >$ .1 compared with saline-incubated control.

erythrocytes (hemolymph of both species contains high levels of hemagglutinins; see Table 14.2) actually caused a significant reduction, rather than an increase, in the number of these particles ingested by the phagocytic plasmatocytes (Table 14.5). This reduction was more noticeable when hemolymph samples with a high agglutinin titer were used, as this led to some agglutination of erythrocytes on the monolayers, leaving fewer test particles available for ingestion. With higher dilutions of hemolymph, this agglutination was all but abolished, and the reduction in phagocytosis also disappeared.

These experiments do not rule out the possibility of participation of agglutinins in recognition processes in insects, as studies to date have been with few insect species, often under suboptimal in vitro conditions, and, more important, without purified molecules. Furthermore, if both phagocyte-associated and humoral agglutinins that bind the target particles are present, then any possible opsonization effect would be masked.

There is, however, some circumstantial evidence for the role of agglutinins in insect recognition processes. The localization of agglutinins on the surfaces of plasmatocytes and granular cells (both types of phagocytic hemocytes, Yeaton, 1980; Komano et al., 1983) but not on other nonphagocytic cells, such as oenocytoids and spherule cells, coincides with a possible role for these molecules in immune recognition. Lackie (1981) has shown that there is a positive correlation between the level of nonself discrimination and the range of specific agglutinins present in different insects. Furthermore, Komano and Natori (1985) have reported the involvement of the agglutinin from *S. peregrina* larvae in the clearance and lysis of ^{51}Cr-labeled sheep erythrocytes following their intrahemocoelic injection. Whether this lysis is humorally induced or requires cellular mediation is still unclear, but phagocytosis apparently plays no role in the process, as evidenced by the lack of radioactivity associated with the blood cell pellet.

Even if some agglutinins lack opsonic activity, they may still play a role by neutralizing (agglutinating) foreign agents so that the hemocytes can more effectively deal with them by phagocytosis, nodule formation, and encapsulation. In addition, lectins found in different regions of the alimentary canal of the bloodsucking bug *Rhodnius prolixus* and the tsetse fly, *Glossina austeni*, with agglutinating activity toward *Trypanosoma cruzi* may control the infectivity of such parasites outside the hemocoel (Pereira et al., 1981; Ibrahim et al., 1984). Lectins found on the peritrophic membrane of the larva of the blowfly *Calliphora erythrocephala* (Peters et al., 1983) may also be important in the regulation of symbiotic and pathogenic microorganisms.

14.4. THE PROPHENOLOXIDASE SYSTEM OF INSECTS

Melanin is a ubiquitous pigment laid down in most arthropods both in the cuticle and also in response to a variety of foreign agents. The enzyme involved in the production of this pigment is phenoloxidase (EC 1.14.18.1), which is activated in both crustaceans and insects from the inactive prophenoloxidase by a serine protease or proteases (Ashida and Dohke, 1980; Ashida, 1981; Söderhäll, 1981, 1982, 1983; Söderhäll and Häll, 1984; Leonard et al., 1985b). Although this melanization response has been well documented, its significance in arthropod immunity has only recently been realized owing to the excellent studies of Kenneth Söderhäll and colleagues with crustaceans (Söderhäll, 1981, 1982; Söderhäll and Smith, 1983, 1984; Smith and Söderhäll, 1983a,b; Söderhäll et al., 1984; see also Chapter 9). These studies have shown that the prophenoloxidase-activating system not only generates melanin but may also have opsonic, coagulating, fungicidal, and bactericidal activity (Söderhäll, 1981; Söderhäll and Ajaxon, 1982; Smith and Söderhäll, 1983a). They have elucidated some of the major steps in the enzymatic cascade (Figure 14.3), which, in the crustaceans examined to date, is localized in fragile, granule-containing hemocytes (Söderhäll and Smith, 1983; Smith and Söderhäll, 1983b). Ashida and coworkers (1982, 1983) have, however, reported that in the silkworm *Bombyx mori* the prophenoloxidase system is primarily present in the plasma, with some activity in oenocytoids. Studies with *G. mellonella* (Pye and Yendol, 1972; Schmit et al., 1977; Ratcliffe et al., 1984; Leonard et al., 1985a) and *B. craniifer* (Leonard et al., 1985b) strongly suggest that at least the major part of the prophenoloxidase system is localized in hemocytes, possibly granular cells and coagulocytes, which degranulate and lyse in vitro unless stabilized before bleeding.

In crustaceans, the prophenoloxidase pathway is activated by β-1,3-glucans (surface components of many microorganisms), such as laminarin, as well as by lipopolysaccharide (isolated from the cell wall of gram-negative bacteria), detergents, and heat (Unestam and Söderhäll, 1977; Söderhäll, 1981, 1982). In *B. mori, G. mellonella,* and *B. craniifer,* although β-1,3-glucans activate the pathway, lipopolysaccharide has no activity (Ashida et al., 1982, 1983; Leonard et al., 1985a,b). Ashida and colleagues (1982, 1983)

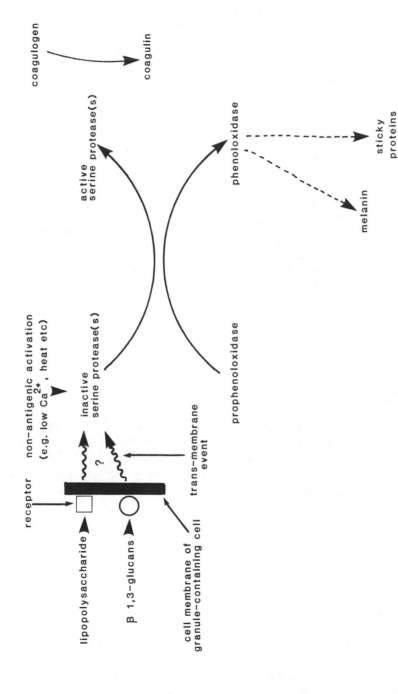

Figure 14.3. Diagrammatic representation of the possible major stages in the prophenoloxidase-activating cascade of insects (largely based on the work of K. Söderhäll and M. Ashida).

TABLE 14.6. ACTIVATION OF *BLABERUS CRANIIFER* PROPHENOLOXIDASE BY BACTERIA AND LIPOPOLYSACCHARIDE

Activator Tested	Phenoloxidase Activity (Δ A490/min/mg protein)	
	Live Bacteria	Heat-Killed Bacteria
Gram-positive Forms		
Bacillus cereus	0.78 ± 0.13[a,b]	0.07 ± 0.03
B. megaterium	0.09 ± 0.02	0.07 ± 0.04
B. subtilis	0.08 ± 0.03	0.08 ± 0.03
B. thuringiensis	0.76 ± 0.12[b]	0.08 ± 0.04
B.1[c]	0.44 ± 0.04[b]	0.06 ± 0.02
Micrococcus luteus	0.08 ± 0.02	0.05 ± 0.02
Staphylococcus aureus	0.06 ± 0.04	0.05 ± 0.03
Gram-negative Forms		
Escherichia coli K12 W6	0.07 ± 0.04	0.07 ± 0.04
Pseudomonas aeruginosa	0.06 ± 0.03	0.04 ± 0.03
Serratia marcescens	0.06 ± 0.03	0.07 ± 0.04
E. coli lipopolysaccharide	0.05 ± 0.03	
Control (buffer only)	0.04 ± 0.03	

Source: From C. Leonard, A. Rowley, and N. Ratcliffe, unpublished.
[a] Mean value ± SD (n = 4).
[b] $P \leq$ .05 compared with control.
[c] B.1 is a natural isolate from an infected cockroach. Preliminary morphologic and biochemical characterizations suggest that it is probably *B. cereus*.

have reported that cell wall extracts from several gram-positive and gram-negative bacteria activate the pathway, but our initial in vitro studies with *B. craniifer* (C. Leonard, A. Rowley, and N. Ratcliffe, unpublished) have shown that only a few species of live bacteria have a stimulatory activity (Table 14.6). This result is somewhat unexpected, because, for the prophenoloxidase-activating system to play a role in immune recognition, it should be activated by most, if not all, nonpathogenic foreign material. However, in vivo experiments suggest that a wider variety of nonself materials can activate the prophenoloxidase cascade, as evidenced by the formation of melanin, than indicated by our in vitro experiments. For example, injection of lipopolysaccharide into *G. mellonella* larvae results in the formation of melanized nodules (Schwalbe and Boush, 1971), and similar experiments with a range of bacteria, some of which fail to activate the prophenoloxidase cascade in our in vitro assay (Table 14.6), also result in the production of this pigment (Ratcliffe and Walters, 1983). Hence, the results of in vitro experiments should be treated with caution because of the complexity of the prophenoloxidase cascade.

14.5. ROLE OF THE PROPHENOLOXIDASE SYSTEM IN IMMUNE RECOGNITION IN INSECTS

It is believed that, following activation of the prophenoloxidase pathway, some components of the system may be released from the granular cells or coagulocytes or other cells by exocytosis, and this leads to the coating and recognition of nonself material (Leonard et al., 1985a) (Fig. 14.3). Furthermore, as both lysozyme (Zachary and Hoffmann, 1984) and agglutinins (Yeaton, 1980) are stored in the granules of such cells, these substances are released too. Hence, degranulation results in the discharge of many factors, some possibly triggering and others effecting both humoral and cellular defense reactions. Evidence for the occurrence of this recognition process in insects comes from experiments in which the effects of laminarin (a β-1,3-glucan), dextran (an αD-glucan, which fails to activate the prophenoloxidase system), and lipopolysaccharide were monitored on the in vitro phagocytosis of heat-killed *Bacillus cereus* by the hemocytes of three insect species (Ratcliffe et al., 1984; Leonard et al., 1985a). Hemocyte monolayers were overlaid with bacteria together with one of these substances and the degree of phagocytosis assessed microscopically. Some experiments were carried out with an inhibitor of serine proteases, *p*-nitrophenyl-*p'*-guanidobenzoate, since these enzymes are involved in the activation of prophenoloxidase (Fig. 14.3). The results demonstrate that the addition of laminarin or lipopolysaccharide enhances the phagocytic activity of the hemocytes towards the test bacteria (Table 14.7). More important, the serine protease inhibitor abolishes the effect, and dextran fails to stimulate phagocytosis at all. Spectrophotometric assays for the production of phenoloxidase in hemocyte lysates revealed that laminarin, but not dextran, activated the cascade. These and other results point to the involvement of the prophenoloxidase system in immune recognition. In crustaceans, Smith and Söderhäll (1983a,b) have suggested that the opsoninlike activity generated is associated with sticky proteins produced by the fragile granule-containing cells (Fig. 14.3). In in-

TABLE 14.7. EFFECT OF LAMINARIN ON THE PHAGOCYTOSIS OF *BACILLUS CEREUS* BY *GALLERIA MELLONELLA* HEMOCYTES

Treatment	Percent Phagocytic Hemocytes	Number of Bacteria per 100 Hemocytes
Laminarin (β-1,3-glucan)	8.0 ± 1.8[a,b]	17.4 ± 3.1[b]
Culture medium only (control)	1.2 ± 0.7	1.8 ± 0.9
pNPGB[c]	1.3 ± 0.5	2.2 ± 1.1
pNPGB + laminarin	1.5 ± 0.6	2.2 ± 0.8
Dextran (α-D-glucan)	1.9 ± 0.6	2.6 ± 0.8

Source: From Leonard et al., 1985a.
[a] Mean value ± SD (*n* = 20).
[b] $P \leq .05$ compared with controls.
[c] *p*-nitrophenyl-*p'*-guanidobenzoate, an inhibitor of serine proteases.

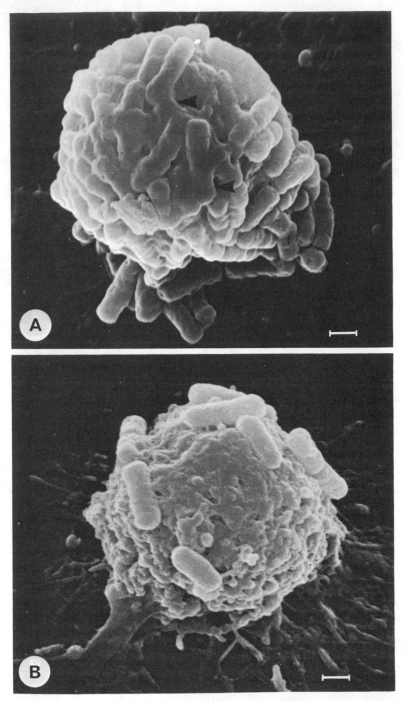

Figure 14.4. (A) Scanning electron micrograph of a granular hemocyte from *Galleria mellonella* incubated with *Bacillus cereus* and laminarin. Note the large number of attached bacteria enmeshed in an amorphous material (unlabeled arrows). (B) Granular hemocyte incubated with *B. cereus* alone. Note that there are fewer attached bacteria and an apparent lack of amorphous material. (Scale bars = 1 μm.)

sects also, there is evidence of the existence of such proteins. Monolayers of *Galleria* hemocytes incubated with *B. cereus* in either insect medium alone or medium containing laminarin were fixed and examined by scanning electron microscopy (Leonard et al., 1985a). Granular cells from monolayers incubated with bacteria and laminarin were found to have numerous bacteria on their surfaces embedded in an amorphous matrix (Fig. 14.4A); this material presumably corresponds to the sticky proteins reported to be generated in crustaceans. In the case of the control monolayers, although a few bacteria were found attached to the granular cells, the amorphous matrix was much less evident (Fig. 14.4B). Bacteria enmeshed in the sticky proteins are therefore more likely to become attached to the phagocytic plasmatocytes, which may account for the enhanced uptake recorded.

In vivo studies also point to a role for the prophenoloxidase system in immune recognition. In crustaceans, fungal spores coated with products of prophenoloxidase activation before their injection have been shown to be more readily encapsulated than saline-incubated spores (Söderhäll et al., 1984), while parasitoids in their habitual hosts may inhibit the activation and/or activity of phenoloxidase and hence abrogate the immune recognition system of these insects (Stoltz and Cook, 1983).

14.6. RECOGNITION OF FOREIGNNESS IN INSECTS: A FINAL NOTE

In this review, although we have considered the agglutinins and components of the prophenoloxidase system as independently aiding in nonself recognition, a syngeneic association may well exist (Ratcliffe et al., 1985). One such possible association is illustrated in Figure 14.5, where the first stage in the recognition process may involve binding of soluble or particulate material via humoral or membrane-bound agglutinins to the granular cells or coagulocytes. This binding may bring about degranulation, with the release of the prophenoloxidase cascade, coagulins (Rowley, 1977), lysozyme (Zachary and Hoffmann, 1984), and agglutinins (Yeaton, 1980). This complex array of hemostatic and antimicrobial factors emphasizes the importance and multipotentiality of the coagulocyte or granular cell type. No doubt other factors will be added to this list as we learn more about the roles of degranulation of granular cells or coagulocytes in the immune defenses of insects.

14.7. SUMMARY

Though insects lack immunoglobulins and complement, this does not imply that they have an ineffectual immunorecognition system. Indeed, it seems possible that the phagocytic hemocytes of insects recognize foreignness both by direct interaction with nonself material and with the aid of humoral fac-

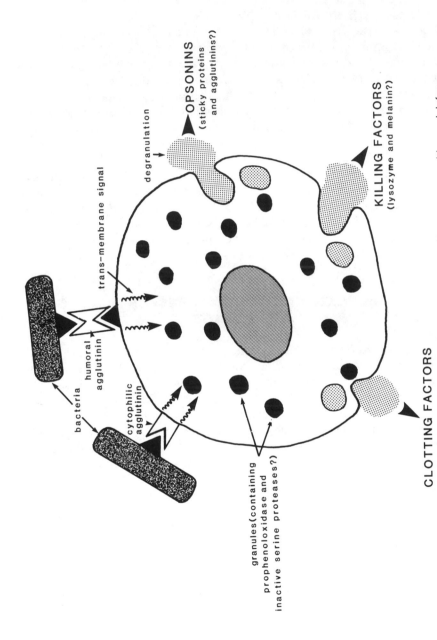

Figure 14.5. Multipotentiality of granular hemocytes in insect immunorecognition and defense reactions.

tors. These factors include components generated by the prophenoloxidase cascade, and the agglutinins. Agglutinins are widely distributed within the Insecta, and they react with a variety of agents, such as bacteria and vertebrate erythrocytes, by binding the carbohydrate residues present on these particles. In insects, the agglutinins are synthesized in both granular hemocytes and fat body cells, but the mechanisms involved in the triggering of this process have yet to be elucidated. To date, although there is no definitive proof that insect agglutinins act either as opsonins or as cytophilic receptors on phagocytes, various reports point to such a role. Proof of such a function can be gained only by using purified agglutinins and suitable in vitro assays.

The prophenoloxidase system, which generates melanin and sticky proteins, has been carefully studied in both insects and crustaceans. The pathway can be activated by a number of factors, including β-1,3-glucans, such as laminarin. The sticky proteins are released from the granular hemocytes, and they coat nonself material. As a result, this material is more readily internalized by the phagocytic plasmatocytes. As with agglutinins, the full potential of the prophenoloxidase system will be realized only when purified components and separated populations of insect hemocytes are available.

ACKNOWLEDGMENTS

The original work quoted in this chapter was supported by the Science and Engineering Research Council (grants GR/D/21684, GR/D/54644 and GR/B/60958 to N. A. R. and A. F. R.), the Royal Society (A. F. R.), the Leverhulme Trust, the British Council (N. A. R.), and the Deutsche Forschungsgemeinschaft (L. R.).

REFERENCES

Amirante, G.A. 1976. Production of heteroagglutinins in haemocytes of *Leucophaea maderae* L. *Experientia* **32:** 526–528.

Amirante, G.A. and F.G. Mazzalai. 1978. Synthesis and localization of hemoagglutinins in hemocytes of the cockroach *Leucophaea maderae* L. *Dev. Comp. Immunol.* 2(4): 735–740.

Anderson, R.S. and R.A. Good. 1976. Opsonic involvement in phagocytosis by mollusk hemocytes. *J. Invertebr. Pathol.* **27:** 57–64.

Anderson, R.S., B. Holmes, and R.A. Good. 1973. *In vitro* bactericidal capacity of *Blaberus craniifer* hemocytes. *J. Invertebr. Pathol.* **22:** 127–135.

Ashida, M. 1981. A cane sugar factor suppressing activation of prophenoloxidase in haemolymph of the silkworm *Bombyx mori. Insect Biochem.* **11:** 57–65.

Ashida, M. and K. Dohke. 1980. Activation of pro-phenoloxidase by the activating enzyme of the silkworm *Bombyx mori. Insect Biochem.* **10:** 37–47.

Ashida, M., Y. Ishizaki, and H. Iwahana. 1983. Activation of prophenoloxidase by bacterial cell walls or β-1,3 glucans in plasma of the silkworm *Bombyx mori. Biochem. Biophys. Res. Commun.* **113:** 562–568.

Ashida, M., R. Iwama, and H. Yoshida. 1982. Control and function of the prophenoloxidase activating system, pp. 81–86. *Proceedings of the Third International Colloquium on Invertebrate Pathology,* University of Sussex.

Bernheimer, A.W. 1952. Hemagglutinins in caterpillar bloods. *Science* (Washington, DC) **115:** 150–151.

Brown, R.C., H. Bass, and J.P. Coombs. 1975. Carbohydrate binding proteins involved in phagocytosis by *Acanthamoeba. Nature* (London) **254:** 434–435.

Ceri, H. 1984. Lectin activity in adult and larval *Drosophila melanogaster. Insect Biochem.* **14:** 547–549.

Coombe, D.R., S.F. Schluter, P.L. Ey, and C.R. Jenkin. 1982. Identification of the HA-2 agglutinin in the haemolymph of the ascidian *Botrylloides leachii* as the factor promoting adhesion of sheep erythrocytes to mouse macrophages. *Dev. Comp. Immunol.* **6**(1): 65–74.

Ey, P.L. and C.R. Jenkin. 1982. Molecular basis of self/non-self discrimination in the Invertebrata, pp. 321–391. In N. Cohen and M.M. Sigel (eds.) *The Reticuloendothelial System: A Comprehensive Treatise,* vol. 3, *Phylogeny and Ontogeny.* Plenum Press, New York.

Gilliam, M. and W.S. Jeter. 1970. Synthesis of agglutinating substances in adult honeybees against *Bacillus larvae. J. Invertebr. Pathol.* **16:** 69–70.

Gold, E.R. and P. Balding 1975. *Receptor-Specific Proteins: Plant and Animal Lectins.* Excerpta Medica, Amsterdam.

Goldenberg, P.Z. and A.H. Greenberg. 1983. Functional heterogeneity of carbohydrate-binding hemolymph proteins: Evidence of a nonagglutinating opsonin in *Homarus americanus. J. Invertebr. Pathol.* **42:** 33–41.

Götz, P. and H.G. Boman. 1985. Cellular and humoral immunity in insects, pp. 453–485. In C.A. Kerkut and L.I. Gilbert (eds.) *Comprehensive Insect Physiology, Biochemistry and Pharmacology,* vol. 3. Pergamon Press, Oxford.

Hapner, K.D. and M.A. Jermyn. 1981. Haemagglutinin activity in the haemolymph of *Teleogryllus commodus* (Walker). *Insect Biochem.* **11:** 287–295.

Hardy, S.W., T.C. Fletcher, and J.A. Olafsen. 1977. Aspects of cellular and humoral defense mechanisms in the Pacific oyster, *Crassostrea gigas,* pp. 59–66. In J.B. Solomon and J.D. Horton (eds.) *Developmental Immunobiology.* Elsevier North Holland, Amsterdam.

Ibrahim, E.A.R., G.A. Ingram, and D.H. Molyneux. 1984. Haemagglutinins and parasite agglutinins in haemolymph and gut of *Glossina. Tropenmed. Parasitol.* **35:** 151–156.

Ingram, G.A., J. East, and D.H. Molyneux. 1983. Agglutinins of *Trypanosoma, Leishmania* and *Crithidia* in insect haemolymph. *Dev. Comp. Immunol.* **7:** 649–652.

Ingram, G.A., J. East, and D.H. Molyneux. 1984. Naturally occurring agglutinins against trypanosomatid flagellates in the haemolymph of insects. *Parasitology* **89:** 435–451.

Jurenka, R., K. Manfredi, and K.D. Hapner. 1982. Haemagglutinin activity in Acrididae (grasshopper) haemolymph. *J. Insect Physiol.* **28**(2): 177–181.

Karp, R.D. and L.A. Rheins. 1980a. Induction of specific humoral immunity to soluble proteins in the American cockroach (*Periplaneta americana*). 2. Nature of the secondary response. *Dev. Comp. Immunol.* **4**: 629–639.

Karp, R.D. and L.A. Rheins. 1980b. A humoral response of the American cockroach to honeybee toxin demonstrating specificity and memory, pp. 65–76. In M.J. Manning (ed.) *Phylogeny of Immunological Memory*. Elsevier North Holland, Amsterdam.

Komano, H. and S. Natori, 1985. Participation of *Sarcophaga peregrina* humoral lectin in the lysis of sheep red blood cells injected into the abdominal cavity of larvae. *Dev. Comp. Immunol.* **9**: 31–40.

Komano, H., D. Mizuno, and S. Natori. 1980. Purification of lectin induced in the hemolymph of *Sarcophaga peregrina* larvae on injury. *J. Biol. Chem.* **255**: 2919–2924.

Komano, H., D. Mizuno, and S. Natori. 1981. A possible mechanism of induction of insect lectin. *J. Biol. Chem.* **256**: 7087–7089.

Komano, H., R. Nozawa, D. Mizuno, and S. Natori. 1983. Measurement of *Sarcophaga peregrina* lectin under various physiological conditions by radioimmunoassay, *J. Biol. Chem.* **258**: 2143–2147.

Kubo, T., H. Komano, M. Okada, and S. Natori. 1984. Identification of hemagglutinating protein and bactericidal activity in the hemolymph of adult *Sarcophaga peregrina* on injury of the body wall. *Dev. Comp. Immunol.* **8**: 283–291.

Lackie, A.M. 1981. The specificity of the serum agglutinins of *Periplaneta americana* and *Schistocerca gregaria* and its relationship to the insects' immune response. *J. Insect Physiol.* **27**: 139–143.

Leonard, C.M., N.A. Ratcliffe, and A.F. Rowley. 1985a. The role of prophenoloxidase activation in non-self recognition and phagocytosis by insect blood cells. *J. Insect Physiol.,* **31**: 789–799.

Leonard, C.M., K. Söderhäll, and N.A. Ratcliffe. 1985b. Studies on prophenoloxidase and protease activity of *Blaberus craniifer. Insect Biochem.,* **15**: 803–810.

Peferoen, M., D. Stynen, and A. De Loof. 1982. A re-examination of the protein pattern of the hemolymph of *Leptinotarsa decemlineata,* with specific reference to the vitellogenins and diapause proteins. *Comp. Biochem. Physiol.* **72B**: 345–351.

Pendland, J.C. and D.G. Boucias. 1985. Hemagglutinin activity in the hemolymph of *Anticarsia gemmatalis* larvae infected with the fungus *Nomuraea rileyi. Dev. Comp. Immunol.* **9**: 21–30.

Pereira, M.E.A., A.F.B. Andrade, and J.M.C. Ribeiro. 1981. Lectins of distinct specificity in *Rhodnius prolixus* interact selectively with *Trypanosoma cruzi. Science* (Washington, DC) **211**: 597–600.

Peters, W., H. Kolb, and V. Kolb-Bachofen. 1983. Evidence for a sugar receptor (lectin) in the peritrophic membrane of the blowfly larva, *Calliphora erythrocephala* Mg. (Diptera). *J. Insect Physiol.* **29**: 275–280.

Price, C.D. and N.A. Ratcliffe. 1974. A reappraisal of insect haemocyte classification by the examination of blood from fifteen insect orders. *Z. Zellforsch. Mikrosk. Anat.* **147**: 537–549.

Pye, A.E. and W.G. Yendol. 1972. Hemocytes containing polyphenoloxidase in *Galleria* larvae after injections of bacteria. *J. Invertebr. Pathol.* **19**: 166–170.

Rabinovitch, M. 1968. Effect of antiserum on the attachment of modified erythrocytes to normal or to trypsinized macrophages. *Proc. Soc. Exp. Biol. Med.* **127:** 351–355.

Rabinovitch, M. and M.-J. De Stefano. 1970. Interactions of red cells with phagocytes of the wax-moth (*Galleria mellonella,* L.) and mouse. *Exp. Cell Res.* **59:** 272–282.

Ratcliffe, N.A. and A.F. Rowley, 1979. Role of hemocytes in defense against biological agents, pp. 331–414. In A.P. Gupta (ed.). *Insect Hemocytes.* Cambridge University Press, Cambridge.

Ratcliffe, N.A. and A.F. Rowley. 1983. Recognition factors in insect hemolymph. *Dev. Comp. Immunol.* **7:** 653–656.

Ratcliffe, N.A. and A.F. Rowley. 1984. Opsonic activity in insect hemolymph, pp. 187–204. In T.C. Cheng (ed.) *Comparative Pathobiology,* vol. 6, *Cells and Serum Factors.* Plenum Press, New York.

Ratcliffe, N.A. and A.F. Rowley. 1986. Insect responses to parasites and other pathogens. In E.J.L. Soulsby (ed.) *Immunology, Immunoprophylaxis and Immunotherapy of Parasitic Infections.* C.R.C. Press, Boca Raton, FL. (in press)

Ratcliffe, N.A. and J.B. Walters. 1983. Studies on the *in vivo* cellular reactions of insects: Clearance of pathogenic and non-pathogenic bacteria in *Galleria mellonella* larvae. *J. Insect Physiol.* **29:** 407–415.

Ratcliffe, N.A., C.M. Leonard, and A.F. Rowley. 1984. Prophenoloxidase activation: Nonself recognition and cell cooperation in insect immunity. *Science* (Washington, DC) **226:** 557–559.

Ratcliffe, N.A., A.F. Rowley, S.W. Fitzgerald, and C.P. Rhodes. 1985. Invertebrate immunity: Basic concepts and recent advances. *Int. Rev. Cytol.* **97:** 183–350.

Renwrantz, L. 1983. Involvement of agglutinins (lectins) in invertebrate defense reactions: The immuno-biological importance of carbohydrate-specific binding molecules. *Dev. Comp. Immunol.* **7**(4): 603–608.

Renwrantz, L. and A. Stahmer. 1983. Opsonizing properties of an isolated hemolymph agglutinin and demonstration of lectin-like recognition molecules at the surface of hemocytes from *Mytilus edulis. J. Comp. Physiol.* **149:** 535–556.

Rheins, L.A. and R.D. Karp. 1982. An inducible humoral factor in the American cockroach (*Periplaneta americana*): Precipitin activity that is sensitive to a proteolytic enzyme. *J. Invertebr. Pathol.* **40:** 190–196.

Rheins, L.A., R.D. Karp, and A. Butz. 1980. Induction of specific humoral immunity to soluble proteins in the American cockroach (*Periplaneta americana*). 1. Nature of the primary response. *Dev. Comp. Immunol.* **4:** 447–458.

Rowley, A.F. 1977. The role of the haemocytes of *Clitumnus extradentatus* in haemolymph coagulation. *Cell Tissue Res.* **182:** 513–524.

Rowley, A.F. and N.A. Ratcliffe. 1980. Insect erythrocyte agglutinins: *In vitro* opsonization experiments with *Clitumnus extradentatus* and *Periplaneta americana* haemocytes. *Immunology* **40:** 483–492.

Rowley, A.F. and N.A. Ratcliffe. 1981. Insects, pp. 421–488. In N.A. Ratcliffe and A.F. Rowley (eds.) *Invertebrate Blood Cells,* vol. 2. Academic Press, New York.

Schluter, S.F., P.L. Ey, D.R. Keough, and C.R. Jenkin. 1981. Identification of two carbohydrate-specific erythrocytes agglutinins in the haemolymph of the proto-chordate *Botrylloides leachii. Immunology* **42:** 241–250.

Schmit, A.R., A.F. Rowley, and N.A. Ratcliffe. 1977. The role of *Galleria mellonella* hemocytes in melanin formation. *J. Invertebr. Pathol.* **29:** 232–234.

Schwalbe, C.P. and G.M. Boush. 1971. Clearance of ^{51}Cr-labeled endotoxin from hemolymph of actively immunized *Galleria mellonella. J. Invertebr. Pathol.* **18:** 85–88.

Scott, M.T. 1971a. A naturally occurring hemagglutinin in the hemolymph of the American cockroach. *Arch. Zool. Exp. Gen.* **112:** 73–80.

Scott, M.T. 1971b. Recognition of foreignness in invertebrates. 2. *In vitro* studies of cockroach phagocytic haemocytes. *Immunology* **21:** 817–827.

Scott, M.T. 1972. Partial characterization of the hemagglutinating activity in hem-olymph of the American cockroach (*Periplaneta americana*). *J. Invertebr. Pathol.* **19:** 66–71.

Sharon, N. 1984. Surface carbohydrates and surface lectins are recognition deter-minants in phagocytosis. *Immunology Today* **5**(5): 143–147.

Smith, V.J. and K. Söderhäll. 1983a. β-1,3-glucan activation of crustacean hemo-cytes *in vitro* and *in vivo. Biol. Bull.* (Woods Hole) **164:** 299–314.

Smith, V.J. and K. Söderhäll. 1983b. Induction of degranulation and lysis of *Astacus astacus* hemocytes by components of the prophenoloxidase activating system *in vitro. Cell Tissue Res.* **233:** 295–303.

Smith, V.J., K. Söderhäll, and M. Hamilton. 1984. β-1,3-glucan induced cellular defence reactions in the shore crab, *Carcinus maenas. Comp. Biochem. Physiol.* **77**A: 635–639.

Söderhäll, K. 1981. Fungal cell wall β-1,3-glucans induce clotting and phenoloxidase attachment to foreign surfaces of crayfish hemocyte lysate. *Dev. Comp. Immunol.* **5:** 565–573.

Söderhäll, K. 1982. Prophenoloxidase activating system and melanization—a rec-ognition mechanism of arthropods? A review. *Dev. Comp. Immunol.* **6:** 601–611.

Söderhäll, K. 1983. β-1,3-glucans enhancement of protease activity in crayfish hem-ocyte lysate. *Comp. Biochem. Physiol.* **74**B: 221–224.

Söderhäll, K. and R. Ajaxon. 1982. Effect of quinones and melanin on mycelial growth of *Aphanomyces* spp. and extracellular protease of *Aphanomyces astaci,* a parasite on crayfish. *J. Invertebr. Pathol.* **39:** 105–109.

Söderhäll, K. and L. Häll. 1984. Lipopolysaccharide-induced activation of pro-phenoloxidase activating system in crayfish haemocyte lysate. *Biochim. Biophys. Acta* **797:** 99–104.

Söderhäll, K. and V.J. Smith. 1983. Separation of the haemocyte populations of *Carcinus maenas* and other marine decapods and prophenoloxidase distribution. *Dev. Comp. Immunol.* **7**(2): 229–239.

Söderhäll, K. and V.J. Smith. 1984. The prophenoloxidase activating system: A complement like pathway in arthropods? pp. 160–167. In D.W. Roberts and J.R. Aist (eds.) *Infection Processes of Fungi,* Rockfeller Foundation, New York.

Söderhäll, K., A. Vey, and M. Ramstedt. 1984. Hemocyte lysate enhancement of fungal spore encapsulation by crayfish hemocytes. *Dev. Comp. Immunol.* **8:** 23–29.

Stein, E. and E.L. Cooper. 1981. Agglutinins as receptor molecules: A phylogenetic approach, pp. 85–98. In E.L. Cooper and M. Brazier (eds.) *Developmental Immunology: Clinical Problems and Aging.* Academic Press, New York.

Stoltz, D.B. and D.I. Cook. 1983. Inhibition of host phenoloxidase activity by parasitoid Hymenoptera. *Experientia* **39:** 1022–1024.

Stynen, D., M. Peferoen, and A. De Loof (1982). Proteins with haemagglutinin activity in larvae of the Colorado beetle, *Leptinotarsa decemlineata. J. Insect Physiol.* **28:**(5): 465–470.

Tripp, M.R. 1966. Hemagglutinin in the blood of the oyster, *Crassostrea virginica. J. Invertebr. Pathol.* **8:** 478–484.

Tripp, M.R. and V.E. Kent. 1967. Studies on oyster cellular immunity. *In Vitro* **3:** 129–135.

Umetsu, K., S. Kosaka, and T. Suzuki. 1984. Purification and characterization of a lectin from the beetle *Allomyrina dichotoma. J. Biochem.* **95:** 239–245.

Unestam, T. and K. Söderhäll. 1977. Soluble fragments from fungal cell walls elicit defence reactions in crayfish. *Nature* (London) **267:** 45–46.

Vasta, G.R. and J.J. Marchalonis. 1983. Humoral recognition factors in the Arthropoda: The specificity of Chelicerata serum lectins. *Am. Zool.* **23:** 157–171.

Weir, D.M. 1980. Surface carbohydrates and lectins in cellular recognition. *Immunol. Today* **1**(2): 45–51.

Weir, D.M. and H.M. Ögmundsdóttir. 1977. Non-specific recognition mechanisms of mononuclear phagocytes. *Clin. Exp. Immunol.* **30:** 323–329.

Whitcomb, R.F., M. Shapiro, and R.R. Granados. 1974. Insect defense mechanisms against microorganisms and parasitoids, pp. 447–536. In M. Rockstein (ed.) *The Physiology of Insecta,* vol. 5, 2nd ed. Academic Press, New York.

Yeaton, R.W. 1980. Lectins of a North American silk moth (*Hyalophora cecropia*): Their molecular characterization and developmental biology. Doctoral dissertation, University of Pennyslvania, Philadelphia.

Yeaton, R.W. 1981a. Invertebrate lectins. 1. Occurrence. *Dev. Comp. Immunol.* **5:** 391–402.

Yeaton, R.W. 1981b. Invertebrate lectins. 2. Diversity of specificity, biological synthesis and function in recognition. *Dev. Comp. Immunol.* **5:** 535–545.

Yeaton, R.W. 1983. Wound responses in insects. *Am. Zool.* **23:** 195–203.

Zachary, D. and D. Hoffmann. 1984. Lysozyme is stored on the granules of certain haemocyte types of *Locusta. J. Insect Physiol.* **30:** 405–413.

Zachary, D. and J.A. Hoffman. 1973. The haemocytes of *Calliphora erythrocephala* (Meig.) (Diptera). *Z. Zellforsch Mikrosk. Anat.* **141:** 55–73.

CHAPTER 15

Humoral Encapsulation in Insects

Peter Götz

Institute for General Zoology
Free University of Berlin
Berlin, West Germany

Alain Vey

Research Station of Comparative Pathology
Institute of Agronomical Research
St. Christol, France

15.1. INTRODUCTION

Cellular encapsulation is the common defense reaction of arthropods against microorganisms and parasites invading the hemocoel (Salt, 1963). Blood cells

407

aggregate around the foreign objects, forming a multicellular envelope of considerable tightness. Melanization of at least the innermost parts of the hemocytic capsule contributes to the resistance and efficiency of the envelope, which is often strong enough to prevent further development of invading organisms. Humoral encapsulation, that is, the formation of a melanotic capsule around foreign material without the visible participation of hemocytes, has until now been found only in certain dipteran species, in which it is effective against invading microorganisms and parasites and certain types of foreign materials. Its occurrence is of special interest in the investigation of the biochemical mechanisms involved in the encapsulation reaction (Götz and Boman, 1985).

15.2. OCCURRENCE OF HUMORAL ENCAPSULATION IN ARTHROPODS: RELATIONSHIP TO BLOOD CELL NUMBERS

Humoral encapsulation was first reported from mosquito and chironomid larvae infected by parasitic nematodes (Wülker, 1961; Bronskill, 1962; Esslinger, 1962). The authors described the deposition of a brownish material on the surface of the nematodes, occurring without visible hemocytic involvement. Among insects, humoral encapsulation was found only in certain species of the families of Culicidae, Chironomidae, Psychodidae, Syrphidae, and Stratiomyidae of Diptera. This finding was established by a comparative study in which larvae and adults of 38 species from 12 insect orders were tested for the type of capsules they formed against nematodes and foreign materials (Götz et al., 1977). The occurrence of humoral encapsulation coincided with low blood cell counts (<6000 cells/mm^3; average, 2000 cells/mm^3). Dipteran species with high blood cell numbers (>6000 cells/mm^3; average, 20,000 cells/mm^3) reacted with cellular encapsulation.

Other arthropods, excluding insects, have been less extensively investigated with regard to their defense reactions. Detailed information is available for crayfish (Unestam and Nylund, 1972; Unestam and Nylhén, 1974; Söderhäll, 1982) and for the crab *Carcinus maenas* (Smith and Söderhäll, 1983a,b). It is surprising that the reaction of *Astacus* against the crayfish plague fungus *Aphanomyces astaci* appears to be related to humoral encapsulation in Diptera.

15.3. HUMORAL ENCAPSULATION IN *CHIRONOMUS* HEMOLYMPH

15.3.1. Humoral Encapsulation of Nematodes

Wülker (1961), investigating parasitism of *Chironomus* larvae by mermithids (Nematoda: Mermithidae), noted that infectious nematode larvae were frequently killed after invasion of the host and that the dead nematodes were covered with dark incrustations. This encapsulation process starts within minutes after invasion of the hemocoel of the host larvae (Fig. 15.1A,B). The intensity with which humoral encapsulation occurs varies from one *Chironomus* larva to another and is dependent on the age of the host and the host and parasite species involved. In nontypical host species, all the invading parasites died after heavy encapsulation, the encapsulated dead nematodes remaining within the hemocoel without initiating further reactions (Fig. 15.2).

Mermithids in adapted host species are less often killed, and a certain percentage of them are able to develop. Free-living populations of *C. riparius* exhibit natural infection rates of between 0.1% and 10% (Götz, 1969). Wülker (1961), studying mermithid parasitism in a local population of *Tanytarsus gregarius*, noticed that the infection rate increased to 100% and that this led to the total collapse of the population. After several years, a new population of *T. gregarius* built up, with new infection rates below 10%. This example demonstrates that it is not advantageous for a parasite to be too successful; moderate success is a better guarantee of a balanced host-parasite relationship and increases the survival chances for both the host and the parasite.

15.3.2. Humoral Encapsulation of Fungi

Humoral encapsulation also proved to be effective against fungi (Götz and Vey, 1974). Spores of *Aspergillus niger, Mucor hiemalis,* and *Beauveria bassiana* injected into *Chironomus* larvae provoked heavy encapsulation (Table 15.1). Nevertheless, most of the encapsulated spores germinated (59–100%). Immediately upon contact with the hemolymph, the outgrowing fungal filaments provoked additional encapsulation. The further development of the fungi varied from species to species. *B. bassiana* was the fastest growing of the fungi tested, and its hyphae were repeatedly able to break the capsular envelope, thus finally exhausting the encapsulation capacity of its host. Injection with *B. bassiana* spores resulted in the development of a fungal mycelium within the hemocoel of the injected larva, and this led to

TABLE 15.1. INJECTION OF FUNGAL SPORES (10^4 PER LARVA) INTO FOURTH-INSTAR LARVAE OF *CHIRONOMUS LURIDUS*

Injected Fungus	Number of Test Animals	Encapsulation of Spores (%)	Germination in Spite of Encapsulation (%)	Development of Fungal Mycellium (%)
Aspergillus niger	17	100	59	0
Mucor hiemalis	14	100	71	71
Beauveria bassiana	6	100	100	100

Source: From Götz and Vey, 1974.

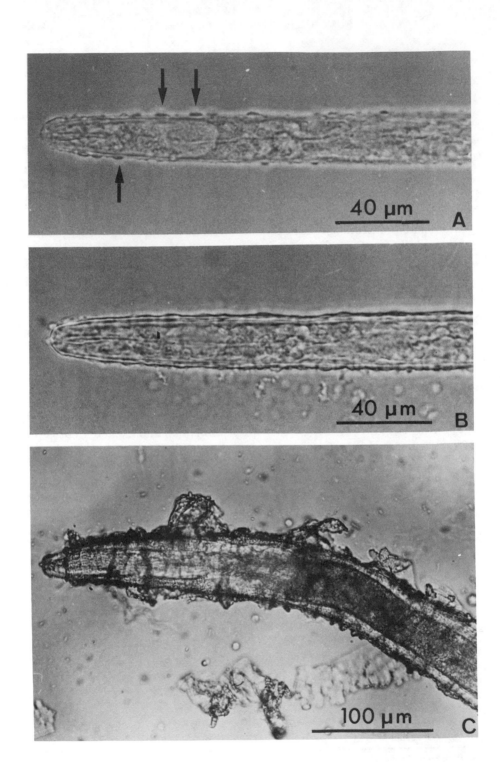

its death within 2–5 days (at a rearing temperature of 22°C). In the case of *M. hiemalis,* 71% of the injected *Chironomus* larvae were killed by the developing fungus, whereas with *A. niger,* none of the germinating spores was able to survive, even though the germination rate was 59%. The development of its mycelium was successfully suppressed by humoral encapsulation.

Of all the fungal species tested, only *B. bassiana* had the capacity to actively invade *Chironomus* larvae. Spores of all three fungal species germinated on the cuticle of *Chironomus* larvae, but only the hyphae of *B. bassiana* succeeded in penetrating the cuticle. The penetrating hyphae provoked heavy defense reactions even within the cuticle and the hypodermal cells (Fig. 15.3). The deposition of a dark, electron-dense material near the surface of the fungal filaments was similar to that found after humoral encapsulation in the hemocoel. Therefore, compounds necessary for the formation of the capsule material must be present in the hemolymph as well as in the cuticle (Götz and Vey, 1974). Similar observations were made with crayfish (Unestan and Nylund, 1972).

All three fungal species were also tested with larvae of *Galleria mellonella,* which react by cellular encapsulation and therefore are suitable as control animals. All fungi tested proved to be pathogenic for *G. mellonella* larvae, and even *A. niger* caused lethal infections in 70% of the test animals at a temperature of 18°C and a mortality of 100% at 35°C after injection of 10^4 spores per larva (Table 15.1).

15.3.3. Humoral Encapsulation of Bacteria

Humoral encapsulation is the only defense reaction of *Chironomus* larvae against bacteria (P. Götz, G. Euderlein and J. Roettgen, in preparation). Twenty-six bacterial strains have been tested by injecting up to 10^6 bacterial cells per larva. Both pathogenic and apathogenic strains of the injected bacteria were immediately encapsulated. Increased mortality of infected larvae by bacteremia occurred only when extremely high injection doses were applied (10^5 or more bacterial cells per *Chironomus* larva). Injection was performed by inserting a glass micropipette into one of the hind prolegs of the larvae and closing the wound with a nylon loop tightened around the injured proleg or by direct perforation of the dorsal abdominal cuticle with especially fine glass pipettes. With injection doses below 10^5, humoral encapsulation prevented any bacterial growth. By contrast, injection of the same bacterial

---

Figure 15.1. Humoral encapsulation of nematodes in isolated hemolymph of *Chironomus riparius.* Before incubation into hemolymph, the nematodes were anaesthetized by addition of Nembutal to the culture medium (1:10, v/v). (A) Infective larva of *Hydromermis contorta* after 5 min-incubation with *Chironomus* hemolmph. Droplets of capsule material (arrows) attach to the surface of the nematode. (B) Five minutes later, the deposited capsule material forms a thin but complete cover around the nematode. (C) Encapsulated larva of *Turbatrix aceti* within its capsule.

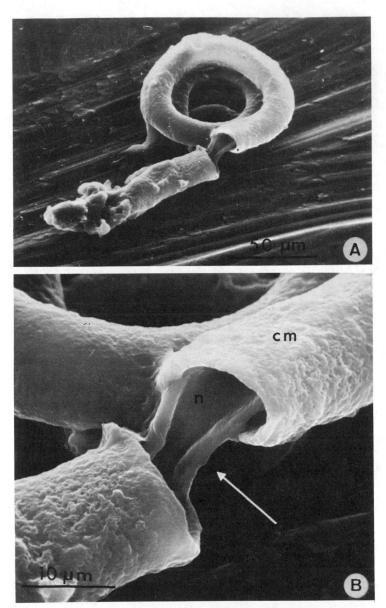

Figure 15.2. Encapsulated nematode (*Hydromermis contorta*) from the hemocoel of a *Chironomus riparius* larva. (A) The nematode is enclosed in a solid tube of capsular material (cm). (B) Where the capsule is broken (arrow), the nematode (n) can be seen within the tubelike envelope. The shrinking of the nematode is an artifact caused during preparation of the material for scanning microscopy. (Courtesy of Ciba Geigy Laboratories, Basel, Switzerland.)

TABLE 15.2. MORTALITY (%)OF *GALLERIA MELLONELLA* AND *CHIVONOMUS TENTANS* LAST-INSTAR LARVAE UNTIL DAY 3 AFTER INJECTION OF 10^4 LOG PHASE CELLS OF VARIOUS BACTERIAL STRAINS

Bacterial Strains	G. mellonella	C. tentans
Pathogenic strains		
Xenorhabdus nematophilus	100	0
Serratia marcescens	100	5
Pseudomonas aeruginosa	100	5
Bacillus thuringiensis	80	0
Apathogenic strains		
Escherichia coli K 12 D 21	0	0
E. coli K 12 D 31	0	0
Isolate water pond II (22°C)	0	0
Isolate water pond II (10°C)	0	5

strains into *G. mellonella* larvae produced different results, depending on the pathogenicity of the bacterial strain used (Table 15.2). Highly pathogenic bacteria (e.g., *Xenorhabdus nematophilus, Serratia marcescens, Pseudomonas aeruginosa,* and *Bacillus thuringiensis*) caused 100% mortality in *G. mellonella.* However, when apathogenic bacteria were injected (e.g., *Escherichia coli* or isolates from freshwater ponds), even doses of 10^6 bacteria per larva were not lethal to last-instar larvae of *G. mellonella.* The success of humoral encapsulation against all types of bacteria must be explained by the rapidity with which this defense reaction occurs. The fast sealing of the bacterial surface prevents the release of toxins or enzymes by which certain bacteria (the pathogenic strains) damage insect hosts and weaken their defense mechanisms.

Besides humoral encapsulation, no other humoral antibacterial activity was detectable in the hemolymph of *Chironomus* larvae. Results of tests for neosynthesis of lysozyme and cecropins (Boman and Hultmark, 1981) in *Chironomus* larvae after bacterial injections were negative. Phagocytic activity was also very low, and cellular encapsulation or nodule formation was completely absent.

15.3.4. In Vitro Experiments with Isolated Hemolymph of *Chironomus* Larvae

Humoral encapsulation has a special advantage in that it occurs in vivo as well as in vitro (Götz, 1969; Götz, 1973; Vey and Götz, 1975). Living or nonliving foreign materials can be incubated with isolated hemolymph and the defense reaction observed directly under the microscope. Since blood cells do not participate in the reaction, observation is easy, and the time until humoral encapsulation becomes obvious is only a few minutes. The addition of inhibitors or substrates allows investigations to be made on the

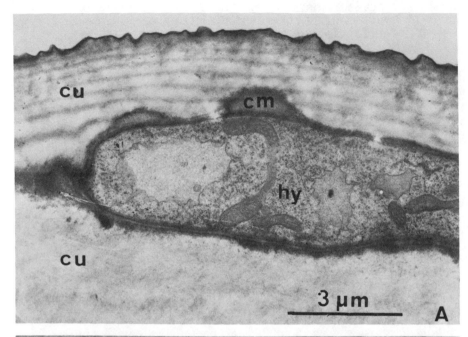

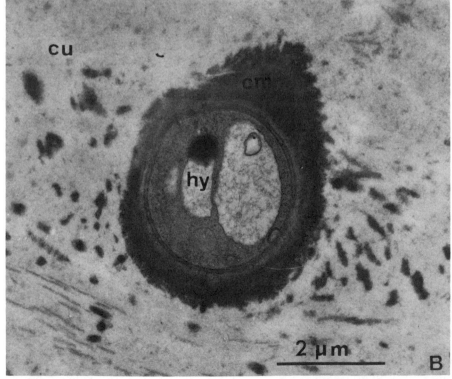

influence of such parameters on the encapsulation reaction. The problem with using *Chironomus* larvae for experimental investigations is their relatively small size and the sensitivity of their hemolymph. Only a few microliters of hemolymph can be withdrawn from one full-grown larva. At room temperature, isolated hemolymph reacts spontaneously. Even without the addition of provoking agents, flocculent material appears in the hemolymph. The encapsulation capacity of *Chironomus* hemolymph is exhausted within about 30 min. However, low temperature extends the encapsulation capacity of *Chironomus* hemolymph samples for a longer period of time. It is therefore advisable to perform in vitro experiments with *Chironomus* hemolymph immediately after bleeding or to collect the hemolymph under low temperature conditions and store it at $-20°C$. To increase the hemolymph yield, we preferred to work with larvae of large *Chironomus* species, such as *C. plumosus,* collected from lakes, or *C. tentans,* which can be easily reared in the laboratory. Last-instar larvae of *C. tentans* usually give about 8–10 µl of hemolymph. For microscopic observation, hemolymph drops were placed in sterile microchambers, and bacteria, fungi, nematodes, or inert materials were added as provocaters (Vey and Götz, 1975). Simple microchambers can be constructed using a glass slide on which thin stripes of petroleum jelly are arranged in a square of about 1.6 cm side length. A drop of Ringer's solution with suspended test organisms or foreign bodies is deposited in the middle of this square, followed by a drop of hemolymph drawn from the larva by severing a leg or puncturing the integument. Dilution of the hemolymph with Ringer's solution must be kept to a minimum (not more than 1:1). Care must also be taken not to contaminate the hemolymph with material adhering to the insect cuticle or with pieces of injured tissue or gut content. The chamber is then quickly closed with a cover slip and sealed by exerting slight pressure on it so as to flatten the petroleum jelly wall. The wall should be flat enough to bring the hemolymph drop in contact with the cover slip, thus offering good optical conditions for microscopy at high magnifications. If sterilized glassware and instruments are used, these hemolymph preparations can be kept for several days.

Deposition of capsule material becomes visible within 2–5 min of incubating foreign objects with the hemolymph and appears as droplets on the provoking surface (Fig. 15.1). The droplets quickly increase in number and size and congregate to form a complete cover. At the beginning of encapsulation, the capsule material is soft and gluey, and motile organisms, such as living nematodes, are still able to move. However, because of this adhesive coating of capsule material, such organisms show an increasing tendency to stick to glass or other surfaces with which they come into contact.

Figure 15.3. Percutaneous infection of *Chironomus luridus* larva by the fungus *Beauveria bassiana.* Ultrathin sections through host cuticle (cu) with penetrating fungal filaments. Note the deposition of capsule material (cm) adjacent to the growing hyphae (hy). (A) Longitudinal section. (B) Cross section through filament.

The capsule itself is still thin and flexible and therefore not easily recognized. The capsule becomes more obvious when the moving nematode bends its body and causes creases in the capsular envelope. A few minutes after deposition, the capsular material hardens, and the nematode becomes fixed within the rigid envelope. Under in vitro conditions, nematodes often succeed in breaking the envelope and escaping, but deposition of further capsule material continues for at least the first 30 min of the experiment. Capsules formed under in vitro conditions reach a thickness of about 0.5–1.0 μm, and after 1–2 hr change color from yellow to brown (bright orange under phase-contrast microscopy). In the vicinity of encapsulated bodies, an accumulation of granular material is commonly observed. This material falls to the bottom of the preparation chamber, forming islands of granular sediments. The granules probably arise from the encapsulation process and may be the result of an overproduction of capsule material that was not deposited on the foreign surface.

15.3.5. Provocation of Humoral Encapsulation by Various Materials

In the search for the mechanism that triggers humoral encapsulation, a variety of organisms and foreign materials have been tested in vivo and in vitro for their capacity to provoke humoral encapsulation (Götz, 1969, 1986a,b; Vey and Götz, 1975; Wilke, 1979; U. Wilke unpublished). These objects can be grouped into three classes: (1) strong provocators, which stimulate a fast and heavy encapsulation in more than 80% of tests carried out; (2) weak provocators, which provoke a delayed reaction in 30%–80% cases; and (3) nonprovocators, which never or only occasionally provoke deposition of small amounts of capsule material (Table 15.3). Strong provocators include living organisms, injured homologous tissues, heterologous tissues, and some organic polymers (Fig. 15.4), while weak provocators are mainly specific organic compounds. Inorganic materials and a third group of organic compounds do not provoke any significant encapsulation. All of the above-mentioned objects have been tested in vitro, whereas the following have been tested in vivo as well as in vitro: injured homologous tissue; bacteria; fungi; mermithids; Sephadex A, G, and C; latex; Dowex; agar; epoxy resin; dental wax; liquid paraffin; glass; activated charcoal; and colloid gold. No defense reaction has been observed in *Chironomus* hemolymph after incubation with *Tetrahymena* sp. (endoparasites of *C. anthracinus*), an undetermined rodlike bacterium from *C. riparius,* and the eggs of the ichneumonid wasp *Pimpla turionella* (Lingg, 1976). Dead organisms (nematodes and spores or hyphae of fungi) provoke a much weaker defense rection than living organisms.

15.3.6. Ultrastructure of Capsule Material

On electron micrographs, the capsule material has an electron-dense appearance, and under high magnifications an irregular fibrillar structure is

TABLE 15.3. PROVOCATION OF HUMORAL ENCAPSULATION BY VARIOUS ORGANISMS AND MATERIALS

Classification of Provocators	Group of Organisms or Materials Tested	Details
Strong provocators	Bacteria	26 strains tested
	Blue-green algae	Undetermined freshwater species
	Green algae	Undetermined freshwater species
	Diatoms	Undetermined freshwater species
	Fungi	*Aspergillus niger, Beauveria bassiana, B. tenella, Mucor hiemalis, Metarhizium anisopliae*
	Sporozoans	*Eimeria tenella* (oocysts)
	Ciliates	*Blepharisma* sp., *Paramecium* sp.
	Nematodes	*Hydromermis* spp., *Neoaplectana* spp., *Turbatrix aceti, Rhabditis* sp.
	Crustaceans	*Daphnia* (eggs), *Cyclops* (eggs), *Artemia salina* (eggs)
	Injured homologous tissue	Fat body of *Chironomus* larvae
	Heterologous tissue	*Tenebrio molitor*
	Erythrocytes	Chicken
	Hydrophilic organic compounds	Silk, cellulose (cotton), agar, latex, Dowex 50 WX 4, polyacrylamide
	Negatively charged or neutral polydextrans	Sephadex C-50, Sephadex G-50
Weak provocators	Organic materials	Hairs, feathers, metacrylate, epoxy resin (araldite), cationic polydextrane (Sephadex A-50), celluloid
Nonprovocators	Hydrophobic organic compounds	Nylon, Perlon, paraffin oil, polyethylene, polystyrol, PVC, dental wax, petroleum jelly
	Inorganic materials	Glass powder, activated charcoal, iron powder, $CaSO_4$ powder, $CaCO_3$ powder, gold colloid

recognized (Fig. 15.5A). The affinity of the capsule material to the foreign surface is very precise, with the capsule forming an accurate template of all the surface structures of the foreign material (e.g., the pattern of the cuticle of a nematode; Fig. 15.5B). At the beginning of humoral encapsulation (about 2–15 min after injection of foreign material into *Chironomus* larva), a loose aggregate of fibrillar capsule material accumulates around the provoking agent, and during the next 15–30 min, this aggregate becomes con-

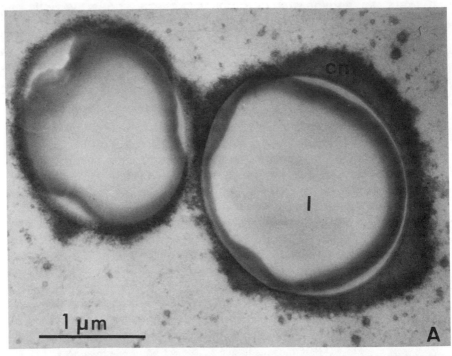

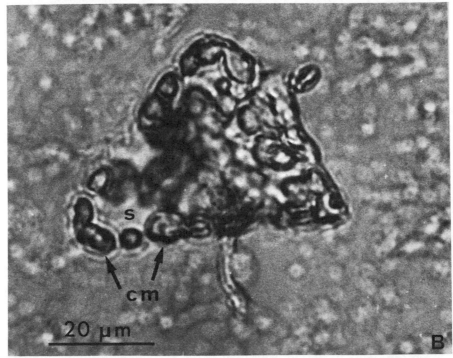

siderably tighter. Within a few hours, the capsular wall reaches its definite electron density and a thickness of 1 to several micrometers (Götz and Vey, 1974).

15.3.7. The Role of Blood Cells during Humoral Encapsulation

In comparison with that of most other insects, the hemolymph of *Chironomus* larvae contains strikingly few blood cells (e.g., *C. riparius* contains 1000–3000 cells/mm^3). Following the nomenclature of Jones (1962), Maier (1969) described the following blood cell types for larvae of *C. riparius:* plasmatocytes (80%), granular cells (18%), adipohemocytes (1.5%), and oenocytoids and others (0.3%). Injection experiments with bacteria, fungal spores, and particles of certain organic compounds (latex, Sephadex) led to the encapsulation of all foreign bodies, but only a small percentage of the encapsulated particles were also phagocytized (Götz and Vey, 1974), demonstrating the much greater speed and higher efficiency of the encapsulation reaction compared with phagocytosis.

15.3.8. Factors Affecting Humoral Encapsulation in *Chironomus*

Humoral encapsulation occurs immediately after contact of hemolymph with foreign organisms. Mermithids, which have not been successfully encapsulated within the first hours of invasion, are able to develop without provoking further defense reactions. Hemolymph from *C. riparius* infected with *Hydromermis contorta* exhibited in vitro a significantly weaker and delayed defense reaction compared to that of hemolymph from noninfected larvae. Encapsulation in the noninfected hemolymph started in vitro in 3.5 min (± 0.5 min) and led to the formation of a complete capsule around the test organism. With hemolymph from infected *Chironomus* larvae, the reaction began after 7 min (± 2 min) and resulted in the production of only a few scattered deposits (Vey and Götz, 1975). Maier (1973) reported that phenoloxidase activity in *C. riparius* larvae parasitized with *H. contorta* was only one-third of its normal value. In addition, it was also found that developing parasitic nematodes were less provocative than freshly invading nematodes. This fact was dicovered by Lingg (1976), who exposed preparasitic and de-

---

Figure 15.4. Humoral encapsulation of nonliving materials in *Chironomus* hemolymph. (A) Ultrathin section through Latex beads (l) encapsulated after injection into *C. luridus* larva. The capsule material (cm) has formed a solid spherule around the bead. Shrinking of the Latex material occurred after encapsulation and was caused by preparation for electron microscopy. (B) Photomicrograph of in vitro encapsulation in isolated hemolymph of *C. riparius.* Large scales of capsule material (cm) were deposited on a Sephadex G particle (s).

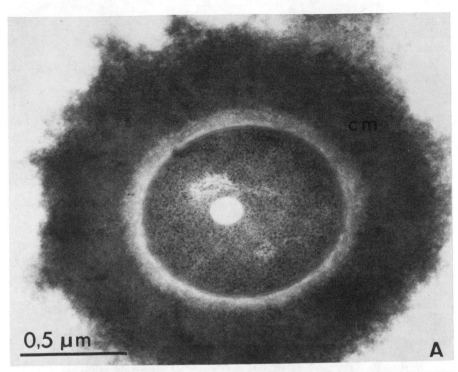

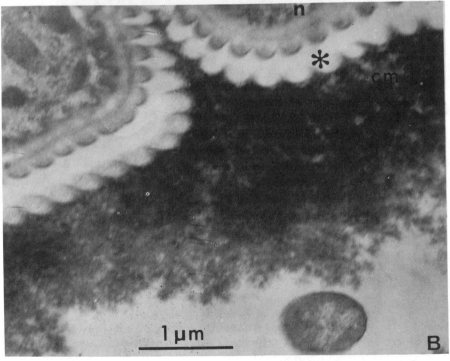

veloping parasitic stages of the mermithid *H. contorta* to isolated hemolymph from noninfected *C. riparius*. Only the preparasitic nematodes were encapsulated, whereas the parasitic stages did not provoke deposition of capsule material. The following observations may be relevant in explaining this phenomenon. When larvae of the nematode *H. rosea* are injected into the hemocoel of *Chironomus* larvae, all the larvae are encapsulated and killed. However, when such larvae enter the hemocoel by penetrating the cuticle, the encapsulation reaction is weaker, and some of the invading parasites are able to develop into adult nematodes. Why do naturally invading nematodes provoke a weaker reaction than do injected nematodes? The answer may lie in the events that occur during active invasion. Before penetration, the nematodes produce an adhesive surface coat that enables them to stick to the surface of the *Chironomus* larva and prepare an invasion hole. When the nematode enters through the narrow entrance, this sticky surface coat is stripped off and remains outside, plugging the entrance behind the invading parasite. After passing the cuticular wall, the nematode does not directly reach the hemocoel, but stays for a while in a hypodermal pocket (15–20 min in an older host), where it is separated from the hemolymph by the basement lamella of the hypodermis (Götz, 1976). It is conceivable that, during this penetration procedure and the following stay within the hypodermal pocket, the parasite undergoes changes in the structure of its surface antigens. Related mechanisms that help parasites avoid the defense reactions of their hosts are known in many insects with cellular encapsulation (Salt, 1968). One of the most spectacular examples of parasite resistance to host defense reactions is the invasion of host tissues by the parasite in order to protect itself against attacking blood cells and humoral factors present in the hemolymph. The mermithid *Filipjevimermis leipsandra,* a parasite of cucumber beetle larva (*Diabrotica undecimpunctata*) takes shelter in the host's nervous system. Here, the infective nematodes remain within the ganglia for 2–4 days, where they rapidly grow until the ganglia become too small for them. They then return to the hemocoel, where they are either too large or no longer "foreign" enough to be successfully attacked by the blood cells. However, every nematode that does not succeed in entering the nervous system within 30 min of invasion of the hemocoel is immediately encapsulated by blood cells (Götz and Poinar, 1968).

Figure 15.5. Ultrastructure of capsule material (cm). (A) State of humoral encapsulation 10 min after injection of a bacterium (*Bacillus thuringiensis*) into a larva of *Chironomus riparius*. (B) In vitro encapsulation of *Hydromermis contorta* in isolated hemolymph of *C. luridus*. Because of the close attachment of capsule material, the capsule represents a template of the surface structures of the nematodes cuticle. The space (∗) between the nematode (n) and capsule (cm) is an artifact resulting from preparation for electron microscopy.

TABLE 15.4. TREATMENT OF CAPSULE MATERIAL WITH VARIOUS CHEMICALS

Chemicals	Solubility	Bleaching
Concentrated H_2SO_4 (22°C, 24 hr)	−	−
40% KOH	+	+
40% NaOH	+	+
KOCl	+ +	+ + +
10% H_2O_2 (22°C, 72 hr)	−	+
Peracetic acid (22°C, 16 hr)	−	+ +
Pyridine (60°C)	−	−

Source: From Götz and Vey, 1974.
Note: Encapsulated nematodes and fungi were dissected from *Chironomus* larvae. After thoroughly washing in Ringer's solution, the capsules were incubated in the indicated chemicals (strong acids, strong bases, oxidizing agents, and organic solvents) to test for solubility and bleaching. Identical results were obtained from in vivo and in vitro encapsulation.

15.3.9. Evidence of the Melanotic Nature of the Capsule Material

Owing to their brownish or black appearance, the capsules from humoral encapsulation have always been assumed to be of a melanotic nature (Bronskill, 1962; Götz, 1969; Poinar and Leutenegger, 1971). This assumption was supported by investigations of the behavior of the capsule material toward organic and inorganic solvents and bleaching agents (Table 15.4), by histochemical tests, by enzymes experiments (Table 15.5), and by the effect of incubating with substrates and inhibitors of phenoloxidase (Maier, 1973; Götz and Vey, 1974; Vey and Götz, 1975). From these results, it was finally concluded that the capsule material consists of a protein-polyquinone com-

TABLE 15.5. TREATMENT OF CAPSULE MATERIAL WITH VARIOUS ENZYMES

Enzymes Tested	Effect on Capsule Material	Effect on Nematodes (control)
α-Glucosidase	−	+
Hyaluronidase	−	+
Neuraminidase	−	+
Chitinase	−	+
Lysozyme	−	+
Trypsin	−	+
Chymotrypsin	−	+
Pronase	−	+
Papain	−	+
Pepsin	−	+

Source: From Albert, 1974.
Note: Incubation time 24 hr at room temperature.

TABLE 15.6. CHARACTERIZATION OF THE TWO PHASES OF HUMORAL ENCAPSULATION

Phase 1: Deposition of Capsule Material	Phase 2: Solidification and Melanization of Capsule Material
During the first 10–15 min	Beginning after 10–15 min
Ca^{2+} dependent	Ca^{2+} independent
Tyrosine independent	Tyrosine consuming
Delayed by PTU	Delayed by PTU
Disintegrated by protease	Protease resistant
Consistency: soft, sticky	Consistency: increasingly solid
Coloration: translucent	Coloration: brown to black

Source: From U. Wilke, in preparation.

plex (Götz and Vey, 1974) and not pure melanin, as was initially thought. Using electronmicroscopical and biochemical techniques, U. Wilke (in preparation) affirmed humoral encapsulation to be a biphasic process that starts with the deposition of a soft, translucent layer during the first 10–15 min and later leads to the formation of a solid, dark brown crust (Table 15.6). Newly deposited capsule material exhibits phenoloxidase activity; black pigments are formed after the addition of tyrosine, DOPA, or other substrates to the capsule material. During the second phase of humoral encapsulation, the hemolymph tyrosine pool is completely exhausted, and phenoloxidase activity stops.

15.4. CELLULAR VERSUS HUMORAL ENCAPSULATION

Cellular encapsulation is the typical defense reaction of arthropods against the invasion of the hemocoel by microorganisms and metazoan parasites. The invaders are surrounded by aggregating blood cells that form an envelope of increasing tightness. During capsule formation, the innermost cells of the hemocytic envelope disintegrate and produce a layer of electron-dense material directly on the foreign surface. This material takes on a brown to black coloration and is considered to be melanin (Salt, 1963; Poinar et al., 1968; Nappi, 1973, 1975; Rowley and Rattcliffe, 1981). Such melanization of at least the inner region of the capsule occurs always when living organisms are encapsulated and occasionally after the encapsulation of inorganic compounds (Table 15.7). It is conceivable that the production of melanin in the course of cellular encapsulation contributes essentially to the effectiveness of the hemocytic capsule. The melanotic material attaches closely to the foreign object and fills all the irregularities on its surface (Poinar et al., 1968). Humoral encapsulation ends in a similar situation, since the surface of the foreign organism is again sealed with a melanized material. Thus,

TABLE 15.7. AGENTS PROVOKING HUMORAL AND/OR CELLULAR ENCAPSULATION

Provocator	Humoral Encapsulation (*Chironomus* sp.)	Cellular Encapsulation (*Galleria mellonella*)	
		Capsule Formation	Melanization
Aspergillus niger	+ +	+ +	+
Beauveria bassiana	+ +	+ +	+
Mucor anisopliae	+ +	+ +	+
Neoaplectana carpocapsae	+ +	−	−
Turbatrix aceti	+ +	(+)	−
Dowex 50 W X4	+ +	+	−
Latex grains	+ +	+	−
Sephadex G	+ +	+	−
Sephadex C	+ +	+	−
Sephadex A	+	+	−
Araldite	+	+	−
Liquid paraffin	−	+	−
Petroleum jelly	−	(+)	−
Dental wax	−	+	−
Glass powder	−	+	−
Activated charcoal	−	+	−
Iron powder	−	+	(+)
CaSO$_4$ powder	−	+	+
CaCO$_3$ powder	−	+	+
Gold colloid	−	+	+

Source: From Wilke, 1979, and Burghause, 1981.
Note: − no reaction; (+) very weak reaction; + good reaction; + + heavy reaction.

humoral and cellular encapsulation have on principal event in common: the process of melanization.

Biochemically, melanization is not a clearly defined reaction. The formation of brown to black pigments generally occurs during quinone sclerotization, which is responsible for the hardening and tanning of the arthropod cuticle. The enzymes controlling quinone sclerotization, phenoloxidases, are present in the cuticle, the hemolymph, and certain hemocytes of arthropods. They are able to oxidize phenol derivatives to quinones, which quickly bind to secondary and terminal amino groups of proteins, thus cross-linking protein chains in a three-dimensional network. Autopolymerization of excess quinone produces dark pigments (pure melanin) that are soluble in strong alkali. The pure melanin seems to be a filler that does not significantly contribute to the number of cross-links between the protein chains (Richards, 1978). β-sclerotization, which is also mediated by phenoloxidases, does not contribute to the coloration of the cross-linked proteins. Quinone sclerotization and β-sclerotization may compete for the same substrate molecules.

Activated hemolymph phenoloxidase is known to be sticky and to attach to available surfaces. Since the quinoid intermediates readily combine with proteins, including phenoloxidase itself, the activated phenoloxidase gradually loses its activity. The end products of phenoloxidase activity, the protein-polyquinone complexes, are mechanically stable and resistant to a broad range of chemicals and enzymes. These characteristics of phenoloxidase indicate its suitability as a component of the defense mechanisms of arthropods. However, since an efficient defense mechanism depends on the ability to discriminate self from nonself, it was necessary for a special phenoloxidase-activating system that reacts only with foreign materials to evolve in the hemolymph of arthropods.

Evidence of such a phenoloxidase-activating cascade came from Dohke (1973a,b), Ashida and Dohke (1980, 1982), Ashida and Söderhäll (1984), and Smith and Söderhäll (1983a,b). Investigating phenoloxidase activation in *Bombyx mori, Astacus* sp., and *Carcinus maenas,* these authors demonstrated several components of the biochemical pathway that leads to active phenoloxidase. Unestam and Söderhäll (1977) described soluble β-1,3-glucans, which commonly occur in fungal and bacterial cell walls, as inducing phenoloxidase activation. Other established components of the activating cascade are bivalent ions and a serine protease (see Chapter 9). Many details of this activating cascade are still unclear, but it is a stimulating concept that may encourage further studies of the mechanisms that control cellular encapsulation. Some authors suggest that the prophenoloxidase-activating system may not only be responsible for controlling cellular encapsulation but may also serve as a general recognition mechanism that also initiates phagocytosis and the neosynthesis of humoral factors (Taylor, 1969; Lackie, 1981; Söderhäll, 1982).

The process of cellular encapsulation requires a system of various types of blood cells, including some that react upon contact with foreign surfaces by disintegration and attachment and others that are attracted by already attached hemocytes. The first visible step in cellular encapsulation is the activation of granular hemocytes, which leads to their degranulation. The material released from the granular cells is thought to be prophenoloxidase, which after activation attaches to foreign objects, such as bacteria, foreign tissues, and inanimate materials (Gagen and Ratcliffe, 1977; Ratcliffe and Gagen, 1977; Schmit and Ratcliffe, 1977, 1978). Factors responsible for the activation of the released prophenoloxidase must originate from the degranulating cells, the hemolymph, and/or the surface of the provoking agent. The substrates for phenoloxidase activity (phenolic compounds) may be released from blood cells or are already present in the serum.

In humoral encapsulation, phenoloxidase in its inactive form (as prophenoloxidase) is present in the hemolymph. A mechanism of blood cell activation is therefore not necessary. Activation of humoral prophenoloxidase occurs upon contact with provoking materials, which include living organisms and certain organic polymers. However, humoral encapsulation is not triggered by inorganic substances and the majority of organic compounds. It is thus more specific than cellular encapsulation, which was provoked by all of the foreign materials tested. Furthermore, cellular encapsulation may or may not include melanization of the innermost layer of disintegrating blood cells. There was no correlation between the provocation of melanization during cellular encapsulation and the initiation of humoral encapsulation in *Chironomus* larvae (Table 15.7).

Recent investigations have shown that the initiation of humoral encapsulation is in agreement with the concept of a phenoloxidase-activating cascade, as has been discussed for cellular encapsulation in insects and crustaceans (Götz, 1986b). As in cellular encapsulation, *Chironomus* hemolymph prophenoloxidase can be activated by incubation with β-1,3-glucans (B. Harmstorf, unpublished). A lack of bivalent ions (e.g., after the addition of EDTA) and the presence of inhibitors of phenoloxidase activity (e.g., *p*-NPGB,*p*-nitrophenyl-*p'*-guanidinobenzoate) prevent the deposition of capsule material on provoking surfaces. Newly deposited capsule material exhibits phenoloxidase activity; the formation of black pigments after the addition of tyrosine, DOPA, or related substrates to fresh capsule material has been demonstrated by U. Wilke (unpublished). Further research is necessary to enable a total understanding of the control mechanisms that govern the initiation and course of humoral encapsulation.

At present, it is not possible to explain the occurrence of humoral encapsulation in certain dipteran groups. Its obvious correlation with low blood cell density provides no clues as to why cellular defense mechanisms have been replaced by humoral encapsulation. Also somewhat conspicuous is the fact that all dipteran larvae that react by humoral encapsulation are aquatic. However, other aquatic groups of Diptera (Simuliidae, Blepharoceridae, and some genera of Culicidae and Chironomidae) have not changed to the humoral type of encapsulation. Compared with cellular encapsulation, humoral encapsulation is more effective against bacteria and fungi, but it is unlikely that these pathogens should represent an increased threat under aquatic conditions. The specific efficiency of humoral encapsulation is noticeable by the scarcity of bacterial and fungal epizootics in natural populations of *Chironomus* larvae. In contrast, intracellular parasites, such as viruses, rickettsiae, and microsporidea, that are not affected by humoral encapsulation are quite common in these insects.

15.5. SUMMARY

Humoral encapsulation has been found to be the principal defense reaction of certain dipteran species, in which it replaces cellular encapsulation. Triggered by contact with foreign surfaces, the hemolymph produces a cell-free capsule around foreign objects. Humoral encapsulation was mainly investigated in vitro and in vivo using *Chironomus* larvae. Humoral encapsulation has proven very efficient against microorganisms and nematodes, and as a defense reaction against bacteria, it is even more successful than the cellular encapsulation of other insects (e.g., larvae of *G. mellonella*). Injection of bacteria that are known to be highly pathogenic to many other insects does not significantly increase mortality of *Chironomus* larvae. The bacteria are rapidly surrounded by capsule material and thus prevented from releasing toxic substances that would otherwise damage their hosts.

Humoral encapsulation is provoked by various kinds of living organisms and by certain organic polymers (hydrophilic and positively charged compounds). In contrast to cellular encapsulation, humoral encapsulation is not provoked by inorganic materials and some hydrophobic organic polymers. The capsule formation occurs in two steps: first the freshly attached capsule material is soft and colorless, but then it solidifies and changes to a brown to black color. Capsule material of the first phase contains activated phenoloxidase, but during solidification the capsule gradually loses its enzyme activity. Concomitantly, the electron density of the capsule material increases, and tyrosine disappears from the hemolymph. Humoral encapsulation is controlled by various factors, including β-1,3-glucans, bivalent ions, and a serine protease. In this respect, humoral encapsulation in *Chironomus* hemolymph follows the activation pathway already demonstrated for cellular encapsulation in crustaceans.

REFERENCES

Albert, H. 1974. Experimente zur humoralen Einkapselung bei *Chironomus*. Staatsexamensarbeit, University of Freiburg, West Germany.

Ashida, M. and K. Dohke, 1980. Activation of prophenoloxidase by the activating enzyme of the silkworm *Bombyx mori*. *Insect Biochem*. **10**: 37–47.

Ashida, M. and K. Söderhäll. 1984. The prophenoloxidase activating system in crayfish. *Comp. Biochem. Physiol*. **77**B: 21–26.

Boman, H.G. and D. Hultmark. 1981. Cell-free immunity in insects. *Trends Biochem. Sci*. **6**: 306–309.

Bronskill, J.F. 1962. Encapsulation of rhabditoid nematodes in mosquitoes. *Can. J. Zool*. **40**: 1269–1275.

Burghause, F. 1980. Die zelluläre Einkapselung von körperfremden Stoffen bei *Gryllus*, p. 329. In W. Rathmayer (ed.) *Verhandlungen der Deutschen Zoologischen Gesellschaft*. Gustav Fischer, Stuttgart, West Germany.

Dohke, K. 1973a. Studies on prophenoloxidase-activating enzyme from the cuticle of the silkworm *Bombyx mori*. 1. Activation reaction by the enzyme. *Arch. Biochem. Biophys*. **157**: 203–209.

Dohke, K. 1973b. Studies on prophenoloxidase-activating enzyme from the cuticle of the silkworm *Bombyx mori*. 2. Purification and characterisation of the enzyme. *Arch. Biochem. Biophys*. **157**: 210–221.

Esslinger, J.H.Y. 1962. Behaviour of microfilariae of *Bruggia pahangi* in *Anopheles quadrimaculatus*. *Am. J. Trop. Med. Hyg*. **1**: 749–758.

Gagen, S.J. and N.A. Ratcliffe, 1976. Studies on the in vivo cellular reactions and fate of injected bacteria in *Galleria mellonella* and *Pieris brassicae* larvae. *J. Invertebr. Pathol*. **28**: 17–24.

Götz, P. 1969. Die Einkapselung von Parasiten in der Hämolymphe von *Chironomus* Larven (Diptera). *Zool. Anz.* **33:** 610–617, supplement.

Götz, P. 1973. Immunreaktionen bei Insekten. *Naturwiss. Rundschau.* **26:** 367–375.

Götz, P. 1976. Parasit-Wirt-Beziehungen zwischen dem Nematoden *Hydromermis contorta* und der Zuckmücke *Chironomus thummi. Publ. Wiss. Filmen* **9:** 67–87.

Götz, P. 1986a. Encapsulation in Arthropods, pp. 153–170. In M. Brehélin (ed.) *Immunity in Invertebrates*, Springer-Verlag, Berlin Heidelberg.

Götz, P. 1986b. Mechanisms of Encapsulation in Dipteran Hosts. In A. Lackie (ed.) *Immune Mechanisms in Invertebrate Vectors*, Oxford University Press.

Götz, P. and H.G. Boman. 1985. Insect Immunity, pp. 453–485. In G.A. Kerkut and L.I. Gilbert (eds.) *Comprehensive Insect Physiology, Biochemistry and Pharmacology,* vol. 3. Pergamon Press, Oxford.

Götz, P. and G.O. Poinar, Jr. 1968. Entwicklung einer Mermithide (Nematode) im Nervensystem von Insekten. *Z. Parasitenkd.* **31:** 10.

Götz, P. and A. Vey. 1974. Humoral encapsulation in Diptera (Insecta): Defence reactions of *Chironomus* larvae against fungi. *Parasitology* **68:** 193–205.

Götz, P., I. Roettgen, and W. Lingg. 1977. Encapsulement humoral en tant que réaction de défense chez les Diptères. *Ann. Parasitol. Hum. Comp.* **52:** 95–97.

Götz, P., G. Euderlein, and J. Roettgen. Immune reaction of *Chironomus* larvae. (Insecta: Diptera) against bacteria (in preparation).

Jones, J.C. 1962. Current concepts concerning insect haemocytes. *Am. Zool.* **2:** 209–246.

Lackie, A.M. 1981. Immune recognition in insects. *Dev. Comp. Immunol.* **5:** 191–204.

Lingg, A. 1976. Experimentelle Untersuchungen zur humoralen Einkapselung. Staatsexamensarbeit, University of Freiburg, West Germany.

Maier, W. 1969. Die Hämocyten der Larven von *Chironomus thummi* (Dipt.). *Z. Zellforsch.* **99:** 54–63.

Maier, A.W. 1973. Die Phenoloxidase von *Chironomus thummi* und ihre Beeinflussung durch parasitäre Mermithiden. *J. Insect Physiol.* **19:** 85–95.

Nappi, A.J. 1973. The role of melanization in the immune reaction of larvae of *Drosophila algonquin* against *Pseudeucoila bochei. Parasitology* **66:** 23–32.

Nappi, A.J. 1975. Parasite encapsulation in insects, pp. 293–326. In K. Maramorosch and R.E. Shope (eds.) *Invertebrate Immunology.* Academic Press, New York.

Poinar, G.O., Jr., and R. Leutenegger. 1971. Ultrastructural investigation of the melanization process in *Culex pipiens* (Culicidae) in response to a nematode. *J. Ultrastruct. Res.* **36:** 149–158.

Poinar, G.O., Jr., R. Leutenegger, and P. Götz. 1968. Ultrastructure of the formation of a melanotic capsule in *Diabrotica* (Coleoptera) in response to a parasitic nematode (Mermithidae). *J. Ultrastruct. Res.* **25:** 293–306.

Ratcliffe, N.A. and S.J. Gagen. 1977. Studies on the *in vivo* cellular reactions of insects; An ultrastructural analysis of nodule formation in *Galleria mellonella. Tissue Cell* **9:** 73–85.

Richards, A.G. 1978. The chemistry of the insect cuticle, pp. 205–232. In M. Rockstein (ed.) *Biochemistry of Insects.* Academic Press, New York.

Rowley, A.F. and N.A. Ratcliffe. 1981. Insects, pp. 421–488. In N.A. Ratcliffe and N.F. Rowley (eds.) *Invertebrate Blood Cells,* vol. 2. Academic Press, New York.

Salt, G. 1963. The defense reactions of insects to metazoan parasites. *Parasitology* **53:** 527–642.

Salt, G. 1968. The resistance of insect parasitoids to the defence reactions of their hosts. *Biol. Rev.* (Cambridge) **43:** 200–232.

Schmit, A.R. and N.A. Ratcliffe. 1977. The encapsulation of foreign tissue implants in *Galleria mellonella* larvae. *J. Insect Physiol.* **23:** 175–184.

Schmit, A.R. and N.A. Ratcliffe. 1978. The encapsulation of Araldite implants and recognition of foreignness in *Clitumnus extradentatus. J. Insect Physiol.* **24:** 511–521.

Smith, V.J. and K. Söderhäll. 1983a. Induction of degranulation and lysis of haemocytes in the freshwater crayfish *Astacus astacus* by components of the prophenoloxidase activating system *in vitro. Cell Tissue Res.* **233:** 295–303.

Smith, V.J. and K. Söderhäll. 1983b. β-1,3-glucan activation of crustacean hemocytes *in vitro* and *in vivo. Biol. Bull.* (Woods Hole) **164:** 299–314.

Söderhäll, K. 1982. Prophenoloxidase activating system and melanization—a recognition mechanism of arthropods? A review. *Dev. Comp. Immunol.* **6:** 601–611.

Taylor, R.L. 1969. A suggested role for the polyphenol-phenoloxidase system in invertebrate immunity. *J. Invertebr. Pathol.* **14:** 427–428.

Unestam, T. and L. Nylhén. 1974. Cellular and noncellular recognition of and reactions to fungi in crayfish, pp. 189–206. In E.L. Cooper (ed.) *Contemporary Topics in Immunology,* vol. 4. Plenum Press, New York.

Unestam, T. and J.E. Nylund. 1972. Blood reactions *in vitro* in crayfish against a fungal parasite, *Aphanomyces astaci. J. Invertebr. Pathol.* **19:** 94–106.

Unestam, T. and K. Söderhäll. 1977. Soluble fragments from fungal cell walls elicit defence reactions in crayfish. *Nature* (London) **267:** 45–46.

Vey, A. and P. Götz. 1975. Humoral encapsulation in Diptera (Insecta): Comparative studies *in vitro. Parasitology* **70:** 77–86.

Wilke, U. 1979. Humorale Infektabwehr bei *Chironomus*-Larven, p. 315. In W. Rathmayer (ed.) *Verhandlugen der Deutschen Zoologischen Gesellschaft.* Gustav Fischer, Stuttgart, West Germany.

Wülker, W. 1961. Untersuchungen über die Intersexualitat der Chironomiden (Dipt.) nach *Paramermis* Infektion. *Arch. Hydrobiol.* **25:** 127–181, supplement.

CHAPTER 16

Mediators in
Insect Immunity

Werner Mohrig
Dietmar Schittek
Dieter Ehlers
Section of Biology
Ernst-Moritz-Arndt University
Greifswald, East Germany

16.1. INTRODUCTION

All metazoa have developed systems to distinguish between self and nonself tissue and to neutralize and eliminate foreign material that invades the body cavity. Such recognition is accomplished by means of reticulate tissue or body fluids with specific cellular components that are distributed within the whole body. Humoral factors act as mediators between cells and tissues. On the basis of their functions and origin, two groups of mediators can be distinguished. The first group may be called induced messenger, or signal, substances, which are released or activated in response to a key stimulus, for example, invasion by nonself tissue. At their target sites, these signal substances initiate one or a series of characteristic reactions by interacting with a specific receptor. Such signal substances inform the host's immune system about a disturbance of its homeostasis or its integrity and initiate appropriate counterregulation. Since 1966, a large variety of mediators formed by immune-competent cells or infected body cells after antigen contact has been demonstrated among vertebrates. The lymphokines, which are released through antigen or mitogen stimulation by lymphocytes and modulate manifold activities of immunologically unspecific effector cells, were the first to be discovered (Bloom and Bennett, 1966). The release of lymphokinelike substances after the grafting of foreign tissue was postulated also for invertebrates (annelids) by Hostetter and Cooper (1974), although, of course, any phylogenetic relationship to the vertebrate line must be excluded.

Signal substances in defense systems of metazoan organisms can be released or activated by injury, microbial invasion, or parasitization. They stimulate cell proliferation in hemopoietic organs, mitosis, adherence and migration behavior of circulating cells, metabolic response capability of cells integrated into defense systems, and degranulation and release of lysosomal enzymes with antibacterial or anti-inflammatory effect (see also Chapter 10).

The second group of mediators includes circulating molecules in the body fluid of animals. They promote adhesion of phagocytic cells to target cells or nonself material, such as bacteria, yeast, viruses, parasites, and soluble material by means of bi- or multivalent molecular structures. As a rule, they react primarily with surface molecules of foreign material and establish the foreign substrate–phagocyte contact. In general, these mediators are characterized as recognition molecules that promote various kinds of cell-"substrate" contacts, mainly defined by the type of the target materials, as well as by the genetically determined function of the effector cell. Such contacts include simple adhesion (e.g., rosette formation); release of Na^+ and Ca^{2+} with an intracellular metabolic change brought about by receptor occupation and activation of the cytoskeleton, as well as the ingestion (phagocytosis) of the fixed material, degranulation, and formation of O_2 metabolites with an enhanced killing effect; and morphologic changes, such as spreading of cells and encapsulation of foreign material.

Since the observations of Wright and Douglas (1903), such mediators have been known as opsonins; the immunoglobulins and the complement factor C3b are well-known examples of such substances among vertebrates. An exclusively phagocytosis-promoting effect is connected with opsonins. In light of more recent studies, the scope of opsonins as primarily effective on foreign bodies must be extended because they represent mediators that initiate such activities as activation of killer cells, lysis of foreign cells, release of active substances, and activation of the hexose monophosphate shunt.

In addition to the two groups of mediators mentioned above, another class of mediators (or activators) was detected with the discovery in mammals of the protein leucokinine (Najjar et al., 1968) and the tetrapeptide tuftsin (Najjar and Schmidt, 1980). These mediators are not induced primarily by a foreign stimulus, nor do they react with the nonself component, as opsonin does; although they act directly on the cell, their mode of action is unknown. In addition to tuftsin, this group of mediators includes factors that perhaps promote phagocytosis only by physicochemical changes in the effector and target cells (surface charges and hydrophobicity). This chapter provides a survey of cellular reactions in insects, given the foregoing classification of mediator substances as background.

16.2. ADHERENCE-INFLUENCING FACTORS

Changed adherence behavior of the hemocytes (plasmatocytes and/or granulocytes) may be detected as the first general reaction, resulting first in a settling of the hemocytes: a sudden disappearance from the circulating hemolymph after injury or injection. The intensity of the reaction is known to depend on the type, volume, and number of injected materials (Gagen and Ratcliffe,1976; Hanschke et al., 1980; Chain and Anderson, 1982). Accordingly, the resulting hemocyte reactions are also different. When a foreign material is injected, phagocytosis occurs with the disappearance of the injected material from the hemolymph. When the host is overloaded with the foreign material or uningestible inert particles, the hemocytes form aggregates that degenerate and are, in a secondary phagocytic reaction, enclosed by intact hemocytes and deposited in the body as melanizing nodules. When larger implants are used, or during invasion by metazoan parasites, encapsulation occurs, resulting in a multistratified capsule of hemocytes around the foreign body. One fundamental prerequisite for subsequent reactions in all these cellular activities is the ability of circulating hemocytes to attach themselves to substrata. The various reactions connected with the changed adhesiveness of hemocytes during phagocytosis, encapsulation, and wound healing are well known, but only in some cases have the molecular fundamentals of these reactions in insects been explained. The significance of adherence-modifying factors for the defense reactions is, however, well recognized. According to their sites of action, adherence-influencing substances can be divided into factors that (1) directly act on the phagocyte (plasmatocyte and/or granulocyte) and (2) change the foreign material, thus causing a changed adherence behavior.

16.2.1. Factors Causing Changes in Hemocyte Activity

Adherence influence due to the change in hemocyte activity can be observed under experimental conditions by means of injury and injection of antigen material. The total hemocyte count is often dramatically altered (Mohrig and Messner, 1970; Gagen and Ratcliffe, 1976; Hanschke et al. 1980). From her experiments on saturniid pupae, Cherbas (1973), considered it possible to deduce the release of an *injury factor*. This factor, with a molecular weight in the range of 50,000 and known as hemokinin, is said to be released from the injured epidermis into the hemolymph and cause changes in the mobility and adhesiveness of hemocytes. It is unclear to what extent these factors are responsible for the formation of wound thrombus and hemolymph co-agulation. It is known that many granulocytes degenerate during hemolymph coagulation, formation of wound thrombus, and removal of foreign bodies. Faye (1978) believes that many of the hemocytes probably stick to the host tissue after the ingestion process. In general, this phenomenon is analogous to the behavior of the phagocytic leucocytes of vertebrates during contact with bacteria. Phagocytic granulocytes form arachidonic acid metabolites (especially Leukotrien B_4 and Thromboxan) from the phospholipids of their membrane (Powell, 1984). It is widely known that these compounds promote adhesiveness, aggregation, and degranulation of leucocytes and act chem-otactically (Naccache and Sha'afi, 1982). Because insects also have enzyme systems that oxidize arachidonic acid, it is likely that the adhesiveness of insect hemocytes is also modified by leukotriens and/or thromboxans. Re-cent observations by Rich and coworkers (1984) indicate that aggregation of marine sponge cells is influenced by leukotriens.

A special aspect of the immediate reaction after injury is stressed by Chain and Anderson (1982, 1983). These authors report on the selective disap-pearance of plasmatocytes (PLs) as a result of the injection of bacteria into the larvae of *Galleria mellonella*. The selectivity is said to be caused by a *PL depletion factor* that can also be produced in vitro. Because this factor is blocked at 4°C, active metabolic processes must be involved in its release. The heating of the factor up to 56°C for 15 min leads to its inactivation. Varying adhesiveness of hemocytes was also found in *Leucophaea maderae*. Bohn (1977a) observed that, in in vitro experiments, PLs attach to glass surface far more rapidly than do granulocytes (GRs), which quickly degen-erate under such conditions. These results are contrary to our own inves-tigations (Hanschke et al., 1980) on the larvae of *G. mellonella* and to those of Horohov and Dunn (1982), who injected *Pseudomonas aeruginosa* into *Manduca sexta*. Furthermore, in in vitro experiments (Ehlers et al., 1987), we have never observed a varying influence on the adhesiveness of GRs and PLs. In the hemolymph of the cockroach *Periplaneta americana*, we were able to demonstrate a factor that reduces the adhesiveness of the hem-ocytes of *G. mellonella* to glass, depending on the concentration (Fig. 16.1). This factor cannot be bound to the glass surface, and therefore it may be concluded that it acts directly on the hemocytes. The availability of hem-ocytes at the reaction site (wound, implant, etc.) is not considered in these observations. Chemotactic factors released from degenerating cells are said to be responsible for the second phase of encapsulation: the attachment of PLs to the implant previously changed by the GRs (Ratcliffe and Gagen, 1977; Schmit et al., 1977; Ratner and Vinson, 1983). The migration of PLs

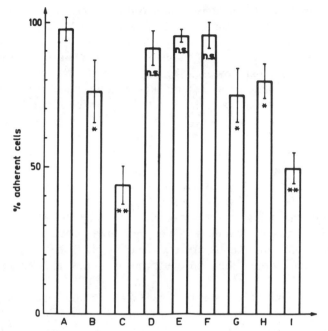

Figure 16.1. Adhesiveness of hemocytes of *Galleria mellonella* on glass in the presence of hemolymph of *Periplaneta americana* and guinea pig serum. (A) Control. (B) With addition of 10% hemolymph. (C) 20% hemolymph. (D) 20% twice frozen and thawed hemolymph. (F) 10% native guinea pig serum. (G) 10% heat-inactivated guinea pig serum. (E) On glass coated with 20% hemolymph of *P. americana*. (H) On glass coated with 10% heat-inactivated guinea pig serum. (I) On glass coated with 10% heat-inactivated guinea pig serum and supplementary addition of 10% heat-inactivated guinea pig serum. Results are the mean of four experiments, each with three parallels. *, significant difference from A, $P = 5\%$; **, significant difference from A, $P = 1\%$; n.s., not significant difference from A, Student's *t* test for paired values.)

to the wound site is also due to chemotaxis, according to Rowley and Ratcliffe (1981). Bohn (1977b) pointed out that the formation of a new epidermis in the course of wound healing presupposes earlier migration of PLs. He states that the PL may produce a factor or factors that control the rearrangement of the epidermal cells into a completely new layer. This so-called *conditioning factor* causes the epidermal cells to attach and flatten over various substrates. It is heat resistant and not inactivated by lipase, RNase, DNase, and various polysaccharide-decomposing enzymes. These results suggest that the conditioning factor is a protein or a proteid.

16.2.2. Factors Causing Changes in Foreign Body Surfaces

Ratner and Vinson (1983) drew attention to the fact that, in in vitro encapsulation reactions with hemocytes of *Heliothis virescens*, the GRs were extremely unstable and formed basophile stromata, with accompanying release of eosinophile granula within 5 min. This material is the prerequisite for the subsequent attachment of the PLs and is therefore designated the *encapsulation-promoting factor*. The change of the implant surface either by components of attached hemocytes or by the lysing of GRs connected with the deposition of coagulated hemolymph seems to be the starting point for encapsulation (Schmit and Ratcliffe, 1977). It is also possible that the attachment of hemocytes does not occur if the surface of an implant or a parasite is altered in a specific way. Rotheram and Crompton (1972) noticed in the helminth *Moniliformis dubius* that the larva becomes surrounded by a membranous coat, or envelope, to which hemocytes do not adhere. But physicochemical changes in the surface of implants also influence the ability of hemocytes to attach to them. Lackie (1981) considers the occurrence of the encapsulation an indicator of surface similarity. The charge of the implanted material exerts an essential influence on the adhesiveness of the hemocytes. While weakly alkaline beads are enclosed by thick hemocyte capsule in *Heliothis* species, only a thin capsule forms around weakly acid beads (Vinson, 1974).

In investigations of silkworm larvae (Walters and Williams, 1967) and spruce budworm (*Choristoneura fumiferana*) larvae (Dunphy and Nolan, 1982), qualitatively similar results were obtained. But the positive surface charge does not, in all cases, seem to be a necessary precondition for the attachment of hemocytes. A similar proportion of hemocytes of the locust *Schistocerca gregaria* adheres to polystyrene plates in vitro, irrespective of the changes in charge and wettability, but the adhesion of hemocytes of the cockroach *P. americana* is proportional to the increase in both parameters. Similarly, the capsules of cockroach hemocytes formed in vivo around more negatively charged polystyrene beads are thicker (Lackie, 1981, 1983). Ratner and Vinson (1983) also observed that surface changes did not always cause modified capsule formation. Thus, coating with silicone, heparin, gelatin, agar, and collagen had no influence on the course of the reaction. These observations agree with the conclusions drawn by Weiss and Blumenson (1967) from their investigations of the dynamics of the adhesion of cells derived from a human osteogenic sarcoma. Both the electrophoretic mobility, as an expression of the surface charge, and the zeta potential are reduced by the stratified deposition of serum to glass beads.

We have examined the influence of surface structure on the adhesiveness of PLs and GRs of *G. mellonella* (Ehlers et al., 1987). On untreated glass surfaces, these cells show a distinct tendency to adhere, and no variation in the adhesiveness of GRs and PLs was observed. Heat-inactivated serum of guinea pigs (GP) reduces the adhesiveness of the hemocytes of *G. mellonella* on glass. This adherence reduction can also be observed when the glass is coated with inactivated GP serum. The reduction of the adhesiveness of hemocytes is especially obvious when they are incubated on coated glass surfaces in the presence of inactivated GP serum (Fig. 16.1). But even under these conditions, the composition of the adherent population remains unchanged. It is obvious that the stratified deposition of the components of mammalian serum to the glass affects the physicochemical conditions (surface charge and wettability) that cause the change in the adhesiveness of insect hemocytes. In a comparative study, we were able to show that the hemolymph of *P. americana* and *Protophormia terra-novae* also reduced the adhesion of mammalian granulocytes (Mohrig et al., 1986). This activity is heat stable at 56°C for 30 min, prevents adsorption to glass, and can be reversed by native GP serum. From this it can be concluded that the adherence-reducing effect acts directly on the GP granulocytes and influences their adherence behavior. The nature of this hemolymph component is as yet unknown.

16.3. PHAGOCYTOSIS-INFLUENCING FACTORS

To accomplish phagocytosis, the phagocytic cells must first recognize the foreign material. In vertebrates, such recognition is mediated by serum factors (opsonins) consisting of two classes of proteins: antibodies and complement components. Both immunglobulins and the C3b component of the complement system form bridges between the foreign element and the membrane of the phagocytic cells. The bridges formed are the result of the existence of specific receptors for the *Fc* region of antibodies and for the C3b component (see also Chapter 1).

16.3.1. Complementlike Factors

Humoral recognition molecules on the basis of γ-globulins are generally absent in insects (Marchalonis, 1977). In addition to the antibodies, the other major opsonic component in vertebrates is the complement system. Some investigations support the conclusion that complement components occur in invertebrate body fluids, especially in insect hemolymph. However, all attempts to isolate complement components from invertebrates have so far failed. Therefore, all attempts to find complement or complementlike factors in invertebrates are directed toward demonstrating functional evidence of these factors (Gewurz et al., 1966; Gigli and Austin, 1971) or, in suitable experiments, proving the existence of factors that can perform the functions of single complement components. Many authors make use of the capability of cobra venom factor (CVF) with C3 proactivator, so-called factor B, to form a stable complex. This CVF factor B complex stimulates the activation of the complement components C5–C9 (late-acting components). On this basis, Day and associates (1970, 1972) have supposedly demonstrated complement components in arthropods (horseshoe crab) and in a sipunculid (*Golfingia* spp.). Anderson and colleagues (1972) showed by means of CVF that C3 proactivator activity is present in the hemolymph of *Blaberus craniifer.*

CVF was also used for the analysis of protective immunity in *G. mellonella* in response to various bacteria (Chadwick et al., 1980). These workers concluded that *C. mellonella* might have such a complementlike effector system, with which the CVF interferes (see also Chapter 10). D'Cruz (1983) reported a peptide of MW 1400 that appears in the hemolymph of *Bombyx mori* and activates the lysis of rabbit red cells by human serum. Moreover, this hemolymph complement activator acts as a specific opsonin for protein A–positive *Staphylococcus aureus*. Modulators of the hemolytic activity of mammalian serum were also described for the hemolymph of larvae of the armyworm *Spodoptera frugiperda* (D'Cruz and Day, 1984). These authors concluded that, in the hemolymph of insects, reactions corresponding to complement activation in mammals occur, but good documentation of their protective role in the defense system was not provided.

16.3.2. Melanization-Related Factors

Melanization occurs as a result of phenoloxidase activity in the hemolymph and/or hemocytes. It is well known that melanization is brought about by various enzymes or enzyme systems that oxidize tyrosine or dopamine. In the hemolymph, such enzymes occur in an inactive form and can be activated by bacteria or microbial products (see also Chapter 1). In addition to studies in which melanization is regarded as a prerequisite for capsule formation (Brewer and Vinson, 1971), other reports show that this process begins only after the attachment of the cells or that it can be inhibited by appropriate inhibitors without effecting capsule formation (Poinar and Leutenegger, 1971; Schmit et al., 1977). Other authors ascribe a protective function to the toxic by-products, such as quinones, that originate during melanization (Taylor, 1969; Walters and Ratcliffe, 1983). The phenoloxidase of insects is said to have an additional function as a nonself recognition factor (Pye and Yendol, 1972). Ratcliffe and coworkers (1984) consider the phenoloxidase system to operate according to a selectively acting recognition principle that generally enhances cellular defenses (see also Chapter 9).

16.3.3. Lectins and Agglutinins

Invertebrate phagocytes have the ability to distinguish self from nonself despite the lack of classic opsonins previously mentioned. The hypothetical model is the self-recognition concept (Rothenberg, 1978) or the cell-cell communication system (Glaser, 1976), based on the existence of protein binding to a specific receptor. The chemical bases of cell-cell recognition are membrane-bound or circulating glycoproteins or glycolipids (lectins) with oligosaccharide binding units. As far as mammals are concerned, there is no doubt that lectin-carbohydrate interaction has an additional function in the recognition mechanisms of the so-called nonimmune phagocytosis (independent of antibodies and complement activation) and that lectins act as mediators in the process of phagocyte-particle attachment (Sharon, 1984). Mammalian phagocytes are able to bind soluble lectins as well as cell surface lectins on bacteria, yeast, and other target cells. As the second step in phagocytosis, ingestion is in many cases still dependent on further surface properties, especially hydrophobicity and electric charge.

Many reports in recent years have indicated the occurrence of lectinlike hemagglutinins in the body fluid of invertebrates, including insects (McKay et al., 1969; Yeaton, 1981; Ceri, 1984). The observation that insect hemolymph contains molecules that act as agglutinins for a wide range of mammalian erythrocytes has led to the opinion that they function as recognition factors (Seaman and Robert, 1968; Anderson et al., 1972; Ratcliffe and Rowley, 1979). Most experiments, however, have shown that the hemocytes do not interact more readily with the erythrocytes or other particles. It can be concluded that hemagglutinins in insects are not involved in the recognition process or in promoting phagocytic activity (Scott, 1971a,b; Anderson et al., 1972; Ratcliffe and Rowley, 1983).

In contrast to the demonstration of hemagglutinating factors, other agglutinating activities in insect hemolymph have rarely been looked into. Scott (1971a) has tested the agglutinating activity in the hemolymph of *P. americana* with four strains of yeast and ten strains of bacteria. Agglutination could be proved only for *Aerobacter aerogens*, which, however, did not have anything to do with the promotion of phagocytosis (see also Ratcliffe and Rowley, 1983). We were able to demonstrate various natural bacterioagglutinins (in addition to hemagglutinins) in *P. terra-novae* and *P. americana* (Mohrig and Kauschke, 1987; Mohrig et al., 1987). Considering the pathogenic effect of the bacteria used, it is questionable to ascribe to the bacterioagglutinins any significance in the defense system of insects because the easily agglutinable strains—*Bacillus thuringiensis subtoxicus* (BTS), *B. megaterium*, and *Proteus vulgaris*—are extraordinarily effective insect pathogens.

It is interesting, however, that the hemolymph of *P. terra-novae* clearly promotes the ingestion of yeast cells by guinea pig polymorphonuclear leucocytes (Fig. 16.2). One part of the phagocytosis-promoting factor is absorbed by yeast cells and is probably identical to the yeast-agglutinating factor. Yeast cells precoated with hemolymph were phagocytized in large numbers. We suppose that in this case a true mediator molecule exists for which both the yeast cells and the GRs have complementary surface receptors. This observation agrees with those of Lackie (1981). Larvae of the tapeworm *Hymenolepis diminuta* are strongly agglutinated in vitro by the hemolymph of *P. americana* but not by that of the locust *Schistocerca*. Consequently, the larvae are encapsulated in vivo by the hemocytes of the cockroach but not by those of the locust. Furthermore, Yamazaki and associates (1983) show that a lectin from the hemolymph of the fly *Sarcophaga peregrina* mediates the binding of mouse macrophages to a target tumor cell. This lectin induces the killing of the tumor cell, which indicates genuine

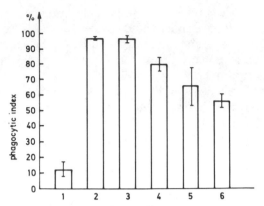

Figure 16.2. Phagocytosis on yeast cells by guinea pig granulocytes under the influence of hemolymph of *Protophromia terra-novae* in a monolayer assay. (1) Phagocytosis in pure Eagle culture medium. (2) With an addition of 10% guinea pig serum. (3) With an addition of 10% native hemolymph of *P. terra-novae*. (4) With 10% heat-treated (56°C, 30 min) hemolymph. (5) With yeast cells precoated in heat-treated hemolymph of *P. terra-novae* without any addition in the culture medium. (6) With yeast cells precoated in native hemolymph of *P. terra-novae*.

activation by a relevant receptor (see also Chapter 18). A mitogenic effect of hemagglutinating component derived from the hemolymph of scorpion *Androctonus australis* on mammalian lymphocytes was demonstrated by Brahmi and Cooper (1980). Plant lectin binding (wheat germ agglutinin) properties of insect hemocytes in connection with the identification of different hemocyte populations in the encapsulation response to the parasite wasp *Leptopilina heterotoma* were reported by Nappi and Silvers (1984). But in this case, lectin binding may function, not just as a mediator, but also as an indicator of the expression of specific carbohydrate moieties of some hemocytes that participate in capsule formation.

16.3.4. Other Phagocytosis-Promoting Factors

In the course of our observations, we found that yeast-absorbed hemolymph of *P. terra-novae* retains one aspect of the phagocytosis-promoting activity. This activity can be ascribed to neither yeast, bacteria, nor abiotic particles. In experiments with albumin-coated cadmium microcrystals, this nonopsonic activity can be easily demonstrated (Fig. 16.3). A similar activity can be found in the hemolymph of *P. americana* and *G. mellonella* (Schittek, 1984; Ehlers and Mohrig, 1986; Mohrig and Kauschke, 1986; Mohrig et al., 1986). The phagocytosis-promoting activity in the cockroach is heat stable (56°C for 30 min), nonopsonic for cells of *B. thuringiensis, Saccharomyces cerevisiae*, and Cd microcrystals and is eluated on Sephadex G75 in the outer volume (Fig. 16.3). The stimulation of the cadmium phagocytosis of

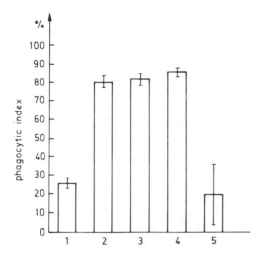

Figure 16.3. Phagocytosis of cadmium microcrystals by guinea pig granulocytes under the influence of hemolymph of *P. americana* in a monolayer assay. (1) Phagocytosis in pure Eagle culture medium. (2) With addition of 10% native hemolymph of *P. americana*. (3) With heat-treated (56°C, 30 min) hemolymph. (4) With heat-treated hemolymph absorbed with cells of *Bacillus thuringiensis subtoxicus*. (5) With 10% native hemolymph and addition of 1 mM trypan blue.

GP granulocytes is inhibited in the presence of trypan blue. If it is correct to assume that trypan blue is a specific inhibitor for complement-mediated phagocytosis (Guckian et al., 1978), then this observation may be a further hint of the existence of complementlike components in insects. The molecular mechanism of these heat-stable insect factors to promote phagocytosis of GP granulocyte remains unclear. As they do not bind on the surface of particles, we assumed that they act directly on the phagocytizing cell and change their reactivity. A comparable nonimmunologic serum activity is known from mammals as the tuftsin, mentioned above, and a recently discovered heat-stable factor in guinea pig serum that is unrelated to antibodies and complement (Schittek and Mohrig, 1985). The possibility that all these factors change the surface charge or the hydrophobicity of a given particle cannot be excluded. It has been suggested that the surface charge and hydrophobicity determine whether a particle can be recognized in the absence of antibodies, complement, or lectins (Griffin, 1982), but experimental evidence is difficult to obtain. Sharon (1984) suggested that in the recognition of biotic material, more specific mechanisms must be involved than merely nonspecific surface properties.

Our own investigations of in vivo phagocytosis of BTS in *G. mellonella* support the idea that particle ingestion by insect hemocytes must also be the result of the action of inducible humoral factors (Mohrig and Schittek, 1979; Mohrig et al. 1979a,b; Schittek, 1984). The BTS ingestion-promoting effect is induced by easily phagocytable particles and can be transferred to untreated larvae of *G. mellonella* by means of particle and cell-free hemolymph. The factor formation and/or release seems to be a result of a successful-recognition of nonself. The phagocytosis-promoting factor does not react with the surface of the BTS as an opsonin does. The inability to adsorb the activity by means of BTS, as well as the absence of preopsonic prop-

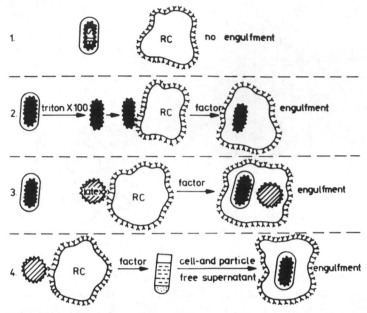

Figure 16.4. Schematic representation of the speculative mode of action of the phagocytosis-promoting factor in *G. mellonella*. (1) Foreign determinant on the surface of *B. thuringiensis subtoxicus* (BTS) is masked. No recognition takes place. (2) Changes of surface properties (demasking of foreign determinants) cause factor release and phagocytosis. (3) Simultaneous injection with easily phagocytable particle (latex beads) produces factor release by latex and consequent ingestion of BTS. (4) Factor released by latex or other phagocytable particles is transmissible to untreated animals and causes phagocytosis of BTS. (RC, hypothetical recognition cell.)

erties, leads to the assumption that the factor acts directly on the phagocyte. Figure 16.4 illustrates a speculative hypothesis to elucidate the observations discussed above, including the assumption of the existence of functionally distinct hemocyte populations.

16.4. SUMMARY

The spatial dispersion of the components of the defense systems, together with the tendency in the animal kingdom to evolve different cell populations with special reactivities in the immune response, requires an adequate humoral control. Such humoral factors are well known as mediator substances (lymphokine, interleukine, complement, etc.) between cells and tissues in vertebrates. Also to be noted are reports in recent years about the activity of mediators in defense reactions of insects or in influencing the reactivity of phagocytic cells. In particular, the change in the adherence behavior of phagocytic cells in the circulating hemolymph after injury and/or invasion of foreign material seems to be dependent on the synthesis or release of signal substances. A prerequisite for successful phagocytosis is the recognition of foreignness and the attachment of the particle to the surface of a given phagocyte. Although in a few cases and for single particles phagocytosis-promoting properties of opsonins have been proved, it must be assumed that they play only a limited and more random role.

As a class of chemically and functionally characterized substances, lectins seem to be a normal product of insect metabolism. They do not represent a defined system of defense substances, as do complement and γ-globulins in vertebrates. Furthermore, some examples have been given of the occurrence of complement factors or complementlike activities in insects and of the possible additional function of the prophenoloxidase system nonself recognition. Some new investigations on phagocytosis-promoting activities have been presented that support the idea that particle engulfment by insect hemocytes can also be a result of the action of inducible humoral factors. The hemolymph of various insects also contains factors favoring the phagocytic activity of mammalian phagocytes. Beside this nonopsonic phagocytosis-promoting effect, it has also been shown that insect hemolymph influences the adhesiveness of mammalian granulocytes and that components of mammalian serum influence the adhesion behavior of insect hemocytes. These findings may have some bearing on a phylogenetic view of the regulatory mechanisms of phagocytic cells in the animal kingdom.

REFERENCES

Anderson, R.S., N.K.B. Day, and R.A. Good. 1972. Specific hemagglutinin and a modulator of complement in cockroach hemolymph. *Infect. Immun.* **5:** 55–59.

Bloom, B.R. and B. Bennett. 1966. Mechanism of a reaction *in vitro* associated with delayed-type hypersensitivity. *Science* (Washington, DC) **153:** 80.

Bohn, H. 1977a. Differential adhesion of the hemocytes of *Leucophaea maderae* (Blattaria) to a glass surface. *J. Insect Physiol.* **23:** 185–194.

Bohn, H. 1977b. Enzymatic and immunological characterization of the conditioning factor for epidermal outgrowth in the cockroach *Leucophaea maderae. J. Insect Physiol.* **23:** 1063–1073.

Brahmi, Z. and E.L. Cooper. 1980. Activation of mammalian lymphocytes by a partially purified fraction of scorpion hemolymph. *Dev. Comp. Immunol.* **4:** 433–446.

Brewer, F.D. and S.B. Vinson. 1971. Chemicals affecting the encapsulation of foreign material in an insect. *J. Invertebr. Pathol.* **18:** 287–289.

Ceri, H. 1984. Lectin activity in adult and larval *Drosophila melanogaster. Insect Biochem.* **14:** 547–549.

Chadwick, J.S., W.P. Aston, and J.R. Ricketson. 1980. Further studies on the effect and role of cobra venom factor on protective immunity in *Galleria mellonella*:

Activity in the response against *Proteus mirabilis. Dev. Comp. Immunol.* **4:** 223–232.

Chain, B.M. and R.S. Anderson. 1982. Selective depletion of the plasmatocytes in *Galleria mellonella* following injection of bacteria. *J. Insect Physiol.* **28:** 377–384.

Chain, B.M. and R.S. Anderson. 1983. Inflammation in insects: The release of a plasmatocyte depletion factor following interaction between bacteria and hemocytes. *J. Insect Physiol.* **29:** 1–4.

Cherbas, L. 1973. The induction of an injury reaction in cultured hemocytes from saturniid pupae. *J. Insect Physiol.* **19:** 2011–2023.

Day, N.K.B., H. Geiger, J. Finstad, and R.A. Good. 1972. A starfish hemolymph factor which activates vertebrate complement in the presence of cobra venom factor. *J. Immunol.* **109:** 164–167.

Day, N.K.B., H. Gewurz, R. Johansen, J. Finstad, and R.A. Good. 1970. Complement and complement like activity in lower vertebrates and invertebrates. *J. Exp. Med.* **132:** 941–950.

D'Cruz, O.J.M. 1983. Interaction of silkworm hemolymph components with human complement. *Fed. Proc.* **42:** 1237.

D'Cruz, O.J.M. and N.K.B. Day. 1984. Purification of a human alternative complement pathway inhibitor from hemolymph of larval fall armyworm (*Spodoptera frugiperda*). *Biochem. Biophys. Res. Commun.* **122:** 1426–1433.

Dunphy, G.B. and R.A. Nolan. 1982. Cellular immune response of spruce budworm larvae to *Entomophthora egressa* protoplast and other test particles. *J. Invertebr. Pathol.* **39:** 81–92.

Ehlers, D., M. Quast, and W. Mohrig. 1987. Adhesivity of haemocytes of *Galleria mellonella. J. Insect Physiol.,* **33,** in press.

Faye, J. 1978. Insect immunity: Early fate of bacteria injected in saturniid pupae. *J. Invertebr. Pathol.* **31:** 19–26.

Gagen, S.J. and N.A. Ratcliffe. 1976. Studies on the *in vivo* cellular reactions and fate of injected bacteria in *Galleria mellonella* and *Pieris brassicae* larvae. *J. Invertebr. Pathol.* **28:** 17–24.

Gewurz, H., J. Finstad, L.H. Muschel, and R.A. Good. 1966. Phylogenetic inquiry into the origins of the complement system, pp. 105–116. In R.T. Smith, P.A. Miescher, and R.A. Good (eds.) *Phylogeny of Immunity.* University of Florida Press, Gainesville, FL.

Gigli, J. and K.F. Austin. 1971. Phylogeny and function of the complement system. *Annu. Rev. Microbiol.* **25:** 309–332.

Glaser, L. 1976. Cell-cell recognition. *TIBS* **4:** 84–86.

Griffin, F.M., Jr. 1982. Mononuclear cell phagocytic mechanism and host defense, pp. 31–55. In J.I. Gallin and A. Fanci (eds.) *Advances in Host Defense Mechanisms.* Raven Press, New York.

Guckian, J.C., W.D. Christensen, and D.P. Fine. 1978. Trypan blue inhibits complement-mediated phagocytosis by human polymorphonuclear leucocytes. *J. Immunol.* **120:** 1580–1586.

Hanschke, R., W. Mohrig, and I. Groth. 1980. Influence of injections of particular suspensions to immediate and late reactions of hemocytes in larvae of the greater wax moth (*Galleria mellonella* L.). *Zool. Jahrb. Physiol.* **84:** 181–197.

Horohov, D.W. and P.E. Dunn. 1982. Changes in the circulating hemocyte population of *Manduca sexta* larvae following injections of *Pseudomonas aeruginosa*. *J. Invertebr. Pathol.* **41:** 203–213.

Hostetter, R.K. and E.L. Cooper. 1974. Earthworm coelomocyte immunity, pp. 91–107. In E.L. Cooper (ed.) *Contemporary Topics in Immunobiology*, vol. 4, *Invertebrate Immunology*. Plenum Press, New York.

Lackie, A.M. 1981. Immune recognition in insects. *Dev. Comp. Immunol.* **5:** 191–204.

Lackie, A.M. 1983. Effect of substratum wettability and charge on adhesion *in vitro* and encapsulation *in vivo* by insect haemocytes. *J. Cell Sci.* **63:** 181–190.

Marchalonis, J.J. 1977. Phylogeny of complement, pp. 203–210. In J.J. Marchalonis (ed.) *Immunity in Evolution*. Harvard University Press, Cambridge, MA.

McKay, D., C.R. Jenkin, and D. Rowley. 1969. Immunity in the invertebrates. 1. Studies on the naturally occurring haemagglutinins in the fluid from invertebrates. *Aus. J. Exp. Biol. Med.* **47:** 125–134.

Mohrig, W. and E. Kauschke. 1987. Lectin-like substances in the hemolymph of the blowfly *Protophormia terra-novae* (Insecta, Diptera) activating phagocytosis in guinea pig polymorphonuclear leukocytes. *Dev. Comp. Immunol.*, **11** in press.

Mohrig, W. and B. Messner. 1970. Immunrektionen bei Insekten. 3. Haemozytenreaktionen und Lysozymverhalten bei *Galleria mellonella* L. *Biol. Zentralbl.* **89:** 611–639.

Mohrig, W. and D. Schittek. 1979. Phagocytosis-stimulating mediators in insects. *Acta Biol. Med. Germ.* **38:** 953–958.

Mohrig, W., D. Schittek, and R. Hanschke. 1979a. Investigations on cellular defence reactions with *Galleria mellonella* against *Bacillus thuringiensis*. *J. Invertebr. Pathol.* **34:** 207–212.

Mohrig, W., D. Schittek, and R. Hanschke. 1979b. Immunological activation of phagocytic cells in *Galleria mellonella*. *J. Invertebr. Pathol.* **34:** 84–87.

Mohrig, W., E. Kauschke, D. Ehlers, and D. Schittek. 1987. Bacterioagglutinins and phagocytosis stimulating activity in the hemolymph of the American cockroach *Periplaneta americana* L. *Dev. Comp. Immunol.*, **11** in press.

Naccache, P.H. and R.I. Sha'afi. 1982. Arachidonic acid, leukotriene B_4, and neutrophil activation. *Ann. N.Y. Acad. Sci.* **414:** 125–130.

Najjar, V.A. and J.J. Schmidt. 1980. The chemistry and biology of tuftsin. *Lymphok. Rept.* **1:** 157–180.

Najjar, V.A., B.V. Fidalgo, and A. Stitt. 1968. The physiological role of the lymphoid system. 7. The disappearance of leucokinin activity following spleenectomy. *Biochemistry* **7:** 2376–2379.

Nappi, A.J. and M. Silvers. 1984. Cell surface changes associated with cellular immune reactions in *Drosophila*. *Science* (Washington, DC) **225:** 1166.

Poinar, G.O., Jr., and R. Leutenegger. 1971. Ultrastructural investigation of the melanization process in *Culex pipiens* (Culicidae) in response to a nematode. *J. Ultrastruct. Res.* **36:** 149–158.

Powell, W.S. 1984. Properties of leukotriene B_4 20-hydroxylase from polymorphonuclear leukocytes. *J. Biol. Chem.* **259:** 3082–3089.

Pye, A.A. and W.G. Yendol. 1972. Hemocytes containing polyphenoloxidase in *Galleria mellonella* larvae after injections of bacteria. *J. Invertebr. Pathol.* **19:** 166–170.

Ratcliffe, N.A. and S.J. Gagen. 1977. Studies on the *in vivo* cellular reaction of insects: An ultrastructural analysis of nodule formation in *Galleria mellonella*. *Tissue Cell* **9:** 73–85.

Ratcliffe, N.A. and A.F. Rowley. 1979. Role of hemocytes in defense against biological agents, pp. 331–414. In A.P. Gupta (ed.) *Insect Hemocytes*. Cambridge University Press, Cambridge.

Ratcliffe, N.A. and A.F. Rowley. 1983. Recognition factors in insect hemolymph. *Dev. Comp. Immunol.* **7:** 653–656.

Ratcliffe, N.A., C. Leonard, and A.F. Rowley. 1984. Prophenoloxidase activation: Nonself recognition and cell cooperation in insect immunity. Science (Washington, DC) **226:** 557–559.

Ratner, S. and S.B. Vinson. 1983. Encapsulation reactions *in vitro* in haemocytes of *Heliothis virescens*. *J. Insect Physiol.* **29:** 855–863.

Rich, A.M., G. Weissman, C. Anderson, L. Vosshall, K.A. Haines, T. Humphreys, and P. Dunham. 1984. Calcium dependent aggregation of marine sponge cells is provoked by leukotriene B_4 and inhibited by inhibitors of arachidonic acid oxidation. *Biochem. Biophys. Res. Commun.* **121:** 863–870.

Rothenberg, B.E. 1978. The self recognition concept: An active function for the molecules of the major histocompatibility complex based on the complementary interaction of protein and carbohydrate. *Dev. Comp. Immunol.* **2:** 23–38.

Rotheram, S. and D.W.T. Crompton. 1972. Observations on the early relationship between *Moniliformis dubius* (Acanthocephala) and the haemocytes of its intermediate host, *Periplaneta americana*. *Parasitology* **64:** 15.

Rowley, A.F. and N.A. Ratcliffe. 1981. Insects, pp. 421–488. In N.A. Ratcliffe and A.F. Rowley (eds.) *Invertebrate Blood Cells*, vol. 2. Academic Press, New York.

Schittek, D. 1984. Beeinflussung der Aktivität von Phagocyten durch Faktoren der Körperfluüssigkeiten bei Everebraten (Insecta) und Vertebraten (Mammalian). Doctoral dissertation, Ernst-Moritz-Arndt University, Greifswald, East Germany.

Schittek, D. and W. Mohrig. 1985. Komplement und Antikörper unabhängige phagozytosestimulierende Aktivität im Meerschweinchenserum. Wiss. Z. Humboldt University, Berlin.

Schmit, A.R. and N.A. Ratcliffe. 1977. The encapsulation of foreign tissue implants in *Galleria mellonella* larvae. *J. Insect Physiol.* **23:** 175–184.

Schmit, A.R., A.F. Rowley, and N.A. Ratcliffe. 1977. The role of *Galleria mellonella* hemocytes in melanin formation. *J. Invertebr. Pathol.* **29:** 232–234.

Scott, M.T. 1971a. A naturally occurring haemagglutinin in the haemolymph of the American cockroach. *Arch. Zool. Exp. Gen.* **112:** 73–80.

Scott, M.T. 1971b. Recognition of foreignness in invertebrates. 2. *In vitro* studies of cockroach phagocytic haemocytes. *Immunology* **21:** 817–828.

Seaman, R. and N.L. Robert. 1968. Immunological response of male cockroaches to injection of *Tetrahymena pyriformis*. *Science* (Washington, DC) **161:** 1359.

Sharon, N. 1984. Carbohydrates as recognition determinants in phagocytosis and in lectin-mediated killing of target cells. *Biol. Cell.* **51**: 239–246.

Taylor, R.L.A. 1969. A suggested role for the polyphenoloxidase system in invertebrate immunology. *J. Invertebr. Pathol.* **14**: 427–428.

Vinson, S.B. 1974. The role of the foreign surface and female parasitoid secretion on the immune response of an insect. *Parasitology* **68**: 27.

Walters, D.R. and C.M. Williams. 1967. Reaggregation of insect cells as studied by a new method of tissue and organ culture. *Science* (Washington, DC) **154**: 516.

Walters, J.R. and N.A. Ratcliffe, 1983. Variable cellular and humoral defense reactivity of *Galleria mellonella* larvae to bacteria of differing pathogenicities. *Dev. Comp. Immunol.* **7**: 661.

Weiss, L. and L.E. Blumenson. 1967. Dynamic adhesion and separation of cells *in vitro*. 2. Interactions of cells with hydrophilic and hydrophobic surfaces. *J. Cell. Physiol.* **70**: 23–32.

Wright, A.E. and S.R. Douglas. 1903. An experimental investigation of the role of blood fluids in connection with phagocytosis. *Proc. R. Soc. Lond.*(Biol.) **72**: 357.

Yamazaki, M., M. Ikenami, H. Komano, S. Tsunawaki, H. Kamiya, S. Natori, and D. Mizuno. 1983. Polymorphonuclear leukocyte-mediated cytolysis induced by animal lectin. *Gann* **74**: 576–783.

Yeaton, R.W. 1981. Invertebrate lectins. 1. Occurrence. *Dev. Comp. Immunol.* **5**: 391–402.

CHAPTER 17

Role of Insect Lysozymes in Endocytobiosis and Immunity of Leafhoppers

Werner Schwemmler
Herold Müller
Institute for Plant Physiology, Cell Biology, and Microbiology
Free University of Berlin
Berlin, West Germany

17.1. INTRODUCTION

The role of lysozymes as antibacterial agents in the immune system of insects was first recognized by Malke (1965) and later by Mohrig and Messner (1967, 1968a,b,c; 1970); however, the general significance of lysozymes as basic antibacterial factors has since been judged more cautiously by other authors (for a summary, see Götz, 1973; Hultmark et al., 1980). Malke (1964a,b) and Malke and Schwartz (1966a,b) also pointed out the influence of lysozymes on the control of intracellular symbiosis or endocytobiosis. The latter authors also used lysozymes to disperse endocytobiosis of cockroaches (see also Wharton and Cola, 1969; Daniel and Brooks, 1972). Ehrhardt (1966) and Hinde (1971) were the first to use lysozymes to eliminate intracellular sym-

449

bionts or endocytobionts, in aphids, Nogge (1976) did the same for tsetse fly symbionts, and Schwemmler (1973a,b; 1974a; 1984) did so for leafhopper endocytobionts. Such studies have yielded general information on the control of endocytobiosis by the insect host. This chapter briefly describes these control mechanisms, using leafhopper endocytobiosis as an example.

17.2. PROCESS OF ENDOCYTOBIOSIS

As with many other insects, the survival of the small leafhoppers *Euscelis incisus* and *Euscelidius variegatus* (Homoptera: Cicadina) depends on endocytobionts. Two types of essential endocytobionts can be distinguished: primary and auxiliary, designated *a* and *t* respectively (Müller, 1949). Type *a*, which is found in primitive lice and leafhoppers, can occur alone. However, type *t*, when present, always appears with type *a*. Both belong to the newly described prokaryotic group of protoplastoids (Schwemmler, 1971; Schwemmler et al., 1975). Biochemical and electron microscopic analyses have shown that the probable maximum molecular weight of their DNA is 10^8. A third, rickettsialike prokaryotic endocytobiont, KR_E, is unnecessary for the leafhoppers; it is known as an accessory endocytobiont. Two forms of fine structures can be distinguished in every essential endocytobiont: the smaller and optically denser infectious form has a double membrane and is transmitted extracellularly to the next host generation; and the larger, intracellular and optically more transparent reproductive or vegetative form has a triple membrane (Louis and Laporte, 1969).

The process of transmission to the next host generation involves the insertion of the infectious forms of the *a* and *t* endocytobionts between the egg envelope and the egg cell, where they form the symbiont ball (Fig. 17.1). Each then undergoes a different development (Körner 1969, 1972). The *a* infectious forms are soon incorporated by the polyploid host cells, which are designated *a* bacteriocytes.* These cells disintegrate at a later stage, and the free infectious *a* endocytobionts then infect the binucleated, polyploid host cells, the a_2 bacteriocytes. Only at this stage are the infectious *t* endocytobionts incorporated into polyploid host cells, the *t* bacteriocytes. In the final a_2 and *t* bacteriocytes, the infectious symbiotic forms are transformed into their vegetative forms. The a_2 and *t* bacteriocytes combine in a particular way to form a common symbiont organ, the bacteriome.

During the embryogenesis of the host, the bacteriome divides and forms two lateral bacteriomes, located at either side of the abdomen. They each consist of a *t* bacteriome enclosed in an *a* bacteriome. During the larval development, no significant changes in the two larval bacteriomes is de-

---

* The current terms *mycetocyte* and *mycetome* were introduced, respectively, for host cells and organs infected with symbiotic yeasts. However, electron microscopic studies show that the leafhopper endocytobionts *a* and *t* are bacterialike cells without nuclei. Therefore, it is correct to use only the terms *bacteriocyte* and *bacteriome* here.

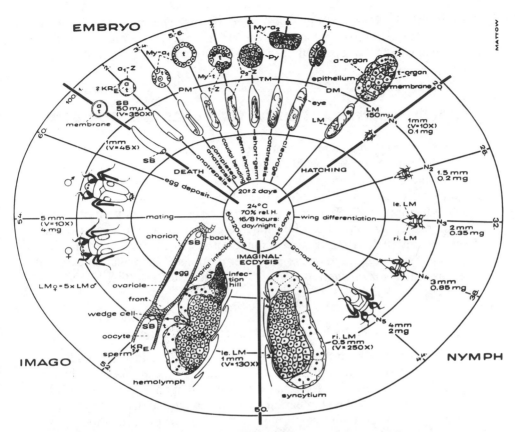

Figure 17.1. Analysis of the development of host cycle and endocytobiont cycle of cicada. The developmental cycle consists of embryonic, nymphal, and adult stages. The inner circle represents stages of host, and the outer circle the corresponding endocytobiotic structures. The time span refers to given standardized breeding conditions. Data were mainly obtained by histocultural methods. Subsequent histologic treatment of in vivo and in vitro material was carried out by regular, semifine, and ultrafine sections and by life observation and preparation according to standard methods. (a, primary endocytobiont a; a_1-Z, a_1 cell; a_2-Z, a_2 cell; DM, double bacteriome; I, imaginal stage; KR_E, rickettsialike accessory symbiont; LM, lateral bacteriome (ri, right, le, left); My-a_1, a_1 bacteriocyte; My-a_2, a_2 bacteriocyte; My-t, t bacteriocyte; N, nymphal stage; PM, primary bacteriome; PY, pyknosis; SB, symbiont ball; TM, transitory bacteriome; t, auxiliary endocytobiont t; t-Z, t cell; V, enlargement.)(From Schwemmler, 1973a.)

tected. Further differentiation occurs only in the bacteriomes of adult females. Cells of ovarian origin migrate into the *a* bacteriome and form a special structure called the infectious mount, in which the infectious *a* forms grow. In regions where the *t* bacteriomes come into contact with the ovary, infectious *t* forms develop into migratory *t* bacteriocytes. Both infectious forms *a* and *t* are released into the hemolymph of the adult female and infect the egg of the host, as previously described, through specialized follicle cells called wedge cells, from which they enter the egg. Those endocytobionts that remain in the wedge cells are lysed, thus completing the symbiotic development cycle.

17.3. LYSOZYMES AS REGULATORS IN ENDOCYTOBIOSIS

The lytic effect of lysozymes stems from their ability to cleave the β-1,4-glycosidic bond between the N-acetylmuramic acid and N-acetyl-glucosamine of the murein in the cell wall or in the remnants of the cell wall of prokaryotes or prokaryotic endocytobionts (Fig 17.2; Karlson, 1970).

The temporal and local effects of lysozymes on the endocytobiotic complex of leafhoppers have been investigated by in vivo and in vitro methods (Schwemmler, 1972, 1973a,b; Schwemmler et al., 1973). The lysozyme content of ovarioles, eggs, and bacteriocytes or portions of them was determined by measuring lysozyme activity using the plaque test (Ceriotti, 1964; Mohrig and Messner, 1968a–c).

17.3.1. Influence of Humoral Lysozymes

The lysozyme content of hemolymph from newly hatched female adult leafhoppers corresponds to an intermediate value, 200 ± 100 μg/ml, which is within the range observed for other insect hemolymphs (Mohrig and Messner, 1967). During the egg infection by the endocytobionts, lysozyme concentration in female hemolymph increases by up to five times the normal concentration. However, it never reaches the extreme value of approximately 10,000 μg/ml (50 times the normal concentration) measured in insects following microbial infection (Mohrig and Messner, 1967). Generally, the increase of lysozyme concentration in infected insect hemolymph induces lysis of microbes. Similarly, increased lysozyme content of leafhopper female hemolymph during egg infection serves to destroy those endocytobionts not incorporated by wedge cells within 1 week.

If the lysozyme content of hemolymph is increased artificially to 50–10,000-fold (10,000–2,000,000 μg/ml) of the normal content either by injection of 1%–2% lysozymes into the abdomen of the insect or by feeding 0.1% lysozymes in a synthetic diet, the vegetative and infectious symbiotic stages are lysed differently (Schwemmler, 1974a,b). The vegetative endocytobionts of bacteriocytes are only partially lysed and only in some of the developing embryonic and larval bacteriomes. In contrast to the vegetative forms, the infectious stages are lysed in the hemolymph in the a infectious mount and the t migratory bacteriocytes. Thus, egg infection is partially or completely interrupted, and resulting eggs have either a reduced or no symbiont ball (Fig. 17.3A,B,C). Eggs without a symbiont ball develop into embryos lacking an abdomen, or so-called head-thorax embryos. The different lytic effects

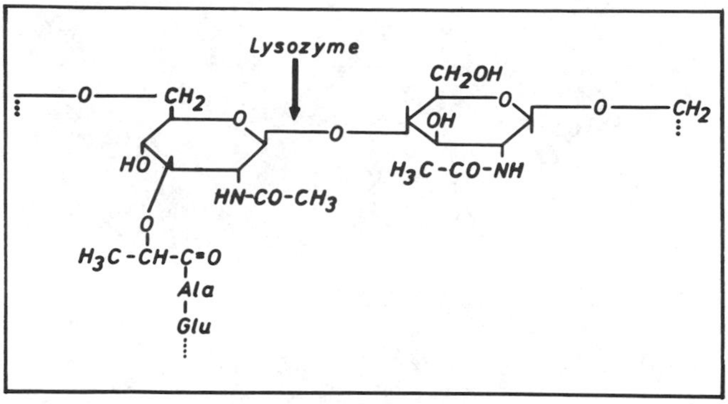

Figure 17.2. Mureine segment.

on the vegetative and infectious stages of endocytobionts can be explained by the fact that the vegetative forms possess three membranes, while the infectious stages possess only two, of which the exterior membrane exhibits activity similar to that of a lysozyme-sensitive cell wall complex (Louis and Laporte, 1969).

17.3.2. Influence of Cellular Lysozymes

The sheath cells of both the ovarioles and the bacteriomes, as well as the oocytes themselves, exhibit high lysozyme activity comparable to that of the hemolymph during the egg infection (Schwemmler, 1974a). Also, in vitro host cells of the fibroblast and epithelial type totally lyse the incorporated endocytobionts with their high level of lysozymes. The symbiont ball and the bacteriocytes of the embryo, nymph, and adult exhibit no lysozyme activity, indicating that no host cells with permanent endocytobiont colonization (e.g., bacteriocytes) possess detectable amounts of lysozymes. In contrast, all cells lacking incorporated endocytobionts (e.g., epithelial cells and oocytes) possess large amounts of lysozymes. The wedge cells, which lyse the remaining symbionts within a week, possess intermediate amounts of lysozymes.

17.3.3. Possible Control Mechanisms

Depending on behavior toward the *a* and *t* endocytobionts, the host cells that participate in the endocytobiotic cycle can be distinguished as follows (Fig. 17.4):

1. Cells that do not incorporate endocytobionts (e.g., epithelial cells of ovarioles and of bacteriomes)
2. Cells that incorporate endocytobionts (e.g., wedge cells and bacteriocytes, the former incorporating both *a* and *t* endocytobionts and the

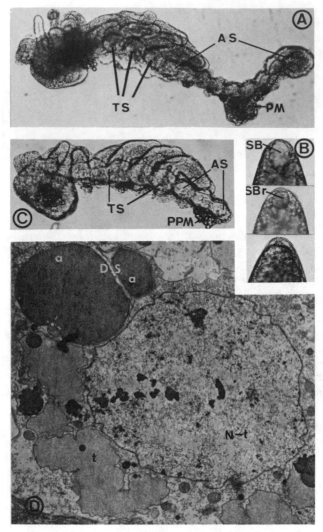

Figure 17.3. (A) Normal 6-day-old leafhopper embryo (×100). (B top) Egg with a normal symbiotic infectious mass (×100). (B, center) Egg with reduced symbiont ball, laid by a 2% lysozyme–injected female (×100). (B, bottom) Posterior egg pole without symbiotic infectious mass, obtained from an 0.1% tetracycline–fed female (×100). (C) Cephalothorax embryo with reduced abdomen (×100). (D) Double infection of embryonic *t* bacteriocyte with *t* and *a* endocytobionts, obtained from an 0.01%–tetracycline fed female (×7500). (a, *a* endocytobiont; AS, abdominal segments; DS, division stage; N-*t* nucleus of bacteriocyte; P(P)M, primary bacteriome (asymbiotic); SB(r), symbiont ball (reduced); *t*, *t* endocytobiont; TS, thoracic segments.)(From Schwemmler, 1974a.)

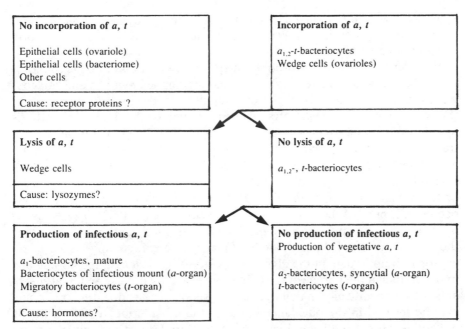

Figure 17.4. Analysis of control mechanisms of leafhopper endocytobiosis. Classification of cells in the endocytobiotic system corresponds with incorporation, lysis, and production of endocytobionts and possible distribution of membrane-bound recognition-specific receptor proteins, lysozymes, and hormones. (From Schwemmler, 1974a.)

latter selectively incorporating a or t endocytobionts), including

a. Incorporating cells that after a certain time lyse the incorporated endocytobionts (e.g., wedge cells)

b. Incorporating cells that do not lyse but integrate the incorporated endocytobionts (e.g., bacteriocytes), including

    (1) Bacteriocytes that produce infectious endocytobionts (e.g., mature a_1 bacteriocytes, a bacteriocytes of the infectious mount, and migratory t bacteriocytes)

    (2) Bacteriocytes that exclusively produce vegetative endocytobionts (e.g., young a_1 bacteriocytes, a_2 bacteriocytes, and t bacteriocytes)

The lysozyme specificity of the host cells was determined by the analysis of the temporal and spatial distribution of cellular and humoral host lysozymes. According to this analysis, all host cells of the endocytobiosis complex that do not incorporate endocytobionts exhibit high lysozyme activity; host cells that temporarily take up endocytobionts have little lysozyme activity; those that permanently shelter endocytobionts have almost no lysozyme activity. This finding explains how the host can channel and control the flow of endocytobionts from the bacteriome to the oocyte, particularly during the egg infection.

In order to study the incorporation specificity of a and t endocytobionts by wedge cells and bacteriocytes, female leafhopper nymphs and imagoes were fed a natural or a synthetic diet supplemented with 0.1% tetracycline (Schwemmler et al., 1973). Tetracycline is known to inhibit protein synthesis in eukaryotic cells (Schlegel, 1974). Embryos that developed from females treated with tetracycline contained, among other forms, t bacteriocytes infected with both t and a endocytobionts (Fig 17.3D; Schwemmler, 1974a). The infection of the epithelial cells of ovarioles and bacteriomes with a endocytobionts has been observed with the light microscope but has not yet been confirmed by electron microscopy (Schwemmler, 1983). These results indicate that normally a specific protein mechanism that specifically inhibits the incorporation of the a endocytobionts, perhaps involving a kind of receptor protein, exists in the glycocalix of the t bacteriocyte membrane and in the epithelial membrane of the bacteriomes and ovarioles. A similar mechanism that specifically inhibits phagocytosis of the t endocytobionts could also be located in the outer membrane of the a bacteriocytes as well as in epithelial cells of the bacteriomes and the ovarioles. At present, however, no known antibiotic is capable of blocking this postulated t-specific phagocytosis mechanism. The supposed distribution of a- and t-specific receptor proteins in the host cell membranes would explain how endocytobionts are specifically incorporated by the host cells and provide regulation of the passage of endocytobionts from the bacteriome to the oocyte in addition to that provided by lysozyme activity.

Finally, the production specificity of infectious endocytobiotic forms takes place only during certain stages of host growth. This specificity is perhaps controlled by the corresponding developmental stage of the host and cannot be influenced by exogenous factors (Schwartz, 1932). Since the production of infectious endocytobionts begins just before egg maturation (approximately 2 days after the hatching of female adults), it is possible that this process is coordinated with changes in the sexual hormones of the host. This assumption is supported by the observation that the infectious mount, where the a infectious stages are differentiated, is formed by immigration of ovary cells and that the migratory t bacteriocytes, which produce the t infectious forms, originate in contact with the ovaries. Addition of a preparation extracted from mature female gonads in vitro produced infectious endocytobionts in bacteriocyte cultures (Schwemmler, 1972, 1974b; Schwemmler et al., 1973). The probable coordination with the hormone system ensures that infectious endocytobionts are produced only at predetermined developmental stages of the host, namely, a short time before egg maturation. Possibly, synthesis in mature a_1 bacteriocytes is controlled by a hormone pool released during oogenesis.

17.4. CONCLUSIONS

The specificity of incorporation of the a and t endocytobionts is probably caused by a- and/or t-specific receptor proteins, but the lysis specificity results from the action of the antimicrobial lysozymes; and the production specificity of infectious endocytobionts is probably determined by coupling with the host's sexual hormone system. Figure 17.5 shows the probable distribution of lysozymes, receptor proteins, and sexual hormones in various cell types participating in the endocytobiotic complex. According to this diagram, epithelial cells of both the ovarioles and the bacteriomes contain a- and t-specific receptor proteins and a high concentration of lysozymes and are not influenced by the hormonal system of the host. Wedge cells probably possess neither a- nor t-specific receptor proteins, have a low concentration of lysozymes, and are free from control by sexual hormones. The t bacteriocytes have an a-specific receptor protein, a correlation with the host hormone system in the migratory t bacteriocytes, and no detectable activity of lysozymes. The a bacteriocytes exhibit a possible t-specific receptor protein; a correlation with the host hormone system in the a_1 bacteriocytes, as in the a bacteriocytes of the infectious mount; and no de-

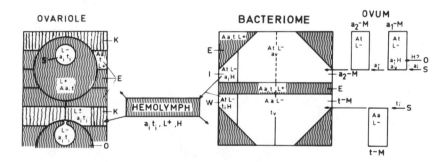

Preliminary model of the endosymbiontic control mechanism in Euscelis plebejus F. according to the possible extra- or intracellular distributions of lysozymes, hormones and receptor proteins

▨ much Lysozyme   ▨ little Lysozyme   ☐ no Lysozyme

Figure 17.5. According to preliminary model, cellular specificity of incorporation of endocytobionts is due to a- and/or t-specific membrane-bound receptor proteins, while lysis specificity results from various actions of antimicrobial lysozymes, and the specificity of infectious endocytobionts production is determined by the host's sexual hormones. Incorporation specificity was tested by antibiotic inhibition; lysozyme activity was determined by immunobiologic plaque tests to measure lysis of membranes from *Micrococcus lysodeikticus*; endocytobiotic production specificity was studied by application of leafhopper gonad extract to bacteriocyte cultures. ($a_{1,2}$-M, $a_{1,2}$ bacteriocyte; $A_{a,t}$, receptor proteins (specifically prohibiting incorporation of a and t); $a_{i,v}$, a endocytobiont, infectious, vegetative stage; E, epithelial cell; H, hormones (including production of a_i and t_i); I, infectious mount cell; K, wedge cell; $L^+, \pm, -$, many, few, or no lysozymes (effecting lysis of a_i and t_i); O, oocyte; S, symbiont ball; $t_{i,v}$, t endocytobiont, infectious, vegetative stage; t-M, t bacteriocyte; W, migratory cell.)(From Schwemmler, 1973b, 1974b.)

tectable lysozyme activity. The extracellular hemolymph system exhibits high lysozyme activity during egg infection and production of sexual hormones. The symbiont balls are free of lysozymes and are not controlled by hormones.

Many questions remain unanswered. For example, the chemical nature of leafhopper lysozymes and the mechanism of hormonal regulation and release of infectious endocytobionts are not known. These and other questions, such as the relationship of lysozyme to insect immunity (see also Chapters 1 and 10) must be solved by further systematic experiments. The model of control mechanisms of leafhopper endocytobiosis represents only the first attempt to evaluate schematically the known data.

17.5. SUMMARY

The small leafhopper species *Euscelis* and *Euscelidius* (Homoptera: Cicadina) maintain intracellular symbiosis (endocytobiosis) with two types of essential intracellular symbionts (endocytobionts) that are selectively harbored in special host cells (bacteriocytes), several of which form symbiont organs (bacteriomes) found on the right and left sides of the abdomen. The endocytobionts are transferred from one generation to the next through special infectious forms.

The leafhopper probably controls this endocytobiosis by means of a system composed of three interdependent mechanisms:

1. With the help of temporal and spatial variation in humoral and cellular lysozyme activity, the host clearly channels the flow of endocytobionts from the bacteriome to the egg.
2. Specific membrane-bound receptor proteins may serve the host by controlling the selective phagocytosis of the two types of endocytobionts.
3. The sexual hormone system of the host may participate in regulating the production of symbiotic infectious forms during egg maturation.

The increase in lysozyme content to about a fivefold level following symbiotic egg infection, compared to an approximately fiftyfold increase following microbial infection, indicates that the host still treats the endocytobionts as pathogenic agents, although in a much weaker form. This finding highlights the importance of lysozymes to the immunity of leafhoppers.

REFERENCES

Ceriotti, G. 1964. Quantitative determination of lysozyme, pp. 22–23. In *Atti III. Symp. Int. Lisozima*. Fleming, Milan.

Daniel, R.S. and M.A. Brooks. 1972. Intracellular bacteroids: Electron microscopy of *Periplaneta americana* injected with lysozyme. *Exp. Parasitol.* **31**: 232–246.

Ehrhardt, P. 1966. Die Wirkung von Lysozyminjektionen auf Aphiden und deren Symbionten. *Z. Vgl. Physiol.* **53**: 130–141.

Götz, P. 1973. Immunreaktionen bei Insekten. *Naturwiss. Rdsch.* **26**: 367–375.

Hinde, R. 1971. The control of the mycetome of the aphids *Brevicoryne brassicae, Myzus persicae,* and *Macrosiphum rosae. J. Insect Physiol.* **17**: 1791–1800.

Hultmark, D., H. Steiner, T. Rasmuson, and H.G. Boman. 1980. Insect immunity: Purification of properties of 3 inducible bacterial proteins from hemolymph of immunized pupae of *Hyalophora cecropia. Eur. J. Biochem.* **106**: 7–16.

Karlson, P. 1970. *Biochemie.* Georg Thieme Verlag, Stuttgart, West Germany.

Körner, H. 1969. Die embryonale Entwicklung der symbiontenführenden Organe von *Eucelis plebejus* (Homopt., Cicad.). *Oecologia* (Berlin) **2**: 319–346.

Körner, H. 1972. Elektronenmikroskopische Untersuchungen am embryonalen Mycetom der Kleinzikade *Euscelis plebejus* FALL, (Homopt., Cicad.). *Z. Parasitenk.* **40**: 203–226.

Louis, C. and M. Laporte. 1969. Caractère ultrastructuraux et différenciation des formes migratrices des symbiotes chez *Euscelis plebejus* (Homoptera, Jassidae). *Ann. Soc. Entomol. Fr.* **5**: 799–809.

Malke, H. 1964a. Wirkung von Lysozym auf die Symbionten der Blattiden. *Z. Allgem. Mikrobiol.* **4**: 88–91.

Malke, H. 1964b. Production of aposymbiontic cockroaches by means of lysozyme. *Nature* (London) **204**: 1223–1224.

Malke, H. 1965. Über das Vorkommen von Lysozym in Insekten. *Z. Allgem. Mikrobiol.* **5**: 42–47.

Malke, H. and W. Schwartz. 1966a. Untersuchungen über die Symbiose von Tieren mit Pilzen und Bakterien. 12. Die Bedeutung der Blattiden-Symbiose. *Z. Allgem. Mikrobiol.* **6**: 34–68.

Malke, H. and W. Schwartz. 1966b. Die Rolle des Wirtslysozyms in der Blattiden-Symbiose. *Arch. Mikrobiol.* **53**: 17–23.

Messner, B. and W. Mohrig. 1970a. Zur Immunität der wirbellosen Tiere: Übersicht, Probleme, Ausblick. *Biol. Rdsch.* **8**: 158–170.

Messner, B. and W. Mohrig. 1970b. Zum gemeinsamen Vorkommen von Hämagglutininen und Lysozymen bei Pflanzen und Tieren. *Acta Biol. Med. Germ.* **25**: 891–903.

Mohrig, W. and B. Messner. 1967. Lysozym im humoralen Abwehrmechanismus spezifisch und unspezifisch immunisierter Insektenlarven. *Biol. Rdsch.* **5**: 181–183.

Mohrig, W. and B. Messner. 1968a. Immunreaktionen bei Insekten. 1. Lysozym als grundlegender antibakterieller Faktor im humoralen Abwehrmechanismus der Insekten. *Biol. Zentralb.* **87**: 439–470.

Mohrig, W. and B. Messner. 1968b. Antibakterielles Abwehrsystem im Darmtrakt von Insekten. *Biol. Rdsch.* **6**: 136–138.

Mohrig, W. and B. Messner. 1968c. Lysozym als antibakterielles Agens im Bienenhonig und Bienengift. *Acta Biol. Med. Germ.* **21**: 85–95.

Mohrig, W., R. Storz, and B. Messner. 1970. Insektenimmunität. 3. Hämocyten-reaktionen und Lysozymänderungen bei *Galleria mellonella*. *Biol. Zentralb.* **89:** 611–639.

Müller, H.J. 1949. Zur Systematik und Phylogenie der Zikadenendosymbiose. *Biol. Zentralb.* **68:** 343–368.

Nogge, G. 1976. Sterility in tsetse flies, *Glossina morsitans,* caused by loss of symbionts. *Experientia* **32:** 995–996.

Schenk, H. and W. Schwemmler 1983. *Endocytobiology* 2. *Intracellular Space as Oligogenetic Ecosystem.* Walter de Gruyter, Berlin.

Schlegel, H.G. 1976. *Allgemeine Mikrobiologie,* Edition 4. Georg Thieme Verlag, Stuttgart, West Germany.

Schwartz, W. 1932. Neue Untersuchungen über die Pilzsymbiose der Schildläuse (Lecaniinen). *Arch. Mikrobiol.* **3:** 446–472.

Schwemmler, W. 1971. Intracellular symbionts: A new type of primitive prokaryote. *Cytobiologie* **3:** 427–429.

Schwemmler, W. 1972. In vivo und in vitro Analysen zur Struktur, Funktion und Evolution der Endosymbiose von *Euscelis plebejus* F. (Hemipt., Homop., Cicad.) Doctoral dissertation, University of Freiburg, West Germany.

Schwemmler, W. 1973a. Beitrag zur Analyse des Endosymbiosezyklus von *Euscelis plebejus* F. (Hemipt., Homop., Cicad.) mittels in vitro Beobachtung. *Biol. Zentralb.* **92:** 749–772.

Schwemmler, W. 1973b. In vitro Vermehrung intrazellulärer Zikaden-Symbionten und Reinfektion asymbiontischer Mycetocyten-Kulturen. *Cytobios* **8:** 63–73.

Schwemmler, W. 1974a. Control mechanisms of leafhopper endosymbiosis, pp. 179–187 In E.L. Cooper (ed.) *Contemporary Topics in Immunbiology,* vol. 4. Plenum Press, New York.

Schwemmler, W. 1974b. Studies on the fine structure of leafhopper intracellular symbionts during their reproductive cycles (Hemiptera, Deltocephalidae). *Appl. Entomol. Zool.* **9:** 205–214.

Schwemmler, W. 1978. *Mechanismen der Zellevolution: Grundriss einer modernen Zelltheorie.* Walter de Gruyter, Berlin.

Schwemmler, W. 1983. Analysis of possible gene transfer between an insect host and its bacteria-like endocytobionts. *Int. Rev. Cytol.* **14:** 247–266.

Schwemmler, W. 1984. *Reconstruction of Cell Evolution: A Periodic System.* CRC Press, Boca Raton, Fl.

Schwemmler, W., J.-M. Quiot, and A. Amargier. 1971. Ètude de la symbiose intracellulaire sur cultures organotypiques et cellulaires de l'homoptère *Euscelis plebejus* F. (Cicad.) en milieux à fraction standard. *Ann. Soc. Entomol. Fr.* **7:** 423–438.

Schwemmler, W., J.-L. Duthoit, G. Kuhl, and C. Vago. 1973. Sprengung der Endosymbiose von *Euscelis plebejus* F. und Ernährung asymbiontischer Tiere mit synthetischer Diät (Hemiptera, Cicadidae). *Z. Morphol. Oekol. Tiere* **74:** 297–322.

Schwemmler, W., G. Hobom, and M. Egel-Mitani. 1975. Isolation and characterization of leafhopper endosymbiont DNA. *Cytobiologie* **10:** 249–259.

Wharton, D.R.A. and J.E. Cola. 1969. Lysozyme action on the cockroach *Periplaneta americana* and its intracellular symbionts. *J. Insect Physiol.* **15:** 1647–1658.

PART III

TECHNIQUES AND BIOMEDICAL APPLICATIONS

CHAPTER 18

Isolation, Characterization, and Inhibition of Arthropod Agglutinins

Mark R. Stebbins

Genetic Systems Corporation
Seattle, Washington

Kenneth D. Hapner

Department of Chemistry
Montana State University
Bozeman, Montana

18.1. INTRODUCTION

Agglutinins or lectins are ubiquitous molecules that occur throughout plants, animals, and microorganisms (Gold and Balding, 1975). They are nonenzymatic proteins or glycoproteins that bind to erythrocyte and other cell surfaces and to glycoconjugates, causing their respective agglutination or precipitation. Their specificity is normally defined by the structures of simple saccharides that inhibit agglutinin-mediated agglutination or precipitation reactions. The broad presence of agglutinins among the Arthropoda has been recorded by Vasta and Marchalonis (1983) and Yeaton (1981). Most arthropodan agglutinins that have been studied are humoral components of the hemolymph. Agglutinins have also been detected in various tissues from the lobster (Hall and Rowlands 1974a), peritrophic membrane (Peters et al., 1983), epidermal cell membranes (Mauchamp, 1982), perivitelline fluid (Shishikura and Sekiguchi 1984a), hemocyte membranes (Amirante, 1976; Amirante and Mazzalai 1978; Amirante and Basso, 1984), fat body (Komano et al., 1983), muscle tissue (Denburg, 1980), and eggs (Stynen et al., 1982). Their concentration may be dependent on age, habitat, or metamorphic stage of the organism.

Although agglutination activity has been detected in many arthropods, only a limited number of agglutinins have been isolated and purified. This chapter focuses on those agglutinins for which some detailed data are available. There are about 20 extensively described agglutinins distributed among four major groups of Arthropoda. The taxa and representative organisms from which agglutinins have been isolated are the Crustacea (lobsters, barnacles, and crabs); Insecta (grasshoppers, crickets, cockroaches, moths, flies, and beetles); Merostomata, or Aquatic Chelicerata (horseshoe crabs); and Arachnida, or Terresterial Chelicerata (scorpions). Isolation, molecular, and inhibitory properties of each arthropodan agglutinin are described in the following sections. Additional discussions, definitions of terms, and comparative analyses of the molecular features of purified arthropodan agglutinins are included in Chapter 8.

18.2. AGGLUTININS FROM THE MEROSTOMATA: AQUATIC CHELICERATA

18.2.1. Xiphosurida: Horseshoe Crabs

Agglutinins have been purified and molecularly characterized from four species of horseshoe crabs: *Limulus polyphemus, Carcinoscorpius rotunda cauda, Tachypleus tridentatus,* and *T. gigas.*

18.2.1.1. Limulus polyphemus. Limulin, the agglutinin present in the hemolymph of *L. polyphemus,* has been isolated by several techniques since its initial purification by Marchalonis and Edelman (1968). Early procedures involving preparative ultracentrifugation (to remove abundant hemocyanin), zone electrophoresis, gel chromatography, and density gradient centrifugation (Finstad et al., 1972) have evolved to include ion exchange (Roche and Monsigny, 1974) and affinity chromatographic procedures (Nowak and Barondes, 1974; Oppenheim et al., 1974; Roche et al., 1975). A single-step procedure for the isolation of limulin that involves affinity chromatography on a column consisting of hog gastric mucin glycopeptide conjugated to Sepharose 4B has recently been reported (Muresan et al., 1982). Each of the various isolation procedures produces apparently pure preparations of limulin. Two procedures, those of Roche and coworkers (1975) and Muresan and coworkers (1982) result in preparations of limulin having extremely high specific activity. In both methods, limulin is eluted from the affinity chromatography column by the use of a series of three Tris·HCl buffers as shown in Figure 18.1. The column containing absorbed limulin is first washed with the sample application buffer that contains 0.1 M NaCl and 0.01 M $CaCl_2$. This solution washes out extraneous materials not retained by the column. A second buffer containing 1.0 M NaCl and 0.01 M $CaCl_2$ releases glycoprotein materials that are apparently associated with limulin and therefore retained by the column but that lack hemagglutinating activity. Finally, the column is eluted with calcium-free buffer that contains 1.0 M NaCl, and highly pure limulin is released from the affinity matrix owing to the depletion of Ca^{2+}, obligatory for binding. Approximately 6–10 mg of limulin may be purified from 1 liter of hemolymph. Greater than quantitative yields (approximately 125%–300%) are realized owing to the presumed presence of inhibitory substances or nonfunctional limulin molecules that are removed

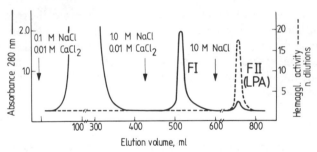

Figure 18.1. Affinity chromatography of limulin on a column (2 × 22 cm) of hog gastric mucin glycopeptide conjugated to Sepharose 4B. Declotted and dialyzed hemolymph (300 ml) is applied, and the column is washed with buffer (0.05 M Tris·HCl, 0.1 M NaCl, 0.01 M CaCl$_2$; pH 7.8) to remove nonbinding materials (first peak). Stepwise elution with 0.05 M Tris·HCl (pH 7.8) containing 1.0 M NaCl and 0.01 M CaCl$_2$ removes a nonactive fraction (FI), and subsequent elution with calcium-free buffer containing 1.0 M NaCl releases the purified limulin (FII) from the column. The FI peak contains heterogeneous glycoproteins unrelated to limulin. Hemagglutinating activity is measured with horse erythrocytes. (From Muresan et al., 1982.)

during the purification steps. These procedures have resulted in 30,000-fold purification of limulin over that present in the original hemolymph. Limulin is the only arthropodan agglutinin that is currently commercially available.

Several laboratories have examined the physicochemical properties of limulin (Marchalonis and Edelman, 1968; Finstad et al., 1972; Finstad et al., 1974; Nowak and Barondes, 1974; Oppenheim et al., 1974; Roche and Monsigny, 1974, 1979; Roche et al., 1975; Muresan et al., 1982). The native structure is a high molecular weight (approximately 400,000) aggregate composed of noncovalently associated glycoprotein subunits of approximate MW 22,500. About 4% carbohydrate is present as glucosamine and neutral sugar. Reports of the subunit molecular weight vary from 19,000 (Roche and Monsigny, 1974) to 29,000 (Muresan et al., 1982), presumably owing to factors related to uncertainties involving polyacrylamide gel electrophoresis of glycoproteins (Furthmayer and Marchesi, 1976) or to variations in the carbohydrate content of limulin (Kehoe et al., 1979). About 18 identical subunits appear to noncovalently assemble into the 400,000-MW aggregate. No interchain disulfide bonds (between subunits) are present, but each subunit contains one intradisulfide bond and one free cysteine (Kaplan et al., 1977). The amino acid composition of limulin (Table 8.2, in Chapter 8) shows typically high amounts of aspartic acid and glutamic acid. Sequence determination indicates leucine to be the only N terminal amino acid and shows no sequence heterogeneity through the first 50 amino acid residues (Kaplan et al., 1977; Kehoe et al., 1979). Lack of sequence heterogeneity throughout this region and the results of peptide mapping (Marchalonis and Edelman, 1968) indicate that the protein subunits are similar or identical. The amino acid sequence of limulin is described elsewhere in this volume (see Chapter 12).

The limulin molecule is labile to heat and EDTA and is dependent on Ca^{2+} for hemagglutination activity. Calcium appears to stabilize the high molecular weight aggregate, since dissociation occurs in its absence. The aggregated structure also dissociates if subjected to high (pH 9.6) or low (pH 3) pH. Hemagglutinating activity and the high molecular weight structural properties are restored by addition of Ca^{2+} but not of Mg^{2+}. Circular dichroism measurements (Finstad et al., 1972; Roche et al., 1978) indicate minimal far ultraviolet ellipticity and little or no ordered secondary structure. Removal of Ca^{2+} induces a slight but uninformative increase in ellipticity. Hence, the role of Ca^{2+} in conformational aspects of the limulin structure is uncertain, but Ca^{2+} does have a major role in hemagglutination and aggregation properties of the molecule.

Limulin is capable of agglutinating many vertebrate erythrocytes and other cell types. Most carbohydrate inhibition studies have used erythrocytes from horse or human sources. Purified limulin agglutinates horse erythrocytes at a minimal concentration of 0.002 µg/ml (Roche et al., 1975). Hemagglutination (horse erythrocytes) by limulin is inhibited by sialic acids and by sialic acid–containing carbohydrates and glycoconjugates. Roche and colleagues (1975) show that 2-0-methyl derivatives of N-acetylneuraminic acid (NANA) and the disaccharide N-acyl-α-neuraminyl(2 → 6)N-acetyl-galactosaminitol are stronger inhibitors than are free NANA and N-glycoylneuraminic acid (NGNA). Inhibition by the neuraminyl disaccharide (0.04 mM) is equal in strength to the normally potent inhibition by sialoglycoproteins. Extensive comparative studies on the inhibition of limulin by structural analogs of sialic acid have been carried out (Roche and Monsigny 1979; Cohen et al., 1983; Cohen et al., 1984). One of the conclusions made is the essentiality of the charged carboxylate group for effective inhibition of hemagglutination. Hemagglutination (chicken erythrocytes) by limulin is also inhibited by 2-keto-3-deoxyoctonate (KDO), a molecular component of bacterial lipopolysaccharide (McSweegan and Pistole, 1982; Rostam-Abadi and Pistole 1982). Antibacterial activity of limulin is discussed in Chapter 11.

18.2.1.2. Carcinoscorpius rotunda cauda. The agglutinin named carcinoscorpin, present in the hemolymph of *Carcinoscorpius rotunda cauda*, has been isolated and characterized by Bishayee and Dorai (1980) and by Dorai and associates (1981). The agglutinin is isolated from dialyzed, declotted hemolymph by affinity chromatography on columns of ovine submaxillary mucin (OSM) conjugated to Sepharose 4B and on columns of fetuin-Sepharose 4B. A DEAE-Sephadex A-50 step interposed between the two affinity columns removes traces of contaminating hemocyanin. Absorbed agglutinin is released from the affinity matrix by elution with Tris·HCl buffer (pH 8.0) that contains 40 mM sodium citrate. The citrate presumably complexes with Ca^{2+}, which is obligatory for agglutinin binding to the affinity

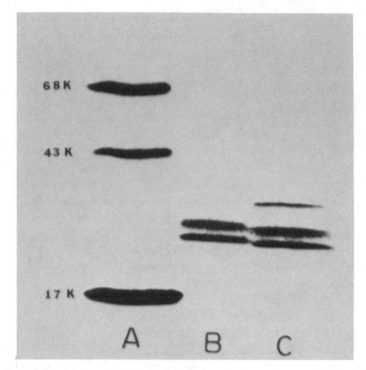

Figure 18.2. SDS polyacrylamide gel electrophoresis of carcinoscorpin. Lane A contains molecular weight markers (17,000, 43,000, and 68,000); lanes B and lane C contain 20 μg of carcinoscorpin and commercial limulin, respectively. Mercaptoethanol is not included. The two carcinoscorpin subunits (MW 27,000 and 28,000) are slightly slower (larger) than those observed in limulin. (From Bishayee and Dorai, 1980).

matrix. This procedure gives homogeneous agglutinin in approximately 60% yield and enables 1.5 mg to be isolated from 100 ml hemolymph.

Carcinoscorpin is a 420,000-MW aggregate composed of 27,000 and 28,000 glycoprotein subunits, as indicated in Figure 18.2. The total carbohydrate content is 5.8%. Carcinoscorpin agglutinates rabbit and horse erythrocytes at minimal concentrations of 0.03 μg/ml and 0.001 μg/ml, respectively. No interchain disulfide bonds are present, and stability toward reducing agents suggests the lack of any intrachain disulfide cross-links. The molecule is stable to 6 M urea but labile to heat, guanidine hydrochloride, and extremes of pH. Calcium ion is required for activity and cannot be replaced by Mg^{2+}, Mn^{2+}, Ba^{2+}, or Sr^{2+}. Reagents that attack tryptophan also destroy agglutination activity. An inhibitory disaccharide, O-(N-acetylneuraminyl)-(2 → 6)-2-acetamido-2-deoxy-D-galactitol (NANG), binds with an association constant of 1.15×10^6 M^{-1} at an inhibitory site involving tryptophan (Mohan et al., 1983). The amino acid composition of carcinoscorpin is known (Table 8.2). The protein subunits are similar or identical, based on identification of a single N terminal amino acid (leucine) and results of peptide mapping. Although similar to limulin in gross structure and binding properties, the two proteins are antigenically unrelated. Like limulin, carcinoscorpin has little or no ordered secondary structure (Mohan et al., 1984).

Carcinoscorpin shows broad binding specificity toward acidic sugars and is inhibited (rabbit erythrocytes) most strongly by sialic acids and glycoconjugates that contain sialic acid. Mohan and colleagues (1982) have analyzed the binding and inhibitory properties of carcinoscorpin with various sugars and glycoproteins. NANA, NGNA, NANG, N-acetylneuraminyl lactose, and sialoglycoproteins are among the strongest inhibitors of both hemagglutination and fetuin binding by carcinoscorpin. It is interesting to note that KDO and glucuronic acid (GlcA), both components of bacterial surfaces, are relatively powerful inhibitors of carcinoscorpin. Binding is dependent on Ca^{2+} and is maximal at 20 mM concentration. The presence of multiple binding sites in carcinoscorpin, characterized by positive cooperativity, may be related to structural changes in the architecture of the molecule and give rise to heterogeneous and multispecific interactions with glycoconjugates (see Section 8.4.1, in Chapter 8).

18.2.1.3. Tachypleus tridentatus. Hemolymph of the Japanese horseshoe crab, *Tachypleus tridentatus*, contains multispecific agglutination activity inhibitable by N-acetyl amino sugars (Shimizu et al., 1977, 1979; Shishikura and Sekiguchi, 1983). Shimizu and coworkers (1977) separated four agglutinin fractions by affinity chromatography on bovine submaxillary mucin (BSM) Sepharose 4B. A major fraction, M3 was purified further on a N-acetylgalactosamine (GalNAc) Sepharose 4B column using 50 mM N-acetylglucosamine (GlcNAc) as the eluant. M3 is isolated in 16% yield, resulting in about 1 mg from 100 ml hemolymph. Alternatively, M3 is isolated in a single step by affinity chromatography of whole hemolymph on a column of GalNAc-Sepharose 4B and elution with 50 mM GlcNAc (Shimizu et al., 1979). Increasing the GlcNAc concentration in the eluant to 500 mM releases a fifth agglutinin, called GN3.

At least three of the *T. tridentatus* agglutinins are large noncovalent protein aggregates, consisting of differently sized subunits of MW 20,000–40,000. Amino terminal sequences of M3 and GN3 agglutinins are Pro-Ile-Pro and Gln-Try-Pro, respectively. This observation, along with differences in amino acid composition, subunit structure, antigenicity, and carbohydrate specificity, indicates the presence of heteroagglutinins in *T. tridentatus* hemolymph (Shimizu et al., 1979). Hemagglutinating activity is destroyed by heat but augmented by $CaCl_2$ in three of the five active fractions. Each heteroagglutinin is multispecific toward animal erythrocytes; however, none agglutinates bovine red blood cells.

Rechromatography of the M3 fractions on GalNAc-Sepharose and stepwise elution with increasing amounts of GalNAc in the elution buffer results in the further separation of M3 into four subfractions. These subgroups are designated G1, G2, G3, and G4, respectively. They are all noncovalently associated aggregates of 65,000–73,000-MW subunits composed of 32,000–38,000-MW polypeptide chains cross-linked by disulfide bonds. The amino acid compositions of G1, G2, G3, and G4 are similar and have typically high contents of glycine and acidic amino acids (see Table 8.2). The GN3 heteroagglutinin shows a different structure that contains subunits near MW 20,000 not stabilized by disulfide bonds.

Inhibition properties of the heteroagglutinins are generally similar. NANA, GlcNAc, GalNAc, N-acetylmannosamine (ManNAc), N-acetyl-muramic acid (NAMA), and sialoglycoproteins inhibit in the 1–5 mM and 1 μg/ml concentration range, but amino sugars, hexoses, and GlcA are non-inhibitory. Lack of inhibition by GlcA is contrary to the strong inhibition of carcinoscorpion by this sugar acid (Bishayee and Dorai, 1980). The inhibition potency of glycoprotein-associated sialic acid is much greater than that of free sialic acid, indicating that the agglutinin binds to structurally associated carbohydrate structures in addition to sialic acid.

Shishikura and Sekiguchi (1983) purified one of the heteroagglutinins from *T. tridentatus* by a combination of affinity chromatography on BSM-Sepharose 4B and gel filtration procedures. The agglutinin, designated TTA-III, exists in multiple high molecular weight polymeric forms (420,000–250,000) composed of identical 42,000-MW subunits. It appears that hemolymph factors and choice of separation procedures may influence the aggregation properties of the agglutinin. TTA-III reacts strongly with horse erythrocytes and is isolated in 44% yield, enabling 0.8 mg of pure agglutinin to be prepared from 100 ml hemolymph. During purification, the total activity increased by tenfold, indicating the possible presence of inhibitors in the initial hemolymph. The agglutinin TTA-III is antigenically distinct from the other isolated *Tachypleus* agglutinins, thus confirming the existence of heteroagglutinins. The *T. tridentatus* heteroagglutinins are also antigenically unrelated to limulin.

18.2.1.4. Tachypleus gigas. A multispecific agglutinin is present in the embryonic perivitelline fluid of the horseshoe crab *Tachypleus gigas* (Shishikura and Sekiguchi, 1984a). The agglutinin is isolated from declotted perivitelline fluid by affinity chromatography on BSM-Sepharose 4B in pH 7.5 Tris·HCl buffer. After nonabsorbed materials are washed through the column, the agglutinin is eluted with 0.5 M GlcNAc. The agglutinin is then purified to homogeneity by gel filtration in the presence of 1 M urea (see Fig. 8.3, in Chapter 8). Two prominent peaks are seen in gel filtration, one containing the pure agglutinin and the other containing glycoprotein materials that may represent inhibitory substances naturally present in the perivitelline fluid (Shishikura and Sekiguchi, 1984b). Recovery of agglutinin applied to the affinity column is 85% and allows isolation of 180 μg of agglutinin from 750 ml perivitelline fluid.

The purified agglutinin is a highly polymerized protein aggregate of

450,000 MW composed of 40,000-MW subunits. Possible disulfide involvement in the aggregated structure is uncertain, since electrophoresis of the denatured molecule was performed only in the presence of mercaptoethanol. The fact that the native molecule is irreversibly inhibited by mercaptoethanol suggests that disulfide bonds have a stabilizing role. Acid and base also cause irreversible inhibition, whereas the effects of urea and guanidine hydrochloride are reversed by dialysis. Dialysis in calcium-free buffers inactivates the agglutinin, and addition of calcium restores activity. Both human and horse erythrocytes are agglutinated by *T. gigas* agglutinin.

The purified agglutinin from the perivitelline fluid is inhibited (horse erythrocytes) by *N*-acetylamino sugars and sialic acid–containing BSM. The order of inhibitory effectiveness is BSM > NANA > GalNAc = GlcNAc > fucose in the concentration range 0.2–50 mM for the carbohydrates. The endogenous inhibitory substance isolated during the gel filtration step is an extremely potent agglutination inhibitor at a concentration of 0.002 μg/ml. Cross-absorption experiments indicate that the agglutinin is capable of being completely absorbed by horse or human erythrocytes, thus demonstrating its multispecificity.

18.3. AGGLUTININS FROM THE CRUSTACEA

18.3.1. Decapoda: Lobsters, Crabs, and Crayfish

18.3.1.1. Homarus americanus. Hemagglutinating activity in the hemolymph of the lobster *Homarus americanus* was first demonstrated by Cornick and Stewart (1968, 1973). Initial isolation attempts by using pevikon block electrophoresis and gel filtration chromatography (Hall and Rowlands, 1974a) resulted in the low yield purification of two structurally different agglutinins. The heteroagglutinins, LAg1 and LAg2, differ in molecular weight, electrophoretic mobility, antigenicity, and erythrocyte agglutination specificity (Hall and Rowlands, 1974b).

More recent isolation methods, based on affinity chromatography, (Hartman et al., 1978; VanderWall et al., 1981) result in better yields of LAg1 and LAg2; however, precautions must be taken to avoid cross-contamination of the heteroagglutinins. In these procedures, the hemolymphatic ammonium sulfate fraction that contains the heteroagglutinins is applied to affinity chromatography columns composed of BSM or fetuin covalently conjugated to Sepharose 4B. Both heteroagglutinins are absorbed onto the glycoprotein affinity matrix, and other, nonabsorbed, extraneous materials are washed from the column with neutral buffer. The agglutinins are desorbed and eluted from the column through application of the appropriate inhibitory carbohydrate. Elution with 0.2 M GalNAc releases LAg2 from the affinity matrix, whereas elution with 0.2 M ManNAc results in release of LAg1. The two agglutinins prepared in this manner, although free of other proteins, are

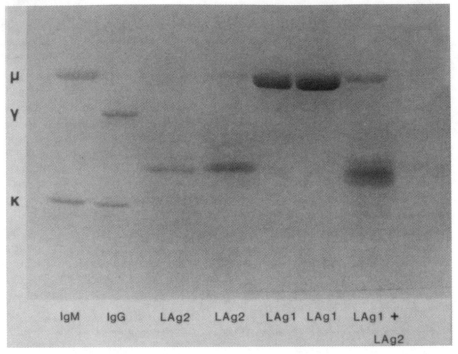

Figure 18.3. SDS-polyacrylamide gel electrophoresis of lobster heteroagglutinins LAg1 and LAg2 prepared by affinity chromatography. Lanes 1 and 2 (from the left) are markers corresponding to MW 22,000 (K), MW 50,000 (Y), and MW 70,000 (u). Lane 3 shows LAg2 that has had contaminating LAg1 removed by treatment with colominic acid–Sepharose. Lane 4 shows LAg2 and contaminating LAg1 resulting from elution of absorbed fetuin-Sepharose 4B with 0.2 M GalNac. Lanes 5 and 6 show LAg1 purified with colominic acid–Sepharose 4B. Lane 7 shows LAg1 and LAg2 eluted from fetuin–Sepharose 4B with 0.2 M both GalNAc and ManNAc. The electrophoretic pattern is the same in the presence or absence of mercaptoethanol. (From VanderWall et al., 1981.)

cross-contaminated. The two agglutinins may be completely purified by application to a second affinity chromatography column, consisting of colominic acid–Sepharose 4B (colominic acid is an α-$(2 \rightarrow 8)$ homopolymer of NANA). LAg1 is quantitatively absorbed to the colominic acid matrix, whereas LAg2 passes through the column and is uncontaminated by LAg1. Pure LAg1 is then eluted from the colominic acid–Sepharose 4B matrix with 0.2 M ManNAc. The ammonium sulfate fractionation step may be omitted from this procedure without risk. In fact, LAg1 may be purified from whole unfractionated hemolymph in a single step by affinity chromatography with colominic acid Sepharose 4B. Figure 18.3 shows the electrophoretic characteristics of the lobster heteroagglutinins purified by these procedures.

The lobster heteroagglutinins have been characterized by Hall and Rowlands (1974a,b), VanderWall and colleagues (1981), and Abel and colleagues

(1984). Both molecules are high molecular weight aggregates of smaller, identical protein subunits. LAg1 is the larger, with a sedimentation constant in excess of 19 S, whereas that of LAg2 is about 11 S. A similar relationship is seen during gel filtration of whole hemolymph, indicating that molecular size is are not influenced by other hemolymphatic components. It is interesting that different molecular weight forms of LAg1 (500,000 or 700,000) are observed depending on the type of affinity matrix used in their isolation (Abel et al., 1984). The molecular weight of the two forms is unaffected by change in pH from 7.8 to 5.5. Immunodiffusion of the respective purified molecules against antiagglutinin rabbit serum gives no common precipitin lines, confirming the heterogeneity between LAg1 and LAg2. LAg1 and LAg2 are, respectively, noncovalent assemblies of 70,000 and 35,000 molecular weight subunits, as determined by electrophoresis (Fig. 18.3). The molecular weight of the subunits is independent of reducing agents, indicating that no interchain disulfide bonds are present and that the 70,000 and 35,000 subunits are each composed of a single polypeptide chain. Both agglutinins are inactivated by heat, trypsin, or EDTA. The effects of EDTA are reversed by dialysis and subsequent addition of $CaCl_2$, showing that agglutination activity is dependent on Ca^{2+}. LAg1 and LAg2 stain positively with periodic acid–Schiff reagent, indicating them to be glycoproteins.

Agglutinating activity is also present in several organs and tissues of *H. americanus*, but the largest amounts of agglutinins are found in the hemolymph and hemocyte extracts, suggesting that lobster hemocytes may synthesize or accumulate them. Both heteroagglutinins are present in the hemolymph of single animals. LAg1 agglutinates human and mouse erythrocytes, whereas LAg2 agglutinates mouse, but not human, red blood cells. This distinction is a useful but insufficient way of distinguishing the two heteroagglutinins during their preparation.

Carbohydrate inhibition properties of LAg1 and LAg2 are described by Hall and Rowlands (1974b) and Abel and associates (1984). Lobster heteroagglutinins are multispecific and agglutinate many cell types, including erythrocytes, lymphocytes, and bacteria. Agglutination of mouse erythrocytes by LAg2 is inhibited by GalNAc and to a lesser extent by ManNAc and NANA. None of these sugars inhibits agglutination of mouse erythrocytes by LAg1. Agglutination of human erythrocytes by LAg1 is inhibited by NANA and less so by NGNA and ManNAc. The LAg1 molecule contains NANA binding sites that react with human erythrocytes, and the LAg2 agglutinin contains GalNAc binding sites that react with mouse erythrocytes. Glycoproteins that inhibit LAg1 (human erythrocytes) are mucin > glycophorin > fetuin, all of which contain terminal sialic acid. Affinity chromatography experiments show that LAg1 can bind to GlcNAc and GalNAc as well, even though these sugars do not effectively inhibit hemagglutination by LAg1. In sum, LAg1 is inhibited primarily by NANA and LAg2 primarily by GalNAc; however, neither binding site shows strict specificity for these carbohydrates.

18.3.1.2. Other Lobsters. Similar, but less well characterized agglutinins are present in the lobsters *Panulirus argus* (Acton et al., 1973; P. Weinheimer, cited in Hall and Rowlands, 1974a), and *P. interruptus* (Tyler and Metz, 1945).

18.3.1.3. Crabs and Crayfish. Agglutinating activity has been described and partly characterized from the hemolymph of *Callinectes sapidus* (Pauley 1974a,b; Vasta and Cassels, 1983) and *Macropipus puber* (Ghidalia et al., 1975). The agglutinins are high molecular weight (approximately 300,000) proteins that show multispecific agglutination of human and other vertebrate erythrocytes. The agglutinating activity is destroyed by heat and extremes of pH. Heteroagglutinins are suggested to be present based on cross-absorption and immunoelectrophoresis experiments. In the case of *C. sapidus*, the agglutinins may be associated with the hemocytic microsomal fraction. The agglutinins are generally inhibited by *N*-acylamino compounds, including sialic acid, NAMA, *N*-acetylglutamic acid (GluNAc), GalNAc, GLcNAc, ManNAc, colominic acid, and glycoproteins that contain these compounds. Ravindranath and Cooper (1984) have extensively described the carbohydrate binding specificity and biomedical applications of crab agglutinins.

The hemolymph of the crayfish *Procambarus clarkii* contains a high molecular weight (> 150,000) agglutinin or agglutinins capable of agglutinating marine bacteria and vertebrate erythrocytes (Miller et al., 1972; Pauley, 1974b). The agglutinating activity is labile to heat and extremes of pH and is resistant to proteolysis. No inhibition studies are reported.

18.3.2. Thoracia: Barnacles

18.3.2.1. Balanus balanoides. The coelomic fluid of certain acorn barnacles contains agglutinating activity toward vertebrate erythrocytes (Kamiya and Shimizu, 1981). An agglutinin present in *Balanus balanoides* has been isolated from pooled and lyopholized coelomic fluid by chromatofocusing procedures (Ogata et al., 1983). Further purification of the fractions containing agglutinating activity (mouse erythrocytes) is achieved by gel filtration followed by chromatography on hydroxyapatite. Elution of the hydroxyapatite column with 0.1 M phosphate buffer results in recovery of the pure molecule, but in low yield. A 400-mg initial sample produces 1.54 mg purified *B. balanoides* agglutinin.

Coelomic fluid from *B. balanoides* shows multispecific agglutinating activity toward animal erythrocytes, marine bacteria, and mouse leukemia cells (Ogata et al., 1983). The purified agglutinin from *B. balanoides* expresses the same characteristics, suggesting that it is responsible for all types of oberved agglutinating activity in whole coelomic fluid. The purified molecule is a nondisulfide protein aggregate of MW 300,000 and sedimentation constant 11 S. The native molecule has one band on polyacrylamide gel electrophoresis and two acidic bands on isoelectric focusing gels. When denatured in sodium dodecyl sulfate (SDS), the agglutinin has one major band at MW 70,000, a minor band at MW 67,000, and a faint band at MW 26,000. The stoichiometric relationship between the various subunits and the 300,000 molecular weight multimer is unknown. The agglutinin is a glycoprotein and contains 4.5% carbohydrate. The amino acid composition of *B. banaloides* agglutinin is distinctive in that serine, glycine, and glutamic acid collectively account for 56% of all amino acid residues, whereas the molecule lacks arginine and half-cystine (see Table 8.1). Lack of half-cystine precludes disulfide bond involvement in the aggregated structure. The minimal concentration of agglutinin necessary to agglutinate rabbit erythrocytes is 6 μg/ml.

The agglutinin from *B. balanoides* is inhibited (rabbit erythrocytes) by NANA, galacturonic acid (GalA), and GlcA. There is no difference in the hemagglutinating activity toward desialylated (neuraminidase-treated) and normal rabbit erythrocytes, suggesting that the binding domain of the agglutinin includes more than just NANA. The relevance of inhibition by GalA and GlcA to binding specificity is uncertain, since these uronic acids are not located on the rabbit erythrocyte membrane.

18.3.2.2. Megabalanus rosa. The coelomic fluid of the barnacle *Megabalanus rosa* contains heteroagglutinins that have been isolated by affinity chromatography on acid-treated Sepharose 4B (Kamiya and Ogata, 1982; Muramoto et al., 1985). The heteroagglutinins are eluted from the affinity matrix with 0.2 M galactose and subsequently resolved by gel filtration on Sephadex G-200 (Fig. 18.4). Three electrophoretically homogeneous agglutinins, active toward mouse erythrocytes and differing in molecular weight, are observed. One liter of coelomic fluid yields approximately 40 mg BRA-1, 120 mg BRA-2, and 60 mg BRA-3. Non–acid-treated Sepharose 4B does not absorb the *M. rosa* heteroagglutinins, and neither form of Sepharose 4B absorbs the agglutinin present in *B. balanoides*.

Heteroagglutinins present in the coelomic fluid of *M. rosa*, (BRA-1, BRA-2, and BRA-3) show multispecific agglutinating activity toward all human ABO and rabbit erythrocytes and several types of tumor cells (Muramoto et al., 1985). The minimal agglutinin concentration required for agglutination of rabbit or human red blood cells is 1–5 μg/ml for all the heteroagglutinins with the exception of BRA-3, which requires 10 μg/ml to agglutinate human erythrocytes. Sheep erythrocytes are not agglutinated at comparable concentration of heteroagglutinins. Each of the *M. rosa* heteroagglutinins is an aggregated structure composed of protein subunits. Native molecular weights of BRA-1, BRA-2, and BRA-3 are 330,000, 140,000, and 64,000, respectively, as determined by high-speed gel filtration at pH 7.0. The electrophoretic analysis of *M. rosa* heteroagglutinins is included in Figure 18.4. BRA-1 is composed primarily of 43,000-MW subunits, but it also contains dimers and higher multimers of the 43,000 subunit that are cross-linked by disulfide bonds. The 43,000 subunit contains two disulfide-linked 22,000-

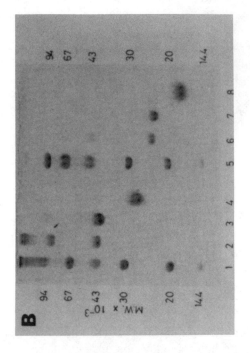

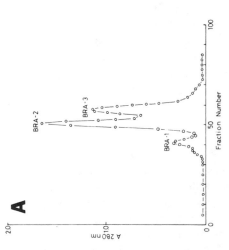

Figure 18.4. Preparation and characterization of *M. rosa* heteroagglutinins. (A) Gel filtration in 50 mM NH₄CO₃ of *M. rosa* heteroagglutinins on a column (3.2 × 100 cm) of Sehpadex G-200. (B) SDS-polyacrylamide electrophoresis of *M. rosa* heteroagglutinins. Samples in lanes 1–4 are nonreduced; those in lanes 5–8 are reduced (mercaptoethanol). Lanes 1 and 5 are protein markers; lanes 2 and 6 are BRA-1; lanes 3 and 7 are BRA-2; lanes 4 and 8 are BRA-3. (From Muramoto et al., 1985).

MW polypeptide chains. BRA-2 is identical to BRA-1 in subunit structure, except that it lacks any disulfide-linked higher molecular weight forms of the principal 43,000 subunit. BRA-3 has a different structure and consists of 28,000-MW subunits that in turn contain two 18,000-MW polypeptide chains cross-linked by disulfide bonds. All three heteroagglutinins are non-covalent assemblies of small subunits; however, BRA-1 does contain some higher molecular weight (> 43,000) disulfide forms. All three proteins are glycoproteins, and the associated carbohydrate may account for the imperfect molecular weight relationships shown in Figure 18.4. Each native heteroagglutinin has a single band on isoelectric focusing in support of its molecular purity.

The amino acid composition of *M. rosa* heteroagglutinins (Table 8.1, in Chapter 8) suggests that BRA-1 and BRA-2 are similar to each other but different from BRA-3. None of the *M. rosa* proteins resembles the *B. balanoides* agglutinin. Compared with *B. balanoides* agglutinin, *M. rosa* agglutinins have high amounts of aspartic acid and glutamic acid but reduced amounts of serine and glycine. The relatedness of BRA-1 and BRA-2 is also seen in their circular dichroism spectra, which suggest a significant amount of β structure. In order to examine similarity of primary structure, the heteroagglutinins were hydrolyzed with proteolytic enzymes and the resulting peptide maps compared by reverse-phase high-performance liquid chromatography. BRA-1 and BRA-2 show similar peptidic profiles, indicating similarity of their amino acid sequence, whereas that of BRA-3 is unrelated. Hence, two of the agglutinins from *M. rosa*—BRA-1 and BRA-2—appear to be similar molecules, differing in degree and nature of aggregation, whereas BRA-3 has a different structure and an unusually low molecular weight compared with other arthropodan agglutinins. This general picture is borne out by the agglutination inhibition patterns exhibited by the heteroagglutinins from *M. rosa*.

The inhibitory character of *M. rosa* heteroagglutinins is similar to that of *B. balanoides*. Hemagglutination (rabbit erythrocytes) by BRA-1, and BRA-2, and BRA-3 is inhibited by NANA, GalA, and GlcA in the concentration range 3–25 mM. BRA-2 appears slightly more sensitive to the inhibitors than do BRA-1 and BRA-3. Binding inhibition was also determined by using individual fluorescein-labeled *M. rosa* heteroagglutinins and rabbit erythrocyte ghosts in the presence of different carbohydrates at 200 mM concentration (Muramoto et al., 1985). These studies reveal similarly broad inhibition patterns for BRA-1 and BRA-2 and a more restrictive pattern for BRA-3. Galactose, galactosamine, NANA, and lactose strongly inhibit all three heteroagglutinins. BRA-1 and BRA-2 are also strongly inhibited by glucose, mannose, GalNAc, xylose, arabinose, maltose, cellobiose, and sucrose. Several other tested monosaccharides were noninhibitory. Binding to the erythrocyte ghosts is dependent on the presence of NaCl or $CaCl_2$. Binding of BRA-3 to the ghosts is uniquely inhibited by $CaCl_2$ above 20 mM concentration.

The carbohydrate inhibitory pattern for the *M. rosa* heteroagglutinins is therefore broad and may be influenced by calcium.

18.3.3. Stomatopoda

18.3.3.1. Squilla mantis. Two human blood group–specific agglutinins have been isolated and partially characterized from the hemolymph of male *Squilla mantis* (Amirante and Basso, 1984). The two agglutinins, anti-H and anti-A, are absorbed from dialyzed hemolymph with their respective target erythrocytes at neutral pH and subsequently desorbed in glycine buffer at pH 2.4 and 37°C. The purified agglutinins give identical single bands at MW 193,000 on SDS polyacrylamide gel electrophoresis under reducing conditions. Molecular weights of the native molecules have not been established.

Anti-H and anti-A agglutinin are inhibited respectively by fucose and GalNAc at 1 mM concentration. The inhibitors correspond to the principal antigenic determinant on the respective erythrocyte surface. Both agglutinins are labile to heat and EDTA, and they reversibly precipitate when dialyzed in distilled water. Monoclonal antibodies that bind to anti-H agglutinin also bind to *S. mantis* granulocytes, demonstrating the presence of agglutinins associated with the cellular membrane. The two agglutinins appear to have at least one common antigenic determinant and may therefore be structurally related.

18.4. AGGLUTININS FROM THE ARACHNIDA: TERRESTRIAL CHELICERATA

18.4.1. Scorpionida: Scorpions

18.4.1.1. Heterometrus bengalensis. Hemolymphatic agglutinins have been partially characterized in the scorpion *Heterometrus bengalensis* (Basu et al., 1984). The *H. bengalensis* agglutinin is unique when compared to other arthropodan agglutinins. The molecule is isolated from an ammonium sulfate fraction of whole dialyzed hemolymph by gel filtration and salt gradient elution from DEAE-Sephacel. The purified molecule has MW 146,000 by gel filtration, and the value is unchanged in the presence of SDS with or without mercaptoethanol, indicating the structure to be monomeric. Dialysis in 0.01 M EDTA has no effect and suggests no requirement for Ca^{2+} or Mg^{2+}. No carbohydrate is present. The agglutinin has activity toward rabbit erythrocytes, but no inhibition of agglutination is observed with 50 mg/ml solutions of numerous sugars and glycoproteins. Those tested carbohydrates and glycoconjugates included hexoses, *N*-acetylhexoses, glycosides, amino sugars, NANA, mucin, fetuin, and thyroglobulin. Similar lack of inhibition is observed with neuraminidase-treated rabbit erythrocytes. This agglutinin appears to have unique molecular and inhibitory properties that are deserving of further study.

18.4.1.2. Centruroides sculpturatus. The agglutinin activity present in the hemolymph of the scorpion *Centruroides sculpturatus* (Vasta and Cohen, 1982) shows characteristics different from those of the *H. bengalensis* agglutinin. In this case, agglutination and inhibitory profiles with various erythrocytes are determined by using whole hemolymph. The *C. sculpturatus* agglutinating activity is destroyed by heat and mercaptoethanol and is reversibly inhibited by EDTA, indicating activity dependence on Ca^{2+}. Cross-absorption experiments indicate multispecific binding and the possible presence of heteroagglutinins. The strongest inhibitors of agglutination (human cells) are NANA, NGNA, GlcA, GalA, GlcNAc, and GalNAc. Treatment of human erythrocytes with pronase or neuraminidase has minimal effect on agglutination by *C. sculpturatus* agglutinin. Although limited molecular detail is available, the agglutinating activity present in *C. sculpturatus* hemolymph is obviously distinct from that reported for *H. bengalensis*.

18.4.2. Araneae: Spiders

No reports concerning the molecular properties of agglutinins from spiders are available. Vasta and Cohen (1984) have carried out extensive agglutination specificity and inhibition experiments on sera of several species of genus *Aphonopelma* and have shown them to exhibit the expected sialic acid–binding properties observed among the Chelicerata (Vasta and Marchalonis, 1983).

18.5. AGGLUTININS FROM THE INSECTA

18.5.1. Diptera: Flies

18.5.1.1. Sarcophaga peregrina. One of the first insect agglutinins to be purified and extensively characterized is present in the hemolymph of injured *Sarcophaga peregrina* larvae (Komano et al., 1980). Diluted and centrifuged hemolymph is applied at 4°C to an affinity chromatography column consisting of Sepharose 4B. Nonabsorbed materials are removed by washing the column contents with pH 7.9 Tris·HCl buffer. The agglutinin is then eluted from the column with 0.2 M galactose, as shown in Figure 18.5. All agglutinating activity is associated with the galactose-eluted peak, showing that the affinity column absorbs all of the agglutinin from the hemolymph. Agglutinin is recovered in 40%–60% yield. No agglutinin can be isolated from the hemolymph of noninjured larvae.

The isolated agglutinin displays one band on polyacrylamide gel electrophoresis, establishing its purity. The molecular weight of the native molecule as determined by gel filtration is 190,000. Electrophoresis in SDS shows two closely spaced bands at MW 32,000 and MW 30,000. Mercaptoethanol has no effect. Since the ratio of the two subunits is 2:1, it is proposed that the

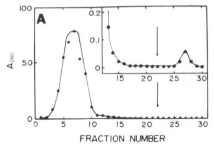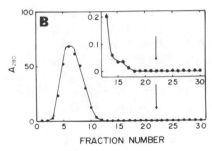

Figure 18.5. Affinity chromatography of hemolymph from (A) injured and (B) noninjured *Sarcophaga peregrina* larvae. About 2.5 ml of hemolymph containing 350 mg protein diluted in buffer (10 mM Tris·HCl, 130 mM NaCl, 5 mM KCl, 1 mM $CaCl_2$; pH 7.9) is applied to the Sepharose 4B column (1.5 × 5 cm) at 4°C. The column is washed extensively, and the absorbed protein is eluted with 0.2 M galactose in the same buffer as indicated by the arrows. The small peak at fractions 25–30 contains the hemagglutinin in A. No hemagglutinin is seen in noninjured larvae (B). The scale of A_{280} is expanded in the inserts. (From Komano et al., 1980.)

native molecule is a noncovalent aggregate consisting of four 32,000-MW subunits and two 30,000-MW subunits of the general formula $(\alpha_2\beta)_2$. The 30,000-MW subunit is derived from the 32,000-MW subunit in response to injury of the larval wall (Komano et al., 1981). A wound-dependent protease presumably converts some of the preexisting inactive 32,000-MW subunits to 30,000-MW units, and the two forms assemble into the active molecule. The two subunits have similar peptide maps, supporting commonality of their amino acid sequence. Fat body apparently synthesizes the agglutinin and, upon injury to the larval wall, secretes it into the hemolymph, where it is sequestered by the hemocytes (Komano et al., 1983). The same agglutinin is present in the hemolymph of *S. peregrina* pupae. The purified agglutinin is stable at room temperature but is destroyed by boiling. However, significant hemagglutinating activity remains after heating for 5 min at 80°C, and the molecule is somewhat resistant to trypsin, indicating considerable stability.

The purified agglutinin from *S. peregrina* agglutinates sheep erythrocytes and various cells from mouse tissues. Hemagglutination is completely inhibited by 50 mM galactose or lactose. Noninhibitory sugars are L-fucose, mannose, glucose, maltose, sucrose, GlcNAc, and α-methylmannoside.

18.5.2. Orthoptera: Crickets, Grasshoppers, and Cockroaches

18.5.2.1. Teleogryllus commodus. Hemolymph from the cricket *Teleogryllus commodus* contains multispecific hemagglutinating activity toward all human ABO and several other vertebrate erythrocyte types (Hapner and Jermyn, 1981). The agglutinin is isolated from centrifuged whole hemolymph by affinity chromatography on fetuin-Sepharose 4B using pH 7.4 phosphate-or Tris-buffered saline solution. The column absorbs 90% or more of the agglutinin activity in the sample. After unbound hemolymphatic components are washed from the column, the agglutinin is eluted from the affinity matrix with 0.1 M GlcNAc or 0.1 M NANA. All absorbed agglutinin is released with NANA, whereas elution with GlcNAc releases only about 30% of the absorbed agglutinin. The remainder may be subsequently eluted with NANA. The agglutinin is isolated in 50%–80% yield, and the minimal agglutinin concentration required to agglutinate human erythrocytes is approximately 0.5 μg/ml.

Molecular weight of the purified molecule is very large (> 10^6), based on exclusion from a column of Sepharose 6B. Incorporation of Triton X-100 detergent in the gel filtration step reduces the apparent molecular weight to near 10^6. The denatured molecule in SDS remains very large and does not enter a 7.5% polyacrylamide gel. In the presence of mercaptoethanol, two subunits are seen, with MW 53,000 and 31,000, respectively. *T. commodus* agglutinin is apparently a high molecular weight covalent aggregate of 53,000-MW and 31,000-MW subunits cross-linked by disulfide bonds. The protein is a glycoprotein based on the presence of amino sugars and is labile to heat, trypsin, and EDTA. The purified molecule is significantly less stable than is the agglutinin when present in hemolymph.

Purified *T. commodus* agglutinin is inhibited (human O erythrocytes) by 10 mM NANA, GlcNAc, and GalNAc and 1 μg/ml fetuin. The unpurified activity is not inhibited by GlcNAc or GalNAc at comparable concentrations, suggesting that materials in the hemolymph may alter the carbohydrate binding site in addition to imparting increased stability to the molecule.

18.5.2.2. Melanoplus sanguinipes and M. differentialis. Hemolymph from *Melanoplus sanquinipes, M. differentialis* and other acridid species exhibits multispecific agglutination activity toward human and numerous other erythrocyte types (Jurenka et al., 1982). The multispecific activity is present in individual insects and is inhibited by many galactosidic and glucosidic carbohydrates (Hapner, 1983). It is interesting that a single agglutinin is responsible for the observed multispecificity and broad carbohydrate inhibitory pattern (Stebbins and Hapner, 1985). The agglutinin present in adult grasshoppers is isolated from centrifuged hemolymph by affinity chromatography on galactose-Sepharose 4B, as shown in Figure 18.6. Hemolymph diluted with Dulbecco's phosphate buffered saline solution is applied to the affinity column and washed with buffer until all nonabsorbed materials are removed. All detectable agglutinin activity present in the hemolymph is absorbed by the affinity matrix and subsequently eluted with 0.2 M galactose. Approximately 350 μg agglutinin is recovered from 35 ml of diluted hemolymph collected from approximately 500 insects. The minimal concentration of purified agglutinin capable of agglutinating human type O asialo eryth-

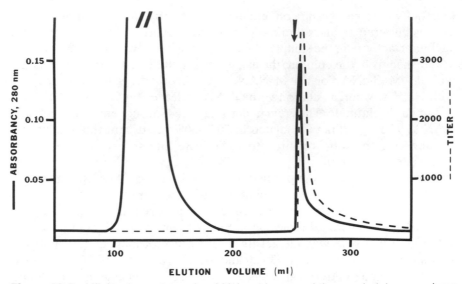

Figure 18.6. Affinity chromatography of *Melanoplus sanquinipes* agglutinin on a column (0.5 × 3 cm) of Sepharose 4B that contains covalently attached galactose. The hemolymph sample (50 ml) diluted in Dulbecco's phosphate buffered saline solution, pH 7.2, is applied to the column at 4°C and washed with Dulbecco's until the absorbancy returns to baseline. Agglutinin is then eluted with 0.2 M galactose in Dulbecco's buffer (arrow). Asialo human erythrocytes are used in detection of the agglutination activity. (From Stebbins and Hapner, 1985.)

rocytes is 0.02 µg/ml, and the approximate hemolymphatic concentration of the agglutinin is calculated as 20 µg/ml.

The molecular weight of native grasshopper agglutinin is estimated to be 590,000 by polyacrylamide electrophoresis and near 700,000 by gel filtration chromatography on Sepharose 6B. The subunit structure of grasshopper agglutinin is indicated in Figure 18.7. Upon SDS polyacrylamide gel electrophoresis, the agglutinin displays one major band at MW 70,000. When reduced with mercaptoethanol, the 70,000-MW band disappears, and two new bands at MW 40,000 and MW 28,000 appear. The native molecule is a noncovalent aggregate of 70,000-MW subunits composed of 40,000- and 28,000-MW polypeptide chains cross-linked by disulfide bonds. Some microheterogeneity may be present in the molecule, as determined by isoelectric focusing in 6 M urea. The protein contains carbohydrate and is a glycoprotein. Amino acid analysis shows relatively high amounts of aspartic acid and glutamic acid and least amount of methionine. The agglutinin from *M. differentialis* has similar or identical amino acid composition (see Table 8.1). Purified grasshopper agglutinin loses activity when heated for 1 min at 56°C, whereas whole hemolymph requires 1 hr for loss of agglutinating activity. Somewhat analogously, the purified molecule shows initial resistance to trypsin but is destroyed after 4 hr exposure, whereas activity in hemo-

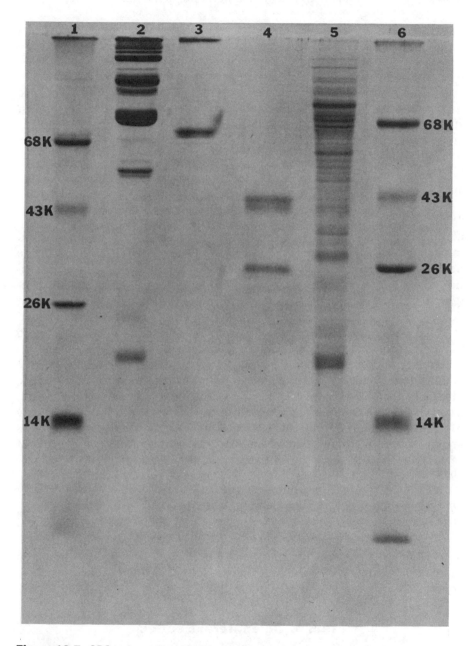

Figure 18.7. SDS polyacrylamide slab gel electrophoresis of purified agglutinin from *M. sanquinipes* (lanes 3 and 4). Lanes 2 and 5 contain whole hemolymph, and lanes 1 and 6 contain molecular weight markers. Samples in lanes 1, 2, and 3 are nonreduced, and samples in lanes 4, 5, and 6 are reduced with mercaptoethanol. (From Stebbins and Hapner, 1985.)

lymph is stable toward trypsin. The molecule is stabilized by galactose and is irreversibly inactivated by 5 mM EDTA.

Purified grasshopper agglutinin is inhibited (asialo human erythrocytes) by numerous glucosidic and galactosidic carbohydrates in the 1–5 mM concentration range (Stebbins and Hapner, 1985). The strongest inhibitors are α-anomers of simple galactosides and galactose-containing oligosaccharides, such as α-methylgalactoside and stachyose, but α-glucosides are closely comparable. Trehalose, the circulating disaccharide in the hemolymph, does not inhibit at similar concentrations. EDTA is among the most potent inhibitors, demonstrating the dependence of agglutinating activity on (presumably) calcium.

18.5.2.3. Leucophaea maderae. Although much work has been done concerning cockroach agglutinins, no extensive reports of their isolation and molecular characterization have appeared. One of two agglutinins in the hemolymph of *Leucophaea maderae* has been isolated by selective absorption to formolated rabbit erythrocytes (Amirante et al., 1976). Whole hemolymph is absorbed with the treated erythrocytes for 1 hr at 37°C or overnight at 4°C. After extensive washing with saline solution to remove loosely bound materials, the agglutinin-erythrocyte complex is dissociated by exposure to pH 2.4 glycine buffer. This procedure releases one of the two agglutinins in apparently pure form. The molecule displays a single precipitin line against antiagglutinin antisera on immunoelectrophoresis and shows a symmetric ultracentrifugation peak at 18.4 S, corresponding to a large molecular size. The second agglutinin is not released from the rabbit erythrocytes at low pH. Antibodies labeled with fluroescence markers show that the agglutinins are associated with both cytoplasmic and outer membrane components of certain *Leucophaea* hemocytes (Amirante, 1976; Amirante and Mazzalai, 1978).

18.5.2.4. Other Cockroaches. Agglutinins in *Periplaneta americana* (Scott, 1971; 1972) and *Blaberus craniifer* (Anderson et al., 1972; Donlon and Wemyss, 1976) are minimally characterized as heat labile, non-dialysable proteins. Agglutinating activity in the hemolymph of *P. americana* is inhibited (human erythrocytes) by glucose, fucose, L-rhamnose, and lactose in the 6–25 mM concentration range (Lackie, 1981).

18.5.3. Coleoptera: Beetles

18.5.3.1. Allomyrina dichotoma. Larval hemolymph from *Allomyrina dichotoma* contains multispecific agglutinating activity toward all types of human erythrocytes (Umetsu et al., 1984). Two similar agglutinins are isolated by affinity chromatography on acid-treated Sepharose 4B using a pH 7.2 phosphate saline buffer. The absorbed agglutinins are eluted from the column with 0.05 M lactose. Further purification is done by DEAE-Cellofine AM ion-exchange chromatography, resulting in separation of the two agglutinins, A-I and A-II. From 250 mg crude hemolymphatic protein, 2 mg A-I and 15 mg A-II are isolated, representing 67% recovery of initial agglutinating activity.

The native agglutinins display one band on polyacrylamide gel electrophoresis and have molecular weights of 65,000 (A-I) and 66,500 (A-II), as determined by gel filtration on Sephadex G-100. SDS polyacrylamide gel electrophoresis in the presence of mercaptoethanol shows that both A-I and A-II contain two sizes of subunits. A-I has subunits of 17,500 MW and 20,000 MW, and A-II contains 19,000- and 20,000-MW subunits. Glutaraldehyde cross-linking experiments indicate that the native molecules exist as dimers of about MW 40,000. The discrepancy between this molecular weight and that arrived at by gel filtration is unexplained. The possible role of disulfide bonds in the molecular structure is uncertain because no molecular weight data were determined on the nonreduced denatured molecule. Isoelectric focusing indicates some ionic heterogeneity as well as different isoelectric pH values for the molecules. The two agglutinins are immunologically indistinguishable and likely represent closely related structural forms of the same molecule, perhaps modified during biosynthetic or preparatory procedures. Inhibitory activity is not affected by 20 mM EDTA.

A. dichotoma agglutinins agglutinate all human ABO blood types equally, and their carbohydrate inhibition properties are identical. The multispecific agglutinins show strict inhibitory specificity for the β-anomeric form of galactosidic carbohydrates. The best inhibitors are lactose > lactulose > β-phenylgalactoside = β-nitrophenylgalactoside at concentrations in the range 3–12 mM. Other galactose derivatives and α-galactosides are noninhibitory. This narrow range of β-galactoside specificity is unusual among the arthropodan agglutinins.

18.5.3.2. Other Beetles. The agglutinating and inhibitory characteristics of two agglutinins present in larval hemolymph of *Leptinotarsa decemlineata* have been described by Stynen and coworkers (1982); however, few molecular data are reported.

18.5.4. Lepidoptera: Butterflies and Moths

18.5.4.1. Pieris brassicae. The agglutinin present in epidermal cell membranes of the cabbage butterfly, *Pieris brassicae*, is the only example of an arthropodan agglutinin to be purified from a tissue source other than hemolymph (Mauchamp, 1982). The agglutinin is isolated from the microsomal fraction of extracts or acetone powders of apolyzed pupal wings. Microsomal fractions are sonicated in phosphate buffer with 0.3% Triton X-100 detergent. This procedure releases activity that agglutinates trypsin-treated and glutaraldehyde-treated rabbit erythrocytes. No agglutinin activity is present in the hemolymph or in hemocyte extracts. The agglutinin is purified by affinity chromatography through application of the microsomal sonicate to a column of immobilized GlcNAc (Selectin 1). Elution with 0.1 M GlcNAc releases purified agglutinin from the column in 50%–60% yield.

The molecular weight of the native molecule is 40,000, as determined by gel filtration. SDS polyacrylamide slab gels in the presence of mercaptoethanol display a single band at 23,000, showing that the molecule is a dimer of 23,000-MW subunits. Whether the subunits are cross-linked by disulfide bonds is unreported. The agglutinin is stable when lyophilized and at 4°C in the presence of 0.1 M GlcNAc but is destroyed by heating at 60°C for 1 hr. At room temperature, the molecule slowly loses activity.

The agglutinin as it exists in the Triton X-100 sonicate is inhibited (treated rabbit erythrocytes) by GlcNAc, di-*N*-acetylchitobiose, and tri-*N*-acetyl-chitotriose at 0.5–1 mM concentration. Other tested monosaccharides, including GalNAc and glucosamine, are noninhibitory.

18.5.4.2. Bombyx mori. Larval hemolymph of the silkworm *Bombyx mori* contains agglutination activity toward trypsinized and glutaraldehyde-treated sheep erythrocytes (Suzuki and Natori, 1983). Although attempts at purification were hampered by general instability and manipulative problems, an agglutinin has been partially purified by gel exclusion from Sephacryl S-300 and identified through radioimmunoprecipitation of the ^{125}I-labeled molecule. The partially purified agglutinin is inactivated by heating at 70°C and by exposure to trypsin. Electrophoretic analysis of the excluded Sephacryl S-300 fraction shows it to contain several protein species in addition to the active substance. The molecular weight of the agglutinin is estimated by SDS electrophoresis of ^{125}I-labeled material that is specifically absorbed by treated sheep erythrocytes. This analysis suggests that a 260,000-MW protein is responsible for agglutinating activity. An antiserum was then produced against the 260,000-MW protein (eluted from a polyacrylamide gel) that is capable of inhibiting agglutination by hemolymph and the partially purified agglutinin. This antibody fraction preferentially precipitates the 260,000-MW protein and supports the conclusion that the 260,000-MW molecule is the active agglutinin.

The partially purified agglutinin is inhibited (treated sheep cells) by GlcA and GalA at 4 mM and 20 mM, respectively. Heparin is also a potent inhibitor of agglutination at 0.4 μg/ml.

18.6. SUMMARY

Agglutinins are ubiquitous carbohydrate-binding protein molecules and are found throughout the Arthropoda. Although their physiologic function or functions are unclear, they are suspected of having recognitory and immunelike roles associated with defense and tissue maintenance systems. Several arthropodan agglutinins have been isolated and purified to homogeneity. About 20 structurally different agglutinins are now characterized from 15 species among 9 orders in 4 classes of the phylum Arthropoda. The agglutinins are generally isolated from the hemolymph, but several are known to be present in other tissues. Affinity chromatography on an agglutinin-specific biomolecular matrix and elution with inhibitory saccharides is generally the most direct and efficient method of purifying agglutinins. Examples of heteroagglutinins, multispecific, and monospecific agglutinins are found among the Arthropoda. Most arthropodan agglutinins are high molecular weight glycoprotein molecules that are nondisulfide aggregates of smaller protein subunits. Agglutinins from the crustaceans, merostomes, and arachnids are commonly inhibited by sialic acids (NANA, NGNA), hexosuronic acids, and *N*-acetylated hexosamines. Certain insect agglutinins show similar carbohydrate inhibition; however, several are inhibited by neutral glucosides and galactosides. More extensive compilations of the physicochemical and inhibitory properties of the agglutinins from the Arthropoda are found in Chapter 8.

ACKNOWLEDGMENT

Some research work described herein and preparation of this chapter were supported by grants from the National Science Foundation MONTS program and NSF-DCB 8510097 and the Montana Agricultural Experiment Station.

REFERENCES

Abel, C.A., P.A. Campbell, J. VanderWall and A.L. Hartman. 1984. Studies on the structure and carbohydrate binding properties of lobster agglutinin, (LAg1), a sialic acid-binding lectin. *Prog. Clin. Biol. Res.* **157:** 103–114.

Acton, R.T., P.F. Weinheimer, and W. Niedermeir. 1973. The carbohydrate composition of invertebrate hemagglutinin subunits isolated from the lobster *Panulirus argus* and the oyster *Crassostrea virginica*. *Comp. Biochem. Physiol.* **44B:** 185–189.

Amirante, G.A. 1976. Production of heteroagglutinins in haemocytes of *Leucophaea maderae* L. *Experientia* **32:** 526–528.

Amirante, G.A. and V. Basso. 1984. Analytical study of lectins in *Squilla mantis* L. (Crustacea: Stomatopoda) using monoclonal antibodies. *Dev. Comp. Immunol.* **8:** 721–726.

Amirante, G.A. and F.G. Mazzalai. 1978. Synthesis and localization of hemagglutinins in hemocytes of the cockroach *Leucophaea maderae* L. *Dev. Comp. Immunol.* **2:** 735–740.

Amirante, G.A., F.L. DeBernardi, and P.C. Magnetti. 1976. Immunochemical studies on heteroagglutinins in the haemolymph of cockroach *Leucophaea maderae* L. (Insecta: Dictyoptera). *Boll. Zool.* **43:** 63–67.

Anderson, R.S., N.K.B. Day, and R.A. Good. 1972. Specific hemagglutinin and a modulator of complement in cockroach hemolymph. *Infect. Immun.* **5**: 55–59.

Basu, P.S., P.K. Datta, P. Agarwal, M.K. Ray, and T.K. Datta. 1984. Purification and partial characterization of an erythroagglutinin from the hemolymph of scorpion, *Heterometrus bengalensis. Biochimie* **66**: 487–491.

Bishayee, S. and D.T. Dorai. 1980. Isolation and characterization of a sialic acid–binding lectin (carcinoscorpin) from Indian horseshoe crab *Carcinoscorpius rotunda cauda. Biochim. Biophys. Acta* **623**: 89–97.

Cohen, E., G.H.V. Ilodi, W. Korytnyk, and M. Sharma. 1983. Inhibition of limulus agglutinins with *N*-acetyl-neuraminic acid and derivatives. *Dev. Comp. Immunol.* **7**: 189–192.

Cohen, E., G.R. Vasta, W. Korytnyk, C.R. Petrie, III, and M. Sharma. 1984. Lectins of the *Limulidae* and hemagglutination-inhibition by sialic acid analogs and derivatives. *Prog. Clin. Biol. Res.* **157**: 55–69.

Cornick, J.W. and J.E. Stewart. 1968. Interaction of the pathogen *Gaffkya homari*, with natural defense mechanisms of *Homarus americanus. J. Fish. Res. Bd. Can.* **25**: 695–709.

Cornick, J.W. and J.E. Stewart. 1973. Partial characterization of a natural agglutinin in the hemolymph of the lobster *Homarus americanus. J. Invertebr. Pathol.* **21**: 255–262.

Denburg, J.L. 1980. Cockroach muscle hemagglutinins: Candidate recognition macromolecules. *Biochem. Biophys. Res. Commun.* **97**: 33–40.

Donlon, W.C. and C.T. Wemyss. 1976. Analysis of the hemagglutinin and general protein element of the hemolymph of the West Indian leaf cockroach *Blaberus craniifer. J. Invertebr. Pathol.* **28**: 191–194.

Dorai, D.T., B.K. Bachhawat, S. Bishayee, K. Kannan, and D.R. Rao. 1981. Further characterization of the sialic acid–binding lectin from the horseshoe crab *Carcinoscorpius rotunda cauda. Arch. Biochem. Biophys.* **209**: 325–333.

Finstad, C.L., R.A. Good, and G.W. Litman. 1974. The erythrocyte agglutinin from *Limulus polyphemus* hemolymph: Molecular structure and biological function. *Ann. N.Y. Acad. Sci.* **234**: 170–180.

Finstad, C.L., G.W. Litman, J. Finstad, and R.A. Good. 1972. The evolution of the immune response. 8. The characterization of purified erythrocyte agglutinins from two invertebrate species *J. Immunol.* **108**: 1704–1711.

Furthmayer, H. and V.T. Marchesi. 1976. Subunit structure of human erythrocyte glycophorin A. *Biochemistry* **15**: 1137–1144.

Ghidalia, W., P. Lambin, and J.M. Fine. 1975. Electrophoretic and immunologic studies of a hemagglutinin in the hemolymph of the decapod *Macropipus puber. J. Invertebr. Pathol.* **25**: 151–157.

Gold, E.R. and P. Balding. 1975. Receptor-specific proteins: Plant and animal lectins, pp. 251–283. In E.R. Gold and P. Balding (eds.) *Excerpta Medica*, American Elsevier, New York.

Hall, J.L. and D.T. Rowlands, Jr. 1974a. Heterogeneity of lobster agglutinins. 1. Purification and physiochemical characterization. *Biochemistry* **13**: 821–827.

Hall, J.L. and D.T. Rowlands, Jr. 1974b. Heterogeneity of lobster agglutinins. 2. Specificity of agglutinin-erythrocyte binding. *Biochemistry* **13**: 828–832.

Hapner, K.D. 1983. Haemagglutinin activity in the haemolymph of individual Acrididae (grasshopper) specimens. *J. Insect Physiol.* **29:** 101–106.

Hapner, K.D. and M.A. Jermyn. 1981. Haemagglutinin activity in the haemolymph of *Teleogryllus commodus* (Walker). *Insect Biochem.* **11:** 287–295.

Hartman, A.L., P.A. Campbell, and C.A. Abel. 1978. An improved method for the isolation of lobster lectins. *Dev. Comp. Immunol.* **2:** 617–625.

Jurenka, R., K. Manfredi, and K.D. Hapner. 1982. Haemagglutinin activity in Acrididae (grasshopper) haemolymph. *J. Insect Physiol.* **28:** 177–181.

Kamiya, H. and K. Ogata. 1982. Hemagglutinins in the acorn barnacle *Balanus* (*Megabalanus*) *roseus*: Purification and partial characterization. *Bull. Jpn. Soc. Sci. Fish.* **48:** 1421–1425.

Kamiya, H. and Y. Shimizu. 1981. Occurrence of hemagglutinin in acorn barnacles. *Bull. Jpn. Soc. Sci. Fish.* **47:** 411–414.

Kaplan, R., S.S.-L. Li, and J.M. Kehoe. 1977. Molecular characterization of limulin, a sialic acid binding lectin from the hemolymph of the horseshoe crab. *Biochemistry* **16:** 4297–4302.

Kehoe, J.M., R. Kaplan, and S.S.-L. Li. 1979. Functional implications of the covalent structure of limulin: An overview. *Prog. Clin. Biol. Res.* **29:** 617–623.

Komano, H., D. Mizuno, and S. Natori. 1980. Purification of lectin induced in the hemolymph of *Sarcophaga peregrina* larvae on injury. *J. Biol. Chem.* **255:** 2919–2924.

Komano, H., P. Mizuno, and S. Natori. 1981. A possible mechanism of induction of insect lectin. *J. Biol. Chem.* **256:** 7087–7089.

Komano, H., R. Nozawa, D. Mizuno, and S. Natori. 1983. Measurement of *Sarcophaga peregrina* lectin under various physiological conditions by radioimmunoassay. *J. Biol. Chem.* **258:** 2143–2147.

Lackie, A.M. 1981. The specificity of the serum agglutinins of *Periplaneta americana* and *Schistocerca gregaria* and its relationship to the insects' immune response. *J. Insect. Physiol.* **27:** 139–143.

Marchalonis, J.J. and G.M. Edelman. 1968. Isolation and characterization of a hemagglutinin from *Limulus polyphemus*. *J. Mol. Biol.* **32:** 453–465.

Mauchamp, B. 1982. Purification of an *N*-acetyl-D-glucosamine specific lectin (P.B.A.) from epidermal cell membranes of *Pieris brassicae* L. *Biochimie* **64:** 1001–1008.

McSweegan, E.F. and T.G. Pistole. 1982. Interaction of the lectin limulin with capsular polysaccharides from *Neisseria meningitidis* and *Escherichia coli*. *Biochem. Biophys. Res. Commun.* **106:** 1390–1397.

Miller, Van H., R.S. Ballback, G.B. Pauley, and S.M. Krassner. 1972. A preliminary physicochemical characterization of an agglutinin found in the hemolymph of the crayfish *Procambarus clarkii*. *J. Invertebr. Pathol.* **19:** 83–93.

Mohan, S., D.T. Dorai, S. Srimal, and B.K. Bachhawat. 1982. Binding studies of a sialic acid-specific lectin from the horseshoe crab *Carcinoscorpius rotunda cauda* with various sialoglycoproteins. *Biochem. J.* **203:** 253–261.

Mohan, S., D.T. Dorai, S. Srimal, B.K. Bachhawat, and M.K. Das. 1983. Fluorescence studies on the interaction of some ligands with carcinoscorpin, the sialic acid specific lectin, from the horseshoe crab *Carcinoscorpius rotunda cauda*. *J. Biosci*. **5**: 155–162.

Mohan, S., D.T. Dorai, S. Srimal, B.K. Bachhawat, and M.K. Das. 1984. Circular dichroism studies on carcinoscorpin, the sialic acid binding lectin of horseshoe crab *Carcinoscorpius rotunda cauda*. *Indian J. Biochem. Biophys*. **21**: 151–154.

Muramoto, K., K. Ogata, and H. Kamiya. 1985. Comparison of the multiple agglutinins of the acorn barnacle *Megabalanus rosa*. *Agric. Biol. Chem*. **49**: 85–93.

Muresan, V., V. Iwanij, Z.D.J. Smith, and J.D. Jamieson. 1982. Purification and use of Limulin: A sialic acid-specific protein. *J. Histochem. Cytochem*. **30**: 938–946.

Nowak, T.P. and S.H. Barondes. 1974. Agglutinin from *Limulus polyphemus*: Purification with formalinized horse erythrocytes as the affinity adsorbent. *Biochim. Biophys. Acta* **393**: 115–123.

Ogata, K., K. Muramoto, M. Yamazaki, and H. Kamiya. 1983. Isolation and characterization of *Balanus balanoides* agglutinin. *Bull. Jap. Soc. Sci. Fish*. **49**: 1371–1375.

Oppenheim, J.D., M.S. Nachbar, M.R.J. Salton, and F. Aull. 1974. Purification of a hemagglutinin from *Limulus polyphemus* by affinity chromatography. *Biochem. Biophys. Res. Commun*. **58**: 1127–1134.

Pauley, G.B. 1974a. Comparison of a natural agglutinin in the hemolymph of the blue crab, *Callinectes sapidus*, with agglutinins of other invertebrates, pp. 241–260. In E.L. Cooper (ed.) *Contemporary Topics in Immunology*, vol. 4. Plenum Press, New York.

Pauley, G.B. 1974b. Physicochemical properties of the natural agglutinins of some mollusks and crustaceans. *Ann. N.Y. Acad. Sci*. **234**: 145–158.

Peters, W., H. Kolb, and V. Kolb-Bachofen. 1983. Evidence for a sugar receptor (lectin) in the peritrophic membrane of the blowfly larva, *Calliphora erythrocephala* Mg. (Diptera). *J. Insect Physiol*. **29**: 275–280.

Ravindranath, M.H. and E.L. Cooper. 1984. Crab lectins: Receptor specificity and biomedical applications. *Prog. Clin. Biol. Res*. **157**: 83–96.

Roche, A.-C. and M. Monsigny. 1974. Purification and properties of limulin: A lectin (agglutinin) from hemolymph of *Limulus polyphemus*. *Biochim. Biophys. Acta* **371**: 242–254.

Roche, A.-C. and M. Monsigny. 1979. Limulin (*Limulus polyphemus* lectin): Isolation, physicochemical properties, sugar specificity and mitogenic activity. *Prog. Clin. Biol. Res*. **29**: 603–161.

Roche, A.-C. R. Schauer, and M. Monsigny. 1975. Protein-sugar interactions: Purification by affinity chromatography of limulin, an *N*-acyl-neuraminidyl–binding protein. *FEBS Lett*. **57**: 245–249.

Roche, A.-C., J. Maurizot, and M. Monsigny. 1978. Circular dichroism of limulin: *Limulus polyphemus* lectin. *FEBS Lett*. **91**(2): 233–236.

Rostam-Abadi, H., and T.G. Pistole. 1982. Lipopolysaccharide-binding lectin from the horseshoe crab *Limulus polyphemus*, with specificity for 2-keto-3-deoxy-octonate (KDO). *Dev. Comp. Immunol*. **6**: 209–218.

Scott, M.T. 1971. A naturally occurring hemagglutinin in the hemolymph of the American cockroach. *Arch. Zool. Exp. Gen*. **112**: 73–80.

Scott, M.T. 1972. Partial characterization of the hemagglutinating activity in hemolymph of the American cockroach (*Periplaneta americana*). *J. Invertebr. Pathol.* **19:** 66–71.

Shimizu, S., M. Ito, and M. Niwa. 1977. Lectins in the hemolymph of Japanese horseshoe crab, *Tachypleus tridentatus*. *Biochim. Biophys. Acta* **500:** 71–79.

Shimizu, S., M. Ito, N. Takahashi, and M. Niwa. 1979. Purification and properties of lectins from the Japanese horseshoe crab, *Tachypleus tridentatus*. *Prog. Clin. Biol. Res.* **29:** 625–639.

Shishikura, F. and K. Sekiguchi. 1983. Agglutinins in the horseshoe crab hemolymph: Purification of a potent agglutinin of horse erythrocytes from the hemolymph of *Tachypleus tridentatus*, the Japanese horseshoe crab. *J. Biochem.* **93:** 1539–1546.

Shishikura, F. and K. Sekiguchi. 1984a. Studies on perivitelline fluid of horseshoe crab embryo. 1. Purification and properties of agglutinin from the perivitelline fluid of *Tachypleus gigas* embryo. *J. Biochem.* **96:** 621–628.

Shishikura, F. and K. Sekiguchi. 1984b. Studies on the perivitelline fluid of horseshoe crab embryo. 2. Purification of agglutinin-binding substance from the perivitelline fluid of *Tachypleus gigas* embryo. *J. Biochem.* **96:** 629–636.

Stebbins, M.R. and K.D. Hapner. 1985. Preparation and properties of haemagglutinin from haemolymph of Acrididae (grasshoppers). *Insect. Biochem.* **15:** 451–462.

Stynen, D., M. Peferson, and A. DeLoof. 1982. Proteins with hemagglutinin activity in larvae of the Colorado potato beetle *Leptinotarsa decemlineata*. *J. Insect. Physiol.* **28:** 465–470.

Suzuki, T. and S. Natori. 1983. Identification of a protein having hemagglutinating activity in the hemolymph of the silkworm *Bombyx mori*. *J. Biochem.* **93:** 583–590.

Tyler, A. and C.B. Metz. 1945. Natural heteroagglutinins in the serum of the spiny lobster *Panulirus interruptus*. *J. Exp. Zool.* **100:** 387–406.

Umetsu, K., S. Kosaka, and T. Suzuki. 1984. Purification and characterization of a lectin from the beetle *Allomyrina dichotoma*. *J. Biol. Chem.* **95:** 239–245.

VanderWall, J., P.A. Campbell, and C.A. Abel. 1981. Isolation of a sialic acid-specific lobster lectin (LAg1) by affinity chromatography on Sepharose-colominic acid beads. *Dev. Comp. Immunol.* **5:** 679–684.

Vasta, G.R. and F.J. Cassels. 1983. *N*-Acylamino-specific lectins in the serum and a hemocyte microsomal fraction from the blue crab *Callinectus sapidus*. *Am. Zool.* **23:** 901.

Vasta, G.R. and E. Cohen. 1982. The specificity of *Centruroides sculpturatus* Ewig (Arizona lethal scorpion) hemolymph agglutinins. *Dev. Comp. Immunol.* **6:** 219–230.

Vasta, G.R. and E. Cohen. 1984. Sialic acid binding lectins in the serum of American spiders of the genus *Aphonopelma*. *Dev. Comp. Immunol.* **8:** 515–522.

Vasta, G.R. and J.J. Marchalonis. 1983. Humoral recognition factors in the Arthropoda: The specificity of Chelicerata serum lectins. *Am. Zool.* **23:** 157–171.

Yeaton, R.W. 1981. Invertebrate lectins: Occurrence. *Dev. Comp. Immunol.* **5:** 391–402.

CHAPTER 19

Detection of Membrane Receptors by Arthropod Agglutinins

Elias Cohen

Department of Laboratory Medicine
Roswell Park Memorial Institute
Departments of Pathology and Microbiology
School of Medicine
State University of New York at Buffalo, New York

19.1. INTRODUCTION

Among arthropods, the horsefoot or horseshoe crab *Limulus polyphemus* belongs to the class *Merostomata* (*Xiphosura*) and the family *Xiphosuridae*. This chapter describes the discovery and application of agglutinins of *Limulus* and related species to the detection of cellular receptors. Particular attention is given to the work that evolved in our laboratory at Roswell Park Memorial Institute.

A remarkable avidity of *Limulus* serum agglutinins for erythrocytes, regardless of species origin, was observed in 1950 while I was a graduate student in the laboratory of Alan A. Boyden at Rutgers University. An effort was made to differentially agglutinate erythrocytes of the blood of cancer patients in order to facilitate cytologic analysis for circulating tumor cells in the supernatant or buffy coat (Watne and Cohen, 1964; Migailo, 1966). This was the first reported biomedical application.

Hemagglutination by *Limulus* serum had been first reported by Noguchi (1903) to differentiate red blood cells of amphibians and reptiles without hemolysis. From 1962 on, I investigated the nature of the *Limulus* agglutinins with human and other vertebrate erythrocytes.

Marchalonis (1964) independently studied hemagglutination by *Limulus* hemolymph serum. We (Cohen et al., 1965) described a *N*-acetyl glucosamine specificity and ultracentrifuge patterns of the hemagglutinins, and I reported the *N*-acetyl-neuraminic acid (NANA) receptor of *Limulus* agglutinins (Cohen, 1968). The most avid hemagglutination was of murine, human, guinea pig, rat, rabbit, and sheep erythrocytes, in that order. Calcium and magnesium cations enhanced hemagglutination, but lithium, potassium, and sodium ions decreased hemagglutination. Marchalonis and Edelman (1968) and Finstad and coworkers (1974) reported an increase in titer when 0.01 M calcium chloride was added to the hemagglutination diluent. Recent reviews by Vasta and Marchalonis (1983) on recognition proteins of *Chelicerata* and Cohen and colleagues (1984) on *Limulidae* cover recent work regarding the specificity of *Limulus* hemolymph agglutinins, as well as the phylogenetic and biochemical basis for agglutination of mammalian erythrocytes and lymphocytes (see also Chapters 8 and 18).

The hemolymph serum of agglutinins (lectins) of invertebrates, as well as plant lectins, are invaluable in the field of immunohematology. They are reactive with erythrocyte and leukocyte receptors. The broad distribution of cellular sialyl receptors is well known. Distinct patterns of hemagglutination of mammalian erythrocytes of different species may be demonstrated by the diversity of receptor display and/or the heterogeneity of the sialyl-agglutinins of invertebrate lectins that are used. Microagglutination techniques have facilitated the study of the nature of cellular agglutination by lymphocyte lectins. Erythrocyte or cell membrane receptor antigens may serve as indicators of the state of health or disease.

This chapter focuses on the agglutinins of *Limulidae* and other selected arthropods, the techniques employed in studying them, and the potential and actual biomedical applications.

19.2. ERYTHROCYTE CELL RECEPTORS

The specificity of *Limulus* agglutinin for terminal *N*-acetylneuraminyl residues was utilized to demonstrate the high density of those residues at human blood group M and N receptor sites (Cohen et al., 1972). Blockade of M or N receptor sites could inhibit absorption of *Limulus* lectins by human erythrocytes. The microagglutination techniques used were comparable to tube test blood grouping and hemagglutination inhibition procedures.

Serial dilutions of *Limulus* hemolymph serum were made in 75×25 mm test tubes by dispensing 0.1-ml volumes into the tubes. The human erythrocytes were obtained from whole blood collected in disodium EDTA (9 mg/7 ml) washed three times with Bacto-hemagglutination buffer, pH 7.2, before preparation of the test suspension of 2% erythrocytes in the buffered saline solution. A volume of 0.1 ml erythrocyte suspension was added to the serum agglutinin dilutions. Incubation was at room temperature (22–25°C) for 15 min, followed by light centrifugation (1000 RPM) in a tabletop Fisher centrifuge for 2 min. Macroscopic reading was used to grade agglutination (0–4 plus). Titers and scores were recorded.

Handa (1980) studied the erythrocytes of 24 breeds of dogs in which the *N*-acetyl-neuraminic acid (NANA) and *N*-glyconeuraminic acid (NGNA) content varied in the different breeds. This kind of variability must be considered in the selection of the mammal species of indicator red blood cells for an anti-NANA lectin.

Shimuzu and associates (1977) and Roche and Monsigny (1979) have used equine and feline erythrocytes, whereas other workers, including myself (1968) have used human red blood cells. Such choices have been a matter of convenience of the source of cells.

Receptor display on the cell membrane can be manipulated by the use of enzymes. Protease or neuraminidase treatment of human, equine, rat, and quail erythrocytes results in characteristic hemagglutination patterns, in contrast to the results with untreated cells (Cohen and Vasta, 1982). *Limulus* hemolymph lectins agglutinate untreated cells but not protease-treated cells. Saharan scorpion (*Androctonus*) lectins exhibit the same titer and scores with untreated and protease-treated human erythrocytes.

Androctonus and *Centruroides* lectins have similar hemagglutination with human erythrocytes but not with equine cells (Cohen and Vasta, 1982). Brahmi and Cooper (1974) were the first to demonstrate *Androctonus* hemolymph hemagglutinins. However, we (Cohen et al., 1979, 1980) showed that *Limulus* and *Androctonus* hemolymph serum agglutinins exhibit similar agglutinin activity, which is decreased by neuraminidase treatment. However, Vasta and coworkers (1982) present data that suggest that *Androctonus* agglutinins have an extra specificity, with higher affinity for an *N*-acylated compound of a neuraminidase-resistant receptor of red blood cells. The occurrence and specificity of the hemagglutinins in the Aquatic and Terrestrial Chelicerata has been described in some detail for the following species; *Limulus polyphemus* in comparison with several species of scorpions, and *Centruroides sculpturatus* (Vasta and Cohen, 1982); *Hadrurus arizonensis* (Vasta and Cohen, 1984a); *Vaejovis confuscius* (Vasta and Cohen, 1984b); *Vaejovis springerus* (Vasta and Cohen, 1984c); *Pauauroctonus mesanensis* (Vasta and Cohen, 1985a); the whip (false) scorpion, *Mastigoproctus gi-*

ganteus (Vasta and Cohen, 1984c), and the tarantula *Aphonopelma chalcodes* (Vasta and Cohen, 1984d).

Hemagglutinins of other arthropods, including the insects and crustacenas, are reviewed by Yeaton (1981a,b). At present, microhemagglutination techniques generally utilize human red blood cells isolated from whole blood, collected with EDTA and ACD (acid citrate dextrose) anticoagulant (4 volumes blood to 1 volume ACD), and washed three times with Tris-buffered saline (TBS) solution (0.05M Tris·HCl, 0.10 M NaCl, 0.01 M $CaCl_2$), pH 7.6. The red blood cells are prepared in 5×10^3 cells/mm^3 cell suspension. Bacto-hemagglutination buffer (0.15 M phosphate-buffered saline solution) has also proved to be a reliable cell washing or suspension diluent for human erythrocytes or lymphocytes. Currently, Terasaki (tissue typing) plastic trays of 62 or 96 wells (Robbins Scientific, Mountainview, CA) are used for hemagglutination or lymphocyte microagglutination. Vasta and colleagues (1982) prepared 0.005 ml of arthropod hemolymph serum or purified lectin solution and placed twofold dilutions of these into the wells and added equal volumes of cell suspension. The cell serum mixture was agitated in a Vortex rotator at speed 1 for 10 sec and incubated at room temperature (22–25°C) for 45 min. The results were read microscopically at 100∈ magnification and graded from 0 to 4 as to the degree of agglutination.

The enzyme treatment of cells is useful for elucidating the nature of cell membrane receptor sites. The method used is that of Uhlenbruck and associates (1968) as applied by Vasta and associates (1982). Lymphocytes are treated with pronase in the following way. One milliliter of 15×10^3 cells/mm suspension in Tris-buffered saline solution is incubated with 0.1 ml of enzyme solution at 37°C for 10 min; the cells are washed three times and resuspended to approximately 5×10^3 cells/mm^3. Neuraminidase teatment of lymphocytes is done by the method of Hellstrom and coworkers (1976). Rh_O (D) human erythrocytes treated with proteolytic enzymes are tested with incomplete anti-D Rh reagent sera to prove that the enzyme is proteolytic. A higher titer with treated cells substantiates the activity of the enzyme being used.

For statistical analysis of means of titers, the log_2 of the reciprocal of the titer is essential. Arithmetic mean values can be calculated for the scores of agglutination. The score of an agglutination test is the summation of all readings of that particular titration.

19.3. LEUKOCYTE RECEPTORS

We (1976) differentiated human leukemic lymphocytes from normal ones by *Limulus* serum agglutination. The human peripheral chronic lymphocytic leukemic (CLL) lymphocytes were agglutinated by *Limulus* serum to a significantly higher titer and score than were peripheral normal human lymphocytes. A lymphoblastoid cell line (B411-4) of cultured human B cells was agglutinated at a higher titer and score than was a cell line (MOLT-4F) of T cells.

Isolation of peripheral blood lymphocytes was by the method of Amos and coworkers (1970). Cultured or peripheral blood lymphocytes were suspended in a phosphate-buffered (pH 7.4) saline solution at a concentration of 4000 cells/mm^3. Microagglutination was done in plastic tissue-typing trays (Falcon). Serial dilutions of *Limulus* serum in buffered saline solution were introduced, in 2-μl volumes, into the conical wells, in addition to equal volumes of cell suspensions. After immediate mixing by horizontal plane agitation, the cell-serum mixtures were incubated at room temperature (22–25°C) for 60 min. Readings at 100× magnification were graded from 0 to 4 as to degree of agglutination; titers and scores were determined.

Purified limulin (*Limulus* agglutinin) was used in followup studies to demonstrate agglutination of human peripheral and cultured lymphocytes by us (Cohen et al., 1979). Microagglutination was performed as described above except that the limulin was diluted in Hanks BSS (once) without calcium and magnesium and 5 lambda volumes were added to each well. Cell suspensions of lymphocytes (4000–6000/mm^3) in 5 lambda volumes were added to each well. The trays were rotated on a Yankee rotator at 120–160 revolutions per 5 min at room temperature (22–25°C). Incubation was for 25 min at room temperature. The agglutination was read microscopically and scored from 1 to 4 plus. Titers and scores were recorded. Calcium chloride was used as an additive at 0.01 M concentration in the diluent for evaluation. The addition of calcium generally gave higher titers and scores. The strongest and most consistent difference was between normal and CLL cells, with CLL cells having a significantly higher mean titer. Cultured B cells gave higher titers than did cultured chronic myelogenous leukemia (CML) lymphocytes and acute myelogenous leukemia (AML) lymphocytes. However, since *Limulus* and related species have heterogenous hemolymph lectins, some consideration must be given to that demonstration by Shimizu and associates (1979).

It was observed in our laboratory by Vasta and colleagues (1981) that phytohemagglutinin (PHA) -stimulated human lymphocytes give higher agglutination titers with *Limulus* lectin. By microagglutination, it was subsequently demonstrated by us (Cohen et al., 1982) that significantly higher agglutination titers and scores are obtained with 6-day allogeneic-stimulated lymphocytes in mixed lymphocyte culture (MLC) than with unstimulated cells. The effect of PHA stimulation on titers and score was confirmed. Further study is in progress toward quantitation of MLC by use of *Limulus* lectin.

Mycosis fungoides is a cutaneous T-cell lymphoma characterized by the appearance of hematologically distinct peripheral blood lymphocytes. The lymphocytes of such patients give distinctly higher titers and scores than do the cells of normal subjects (Gaesser et al., 1982). Also, suspensions of lymphoblastoid cell cultures derived from such patients and cultured B-cell lines yield higher titers and scores than do control (normal) healthy blood donors.

Granulocytes of patients with CML were reported to have increased agglutinability with *Limulus* lectin than did those of normal control subjects (Taub et al., 1980). A mean log 2 reciprocal of titer of 3.35 and a mean score of 10.09 with CML patient granulocytes were reported (Cohen et al., 1984), compared to a mean log 2 reciprocal of titer of 2.75 and a mean score of 7.93 with normal granulocytes.

The significant differences in titers and scores between normal and leukemic (CLL or CML) cells and between B- and T-lymphocytes may be due to a greater availability of cell membrane surface receptors on the more reactive cells. The precise mechanism requires further investigation.

19.4. PLATELET AGGLUTININS

It has been reported that a sialic acid loss from platelets during cell senescence facilitates in vivo clearance by the reticuloendothelial system. Experimental removal of platelet sialic acid by neuraminidase is known to affect platelet function and platelet survival (Greenberg et al., 1975).

Limulus hemolymph serum has been reported to agglutinate *human platelets* (Pardoe et al., 1970). In our laboratory, a *Limulus* microhemagglutination inhibition assay was used to measure sialic acid (NANA) receptor display of human platelets during blood bank storage (Cohen et al., 1983). Paired analyses were made between fresh versus 6- to 7-day-old platelets of the same donors. A mean log 2 titer value of 2.43 and a mean score value of 8.0 were recorded for fresh platelets, compared to a mean titer value of 4.0 and a mean score value of 9.5 for 6- to 7-day-stored platelets. It remains to be proved whether the simple *Limulus* hemagglutination inhibition microassay may be applicable to the evaluation of platelet aging, or senescence, during storage. However, the assay may be useful for the screening of substances that may inhibit the loss of surface NANA from stored platelets. Perhaps other arthropod sera may contain anti-NANA–specific lectins applicable for similar biomedical purposes.

19.5. ONCOLOGY AND SIALIC ACID DERIVATIVES

Evidence has accumulated that cell membrane surface sialic acid may influence malignant cell survival. Therefore, Walter Korytnyk, of the Roswell Park Memorial Institute, developed a program of research on inhibitors of sialic acid biosynthesis.

Sialic acid analogues and their derivatives were tested (Cohen et al., 1983, 1984) by *Limulus* hemagglutination inhibition to determine the minimal number of millimoles (mM) required for at least 50% inhibition of titer. Even laboratory purified *Limulus* lectin could recognize differences in sialic acid analogues and derivatives. NANA, the receptor of *Limulus* lectin, was the best inhibitor. The hydrophobic bonding and negatively charged caroboxylic acid elements of the NANA molecule were the most important. Modification of those biochemical sites could affect the inhibition capacity of test molecules. Blockage of the carboxylic acid group of NANA by a methyl group, as in methyl-*N*-acetyl-neuraminate, would obliterate inhibition by the modified NANA.

19.6. OTHER CELL MEMBRANE RECEPTORS

Nakano and coworkers (1980) demonstrated that murine helper T cells are preferentially agglutinated by *Limulus* lectin but not by peanut agglutinin (PNA), which binds to β-D-Gal-(1,3)-D GalNAac. The stimulatory effect of *Limulus* lectin-agglutinated cells on the murine primary antibody response was abolished by treatment of the cells with anti-Thy-1.2 and complement, suggesting that helper cells induced by *Lens culinarius* (plant lectin) agglutinin were T cells and that they have abundant sialic acid residues on their cell surfaces.

It was established by Bee (1982) that retinal neurons of the chick neural crest have intense staining with fluorescein-labeled *Limulus* lectin, with the lectin binding to the point on the neurite hillock from which the neurite projects.

Mazzuca and colleagues (1977) discovered a differential use of *Limulus* lectin for the identification of cell types in human bronchial gland cells: serous, positive; mucous, negative; and goblet, negative.

Jamieson and associates (1981) demonstrated that rat exocrine pancreas cell were stained by labeled *Limulus* lectin as follows: acinar, strong positive; duct, weak positive.

Hormone-sensitive and -insensitive breast carcinoma cells could be differentiated by labeled peanut lectins (Klein et al., 1981). Similar application may lie ahead for invertebrate arthropod lectins.

19.7. BACTERIAL AGGLUTININS

Pistole and Graf (1984) have described gram-positive bactericidal activity of the hemolymph of *Limulus*. Ravindranath and Cooper (1984) demonstrated that crab (*Cancer*) lectins can be used to differentiate variants of *Escherichia coli*. This was based on the presence or absence of receptor *O*-acetyl groups in the outer capsular polysaccharide (K-antigen). *Limulus* has been reported to possess a *Staphylococcus aureus*–agglutinating lectin. This subject of *Limulus* and other arthropod antibacterial lectins is discussed elsewhere in Chapter 11.

19.8. SUMMARY

Selected arthropod hemolymph lectins have been described from *Limulidae*, scorpions, spiders, and related species of Chelicerata, several with sialic-acid specificities. The cell receptors studied included those of human and other mammalian erythrocytes, the high density of NANA of the human MN blood group; greater reactivity than normal for leukocytes of CLL and CML patients and for mycosis fungoides; lymphoblastoid culture cell lines of B and T cells, with greater reactivity for B cells, PHA-induced T cell blasts, and human allogeneic-stimulated lymphocytes in mixed cell culture. In addition, loss of sialic acid receptor in platelet senescence in storage was semiquantitated by use of an arthropod (*Limulus*) lectin. Sialic acid derivatives were used to analyze the nature of the NANA receptor of erythrocytes with use of an arthropod (*Limulus*) lectin.

The techniques of microhemagglutination and leukocyte agglutination were presented, as was the use of the logarithm to the base 2 determination for statistical analysis of titers.

New perspectives to be considered are the advances in laboratory technology, such as automated cell sorters. These developments should facilitate quantitation of cellular agglutination by lectins and/or the binding of lectins to subpopulations of mammalian hematopoietic cells. Arthropod lectins present unique cell probes for utilization in cell differentiation, pathobiology, oncology, and aging. However, it is hoped that some guidelines may be developed to prevent extinction of arthropod species and their ecology. Advances in microtechnology and the efficient storage of extracted arthropod cell probe lectins may be necessary to protect valuable species of arthropods from overzealous commercial exploitation. Effective vivarium maintenance and successful aquaculture will help to rear species whose existence is threatened by biologic and medical research.

Criteria for use of arthropods as a source of hemolymph agglutinins might include the following: (1) large volume of hemolymph, (2) feasible collection of hemolymph without destruction of animals, (3) preservation of ecology, (4) use of microtechnology to spare reagent hemolymph serum, (5) feasible storage of hemolymph and/or lectin extracts.

The pathophysiology of circulating hemolymph lectin levels may be worthy of investigation from the viewpoint of health and disease of arthropods species under consideration, as well as the quality of biomedically valuable arthropod agglutinins (lectins).

ACKNOWLEDGMENTS

This report includes work done in part with the support of grant HE07728, National Institutes of Health; the Juliette and Israel Cohen Memorial Research Fund, Western New York Chapter of the Arthritis Foundation, and the Sanibel-Captiva Foundation.

REFERENCES

Amos, D.B., H. Bashir, W. Boyle, M. MacQueen, and A. Tilikainen. 1970. A simple microcytoxicity test. *Transplantation* **7**: 220–223.

Bee, J.A. 1981. A cytochemical study of lectin receptors on isolated chick neural-retina neurons *in vitro*. *J. Cell. Sci.* **53**: 1–20.

Brahmi, Z. and E.L. Cooper. 1974. Characteristics of the agglutinin in the scorpion. *Androctonus australis*. *Contemp. Top. Immunol.* **4**: 261–270.

Cohen, E. 1968. Immunologic observations of the agglutinins of the hemolymph of *Limulus polyphemus* and *Birgus latro*. *Trans. N.Y. Acad. Sci.* **30**(3): 427–443.

Cohen, E., S.C. Roberts, S. Nordling, and G. Uhlenbruck. 1972. Specificity of *Limulus polyphemus* agglutinins for erythrocyte receptor sites common to M and N antigenic determinants. *Vox Sang.* **23**: 300–307.

Cohen, E., J. Minowada, M. Pliss, L. Pliss, and L.E. Blumenson. 1976. Differentiation of human leukemic from normal lymphocytes by *Limulus* serum agglutination. *Vox Sang.* **31**: 117–123.

Cohen, E., L.E. Blumenson, M. Pliss, and J. Minowada. 1979. Differentiation of human leukemic from normal lymphocytes by purified *Limulus* agglutinin, pp. 589–600. In E. Cohen (ed.) *Biomedical Applications of the Horseshoe Crab* (Limulidae). Alan R. Liss, New York.

Cohen, E., G.H.V. Ilodi, Z. Brahmi, and J. Minowada. 1979. The nature of cellular agglutinins of *Androctonus australis* (Saharan scorpion) serum. *Dev. Comp. Immunol.* **3**: 429–440.

Cohen, E., G.H.V. Ilodi, E.W. Korytnyk, and M. Sharma. 1979. The nature of cellular agglutinins of *Androctonus australis* (Saharan scorpion) serum. *Dev. Comp. Immunol.* **7**(1): 189–192.

Cohen, E., G.H.V. Ilodi, G.R. Vasta, J. Minowada, and Z. Brahmi. 1980. Lymphocyte activation and hemagglutination by *Androctonus* and *Limulus* agglutinins, pp. 417–422. In H. Peeters (ed.) *Protides Biological Fluids*. Pergamon Press, New York.

Cohen, E., S.G. Gregory, U. Khurana, F. Orsini, J.E. Fitzpatrick, and P. Reese. 1982. *Limulus* lectin microagglutination assay of human lymphocyte stimulation. *Am. Assoc. Clin. Histocompat. Testing, 8th Annual Meeting Abstracts*, A-24.

Cohen, E. and G.R. Vasta. 1982. Immunohematological significance of ubiquitous lectins, pp. 99–106. In E.L. Cooper and M.A.B. Brazier (eds.) *Developmental Immunology: Clinical Problems and Aging*. Academic Press, New York.

Cohen, E., S.G. Gregory, J.E. Fitzpatrick, and P. Reese. 1983. Assay of platelet sialic acid (NANA) receptor display by *Limulus* lectin, pp. 107–112. In E. Cohen and D.P. Singal (eds.) *Non-HLA Antigens in Health, Aging and Malignancy*. Alan R. Liss, New York.

Cohen, E., G.R. Vasta, W. Korytnyk, C.R. Petrie III, and M. Sharma. 1984. Lectins of the *Limulidae* and hemagglutination-inhibition by sialic acid analogs and derivatives, pp. 55–69. In E. Cohen (ed.) *Recognition Proteins, Receptors and Probes: Invertebrates*. Alan R. Liss, New York.

Finstad, C.L., R.A. Good, and G.W. Litman. 1974. The erythrocyte agglutinin from *Limulus polyphemus* hemolymph: Molecular structure and biological function, pp. 170–182. In E. Cohen (ed.) *Biomedical Perspectives of Agglutinins of Invertebrate and Plant Origins*. New York Academy of Science, New York.

Gaesser, G., E. Cohen, T. Helm. O. Holtermann, and J. Fitzpatrick. 1982. Mycosis fungoides, collagen disease and normal lymphocyte agglutinability by *Limulus polyphemus* lectin. *Fed. Proc.* **41**(3): 356, abstract.

Greenberg, J.P., M.A. Packham, J.P. Cazenave, H.J. Reamers, and J.F. Mustard. 1975. Effects on platelet function of removal of platelet sialic acid by neuraminidase. *Lab. Invest.* **32**: 476–479.

Handa, S. 1980. Dog breeds in regard to sialic acid structure. *Seikagaku* **52**: 311–317.

Hellstrom, U.H. Mellstedt, P. Perlmann, G. Holm, and D. Petterson. 1978. Receptors or *Helix pomatia*: A haemagglutinin on leukaemic lymphocytes from patients with leukaemic lymphocytic leukemia (CLL). *Clin. Exp. Immunol.* **26**: 196–203.

Jamieson, J.D., D.E. Ingber, V. Muresan, B.E. Hull, M.P. Sarras, M.F. Maylie-Pfenniger, and V. Iwaniji. 1981. Cell-surface properties of normal, differentiating and neoplastic pancreatic acinar cells. *Cancer* **47**: 1516–1525.

Klein, P.J., M. Vierbuchen, K.D. Schulz, H. Wurz, P. Citoler, G. Uhlenbruck, M. Ortmann, and R. Fisher. 1981. Hormone-dependent lectin binding site. 2. Lectin receptors as an indicator of hormone sensitive mammary carcinomas. *Tumor Diagnostik* **2**: 240–245.

Marchalonis, J.J. 1964. Natural hemagglutinin from *Limulus polyphemus*. *Fed. Proc.* **23**: 1468, abstract.

Marchalonis, J.J. and G.M. Edelman. 1968. Isolation and characterization of a hemagglutinin from *Limulus polyphemus*. *J. Mol. Biol.* **32**: 453–465.

Mazzuca, M, A.C. Roche, M. Lhermitte, and P. Roussel. 1977. *Limulus polyphemus* lectin sites in human bronchial mucosa. *J. Histochem. Cytochem.* **25**(6): 470–473.

Nakamo, T., Y. Oguchi, Y. Imar, and T. Osawa. 1980. Induction and separation of mouse helper thymus derived cells by lectins. *Immunology* **18**: 73–83.

Noguchi, H. 1903. On the multiplicity of the serum haemagglutinins of cold-blooded animals. *Zentralb. Bakterial. Parasitenk. Abt. Orig.* **34**: 286–288.

Pardoe, G., G. Uhlenbruck, and G.W.G. Bird. 1970. Studies on some heterophile receptors of the Burkitt EB2 lymphoma cell. *Immunology* **18**: 73–83.

Pistole, T.D. and S.A. Graf. 1984. Bactericidal activity of *Limulus* lectins and amoebocytes, pp. 71–82. In E. Cohen (ed.) *Recognition Proteins, Receptors, and Probes: Invertebrates*. Alan R. Liss, New York.

Shimizu, S., M. Ito, and M. Niwa. 1977. Lectins in the hemolymph of the Japanese horseshoe crab *Tachypleus tridentatus*. *Biochem. Biophys. Acta* **500**: 71–79.

Taub, R.N., M.A. Baker, and K.R. Madyastha. 1980. Masking of neutrophil surface lectin-binding sites in chronic myelogenous leukemia. *Blood* **55**(2): 294–298.

Uhlenbruck, G., A. Rothe, and G.I. Pardoe. 1968. Bemerkonswerte chemische, serologishe und elektrokinetische. Phänomene be Katzenerythrozyten: Zugleich ein

Beitrag zur Bedeuntung von mucoid-und glyko-lipididgebundener Erythrozyten-neuraminsäure. *Z. Immun. Forsch.* **136:** 79–97.

Vasta, G.R. and E. Cohen. 1982. The specificity of *Centruroides sculpturatus* Ewing (Arizona lethal scorpion) hemolymph agglutinins. *Dev. Comp. Immunol.* **6:** 219–230.

Vasta, G.R. and E. Cohen. 1984a. Characterization of the carbohydrate specificity of serum lectins from the scorpion *Hadrurus arizonensis* Stahnke. *Comp. Biochem. Physiol.* **77**B(4)**:** 721–727.

Vasta, G.R. and E. Cohen. 1984b. Humoral lectins in the scorpion *Vaejovus confuscius*: A serological characterization. *J. Invertebr. Pathol.* **43:** 333–342.

Vasta, G.R. and E. Cohen. 1984c. Sialic acid-binding lectins in the "whip scorpion" (*Mastigoproctus giganteus*) serum. *J. Invertebr. Pathol.* **43:** 333–342.

Vasta, G.R. and E. Cohen. 1984d. Sialic acid specific lectins in the serum of the genus *Aphonopelma. Dev. Comp. Immunol.* **8**(3)**:** 515–522.

Vasta, G.R. and E. Cohen. 1985a. Serum lectins from the scorpion *Vaejovus spinigerus* bind sialic acids. *Experientia* **40:** 485.

Vasta, G.R. and E. Cohen. 1985b. Naturally occurring hemagglutinins in the hemolymph of the scorpion *Paurauronctonus mesanensis* Stahnke. *Experientia* **39:** 721–722.

Vasta, G.R. and J.J. Marchalonis. 1983. Humoral recognition factors in the Arthropoda: The specificity of Chelicerata serum lectins. *Am. Zool.* **23:** 157–171.

Vasta, G.R., E. Cohen, and J. Minowada. 1981. Normal and leukemic lymphocyte cultured lymphoblastoid cell lines agglutinability by *Limulus polyphemus* lectin. *Fed. Proc.* **40:** 1099, abstract.

Watne, A.L. and E. Cohen. 1964. *Limulus* heteroagglutinins for the recovery of tumor cells from the blood. *Proc. Am. Assoc. Cancer Res.* **5**(66)**:** 261, abstract.

Yeaton, R.W. 1981a. Invertebrate lectins. 1. Occurrence. *Dev. Comp. Immunol.* **5:** 391–402.

Yeaton, R.W. 1981b. Invertebrate lectins. 2. Diversity of specificity, biological synthesis and function in recognition. *Dev. Comp. Immunol.* **5:** 535–545.

CHAPTER 20

Occurrence of Nonlymphatic Hemagglutinins in Arthropods and Their Possible Functions

Robin W. Yeaton

Department of Microbiology
Molecular Biology Institute
University of California
Los Angeles, California

20.1. INTRODUCTION

In view of the demonstrated specific affinity of hemagglutinins (lectins) for the oligosaccharides on erythrocyte (RBC) surfaces, it often has been postulated that animal hemagglutinins may function naturally in cellular adhesion during development or the immune response. In order to clarify the function of hemagglutinins, it is important to discover where they occur within the organism. There are many published studies of arthropod hemolymph lectins (Yeaton, 1981a,b; see also Chapters 8 and 18), but to date few studies have proceeded beyond the liquid phase to discover nonlymphatic sites of hemagglutinin concentration (Amirante and Mazzalai, 1978; Pereira et al., 1981). This chapter explains the use of immunocytochemical localization with antibodies produced against hemagglutinins that have been secreted into the hemolymph in conjunction with fluorescein- and peroxidase-labeled antibodies to localize hemagglutinins in various cells and tissues of an organism.

The experimental organism for this example is the North American silkmoth, *Hyalophora cecropia*. Cecropia hemolymph contains high titers of at least three divalent cation–requiring hemagglutinins (Bernheimer, 1952; Yeaton, 1980). A complex metamorphosis—embryonic development in a terrestrial chorionated egg, five larval instars specialized for feeding, diapausing pupa with a 9-month low metabolism dormancy, developing adult, and eclosion to a nonfeeding adult moth that leads a brief life dedicated to reproduction—provides a clear separation of many developmental and physiologic functions in cecropia.

20.2. METHODS

20.2.1. Specimens

Tissue donors should be selected with care, taking into account, in addition to the hypotheses to be tested, experimental results that implicate a particular age or sex. Also, the source of hemagglutinin used for generation of antibodies is critically important in systems containing more than one hemagglutinin or expressing hemagglutinin activity at different times during development. In the experiments described here, a fifth-instar female *H. cecropia* was selected as the tissue donor, because (1) synthesis of hemagglutinins by cecropia hemocytes reaches maximal activity at that stage (Yeaton, 1980), (2) no difference in titer or specificity was observed between male and female cecropia at any stage of development (Yeaton, 1980), (3) protein synthesis and storage in fat body cells reaches a peak during metamorphosis from larva to pupa (Locke and Collins, 1965; Tojo et al., 1978), (4) fat body cells aggregate during this period of development to form fatty tissue (Walters, 1969), and (5) the antibody was generated against pooled cecropia hemolymph hemagglutinins, which adhere to rabbit erythrocytes (Yeaton, 1980). Specific questions that were addressed by this study included:

1. Can hemagglutinins be visualized on and/or in the hemocytes, which are responsible for hemagglutinin synthesis?

2. Are hemagglutinins among the proteins stored in granules by fat body cells?

3. Are hemagglutinins responsible for aggregating fat body cells?

20.2.2. Reagents

Chemicals were purchased from Sigma Chemical Co., St. Louis, MO, to prepare the following reagents: 0.05 M Tris-buffered saline solution (0.9% NaCl), pH 7.6, containing 1% normal sheep serum (TBS); 0.2 M phosphate-buffered saline solution (0.9%), pH 7.4 (PBS); 0.05 M Tris, pH 7.6 (TB); and diaminobenzidine (0.03% in TB), 0.025% hydrogen peroxide (DAB). Instead of phosphate-buffered saline solution, cecropia Ringer's (CR; 0.11 M Tris-succinate, pH 6.2, 0.1 M NaCl, 0.04 M KCl, 0.015 M $MgCl_2$, 0.004 M $CaCl_2$) was used, since it was optimal physiologically for cecropia systems, providing sufficient divalent cations and correct pH to sustain hemagglutinin activity (Yeaton, 1980). Moreover, vertebrate RBC washed and suspended in CR remained round and unlysed for several days at 4°C.

Kodak ASA 160-Tungsten film was used for light and fluorescent photomicrographs. Fluorescent exposures ranged in time from 1–2 min at a camera setting of ASA 200.

20.2.3. Antisera

Primary rabbit antiserum (1° ab) was prepared against male pupal hemolymph hemagglutinins as follows. Two milliliters of rabbit RBC were collected in 3.8% sodium citrate, washed extensively in CR, and incubated with 2 ml pooled, centrifuged (to remove hemocytes) hemolymph of male cecropia diapausing pupae for 4 hr at 4°C. The agglutinated rabbit RBCs were four times alternately centrifuged (3000 RPM, 5 min) and gently washed in 20 volumes of CR. One milliliter of the washed, agglutinated rabbit RBC was emulsified with 2 ml of Freund's complete (FCA) or incomplete (FICA) adjuvant (Difco). One ml of this emulsion was injected subcutaneously into multiple sites (0.1 ml/site) on the shaven nape of the RBC donor rabbit. Injections were made at three to four week intervals, the first with FCA and the two boosters with FICA.

Antibody titers were tested two weeks after the final injection by a modified antibody titration assay of Boyden (1951). The serum to be tested was diluted 1:1 in CR at 56°C for 30 min to inactivate complement. Washed RBC ($\frac{1}{8}$ vol of any species normally agglutinated by the hemagglutinins against which the antibodies have been raised) were then incubated with the cooled serum at 24°C for 20 min, followed by centrifugation at 3000 RPM for 10 min. This incubation was repeated to ensure removal of nonspecific activity against the RBCs. Next, hemolymph was serially diluted in 0.5 ml microfuge tubes; and a 10% volume of absorbed antiserum was added to each tube,

**TABLE 20.1. PARAFFIN REMOVAL AND REHYDRATION OF TISSUE
SECTIONS**

Solution	Number of Changes	Time (min/treatment)
Xylene	3	5, 3, 3
100% ethanol	3	3, 3, 3
95% ethanol	2	3, 3
70% ethanol	1	3
Distilled water	1	3
TBS	1	15–120

incubated at 37°C for 1 hr and centrifuged in a microfuge at 8000 RPM for 2 min. Fifty microliters from each tube were plated into corresponding microtiter wells, and an equal volume of RBCs (diluted to 1% in CR) of the same species used to preabsorb the serum was added. Agglutination titers were observed after 24 hr at 4°C. Controls consisted of hemolymph incubated with normal rabbit serum or without serum. The antiserum used in the present study inhibited cecropia hemagglutinin titer against rabbit RBC 128-fold, from $\frac{1}{5120}$ to $\frac{1}{40}$.

Secondary antisera were FITC-conjugated goat anti-rabbit IgG and sheep anti-rabbit IgG (Antibodies, Inc.) as was normal sheep serum. The tertiary antiserum was rabbit horseradish peroxidase-antiperoxidase (HRP-PAP) complex (Cappel Laboratories, Malvern, PA).

20.2.4. Tissue Preparation

Tissues such as brain, fat body, salivary gland, silk gland, gut (anterior, posterior, and hind), heart and pericardium, integument, malpighian tubules, muscle, and ovary were dissected from an anaesthetized cecropia larva, immediately fixed in 2.5% formalin, embedded in paraffin, and cut in 5 μm sections. Two sections per tissue were affixed to a glass slide, and serial replicates were prepared for different antiserum treatments.

According to the method of Scott (1971), monolayer slides were prepared of hemocytes from a diapausing male pupa whose brain had been removed 6 days earlier to stimulate production of large numbers of hemocytes (Harvey and Williams, 1961). Hemolymph (1 ml) was pipetted onto a slide with a culture chamber affixed to it, aspirated after 1 hr, rinsed with PBS, and fixed in 70% ethanol for 1 min.

20.2.5. Immunocytochemistry

Slides of paraffin-embedded tissue sections were immersed to remove paraffin and rehydrated as described in Table 20.1. Both types of slides were

TABLE 20.2. IMMUNOCYTOCHEMICAL STAINING

Stain	Treatment[a]	Dilution	Solvent[b]	Incubation[c] Time (min)	Temp (°C)	Rinse[d]
FITC	1° ab	1/50	TBS	45	25	PBS
	FITC-conjugated goat anti-rabbit IgG	1/100	PBS	30–60	25	PBS
	Mount coverslip[e]					
PAP	1° ab	As above for FITC				
	Sheep anti-rabbit IgG	1/100	TBS	30–60	25	TBS
	HRP-PAP	1/200–1/300	TBS	30–60	25	TB
	TB soak[f]		TB	<120 or >120	25 / 4	
	DAB-H₂O₂[g]		Distilled water	5–30[h]	25	TB
	Distilled water soak				25	
	Counterstain[i]					
	Mount coverslip[e]					

[a] To improve the signal/noise ratio, slides were carefully dried by wiping around the tissue sections with lint-free absorbent photowipes. This drying technique was repeated after each antibody treatment.

[b] Total applied volume approximately 0.15 ml (2 drops).

[c] Slides were incubated in a humidity chamber (in this case, a covered pan lined with wet paper towels).

[d] All rinse buffers were chilled to 4°C in squirt bottles, and slides were washed by a jet of 5 ml of cold buffer. Care was taken to avoid hitting sections directly.

[e] Coverslips were affixed with Permount for PAP and with PBS 1:1 glycerol + 7.5% gelatin for FITC.

[f] Soak in TB in staining dish.

[g] DAB-H₂O₂ solution should be freshly prepared. To be sure it is active, test with a drop of HRP-PAP.

[h] Time depends on development of stain. Rinse just before stains appears dark enough.

[i] Light counterstains with hematoxylin-eosin, etc., are optional, but should be avoided if the titer of the 1°ab is not known and when FITC is the primary stain.

immunocytochemically stained as described in Table 20.2. Optimal 1° ab dilutions indicated in this protocol were determined by a pilot experiment.

20.3. RESULTS

No positive staining with either HRP-PAP or FITC was observed within or among cells of the following tissues: ventral thoracic fat body, brain, heart, ventral longitudinal muscles, and silk or salivary glands (Yeaton, 1980). However, positive results (described in detail elsewhere; Yeaton, 1980, 1982) were obtained where hemocytes or heavy perfusion of hemolymph occurred, including the lumen of the hindgut, the subvitelline membrane layer and inter-yolk sphere foci of a terminal growth-phase oocyte, and the cuticular basement lamina.

Most strikingly labeled were the hemocytes, which are used here to illustrate the immunocytochemical results. Negative controls treated with normal rabbit serum, instead of antihemagglutinin serum, did not show positive staining of any hemocytes.

Cecropia larva hemocytes were identified according to Lea and Gilbert (1966):

1. Prohemocyte (7–14 μm): round, large nucleus with mitotic figures

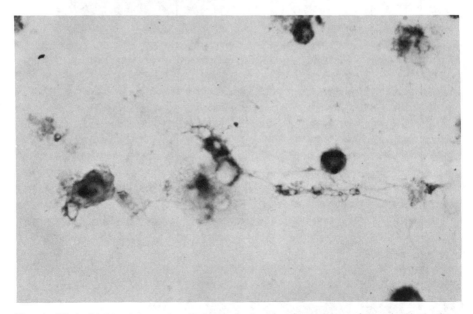

Figure 20.1. Light microscope photograph showing hemocytes of a male *Hyalophora cecropia* pupa labeled immunocytochemically with rabbit antiserum specific for pupal hemolymph hemagglutinins, which bind to rabbit RBC, and horseradish peroxidase (×400).

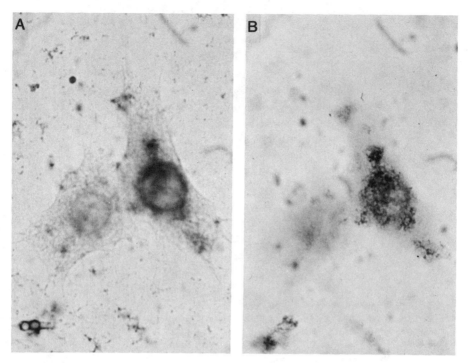

Figure 20.2. Light microscope photograph showing a plasmatocyte of a male *H. cecropia* pupa labeled immunocytochemically with rabbit antiserum specific for pupal hemolymph hemagglutinins, which bind to rabbit RBC, and horseradish peroxidase (×1000). (A) Focused at the interface of slide and cell. (B) Focused to clarify the proximal plasma membrane components.

2. Plasmatocyte (10–138 μm): fusiform to polymorphic, irregular nucleus, copious cytoplasm, many fine lipid granules; phagocytic

3. Spherule cell (4–25 μm): oval, irregular outline, cytoplasm filled with inclusions

4. Oenocytoid (7–47 μm): ellipsoidal, smooth periphery, few inclusions in abundant cytoplasm, small nucleus

5. Adipohemocyte (7–45 μm): granules filled with lipids, small nucleus; numbers increase when pupal fat body dissociates

A conclusive ontogeny of these hemocytes has not been established, although prohemocytes (not observed in this study) are known to be precursors of plasmatocytes (Lea and Gilbert, 1966; see also Chapter 1), and plasmatocytes appear to develop through several transitional stages into granulocytes (spherule cells, oenocytoids and adipohemocytes) (Gupta and Sutherland, 1966; Monpeyssin and Beaulaton, 1978). No immunocytochemical

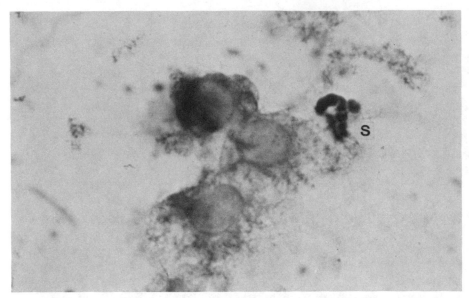

Figure 20.3. Light microscope photograph showing plasmatocytes and a spherule cell (s) of a male *H. cecropia* pupa labeled immunocytochemically with rabbit antiserum specific for pupal hemolymph hemagglutinins, which bind to rabbit RBC, and horseradish peroxidase (×1000).

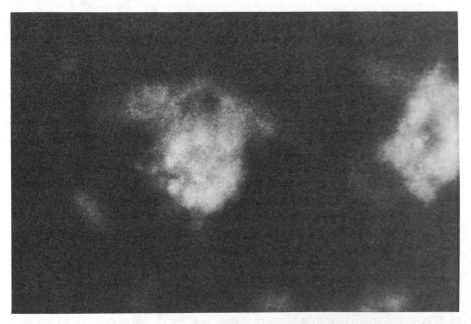

Figure 20.4. Fluorescent microscope photograph showing granulocytes of a male *H. cecropia* pupa labeled immunocytochemically with rabbit antiserum specific for pupal hemolymph hemagglutinins, which bind to rabbit RBC, and fluorescein-isothiocyanate–labeled goat anti–rabbit IgG (×1000).

labeling differences in hemagglutinin concentration or localization were observed among spherule cells, oenocytoids and adipohemocytes.

While not all plasmatocytes were stained above background levels, most revealed positive labeling of the plasma membrane around the body of the cell (Fig. 20.1), occasionally extending into pseudopodia (Fig. 20.2). Granulocytes, whether discrete or aggregated, often surrounded by plasmatocyte processes, contained heavily stained granules indicating high hemagglutinin concentrations in vesicles (Figs. 20.3 and 20.4).

20.4. DISCUSSION

These results confirm and extend previous observations of hemagglutinins on insect hemocytes (Amirante and Mazzalai, 1978; see also Chapter 2). Hemagglutinins have also been observed on the hemocytes of molluscs (van der Knaap et al., 1981; Renwrantz and Stahmer, 1983). Combining these observations with the synthesis of hemagglutinins by hemocytes (Amirante and Mazzalai, 1978; Yeaton, 1980; Chapters 1, 2, and 9), secretion of hemagglutinins into the hemolymph and their exquisite carbohydrate selectivity, the common belief is that hemagglutinins must be involved in the clearance of foreign materials from invertebrate humoral systems (Yeaton, 1981b, 1982, 1983; Renwrantz, 1983).

Specifically, the positive results here implicate granulocytes in the production of hemagglutinins. Van der Knaap and coworkers (1981) observed discretely stained hemagglutinin concentrations in *Lymnaea stagnalis* granulocytes and hypothesized these areas to be the rough endoplasmic reticulum. The positive staining in cecropia granulocytes indicates large vesicles filled with hemagglutinins, perhaps stored for release as needed, upon injury or when a foreign substance is introduced into the hemolymph. The localization of hemagglutinins on the plasma membrane of plasmatocytes increases the probability that lectins there may act as receptors for oligosaccharides on viruses or other cell surfaces. It has also been postulated that the plasmatocytes might work in concert with an opsoninlike activity of humoral hemagglutinins (Tyson and Jenkin, 1973; Yeaton, 1981b).

The absence of positive labeling in fat body cells (neither in the storage granules nor between cells) is particularly interesting. The fat body functions in accumulation, storage, resynthesis, and release of the products of digestion, analogous to the vertebrate liver (Benz, 1963). During the larval-pupal molt, in addition to synthesizing many proteins, fat body cells accumulate large quantities of two major hemolymph proteins that are stored in crystalline membrane-limited granules in the cytoplasm (Tojo et al., 1978). These proteins are accumulated by specific micropinocytosis (Locke and Collins, 1965, 1968). The negative results obtained here suggest that cecropia hemagglutinins are neither of these sequestered proteins (unless they have been modified beyond antigenic recognition), nor are they surface receptors for these proteins.

Moreover, during the first week of adult development, the fat body dissociates into single cells, which are afterward reaggregated by the activity of plasmatocytes (or by artificial agitation in vitro; Walters and Williams, 1966). This aggregation of single cells into the fat body organ at all other times of development is mediated by a protein (present at least in the hemolymph of diapausing pupa and developing adult), which requires the presence of divalent cations for its activity only during the reassociation phenomenon (Walters, 1969). The intact fifth-instar larval fat body tested in this experiment is not stained by antidiapausing pupal hemagglutinin serum, so hemagglutinins are presumably not involved in the intracellular interaction, at least at this stage.

The negative results obtained for the fat body were confirmed by the lack of hemagglutinating activity observed in fat body cell extracts at all stages of development and the failure of fat body cells cultured in vitro to synthesize hemagglutinins (Yeaton, 1980).

Immunocytochemical analysis of hemocytes thus perpetuates the notion that hemagglutinins remain closely associated with the cells which produce them, most likely then to mediate some function (perhaps immunity) that hemocytes perform. The presence of hemagglutinins on the subvitelline layer of the egg, the lining of the hindgut, and beneath the cuticle suggests again a function in protection against pathogens. The absence of positive staining of muscular, glandular, and neural complexes of the late fifth instar larva implies that hemagglutinins may not be involved in integrating enzyme complexes, moving ions within these systems or mediating cellular adhesion during development, as they do in neonatal rat brain (Simpson et al., 1977).

The procedures described in this chapter provide a very high signal to noise ratio of staining to background. The permanence of the HRP-PAP stain, its clarity and resolution make it the preferred technique (compare Figs. 20.3 and 20.4). Definition of different hemagglutinins within an animal can be done in at least two ways. First, an oligosaccharide competitor can be used to prevent some of the polyclonal antibodies from binding to one of the hemagglutinins, a technique used as a negative control in a plant lectin study using natural binding of FITC-globulin to localize concanavalin A in *Canavalia ensiformis* and PHA in *Phaseolus vulgaris* (Clarke et al., 1975). However, without question a second method would be the technique of choice: to develop monoclonal antibodies to define differential expression, localization, function and antigenic similarity of co-occurring hemagglutinins. Higher resolution by electron microscopy could be obtained by coupling ferritin to the antibodies or radiolabeling them.

20.5. SUMMARY

Immunocytochemical localization of hemagglutinins, which naturally occur in arthropods, can be used as a tool to explore the function of hemagglutinins.

This chapter describes the preparation of rabbit polyclonal antisera directed against hemagglutinins in the hemolymph of the North American silkmoth, *H. cecropia*. These hemagglutinins are synthesized and secreted by cecropia hemocytes. Techniques, using both fluorescein-isothiocyanate and horseradish peroxidase, are demonstrated for use of these antisera to immunocytochemically locate hemagglutinins in various tissues and hemocytes of cecropia. Results verify a positive relationship between hemagglutinins and hemocytes; immunocytochemistry showed hemagglutinins on the plasma membrane of plasmatocytes and in granules of granulocytes. No positive staining was observed in the fat body, brain, heart, ventral longitudinal muscles, silk or salivary glands. Implications of these findings are discussed; most probably, hemagglutinins serve some immune function in cecropia.

ACKNOWLEDGMENTS

Special thanks go to William H. Telfer, who directed my dissertation research; John J. Marchalonis, who provided insights on antibody production; Frantisal Sehnak, who trained me in dissection; Denis Baskin, who introduced me to immunocytochemistry; and Ayodhya P. Gupta, who showed great patience and guidance in the preparation of this chapter.

REFERENCES

Amirante, G.A. and F.G. Mazzalai. 1978. Synthesis and localization of hemagglutinins in hemocytes of the cockroach *Leucophaea maderae* L. *Dev. Compl. Immunol.* **2:** 735–740.

Benz, G. 1963. Physiopathology and histochemistry, Chapter 10. pp. 298–338. In E.A. Steinhaus (ed.) *Insect Pathology: An Advanced Treatise*, vol 1. Academic Press, New York.

Bernheimer, A.W. 1952. Hemagglutinins in caterpillar bloods. *Science* (Washington, DC) **115:** 150–151.

Boyden, S.V. 1951. The adsorption of proteins on erythrocytes treated with tannic acid and subsequent hemagglutination by antiprotein sera. *J. Exp. Med.* **93:** 107–120.

Clarke, A.E., R.B. Knox, and M.A. Jermyn. 1975. Localization of lectins in legume cotyledons. *J. Cell Sci.* **19:** 157–167.

Gupta, A.P. and D.J. Sutherland. 1966. *In vitro* transformations of the insect plasmatocyte in some insects. *J. Insect Physiol.* **12:** 1369–1375.

Harvey, W.R. and C.M. Williams. 1961. The injury metabolism of the cecropia silkworm. 1. Biological amplification of the effects of localized injury. *J. Insect Physiol.* **7:** 81–99.

Lea, M.S. and L.I. Gilbert. 1966. The hemocytes of *Hyalophora cecropia* (Lepidoptera). *J. Morphol.* **118:** 197–216.

Locke, M. and J.V. Collins. 1965. The structure and formation of protein granules in the fat body of an insect. *J. Cell Biol.* **36:** 857–884.

Locke, M. and J.V. Collins. 1968. Protein uptake into multivesicular bodies and storage granules in the fat body of an insect. *J. Cell Biol.* **36:** 453–483.

Monpeyssin, M. and J. Beaulaton. 1978. Hemacytopoiesis in the oak silkworm *Antheraea pernyi* and some other Lepidoptera. 1. Ultrastructural study of normal processes. *J. Ultrastruct. Res.* **64:** 35–45.

Pereira, M.E.A., A.F.B. Andrade, and J.M.C. Ribeiro. 1981. Lectins of distinct specificity in *Rhodnius prolixus* interact selectively with *Trypanosoma cruzi*. *Science* (Washington, DC) **211:** 597–600.

Renwrantz, L. 1983. Involvement of agglutinins (lectins) in invertebrate defense reactions: The immuno-biological importance of carbohydrate-specific binding molecules. *Dev. Comp. Immunol.* **7:** 603–608.

Renwrantz, L. and A. Stahmer. 1983. Opsonizing properties of an isolated hemolymph agglutinin and demonstration of lectin-like recognition molecules at the surface of hemocytes from *Mytilis edulis*. *J. Comp. Physiol.* **149B:** 535–546.

Scott, M.T. 1971. Recognition of foreignness in invertebrates. 2. *In vitro* studies of cockroach phagocytic haemocytes. *Immunology* **21:** 817–828.

Simpson, D.L., D.R. Thorne, and H.H. Loh. 1977. Developmentally regulated lectin in neonatal rat brain. *Nature* (London) **266:** 367–369.

Tojo, S., T. Betchaku, V.J. Ziccardi, and G.R. Wyatt. 1978. Fat body protein granules and storage proteins in the silkmoth, *Hyalophora cecropia*. *J. Cell Biol.* **78:** 823–838.

Tyson, C.J. and C.R. Jenkin. 1973. The importance of opsonic factors in the removal of bacteria from the circulation of the crayfish (*Parachaeraps bicarinatus*). *Aust. J. Exp. Biol. Med. Sci.* **51:** 609–615.

van der Knaap, W.P.W., L.H. Boerrigter-Barendsen, D.S.P. van den Hoeven, and T. Sminia. 1981. Immunocytochemical demonstration of a humoral defence factor in blood cells (amoebocytes) of the pond snail, *Lymnaea stagnalis*. *Cell Tissue Res.* **219:** 291–296.

Walters, D.R. 1969. Reaggregation of insect cells *in vitro*. 1. Adhesive properties of dissociated fat-body cells from developing saturniid moths. *Biol. Bull.* (Woods Hole) **137:** 217–227.

Walters, D.R. and C.M. Williams. 1966. Reaggregation of insect cells as studied by a new method of tissue and organ culture. *Science* (Washington, DC) **154:** 516–517.

Yeaton, R.W. 1980. Lectins of a North American Silkmoth (*Hyalophora cecropia*): Their Molecular Characterization and Developmental Biology. Doctoral dissertation, University of Pennsylvania, Philadelphia.

Yeaton, R.W. 1981a. Invertebrate lectins. 1. Occurrence. *Dev. Comp. Immunol.* **5:** 391–402.

Yeaton, R.W. 1981b. Invertebrate lectins. 2. Diversity of specificity, biological synthesis and function in recognition. *Dev. Comp. Immunol.* **5:** 535–545.

Yeaton, R.W. 1982. Are invertebrate lectins primordial receptors? pp. 73–83. in E.L. Cooper and M.A.B. Brazier (eds.) *Developmental Immunology: Clinical Problems and Aging*. Academic Press, New York.

Yeaton, R.W. 1983. Wound responses in insects. *Am. Zool.* **23:** 195–203.

Taxonomic Index

Subject Index